Methods in Enzymology

Volume 390
REGULATORS OF G-PROTEIN SIGNALING
Part B

METHODS IN ENZYMOLOGY

EDITORS-IN-CHIEF

John N. Abelson   Melvin I. Simon

DIVISION OF BIOLOGY
CALIFORNIA INSTITUTE OF TECHNOLOGY
PASADENA, CALIFORNIA

FOUNDING EDITORS

Sidney P. Colowick and Nathan O. Kaplan

Methods in Enzymology

Volume 390

Regulators of G-Protein Signaling

Part B

EDITED BY

David P. Siderovski

ELSEVIER
ACADEMIC
PRESS

AMSTERDAM • BOSTON • HEIDELBERG • LONDON
NEW YORK • OXFORD • PARIS • SAN DIEGO
SAN FRANCISCO • SINGAPORE • SYDNEY • TOKYO

Elsevier Academic Press
525 B Street, Suite 1900, San Diego, California 92101-4495, USA
84 Theobald's Road, London WC1X 8RR, UK

This book is printed on acid-free paper. ∞

For all information on all Academic Press publications
visit our Web site at www.academicpress.com

ISBN: 0-12-182795-X

PRINTED IN THE UNITED STATES OF AMERICA
04 05 06 07 08 9 8 7 6 5 4 3 2 1

Table of Contents

Section I. The RGS Protein Superfamily
Subsection A. RZ Subfamily

Subsection B. R4 Subfamily

Subsection C. R7 Subfamily

v

Contributors to Volume 390

Article numbers are in parentheses and following the names of contributors.
Affiliations listed are current.

NOREEN AKHTAR MOUGHAL (28), *Department of Physiology and Pharmacology, Strathclyde Institute for Biomedical Sciences, University of Strathclyde, Glasgow, G4 ONR, Scotland*

VADIM Y. ARSHAVSKY (13), *Department of Ophthalmology, Harvard Medical School, Boston, Massachusetts 02114*

D. M. BARRACLOUGH (22), *Department of Chemistry and Biochemistry, Howard Hughes Medical Institute, University of California San Diego, La Jolla, California 92093*

JOHN BASILE (17), *Oral and Pharyngeal Cancer Branch, National Institute of Dental and Craniofacial Research, National Institutes of Health, Bethesda, Maryland 20892*

JEFFREY L. BENOVIC (19), *Department of Microbiology and Immunology, Thomas Jefferson University, Philadelphia, Pennsylvania 19107*

ANTONIO DE BLASI (21), *I.N.M. Neuromed, I.R.C.C.S., Pozzilli and Dpt of Human Physiology, University of Rome "La Sapienza" localitá Camerelle, 86077*

L. L. BURNS-HAMURO (22), *Provid Pharmaceuticals, Piscataway, New Jersey 08854*

LOREDANA CAPOBIANCO (21), *I.N.M. Neuromed, I.R.C.C.S., Pozzilli and Dpt of Human Physiology, University of Rome "La Sapienza"*

PATRICK J. CASEY (4), *Departments of Pharmacology and Cancer Biology and of Biochemistry, Duke University Medical Center, Durham, North Carolina 27710*

PATRICE CODOGNO (2), *INSERM U504 Glycobiologie et Signalisation Cellulaire, Institut André Lwoff, Glasgow, 94807 Villejuif, cedex France*

MICHELLE CONNELL (28), *Department of Physiology and Pharmacology, Strathclyde Institute for Biomedical Sciences, University of Strathclyde, G4 ONR, Scotland*

LINA DAGNINO (16), *Department of Physiology and Pharmacology, London, Ontario N6A 5C1, Canada*

PETER W. DAY (19), *Department of Microbiology and Immunology, Thomas Jefferson University, Philadelphia, Pennsylvania 19107*

CARMEN W. DESSAUER (6), *Department of Integrative Biology and Pharmacology, University of Texas Health Science Center at Houston, Houston, Texas 77030*

GURPREET K. DHAMI (20), *Cell Biology Research Group, Robarts Research Institute and Department of Physiology & Pharmacology, The University of Western Ontario, London, Ontario, N6A 5K8, Canada*

MARÍA A. DIVERSÉ-PIERLUISSI (15), *Department of Pharmacology and Biological Chemistry, Mount Sinai School of Medicine, New York, New York 10029*

ROBERT J. DONATI (24), *Departments of Physiology & Biophysics and Psychiatry, University of Illinois, Chicago, Chicago, Illinois 60612*

SUDHIR J. A. D'SOUZA (16), *Department of Physiology and Pharmacology, London, Ontario N6A 5C1, Canada*

KATHLEEN DUNLAP (7), *Department of Neuroscience, Tufts University School of Medicine, Boston Massachusetts 02111*

STEPHEN S. G. FERGUSON (20), *Cell Biology Research Group, Robarts Research Institute and Department of Physiology & Pharmacology, The University of Western Ontario, London, Ontario N6A 5K8, Canada*

JOHN G. FLANAGAN (8), *Department of Cell Biology and Program in Neuroscience, Harvard Medical School, Boston, Massachusetts 02115*

ALFRED G. GILMAN (23), *Department of Pharmacology, University of Texas Southwestern Medical Center, Dallas, Texas 75390*

J. SILVIO GUTKIND (17), *Oral and Pharyngeal Cancer Branch, National Institute of Dental and Craniofacial Research, National Institutes of Health, Bethesda, Maryland 20892*

T. KENDALL HARDEN (11), *Department of Pharmacology, University of North Carolina, Chapel Hill, North Carolina 27599*

SCOTT P. HEXIMER (5), *Department of Physiology, Heart & Stroke/Richard Lewar Centre of Excellence in Cardiovascular Research, University of Toronto, Toronto, Ontario M5S 1A8 Canada*

HANS-JÖRG HIPPE (25), *Innere Medizin III - Kardiologie, Universitätsklinikum, Universität Heidelberg, D-69115 Heidelberg, Germany*

SHELLEY B. HOOKS (11), *Department of Pharmacology, University of North Carolina, Chapel Hill, North Carolina 27599*

GUANG HU (12), *Brigham and Women's Hospital, Department of Medicine, Divisions of Genetics, Boston, Massachusetts 02115*

LUISA IACOVELLI (21), *I.N.M. Neuromed, I.R.C.C.S., Pozzilli and Dpt of Human Physiology, University of Rome "La Sapienza"*

NAOYUKI IIDA (29), *Department of Pharmacology, University of Illinois, Chicago, Illinois*

RANDALL J. KIMPLE (26), *Department of Pharmacology, Lineberger Comprehensive Cancer Center, and UNC Neuroscience Center, The University of North Carolina, Chapel Hill, North Carolina 27599*

TOHRU KOZASA (18,29), *Department of Pharmacology, University of Illinois at Chicago, Chicago, Illinois 60612*

BARRY KREUTZ (18), *Department of Pharmacology, University of Illinois at Chicago, Chicago, Illinois, 60612*

QIANG LU (8), *Department of Cell Biology, Harvard Medical School, Boston, Massachusetts 02115*

SUSANNE LUTZ (25), *Institut für Pharmakologie und Toxikologie, Fakultät für Klinische Medizin Mannheim, Universität Heidelberg, D-68169 Mannheim, Germany*

KIRILL A. MARTEMYANOV (13), *Howe Laboratory of Ophthalmology, Harvard Medical School and Massachusetts Eye and Ear Infirmary Boston, Massachusetts 02114*

LUKE MARTIN-MCCAFFREY (16), *Department of Physiology and Pharmacology, London, Ontario N6A 5C1, Canada*

FERAYDOON NIROOMAND (25), *Institut für Pharmakologie und Toxikologie, Fakultät für Klinische Medizin Mannheim, Universität Heidelberg, D-68169 Mannheim, Germany*

ANDREW B. NIXON (4), *Department of Pharmacology and Cancer Biology, Duke University Medical Center, Durham, North Carolina 27710*

GERRY S. OXFORD (27), *Stark Neurosciences Research Institute, Indiana*

University School of Medicine, Indianapolis, Indiana 46202

AGNIESZKA PAJAK (16), Department of Physiology and Pharmacology, London, Ontario N6A 5C1, Canada

SOPHIE PATTINGRE (2), Division of Infectious Diseases, Southwestern Medical school, Dallas, Texas 75390

ANNE PETIOT (2), INSERM U504 Glycobiologie et Signalisation Cellulaire, Institut André Lwoff, 94807 Villejuif, cedex France

ANTONIETTA PICASCIA (21), I.N.M. Neuromed, I.R.C.C.S., Pozzilli and Dpt of Human Physiology, University of Rome "La Sapienza"

JULIANA S. POPOVA (24), Departments of Physiology & Biophysics and Psychiatry, University of Illinois, Chicago, Chicago, Illinois 60612

NIGEL J. PYNE (28), Department of Physiology and Pharmacology, Strathclyde Institute for Biomedical Sciences, University of Strathclyde, Glasgow G4 ONR, Scotland

SUSAN PYNE (28), Department of Physiology, Strathclyde Institute for Biomedical Sciences, University of Strathclyde, Glasgow G4 ONR, Scotland

MARK M. RASENICK (24), Departments of Physiology & Biophysics and Psychiatry, University of Illinois, Chicago, Chicago, Illinois 60612

RYAN W. RICHMAN (15), Department of Pharmacology and Biological Chemistry Mount Sinai School of Medicine, New York, New York, 10029

OSAMU SAITOH (9), Department of Molecular Cell Signaling, Tokyo Metropolitan Institute for Neuroscience, Fuchu-shi, Tokyo 183–8526, Japan; Department of Bio-Science, Faculty of Bio-Science, Nagahama Institute of Bio-Science and Technology, Nagahama-shi, Shiga 526–0829, Japan

SAMINA SALIM (6), Department of Integrative Biology and Pharmacology, University of Texas Health Science Center at Houston, Houston, Texas 77030

DAVID P. SIDEROVSKI (16, 26), Department of Pharmacology, UNC Neuroscience Center, and Lineberger Comprehensive Cancer Center, University of North Carolina, Chapel Hill, North Carolina 27599

WILLIAM F. SIMONDS (14), Metabolic Diseases Branch/NIDDK, National Institutes of Health, Bethesda, Maryland 20892

VLADLEN Z. SLEPAK (10), Department of Molecular and Cellular Pharmacology, University of Miami School of Medicine, Miami, Florida 33133

BALWINDER SAMBI (28), Department of Physiology and Pharmacology, Strathclyde Institute for Biomedical Sciences, University of Strathclyde, Glasgow G4 ONR, Scotland

RACHEL STERNE-MARR (20), Biology Department, Siena College, Loudonville, New York 12211

EDNA E. SUN (8), Department of Cell Biology and Program in Neuroscience, Harvard Medical School, Boston, 02115

NOBUCHIKA SUZUKI (18), Department of Pharmacology, University of Illinois at Chicago, Chicago, Illinois 60612

GREGORY G. TALL (23), Department of Pharmacology, University of Texas Southwestern Medical Center, Dallas, Texas 75390

SHIHORI TANABE (18), Department of Pharmacology, University of Illinois at Chicago, Chicago, Illinois 60612

S. S. TAYLOR (22), Department of Chemistry and Biochemistry, Howard Hughes Medical Institute, University of California San Diego, La Jolla, California 92093

JOHN J. G. TESMER (20), Institute for Cellular and Molecular Biology, Department of

Chemistry and Biochemistry, University of Texas at Austin, Austin, Texas 78712

PATRIZIA TOSETTI (1, 7), *Institut de Neurobiologie de la Mediterranée, INMED/ INSERM U29, 13273 Marseille Cedex 09, France*

JOSÉ VÁZQUEZ-PRADO (17), *Department of Pharmacology CINVESTAV-IPN, Av. Instituto Politécnico Nacional, 07000 México, D.F. Mexico*

YUREN WANG (3), *Neuroscience Discovery Research, Wyeth Research, Princeton, New Jersey 08543*

CATHERINE WATERS (28), *Department of Physiology and Pharmacology, Strathclyde Institute for Biomedical Sciences, University of Strathclyde, Glasgow G4 ONR, Scotland*

CHRISTINA K. WEBB (27), *The Laboratory of Signal Transduction, National Institute of Environmental Health Sciences, Research Triangle Park, North Carolina 27709*

PHILIP B. WEDEGAERTNER (19), *Department of Microbiology and Immunology, Thomas Jefferson University, Philadelphia, Pennsylvania 19107*

THEODORE G. WENSEL (12), *Department of Biochemistry and Molecular Biology, Baylor College of Medicine, Houston, Texas 77030*

THOMAS WIELAND (25), *Institut für Pharmakologie und Toxikologie, Fakultät für*

Klinische Medizin Mannheim, Universität Heidelberg, D-68169 Mannheim, Germany

FRANCIS S. WILLARD (16, 26), *Department of Pharmacology, UNC Neuroscience Center, and Lineberger Comprehensive Cancer Center, University of North Carolina, Chapel Hill, North Carolina 27599*

D. SCOTT WITHEROW (10), *Howard Hughes Medical Institute, Duke University Medical Center, Durham, North Carolina 27710*

GEOFFREY E. WOODARD (14), *Metabolic Diseases Branch/NIDDK, National Institutes of Health, Bethesda, Maryland 20892*

KUBO YOSHIHIRO (9), *Department of Physiology and Cell Biology, Tokyo Medical and Dental University, Graduate School and Faculty of Medicine, Bunkyo-ku, Tokyo 113–8519, Japan; Division of Biophysics and Neurobiology, National Institute for Physiological Sciences, Okazaki 444–8585, Japan*

KATHLEEN H. YOUNG (3), *Neuroscience Discovery Research, Wyeth Research, Princeton, New Jersey 08543*

JIANG-ZHOU YU (24), *Department of Physiology & Biophysics and Psychiatry, University of Illinois, Chicago, Chicago, Illinois 60612*

JIAN-HUA ZHANG (14), *Metabolic Diseases Branch/NIDDK, National Institutes of Health, Bethesda, Maryland 20892*

Preface

Heterotrimeric G protein signaling has been the subject of several previous volumes of this August series, most recently with the "G Protein Pathways" Volumes 343 to 345. Volume 344, in particular, contains several notable chapters devoted to emergent methods of analysis of the "regulator of G-protein signaling" (RGS) proteins. This superfamily of G-protein modulators first came to prominence in late 1995/early 1996 with a flurry of papers, spanning from yeast and worm genetics to T- and B-lymphocyte signaling, that heralded the discovery of a missing fourth component to the established receptor > G-protein > effector signaling axis. The ability of RGS proteins, via their hallmark RGS domain or "RGS-box", to accelerate the intrinsic guanosine triphosphatase (GTPase) activity of G-protein alpha subunits helped explain the timing paradox between GTPase activity seen with isolated G-alpha subunits *in vitro* and the rapid physiological timing of G protein-coupled receptor (GPCR) signaling observed *in vivo*.

The nomenclature used in the field since the discovery of the RGS protein superfamily has certainly drifted from the first efforts in 1996 of Druey, Horvitz, Kehrl, and Koelle to establish the name "regulators of G-protein signaling" for all proteins containing the RGS-box. Thus, before considering the breadth of content within Volumes 389 (Part A) and 390 (Part B), I will try to clarify some of my own editorial decisions on issues of nomenclature. I greatly prefer "GTPase-*accelerating* protein" rather than "GTPase-*activating* protein" as the definition underlying the acronym "GAP" that describes RGS-box action on G-alpha subunits. This distinction helps highlight that the intrinsic GTPase activity of G-alpha subunits is much greater than that of Ras-superfamily monomeric GTPases – distant cousins of G-alpha whose minimal GTPase activity is truly activated by GAPs (distinct from the RGS proteins) that contribute a key catalytic residue. I also greatly prefer the use of "RGS-box" or "RGS domain" to describe the conserved alpha-helical bundle that comprises the G-alpha binding site and catalytic GAP activity common to the superfamily. Undoubtedly, with such a large superfamily, there will be structural variations on the central theme of the RGS-box nine-helical bundle; however, the proliferation of alternate names for subgroups of RGS domains (such as "rgRGS" for the RhoA guanine nucleotide exchange factors [Rho-GEFs], or "RH" for the N termini of the GPCR kinases [GRKs]) does a disservice to those of you looking in from outside the field.

Clearly, there is now no recourse to the fact that the numbering system for RGS proteins, first implemented in 1996 by Koelle and Horvitz, has gone awry and no longer allows facile cross-species comparisons. For example, three "RGS1" proteins are discussed in volume 389: mouse/human RGS1 (Chapters 2, 6, 8), *Caenorhadbitis elegans* RGS-1 (Chapter 18), and Arabidopsis AtRGS1 (Chapters 19 and 20), yet the mammalian, nematode, and plant "RGS1" proteins share no structural (nor functional) relatedness. Also, whereas *Drosophila* RGS7 (Volume 389, Chapter 21) and mammalian RGS7 proteins (Volume 390, Chapters 10–14) share the same multi-domain architecture of DEP, GGL, and RGS domains, the *C. elegans* RGS-7 does not (yet the *C. elegans* EAT-16 and EGL-10 proteins do; Volume 389, Chapter 18). Therefore, reader beware.

In my choices regarding content for these volumes, I have read the mandate "regulators of G-protein signaling" broadly. Thus, not only does Part B (Volume 390) contain 22 consecutive chapters spanning seven subfamilies of RGS-box proteins, but both Parts A and B also contain chapters devoted to G-protein signaling regulators outside the RGS-box superfamily. For example, four chapters deal with individual components of an emerging, non-GPCR-related nucleotide cycle for G-alpha subunits underlying microtubule dynamics, mitotic spindle organization, and chromosomal segregation: namely, the GoLoco motif-containing proteins Pins and RGS14 (Volume 389, Chapter 22; Volume 390, Chapter 16), the G-alpha exchange factor Ric-8A (Volume 390, Chapter 23), and the microtubule constituent tubulin (Volume 390, Chapter 24). Chasse & Dohlman (Volume 389, Chapter 24) describe their adroit exploitation of the "awesome power of yeast genetics" to identify a multitude of novel modulators of the GPCR signaling archetype: the *Saccharomyces cerevisiae* pheromone response pathway. In the preceding chapter (Volume 389, Chapter 23), Dohlman, Elston, and co-workers describe mathematical modeling of the actions of pheromone signaling components and modulators, like the RGS protein prototype SST2, that are known in great detail. It is my hope, and surely of the authors as well, that the modeling techniques detailed in this chapter will be employed by others to elicit understanding of the interplay between regulators of G-protein signaling and their targets in other systems and organisms.

Indeed, the overall intent of all of the chapters in Parts A and B is to facilitate further investigations of these newly appreciated G-protein signaling modulators in their physiological contexts. My personal belief is that, over time, RGS proteins will prove valuable drug discovery targets. RGS-insensitive ("RGS-i") G-alpha subunits, described most thoroughly here by Neubig and colleagues (Volume 389, Chapter 14), have facilitated several unequivocal demonstrations that endogenous RGS proteins control the dynamics of GPCR signal transduction *in vivo* (e.g., Volume 389, Chapters 10–12 and 14). Chapters 16 and 17 of Volume 389 represent initial forays into developing peptidomimetic and small molecule inhibitors of RGS protein action that should provide "proofs-of-principle" required to energize the actions of "Big Pharma" in considering RGS proteins as legitimate drug targets. Real-time fluorescence and surface plasmon resonance assays of the RGS protein/G-alpha interaction, as described in Chapters 4 and 19 of Volume 389, should help accelerate this pursuit.

I remain ever grateful to my many colleagues in the G-protein signaling community, and members of my own laboratory, who readily contributed chapters for these two volumes. I also thank Ms. Cindy Minor for her organizational assistance, and my family for being patient with me during the completion of this enormous task. Indeed, it must be formally stated to my children that, although I wrote and edited chapter contributions during your ballet, baseball, and basketball practices, I *never* did during your games or performances!

DAVID P. SIDEROVSKI

METHODS IN ENZYMOLOGY

VOLUME 371. RNA Polymerases and Associated Factors (Part D)
Edited by SANKAR L. ADHYA AND SUSAN GARGES

VOLUME 372. Liposomes (Part B)
Edited by NEJAT DÜZGÜNEŞ

VOLUME 373. Liposomes (Part C)
Edited by NEJAT DÜZGÜNEŞ

VOLUME 374. Macromolecular Crystallography (Part D)
Edited by CHARLES W. CARTER, JR., AND ROBERT W. SWEET

VOLUME 375. Chromatin and Chromatin Remodeling Enzymes (Part A)
Edited by C. DAVID ALLIS AND CARL WU

VOLUME 376. Chromatin and Chromatin Remodeling Enzymes (Part B)
Edited by C. DAVID ALLIS AND CARL WU

VOLUME 377. Chromatin and Chromatin Remodeling Enzymes (Part C)
Edited by C. DAVID ALLIS AND CARL WU

VOLUME 378. Quinones and Quinone Enzymes (Part A)
Edited by HELMUT SIES AND LESTER PACKER

VOLUME 379. Energetics of Biological Macromolecules (Part D)
Edited by JO M. HOLT, MICHAEL L. JOHNSON, AND GARY K. ACKERS

VOLUME 380. Energetics of Biological Macromolecules (Part E)
Edited by JO M. HOLT, MICHAEL L. JOHNSON, AND GARY K. ACKERS

VOLUME 381. Oxygen Sensing
Edited by CHANDAN K. SEN AND GREGG L. SEMENZA

VOLUME 382. Quinones and Quinone Enzymes (Part B)
Edited by HELMUT SIES AND LESTER PACKER

VOLUME 383. Numerical Computer Methods (Part D)
Edited by LUDWIG BRAND AND MICHAEL L. JOHNSON

VOLUME 384. Numerical Computer Methods (Part E)
Edited by LUDWIG BRAND AND MICHAEL L. JOHNSON

VOLUME 385. Imaging in Biological Research (Part A)
Edited by P. MICHAEL CONN

VOLUME 386. Imaging in Biological Research (Part B)
Edited by P. MICHAEL CONN

VOLUME 387. Liposomes (Part D)
Edited by NEJAT DÜZGÜNEŞ

VOLUME 388. Protein Engineering
Edited by DAN E. ROBERTSON

VOLUME 389. Regulators of G-Protein Signaling (Part A)
Edited by DAVID P. SIDEROVSKI

Section I

The RGS Protein Superfamily

Subsection A

RZ Subfamily

[1]   Evaluating Chick Gα-Interacting
Protein Selectivity

By PATRIZIA TOSETTI

Abstract

Regulators of G-protein signaling (RGS) proteins constitute a large family of GTPase-accelerating proteins (GAPs) for heterotrimeric G proteins. More than 30 RGS genes have been identified in mammals. One of these, the Gα-interacting protein (GAIP), interacts preferentially with members of the $G_{i/o}$ subfamily of G-protein α subunits in mammalian cells. A unique isoform of GAIP, derived from embryonic chicken dorsal root ganglion neurons, has a short N terminus that is only 41% identical to known mammalian orthologs. Consistent with this unique primary structure, chick GAIP has higher target specificity than its mammalian counterparts. This article describes both *in vitro* and *in vivo* methods used to characterize chick GAIP selectivity.

Introduction

The Gα-interacting protein (GAIP) is a member of the regulator of G-protein signaling (RGS) protein family that opposes cellular signaling via $G_{i/o}$ and $G_{q/11}$ protein-coupled receptors by accelerating the intrinsic GTPase activity of the Gα subunits (De Vries *et al.*, 1995; Ross and Wilkie, 2000). The GTPase-accelerating (or GAP) activity of RGS proteins has been demonstrated to shape the time course of several G-protein-dependent cellular functions, including ion channel modulation (Diversé-Pierluissi *et al.*, 1999; Tosetti *et al.*, 2002, 2003a,b).

Although most RGS protein targets are known, the issue of RGS protein selectivity is still an open one. The abundance of signaling cascades where RGS and G proteins are involved raises the issue of how specificity in their mutual recognition is achieved. It has been difficult to evaluate precisely the selectivity of most of the mammalian RGS proteins among their potential $G_{i/o}$ and $G_{q/11}$ class targets. Independent *in vitro* and *in vivo* studies have sometimes provided contradictory results (Posner *et al.*, 1999;

Shuey *et al.*, 1998; Witherow *et al.*, 2000). As observed for GAPs for monomeric G proteins (Muller *et al.*, 1997; Ridley *et al.*, 1993), some specificity of interaction has been described between RGS proteins and specific heterotrimeric Gα subunits (Berman *et al.*, 1996; Chen *et al.*, 1997; De Vries *et al.*, 1995; Yan *et al.*, 1997). In addition, some selectivity is conferred by restrictive patterns of cellular expression (e.g., RGS9) (Cowan *et al.*, 1998). The N terminus has also been suggested to play a role in the specificity of several RGS proteins, including GAIP, by possibly confining their localization to discrete intracellular compartments (Chatterjee *et al.*, 2003; Zeng *et al.*, 1998). A GAIP mutant lacking the N terminus loses its ability to discriminate between G$_i$- and G$_o$-mediated pathways in chick dorsal root ganglion (DRG) neurons (Diversé-Pierluissi *et al.*, 1999). Similarly, a chick GAIP variant with a structurally unique N terminus shows an enhanced G-protein selectivity (Tosetti *et al.*, 2002).

Most of the insights on RGS protein specificity have been gained by studying RGS/G protein interactions using purified proteins or heterologous expression studies (Ross and Wilkie, 2000). However, *in vitro* assay conditions may give rise to artifactually low specificity, as the ability of RGS proteins to discriminate among G proteins is modulated by regulatory molecules and environmental factors (Martemyanov *et al.*, 2003; Tosetti *et al.*, 2003a). Studies of RGS proteins in primary cells in which G-protein-dependent signaling pathways have been well characterized are therefore a necessary complementary approach to shed light on this issue. The strategy used to characterize chick GAIP selectivity combined both an *in vivo* and an *in vitro* approach. At first, the biologically active, recombinant protein was purified and its GTPase activity was tested on several Gα subunits *in vitro*. Later, *in vitro* data were used to choose the appropriate endogenous G-protein-mediated pathways for *in vivo* testing in its native environment, the DRG neuron.

This article describes two methods employed to determine chick GAIP selectivity. The first part focuses on the *in vitro* characterization of chick GAIP selectivity using a novel, subsecond-resolution GTPase assay. The second part describes how chick GAIP specificity was assessed in its tissue of origin, the DRG neuron, by monitoring its ability to modulate biochemically identified G-protein pathways mediating Ca^{2+} channel inhibition.

Assaying Chick GAIP Specificity *In Vitro* with a Subsecond-Resolution GTPase Assay

The ability to act as GAPs for heterotrimeric Gα subunits is the best-characterized function of RGS proteins (Berman *et al.*, 1996; Ross and Wilkie, 2000). The GTPase activity of a molecule can be assessed via a single turnover GTPase assay, which measures the rate at which a G-protein α subunit completes a whole round of GTP hydrolysis in the presence or absence of the putative GAP molecule (Berman *et al.*, 1996). This test is carried out *in vitro* and requires the presence of both the GAP molecule and the target Gα subunit in a purified, biologically active form. The purified Gα subunit is first activated by binding to [γ-^{32}P]GTP, a molecule having a radioactive γ-phosphate (the leaving group in the hydrolysis reaction). The activation process takes place in ionic conditions where GTP hydrolysis cannot occur [*i.e.*, lack of Mg^{2+} ions (Berman *et al.*, 1996)], allowing loading of the Gα subunit pool with GTP without competing hydrolysis. Mg^{2+} ions are then added to the activated Gα subunits to trigger GTP hydrolysis. The intrinsic hydrolytic activity of the Gα subunit causes the progressive release of ^{32}P$_i$ into the assay solution. The amount of radioactive phosphate released over time determines the rate of hydrolysis. In the presence of a GAP molecule, the intrinsic GTPase activity of the Gα subunit is accelerated and more phosphate is released. GAP activity is measured by the fold change of Gα GTPase rate occurring in the presence of the putative GAP molecule.

Experimental Results

In order to perform a single turnover GTPase assay successfully, the proteins involved must be functional. Biologically active, recombinant chick GAIP can be produced efficiently in an *Escherichia coli* protein expression system and purified using metal chelate affinity chromatography. Following the engineering of a C-terminal sequence encoding six consecutive histidines, chick GAIP cDNA was introduced into a bacterial expression vector under the control of the T7 RNA polymerase promoter (pET11a by Novagen). The construct was then used to transform BL21(DE3) *E. coli* cells, a host strain containing the T7 RNA polymerase gene under the control of the inducible *lacUV5* promoter. Isopropyl-1-thio-β-D-galactopyranoside (IPTG) was used to induce the synthesis of the T7 RNA polymerase that bound the T7 promoter and initiated the transcription of recombinant chick GAIP.

High levels of expression of the target protein in the bacterial system do not always translate in a good protein yield. They may result in insolubility of the product at the concentrations being produced and in the formation of inclusion bodies. Although inclusion bodies consist mainly of the desired recombinant protein (see Fig. 1) and are easy to separate from other cellular constituents, they contain the protein in an insoluble, nonnative state. Purification of proteins from inclusion bodies always requires denaturing conditions, leading to partial or complete loss of function. In contrast, soluble proteins can be purified in conditions preserving their native

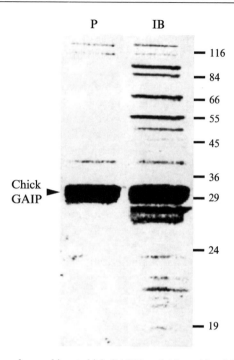

FIG. 1. Distribution of recombinant chick GAIP in soluble and insoluble cellular fractions. Polyacrylamide gel electrophoresis of the crude inclusion body fraction (second lane, IB) or the Ni-NTA-purified soluble cellular fraction (first lane, P) from a recombinant chick GAIP preparation. Molecular mass standards (kDa) are indicated at right. The amounts of recombinant chick GAIP in soluble and insoluble fractions are approximately equivalent. Note that chick GAIP is more than 80% pure after Ni-NTA purification.

biological activity. Furthermore, better yields are usually obtained from processing the soluble fraction, even though it might constitute only a minor portion of the total expressed protein. When optimizing the production of biologically active proteins, priority should therefore be given to conditions reducing the formation of inclusion bodies in favor of the cytoplasmic, soluble form of the recombinant protein. Inducing GAIP expression with a lower IPTG concentration (50 μM), but for a longer period of time (12 h), increased the fraction of chick GAIP that remained soluble. Between 40 and 50% of total chick GAIP was cytoplasmic and could be purified in conditions preserving its biological activity (Fig. 1).

After lysing the bacterial cells and recovering the supernatant (containing the soluble GAIP fraction), recombinant chick GAIP was purified using a resin containing nickel ions immobilized by covalently attached nitrilotriacetic acid (Ni-NTA). The polyhistidine tail of chick GAIP bound selectively to the Ni-NTA resin by coordination bonding of the nonprotonated nitrogen atoms of its imidazole side chains to Ni^{2+} ions. Most contaminating proteins present in the supernatant did not bind to the resin, as expected, and were eliminated by repetitive washing. Purified chick GAIP was then unbound from the resin by the competitive action of high concentrations of free imidazole. GAIP protein preparations were more than 80% pure as assessed by Coomassie Blue staining of conventional polyacrylamide gels (Fig. 1, first lane).

The biological activity of purified, recombinant chick GAIP and its $G_{i/o}\alpha$ subunit selectivity were tested *in vitro* using an automated, fast superfusion system that allowed subsecond resolution of the time course of GTP hydrolysis (Fig. 2) (Tosetti *et al.*, 2002). This method, originally developed to measure neurotransmitter release (Turner *et al.*, 1989), was optimized for the study of GTPase activity. Bovine $[\gamma^{-32}P]$GTP-bound $G\alpha$ subunits, alone or complexed with recombinant chick GAIP, were immobilized on glass fiber filters and positioned in a superfusion chamber accessed by three computer-controlled valves. Each valve was connected to a pressurized reservoir containing a separate solution. N_2 (10 psi) was used to drive the fluid flow. The opening of a valve triggered the rapid delivery of a Mg^{2+}-containing solution to the chamber that initiated GTP hydrolysis and the release of $^{32}P_i$. The pressurized fluid flow passed through the glass fiber filter, washing the hydrolyzed radioactive phosphate off the filter. A rotating 50-vial rack collected the eluted superfusate. The amount of $^{32}P_i$ in each fraction was determined by liquid scintillation counting.

The main novelty of the system resides with a time resolution for GTP hydrolysis of at least 100 ms. This property stems from (1) the possibility of exchanging the superfusing solution rapidly via high-speed, solenoid-driven valves; (2) a relatively high, pressure-driven solution flow rate; and (3) the automated, fast collection of eluted $^{32}P_i$ fractions. Previously reported GTPase assays have a second to minute time resolution, dictated by physical constraints (*i.e.*, most steps are carried out by hand) (Berman *et al.*, 1996; Fischer *et al.*, 1999). However, the time course of RGS protein-mediated termination of the response has been shown to occur on a subsecond time scale for some processes (He *et al.*, 1998; Skiba *et al.*, 2000). Resolving the early phases of the hydrolytic process may therefore provide more precise indications on RGS protein specificity and efficacy.

As shown in Fig. 3, chick GAIP effectively increased the GTPase activity of all $G\alpha$ subunits tested ($G\alpha_{i1}$, $G\alpha_{i3}$, and $G\alpha_o$). Surprisingly, chick GAIP exerted the most significant GAP activity on $G\alpha_{i1}$, in contrast to what has been reported for human GAIP (De Vries *et al.*, 1995).

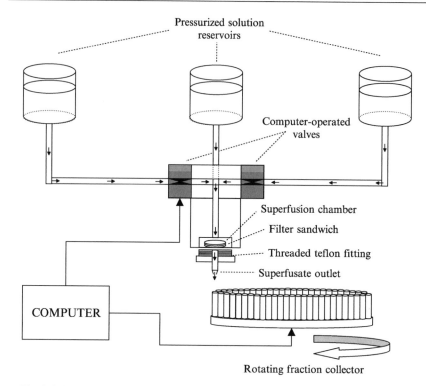

FIG. 2. Schematic diagram of the fast superfusion system used for single turnover GTPase assays. [γ-^{32}P]GTP-bound Gα subunits, alone or complexed with recombinant chick GAIP, are immobilized on a HA/GFF/HA filter sandwich that is positioned in the superfusion chamber and secured into place with a Teflon outlet fitting. A computer program controls the opening of the solenoid-driven valves and the flow of individual solutions through the system. Filters are superfused either with Mg^{2+}-free HDE buffer, to estimate background ^{32}P$_i$ release, or with HDE + 10 mM MgCl$_2$ to trigger GTP hydrolysis. The superfusate, carrying either background or hydrolyzed ^{32}P$_i$, exits as an uninterrupted stream onto the turning fraction collector that separates it into equal volumes. The amount of ^{32}P$_i$ in each fraction is determined by liquid scintillation counting.

Protocols

Engineering a Polyhistidine Tag on the Chick GAIP C Terminus. To allow affinity purification using Ni-NTA resin, a sequence of six histidines (His$_6$) is engineered by polymerase chain reaction (PCR) at the C terminus of chick GAIP. In addition to the His$_6$ tag, PCR primers contain the restriction site sequences (*Nde*I and *Bam*HI; underlined) needed for later in-frame insertion of the PCR product in the pET11a expression vector

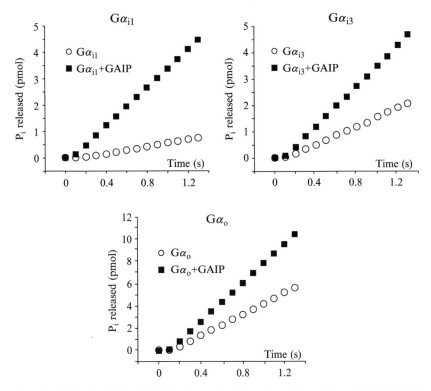

FIG. 3. Chick GAIP is a GAP for $G_{i/o}\alpha$ subunits *in vitro*. The ability of chick GAIP to accelerate the intrinsic GTPase activity of several $G\alpha$ subunits was tested *in vitro* using a subsecond resolution GTPase assay. $^{32}P_i$ release as a function of time was measured for $G\alpha_{i1}$, $G\alpha_{i3}$, and $G\alpha_o$ in the presence and absence of chick GAIP. Chick GAIP acts as GAP for all three tested subunits, but is most effective on $G\alpha_{i1}$. All points are means $\pm$ SEMs from three separate experiments. Data modified from Tosetti *et al.* (2003b), with permission.

(Novagen, Madison, WI). Primer sequences are 5'-ACC CCT TCA TAT GTC CCG GCA TGA GAC TTC TCC G-3' (sense) and 5'-CGT CGG ATC CTA ATG GTG ATG GTG ATG GTG GGA CTC CGA GGA GGA-3' (antisense). PCR is performed using the following cycling protocol: 94° for 90 s (1 cycle), 94° for 30 s, 50° for 30 s, 68° for 1 min (40 cycles), and 72° for 5 min (1 cycle). The PCR product of the expected size (675 bp) is gel purified using the QIAquick gel extraction kit (Qiagen, Valencia, CA) and is subjected to successive *Bam*HI and *Nde*I digestions in parallel with the empty pET11a vector. The digested fragments are then ligated using the Rapid DNA ligation kit from Roche/Boehringer Mannheim (Indianapolis, IN), and the resulting construct is transformed into

TOP10F'-competent cells (Invitrogen, Carlsbad, CA). The restriction enzymes are from New England Biolabs (Beverly, MA). Properly subcloned constructs are identified by PCR and sequencing.

Expression and Purification of Recombinant Chick GAIP. Chick GAIP in pET11a is used to transform *E. coli* BL21(DE3) host cells. This protein expression system by Novagen has the protein of interest under the transcriptional control of the T7 RNA polymerase promoter. One liter of LB medium with ampicillin (100 μg/ml) is inoculated with a 10-ml overnight culture from a single colony at 37°. Induction is performed at OD_{600} = 0.6–0.8 by the addition of IPTG (50 μM). Bacteria are shaken for an additional 12 h prior to harvest, centrifuged to eliminate culture medium, and resuspended in buffer A (50 mM HEPES, pH 8, 20 mM β-mercaptoethanol, 100 mM NaCl, 0.1 mM phenylmethylsulfonyl fluoride). Cell membranes are lysed by the addition of lysozyme (0.2 mg/ml) and sonication. DNase I (5 μg/ml) is added to cut genomic DNA and reduce viscosity in order to facilitate the purification process. The resulting clear lysate is centrifuged (12,000g for 30 min at 4°) to eliminate cellular debris and insoluble proteins. The supernatant fraction is loaded onto a column containing 1 ml of Ni-NTA affinity resin (Qiagen, Hilden, Germany) preequilibrated with buffer A. The column is washed with 50 ml of buffer A + 50 mM imidazole to eliminate unbound and low-affinity-bound contaminating proteins. All eluates from the washing steps are collected and examined with standard polyacrylamide gel electrophoresis (SDS–PAGE) to monitor the progressive elution of unbound proteins. Chick GAIP is eluted with 5 ml of buffer A + 250 mM imidazole. The eluted protein is dialyzed overnight against 1 liter of buffer B [50 mM HEPES, 2 mM dithiothreitol (DTT), pH 8], concentrated using a Centricon 10 column (Millipore, Bedford, MA), checked for purity using SDS–PAGE, and stored at −80°.

Single Turnover GTPase Assay. GTPase assays are performed using an automated fast-superfusion system with subsecond resolution (Turner *et al.*, 1989). A schematic diagram of the system can be seen in Fig. 2. The system consists of three pressurized fluid reservoirs (pressure filtration funnels, Pall/Gelman Science, Ann Arbor, MI), a superfusion device (hand-assembled; see Turner *et al.*, 1989), a computerized control system (Apple-IIe computer with an Applied Engineering digital-to-analog converter), and a rotating fraction collector (hand assembled; see Turner *et al.*, 1989). Protein samples (see later) are immobilized on a glass fiber filter (0.7 μm, Whatman) and positioned in a superfusion chamber connected via solenoid-driven valves (General Valve, Brookshire, TX) to pressurized (10 psi) solution reservoirs. The filter is secured into place in the superfusion chamber with a Teflon outlet fitting. When one valve opens, the reservoir solution flows through the protein complexes immobilized on the filter and exits as an uninterrupted stream onto the fraction collector, a turning circular platform with 50 wells designed to accommodate 15×45-mm glass minivials. Because the superfusate streams into the center of each minivial, the turning of the platform effectively separates it into equal volumes. The solenoid-driven valves allow for the fast switching between reservoir solutions. A computer program controls the timing of valve opening and fraction collection, as well as the desired solution sequence. The minimal dead volume of the chamber and the relatively high solution flow rate permit a time resolution for $^{32}P_i$ release of at least 100 ms. Bovine $G\alpha_{i1}$, $G\alpha_{i3}$, or $G\alpha_o$ (2 μl, 100 $\mu g/ml$, Calbiochem, CA) is incubated with 2 μl $[\gamma^{-32}P]GTP$ (6000 Ci/mM, 10 Ci/l) and 6 μl of Mg^{2+}-free HDE buffer (50 mM HEPES, 5 mM EDTA, 2 mM DTT, pH 8.0) for 30 min at 30° to allow equilibrium binding. Samples are then stored on ice, and His_6-tagged recombinant chick GAIP (3 μM final concentration) is added to half. The labeled $G\alpha$ subunits, with or without GAIP, are retained on a Whatman glass fiber filter (GFF) and placed in the superfusion chamber. Because GFFs tend to disintegrate with high flow rates, a Millipore cellulose ester filter (HA) is added above and below the GFF to provide structural support, creating a 5-mm HA/GFF/HA filter sandwich. Samples are superfused with either Mg^{2+}-free HDE buffer, to measure background $^{32}P_i$ release, or HDE + 10 mM $MgCl_2$ to trigger hydrolysis. Superfusate fractions are collected every 0.1 s for a total time of 1.3 s. One milliliter of scintillation cocktail (Hydrofluor, National Diagnostic, Manville, NJ) is added to each fraction, and the amount of $^{32}P_i$ is determined by liquid scintillation counting. For all experiments, samples are assayed in triplicate, averaged, and $^{32}P_i$ release is expressed as a percentage of total radioactivity added to each filter.

Assaying Chick GAIP Specificity in the Intact Neuron

Results from the *in vitro* GTPase assay suggest that chick GAIP may have a selectivity that differs from that reported for mammalian GAIP (De Vries *et al.*, 1995). *In vitro* data, however, needed to be confirmed in the intact neuron, as (i) selectivity was tested on bovine subunits that are not chick GAIP physiological targets and (ii) selectivity may be modulated by regulatory factors intrinsic to the native neuron.

For this reason, the issue of GAIP target specificity was further explored in the chick DRG neuron, the tissue from which chick GAIP was originally cloned. In the chick DRG neuron, voltage-gated Ca^{2+} channels are inhibited through at least three distinct G-protein-mediated pathways (Fig. 4A).

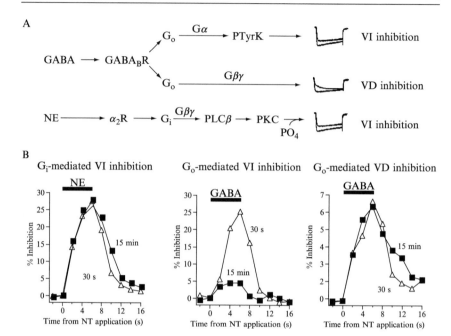

FIG. 4. Chick GAIP selectively reduces G_o-mediated, voltage-independent inhibition of Ca^{2+} currents. (A) Schematic diagram of distinct G-protein-mediated pathways leading to Ca^{2+} channel inhibition in chick DRG neurons. GABA, γ-aminobutyric acid; NE, norepinephrine; $GABA_BR$, $GABA_B$ receptors; α_2R, α_2 adrenergic receptors; PTyrK, protein tyrosine kinase; PKC, protein kinase C; $PLC\beta$, phospholipase $C\beta$. (B) Voltage-independent G_i-mediated (*left*), voltage-independent G_o-mediated (*center*), and voltage-dependent G_o-mediated Ca^{2+} current inhibition (*right*) measured in a single cell after 30 s ($\triangle$) or 15 min (■) of chick GAIP dialysis and plotted as a function of time. Bar shows timing of neurotransmitter application. Recombinant chick GAIP selectively antagonizes the G_o-mediated, voltage-independent component of the inhibition. Data modified from Tosetti *et al.* (2003b), with permission.

One G_o-mediated pathway is linked to $GABA_B$ receptor activation, involves $G_o\alpha$ and a Src-like tyrosine kinase, and reduces Ca^{2+} current in a voltage-independent (VI) manner (Diversé-Pierluissi *et al.*, 1995, 1997). GABA also activates a second G_o-mediated pathway that reduces Ca^{2+} currents in a voltage-dependent (VD) manner via a direct interaction between $G_o\beta\gamma$ and the channel (Diversé-Pierluissi *et al.*, 1995; Lu and Dunlap, 1999). Finally, norepinephrine (NE), acting on α_2-adrenergic receptors, triggers a third, G_i-mediated pathway that involves $G_i\beta\gamma$ and protein kinase C, and induces a VI inhibition of Ca^{2+} currents (Diversé-Pierluissi *et al.*, 1995). These well-characterized G-protein-mediated pathways leading to Ca^{2+} channel inhibition have proven to be a valuable model in which to test specificity of the G protein and RGS protein-dependent signaling (Diversé-Pierluissi *et al.*, 1999; Tosetti *et al.*, 2002).

Experimental Results

To test for possible effects of chick GAIP on G-protein-mediated pathways, Ca^{2+} channel inhibition was induced by the application of neurotransmitters (NT) (100 μM GABA or 100 μM NE) after the recombinant protein was dialyzed into the cells through the patch pipette. NT-induced inhibition of Ca^{2+} channels was measured at two time points: 30 s after break in (prior to significant GAIP diffusion) and 15 min after break in (after equilibration of GAIP with the cell interior). This approach allowed (i) cells not responding to GABA or NE to be eliminated from the analysis and (ii) each cell to be used as its own internal control. GABA inhibited chick DRG Ca^{2+} channels via both a VI and a VD G_o-coupled pathway, whereas NE activated mostly a G_i-coupled pathway leading to VI inhibition. The VD and VI forms of inhibition were separated electrophysiologically (see protocols), allowing the quantification of chick GAIP effects on individual pathways.

When chick GAIP was tested on DRG neurons through intracellular application of the recombinant protein, it was found to antagonize the VI form of G_o-mediated inhibition (Fig. 4B). In contrast, the other two inhibitory pathways present in the same cells (G_o-mediated VD and G_i-mediated VI inhibition) were not affected (Fig. 4B) (Tosetti *et al.*, 2002). Similar results were obtained when comparing Ca^{2+} current inhibition in two distinct pools of neurons dialyzed for 15 min with control and GAIP-containing intracellular solution, respectively (Tosetti *et al.*, 2002). This eliminated the possibility that the reduced inhibition was due to nonspecific washout of intracellular components. This specificity for G_o was somewhat surprising in light of the significant GAP activity of chick GAIP demonstrated *in vitro* against all $G_i\alpha$ and $G_o\alpha$ subunits tested, and notably toward $G\alpha_{i1}$ (Fig. 3).

These results indicate that, in intact systems, chick GAIP is able to distinguish among pathways involving either the same G protein or different isoforms of the same G protein. Therefore, when tested in native chick cells, chick GAIP has a higher degree of specificity than *in vitro*. It also has a G-protein selectivity that differs from that reported for human GAIP in the same cells, a property that can likely be ascribed to the unique N-terminal sequence of chick GAIP (Diversé-Pierluissi *et al.*, 1999; Tosetti *et al.*, 2002). These results further suggest that the native cellular environment can greatly modify the intrinsic properties of RGS proteins measured *in vitro* and underscore the importance of testing selectivity of RGS proteins in their tissue of origin.

Protocols

Preparation and Electrophysiological Recordings of Chick DRG Neurons. Eleven- to 12-day-old chick embryos (Charles River SPAFAS, North Franklin, CT) are decapitated with forceps, and ~30 dorsal root ganglia are dissected out of each embryo [for a detailed description of the dissection procedure, see Anantharam and Diversé-Pierluissi (2002)]. DRGs are incubated for 30 min at 37° in Ca^{2+}-, Mg^{2+}-free saline containing 0.1% collagenase A (Boehringer Mannheim, Indianapolis, IN) to favor cell dissociation. After removing the collagenase solution gently, ganglia are first resuspended in 1 ml of culture medium (DMEM, supplemented with 10% heat-inactivated horse serum, 5% chicken embryo extract, 50 U/ml penicillin, 50 mg/ml streptomycin, 1 mM glutamine, and NGF) and are then dissociated mechanically by trituration through a small-bore, fire-polished Pasteur pipette (see also Richman and Diversé-Pierluissi, 2004). A small drop of the cell suspension (~100 μl) is placed in the center of a 35-mm tissue culture dish and incubated for 1 h at 37° to allow the cells to attach to the substrate. The plated dishes are then filled with culture medium (2 ml/dish).

Cultured DRG neurons are employed for electrophysiological recordings 4 to 48 h after plating. Cells are visualized on the stage of an inverted microscope, and Ca^{2+} currents are measured using standard tight seal, whole cell methods (Hamill *et al.*, 1981). Acutely dissociated DRG neurons express N-type, ω-conotoxin GVIA-sensitive Ca^{2+} channels almost exclusively (>95% of total Ca^{2+} current). Before recording, the cell culture medium is replaced with (in mM) 133 NaCl, 1 $CaCl_2$, 0.8 $MgCl_2$, 25 HEPES, pH 7.4, 12.5 NaOH, 5 glucose, 10 tetraethylammonium Cl, 6×10^{-4} tetrodotoxin, and 10^{-2} bicuculline. Recordings are performed with glass pipettes fabricated from microhematocrit tubing (Fisher), having a resistance between 0.8 and 1.5 MΩ when filled with recording solution. The internal recording solution contains (in mM) 150 CsCl, 5 BAPTA, 5 MgATP, and 10 HEPES, pH 7.4. All reagents are from Sigma Chemical Company (St. Louis, MO). Ca^{2+} currents are recorded using a List Biologic (Campbell, CA) EPC-7 patch-clamp amplifier, filtered at 3 kHz, digitized at 10 kHz using an ITC16 A/D interface (Instrutech Corp., Great Neck, NY), and stored on a Macintosh G3 PowerPC running Pulse software (HEKA Elektronik, Germany). Capacitive transients and series resistances are compensated with the EPC-7 circuitry. Leak currents are subtracted with a standard P/4 protocol. All recordings are performed at room temperature. Igor software (Wavemetrics, Lake Oswego, OR) is employed for data analysis.

In chick DRGs, neurotransmitters (NTs) inhibit N-type Ca^{2+} currents in two biophysically distinct modes: voltage dependent and voltage independent. Depolarizing prepulses at positive potentials reverse VD inhibition, leaving VI inhibition unaffected. The two biophysical components of NT-mediated inhibition of Ca^{2+} channels could therefore be separated using an appropriate voltage protocol. Ca^{2+} currents are evoked by a 20-ms test pulse to 0 mV in the presence or absence of NT to activate G-protein-mediated pathways. While VI inhibition is constant all along the test pulse, VD inhibition is maximal at the beginning of the pulse, then diminishes rapidly due to the depolarization at 0 mV. Both control and inhibited currents are measured at the time that control currents peak (T_p) and at the end of the pulse (T_e). Currents are normalized to the maximal control current and are expressed as a percentage. The difference between control and modulated current at T_p constitutes total inhibition. The difference between the current inhibition at T_p and T_e represents the VD portion of the inhibition; the remaining current is taken as an estimate of VI inhibition.

Ca^{2+} currents are corrected for rundown by measuring Ca^{2+} currents as a function of time in control cells in the absence of NT. Analysis is confined to data from cells that exhibit a rundown of less than 1%/min. All results are expressed as means ± standard errors of the mean. Statistical differences are analyzed using either paired or unpaired Student's t tests, as appropriate, and differences are considered significant when $p < 0.05$.

Solutions and Drug Delivery. Drug and recording solutions are freshly prepared immediately before the experiment. For all experiments, NTs are diluted into the extracellular solution at a working concentration of 100 μM, while GAIP is diluted into the patch pipette solution to a final concentration of 300 nM. A pressurized system of multiple converging quartz tubes (140 μm internal diameter) positioned in the vicinity of the cell of interest allows rapid delivery (<1 s) of the drugs.

Acknowledgment

This work was supported by the Human Frontier Science Program.

References

Anantharam, A., and Diversé-Pierluissi, M. A. (2002). Biochemical approaches to study interaction of calcium channels with RGS12 in primary neuronal cultures. *Methods Enzymol.* **345**, 60–70.

Berman, D. M., Wilkie, T. M., and Gilman, A. G. (1996). GAIP and RGS4 are GTPase-activating proteins for the Gi subfamily of G protein alpha subunits. *Cell* **86**, 445–452.

Chatterjee, T. K., Liu, Z., and Fisher, R. A. (2003). Human RGS6 gene structure, complex alternative splicing, and role of N terminus and G protein γ-subunit-like (GGL) domain in subcellular localization of RGS6 splice variants. *J. Biol. Chem.* **278**, 30261–30271.

Chen, C., Zheng, B., Han, J., and Lin, S. C. (1997). Characterization of a novel mammalian RGS protein that binds to Galpha proteins and inhibits pheromone signaling in yeast. *J. Biol. Chem.* **272,** 8679–8685.

Cowan, C. W., Fariss, R. N., Sokal, I., Palczewski, K., and Wensel, T. G. (1998). High expression levels in cones of RGS9, the predominant GTPase accelerating protein of rods. *Proc. Natl. Acad. Sci. USA* **95,** 5351–5356.

De Vries, L., Mousli, M., Wurmser, A., and Farquhar, M. G. (1995). GAIP, a protein that specifically interacts with the trimeric G protein G alpha i3, is a member of a protein family with a highly conserved core domain. *Proc. Natl. Acad. Sci. USA* **92,** 11916–11920.

Diversé-Pierluissi, M., Goldsmith, P. K., and Dunlap, K. (1995). Transmitter-mediated inhibition of N-type calcium channels in sensory neurons involves multiple GTP-binding proteins and subunits. *Neuron* **14,** 191–200.

Diversé-Pierluissi, M., Remmers, A. E., Neubig, R. R., and Dunlap, K. (1997). Novel form of crosstalk between G protein and tyrosine kinase pathways. *Proc. Natl. Acad. Sci. USA* **94,** 5417–5421.

Diversé-Pierluissi, M. A., Fischer, T., Jordan, J. D., Schiff, M., Ortiz, D. F., Farquhar, M. G., and De Vries, L. (1999). Regulators of G protein signaling proteins as determinants of the rate of desensitization of presynaptic calcium channels. *J. Biol. Chem.* **274,** 14490–14494.

Fischer, T., Elenko, E., McCaffery, J. M., DeVries, L., and Farquhar, M. G. (1999). Clathrin-coated vesicles bearing GAIP possess GTPase-activating protein activity *in vitro. Proc. Natl. Acad. Sci. USA* **96,** 6722–6727.

Hamill, O. P., Marty, A., Neher, E., Sakmann, B., and Sigworth, F. J. (1981). Improved patch-clamp techniques for high-resolution current recording from cells and cell-free membrane patches. *Pflug. Arch.* **391,** 85–100.

He, W., Cowan, C. W., and Wensel, T. G. (1998). RGS9, a GTPase accelerator for phototransduction. *Neuron* **20,** 95–102.

Lu, Q., and Dunlap, K. (1999). Cloning and functional expression of novel N-type Ca(2+) channel variants. *J. Biol. Chem.* **274,** 34566–34575.

Martemyanov, K., Hopp, J., and Arshavsky, V. (2003). Specificity of G protein-RGS protein recognition is regulated by affinity adapters. *Neuron* **38,** 857–862.

Muller, R. T., Honnert, U., Reinhard, J., and Bahler, M. (1997). The rat myosin myr 5 is a GTPase-activating protein for Rho *in vivo*: Essential role of arginine 1695. *Mol. Biol. Cell* **8,** 2039–2053.

Posner, B. A., Gilman, A. G., and Harris, B. A. (1999). Regulators of G protein signaling 6 and 7. Purification of complexes with Gbeta5 and assessment of their effects on g protein-mediated signaling pathways. *J. Biol. Chem.* **274,** 31087–31093.

Richman, R. W., and Diversé-Pierluissi, M. (2004). Mapping of RGS12-Cav2.2 channel interaction. *Methods Enzymol.* **390,** 224–229.

Ridley, A. J., Self, A. J., Kasmi, F., Paterson, H. F., Hall, A., Marshall, C. J., and Ellis, C. (1993). rho family GTPase activating proteins p190, bcr and rhoGAP show distinct specificities *in vitro* and *in vivo. EMBO J.* **12,** 5151–5160.

Ross, E. M., and Wilkie, T. M. (2000). GTPase-activating proteins for heterotrimeric G proteins: Regulators of G protein signaling (RGS) and RGS-like proteins. *Annu. Rev. Biochem.* **69,** 795–827.

Shuey, D. J., Betty, M., Jones, P. G., Khawaja, X. Z., and Cockett, M. I. (1998). RGS7 attenuates signal transduction through the G(alpha q) family of heterotrimeric G proteins in mammalian cells. *J. Neurochem.* **70,** 1964–1972.

Skiba, N. P., Hopp, J. A., and Arshavsky, V. Y. (2000). The effector enzyme regulates the duration of G protein signaling in vertebrate photoreceptors by increasing the affinity between transducin and RGS protein. *J. Biol. Chem.* **275,** 32716–32720.

Tosetti, P., Parente, V., Taglietti, V., Dunlap, K., and Toselli, M. (2003a). Chick RGS2L demonstrates concentration-dependent selectivity for pertussis toxin-sensitive and -insensitive pathways that inhibit L-type Ca^{2+} channels. *J. Physiol.* **549**, 157–169.

Tosetti, P., Pathak, N., Jacob, M. H., and Dunlap, K. (2003b). RGS3 mediates a calcium-dependent termination of G protein signaling in sensory neurons. *Proc. Natl. Acad. Sci. USA* **100**, 7337–7342.

Tosetti, P., Turner, T., Lu, Q., and Dunlap, K. (2002). Unique isoform of Galpha-interacting protein (RGS-GAIP) selectively discriminates between two Go-mediated pathways that inhibit Ca^{2+} channels. *J. Biol. Chem.* **277**, 46001–46009.

Turner, T. J., Pearce, L. B., and Goldin, S. M. (1989). A superfusion system designed to measure release of radiolabeled neurotransmitters on a subsecond time scale. *Anal. Biochem.* **178**, 8–16.

Witherow, D. S., Wang, Q., Levay, K., Cabrera, J. L., Chen, J., Willars, G. B., and Slepak, V. Z. (2000). Complexes of the G protein subunit Gbeta 5 with the regulators of G protein signaling RGS7 and RGS9: Characterization in native tissues and in transfected cells. *J. Biol. Chem.* **275**, 24872–24880.

Yan, Y., Chi, P. P., and Bourne, H. R. (1997). RGS4 inhibits Gq-mediated activation of mitogen-activated protein kinase and phosphoinositide synthesis. *J. Biol. Chem.* **272**, 11924–11927.

Zeng, W., Xu, X., Popov, S., Mukhopadhyay, S., Chidiac, P., Swistok, J., Danho, W., Yagaloff, K. A., Fisher, S. L., Ross, E. M., Muallem, S., and Wilkie, T. M. (1998). The N-terminal domain of RGS4 confers receptor-selective inhibition of G protein signaling. *J. Biol. Chem.* **273**, 34687–34690.

[2] Analyses of Gα-Interacting Protein and Activator of G-Protein-Signaling-3 Functions in Macroautophagy

By Sophie Pattingre, Anne Petiot, and Patrice Codogno

Abstract

Macroautophagy or autophagy is an ubiquitous and conserved degradative pathway of cytosolic components, macromolecules or organelles, into the lysosome. By using biochemical and microscopic methods, which allow one to measure the rate of autophagy, the role of two regulators of G_{i3} protein activity, activator of G-protein-signaling-3 (AGS3) and Gα-interacting protein (GAIP), was studied in the control of autophagy in human colon cancer HT-29 cells. In HT-29 cells, autophagy is under the control of the G_{i3} protein and, when bound to the GTP, the $G\alpha_{i3}$ protein inhibits autophagy, whereas it stimulates autophagy when bound to the GDP. GAIP, which enhances the intrinsic GTPase-activating protein activity of the $G\alpha_{i3}$ protein, stimulates autophagy by favoring the GDP-bound form of

$G\alpha_{i3}$. We showed that GAIP is phosphorylated on its serine 151 and that this phosphorylation is dependent on the presence of amino acids that modulate Raf-1 activity, the kinase upstream of Erk1/2. AGS3, a guanine nucleotide dissociation inhibitor, stimulates autophagy by binding $G\alpha_{i3}$ proteins. The intracellular localization of AGS3 (Golgi apparatus and endoplasmic reticulum, two membranes known to be at the origin of autophagosomes) is consistent with its role in autophagy.

Introduction

Macroautophagy or autophagy is a general and evolutionary conserved catabolic pathway terminating in the lysosomal compartment (Klionsky and Emr, 2000; Klionsky and Ohsumi, 1999; Seglen and Bohley, 1992) and is implicated in development (Melendez et al., 2003), tumorigenesis (Liang et al., 1999), and myopathies (for review, see Nishino, 2003). From a cell biology standpoint, autophagy is characterized by formation of a multimembrane-bound autophagosome that engulfs portions of the cytoplasm (see Fig. 1). The delimiting membrane of the autophagosome is derived from an isolation membrane or phagophore of unknown origin (Fengsrud et al., 1995). In mammalian cells, most of autophagosomes can fuse directly with lysosome or merge with endocytic compartments (Liou et al., 1997; Stromhaug and Seglen, 1993) to form an amphisome where the sequestered material is denatured due to the acidic environment. The final step is the fusion of amphisomes with the lysosomal compartment where the sequestered material is degraded.

The discovery of "ATG" (AuTophaGy) genes (Klionsky et al., 2003) that regulate autophagy in the yeast Saccharomyces cerevisiae, which are almost completely conserved in mammalian cells, has contributed greatly to our understanding of the molecular control of autophagy and its role in multicellular organisms (Klionsky and Ohsumi, 1999; Ohsumi, 2001; Stromhaug and Klionsky, 2001; Thumm, 2000).

Several signaling pathways are able to transduce extracellular or intracellular stimuli to the molecular machinery involved in the formation of the autophagosome (reviewed in Codogno and Meijer, 2003). Autophagy, as with other transport mechanisms in the vacuolar system, is characterized by the involvement of GTPases (Nuoffer and Balch, 1994). The first studies implicating G proteins in the control of autophagy showed that the treatment of hepatocytes with GTPγS, a nonhydrolyzable analog of GTP, inhibits autophagy (Kadowaki et al., 1994). Furthermore, we have shown that the trimeric G_{i3} protein plays a role in the control of autophagy in the human colon cancer HT-29 cell line (Ogier-Denis et al., 1995). The rate of autophagy is low when the $G\alpha_{i3}$ protein is in the GTP-bound form, but it is

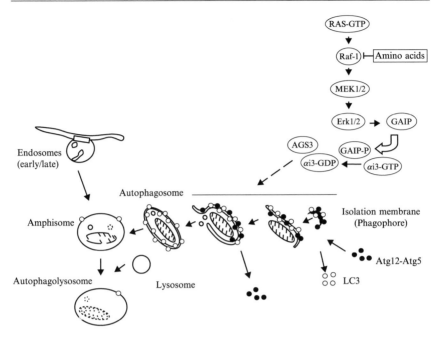

FIG. 1. The delimiting membrane of the autophagosome is derived from an isolation membrane or phagophore of unknown origin (compartments of the vacuolar system, such as the endoplasmic reticulum and the Golgi apparatus, probably contribute to the genesis of the isolation membrane). Atg5–Atg12 and LC3 complexes are recruited to the isolation membrane. Atg5–Atg12 is recycled during formation of the autophagosome, whereas LC3 remains associated to the membrane. During autophagy, the LC3-I form is processed into the LC3-II membrane-bound form (see text for details). The autophagosome is then matured by fusion with endocytic vesicles to form an amphisome, and ultimately with lysosomes to form an autophagolysosome. The quantity of LC3-II associated with autophagic vacuoles decreases during their maturation because of its degradation in the lysosomal compartment. The $G_{\alpha i3}$ protein, GAIP, and AGS3 are tentatively placed upstream of the formation of the autophagosome. Details concerning the MAP-ERK1/2 pathway and the inhibitory effect of amino acids are given in the text.

stimulated when GDP is bound to the $G_{\alpha i3}$ protein (Ogier-Denis *et al.*, 1996). The endoplasmic reticulum (ER)/Golgi localization of the $G_{\alpha i3}$ protein is mandatory to control autophagy, whereas its targeting to the plasma membrane suppresses its role in autophagy (Petiot *et al.*, 1999).

This article presents a technical review of the most used methods to study autophagy in mammalian cells before focusing on the role of G_{α}-interacting protein (RGS-GAIP) and activator of G-protein-signaling-3 (AGS3), two proteins that control the nucleotide cycling and hydrolysis on the $G_{\alpha i3}$ protein, in the regulation of autophagy.

Assays for Autophagy in Cultured Mammalian Cells

Autophagy can be analyzed by a combination of biochemical and microscopic methods. Electron microscopy is the gold standard method used to investigate autophagy by providing ultrastructural images, but also by allowing morphometric quantification of the fractional volume occupied in the cytoplasm by the different categories of autophagic compartments (Pattingre *et al.*, 2003b; Petiot *et al.*, 1999). Readers should consult the following publications for more details on the ultrastructural aspect of autophagy (Dunn, 1990a,b; Fengsrud *et al.*, 1995; Liou *et al.*, 1997; Punnonen *et al.*, 1993; Rabouille *et al.*, 1993).

Analysis of Protein Degradation

The following protocol has been validated in human colon cancer HT-29 cells (Houri *et al.*, 1995; see Fig. 2) and can be optimized depending on the cell system used.

Intracellular proteins are labeled for 18 h at 37° with 0.2 mCi/ml of L-[^{14}C]valine (288.5 mCi/mmol, NEN Dupont de Nemours) in complete medium [Dulbecco's modified Eagle's medium (DMEM, GIBCO), 10% fetal bovine serum (GIBCO), 1% penicillin–streptomycin]. Unincorporated radioactivity is removed by three rinses with phosphate-buffered saline

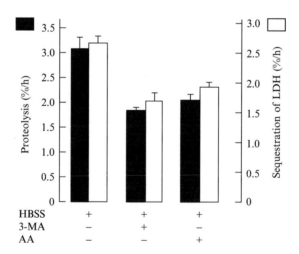

FIG. 2. Measurement of autophagy in HT-29 cells. Analysis of [^{14}C]valine-labeled protein degradation and LDH sequestration. Cells were incubated for 5 h in amino acid and serum-free medium (HBSS). When required, 10 m*M* 3-MA or 4× AA was added at the beginning of the chase period. For the LDH sequestration assay, leupeptin (100 μg/ml) was added 16 h prior to the experiment. Values reported are the mean ± SD of four independent experiments.

(PBS; 137 mM NaCl, 2.7 mM KCl, 4.3 mM Na$_2$HPO$_4$/7H$_2$O, 1.4 mM KH$_2$PO$_4$, pH 7.3). Cells are then incubated in Hanks' balanced salt solution (HBSS, GIBCO), plus 0.1% of bovine serum albumin (BSA), and 10 mM valine (Sigma) to stimulate autophagy. Valine is used because this amino acid does not interfere with autophagy in most cell types. The choice of the amino acid used during the pulse–chase is important because some of them, such as leucine, are potent inhibitors of autophagy (van Sluijters et al., 2000). When required, 10 mM 3-MA (3-methyladenine; Fluka) is added throughout the chase period to inhibit the de novo formation of autophagic vacuoles (Seglen and Gordon, 1982). After the first hour of incubation, at which time short-lived proteins are being degraded, the medium is replaced with the appropriate fresh medium, and incubation is continued for an additional 4-h period. Thereafter, the medium is precipitated with trichloro-acetic acid (TCA) added to a final 10% concentration. After centrifugation for 10 min at 2000 rpm at 4°, the acid-soluble radioactivity is measured by liquid scintillation counting. Cells are washed twice with cold 10% TCA and dissolved at 37° in 0.2 N NaOH. Radioactivity is then measured by liquid scintillation counting. The rate of long-lived protein degradation is calculated from the ratio of the acid-soluble radioactivity in the medium to the cell acid precipitable fraction. As mentioned previously, amino acids are physiological inhibitors of autophagy (van Sluijters et al., 2000). A cocktail of amino acids [the final concentration of amino acids (in μM) is asparagine 60, isoleucine 100, leucine 250, lysine 300, methionine 40, phenylalnine 50, proline 100, threonine 180, tryptophane 70, valine 180, alanine 400, aspartate 30, glutamate 100, glutamine 350, glycine 300, cysteine 60, histidine 60, serine 200, tyrosine 75, ornithine 100] representing four times their concentration (4x) in the portal vein of starved rats is also very potent to inhibit autophagy (Fig. 2).

Regarding the specificity of 3-MA, while investigating a 3-MA-sensitive process, it should be kept in mind that this drug interferes with PI3K-dependent pathways (Blommaart et al., 1997; Petiot et al., 2000) and intracellular trafficking dependent on the activity of PI3K such as endosome-to-lysosome transport (Punnonen et al., 1994). This inhibitory effect is probably responsible for the increase in intralysosomal pH observed in 3-MA-treated cells (Caro et al., 1988).

Analysis of Lactate Dehydrogenase (LDH) Sequestration

This method, described originally in rat liver, is aimed to assay the sequestration of LDH, a cytosolic enzyme, into autophagic vacuoles (Kopitz et al., 1990). This method can be adapted to cultured cells. The following method has been used in cultured HT-29 cells (Houri et al., 1995; Fig. 2).

Cells (1×10^7) are washed gently three times with PBS + 10% sucrose (buffer A) on ice and are then centrifuged for 10 min at 650g at 4°. The pellet is resuspended in 500 μl buffer A + 100 μg/ml BSA and homogenized by 13 strokes in a glass homogenizer with a Teflon pestle (Thomas Scentific) on ice. Five hundred microliters of 100 mM phosphate buffer (KH$_2$PO$_4$/K$_2$HPO$_4$), pH 7.5, 2 mM EDTA, 2 mM dithiothreitol (DTT), 0.01% Tween 20, 100 μg/ml BSA, and 5% sucrose are added to the homogenate before centrifugation at 1000g for 10 min at 4° to eliminate cell debris and nuclei. An aliquot of 100 μl of the supernatant is taken to determine the total LDH activity (see later). Postnuclear material is layered on 3 ml 8% metrizamide (Sigma) dissolved in 50 mM phosphate buffer (KH$_2$PO$_4$/K$_2$HPO$_4$), pH 7.5, 1 mM EDTA, 1 mM DTT, 0.01% Tween 20, and 100 μg/ml BSA and is centrifuged at 7000g for 60 min at 4°. Finally, the pellet is resuspended in 300 μl 50 mM phosphate buffer (KH$_2$PO$_4$/K$_2$HPO$_4$), pH 7.5, 1 mM EDTA, 1 mM DTT, 0.5% Triton X-100, and 100 μg/ml BSA. LDH activity (kit from Sigma) is assayed by measuring the oxidation of NADH with pyruvate as a substrate at 340 nm, according to the manufacturer's instruction, in the pellet containing sedimentable cell components (material enriched in autophagic vacuoles) and in the supernatant (cytosol). Usually, experiments are conducted on cells pretreated overnight with 100 μg/ml leupeptin (Sigma) to limit degradation of the trapped LDH. It is recommended to perform control experiments on 3-MA-treated cells where autophagy is inhibited.

Labeling of Autophagic Vacuoles with Monodansylcadaverine (MDC)

Monodansylcadaverine, which is an autofluorescent compound due to the dansyl residue conjugated to cadaverine, have been shown to accumulate in acidic autophagic vacuoles (Biederbick *et al.*, 1995). The concentration of MDC in autophagic vacuole is the consequence of an ion-trapping mechanism and an interaction with lipids in autophagic vacuoles (autophagic vacuoles are rich in membrane lipids) (Niemann *et al.*, 2000). MDC-labeled vacuoles can be isolated by subcellular fractionation for biochemical analysis (Biederbick *et al.*, 1995). The use of MDC staining is a rapid and convenient approach to assay autophagy, as shown in cultured cells (Munafò and Colombo, 2001; see Fig. 3). Monodansylpentane (MDH), where a neutral group replaces a primary amino group, was shown to accumulate into autophagic vacuoles, including nascent autophagosomes, independently of an ion-trapping mechanism, and this compound still shows a preference for autophagic vacuole membrane lipids (Niemann *et al.*, 2001).

For MDC assay, cells are first washed with HBSS and incubated for 3 h at 37° in HBSS to stimulate autophagy. For staining of autophagic vacuoles, an

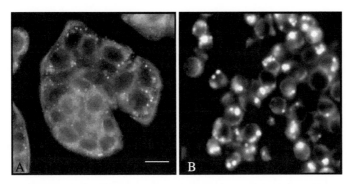

Fig. 3. MDC labeling. HT-29 cells were cultivated for 2 h at 37° in complete medium (DMEM, 1% FBS, 1% penicillin–streptomycin) in the absence (A) or presence (B) of the cyclooxygenase-2 inhibitor celecoxib (100 μM), which induces cell death in HT-29 cells (Arico et al., 2002) and the accumulation of autophagic vacuoles. Bar represents 10 μm.

MDC (Sigma) stock solution [0.1 M in dimethyl sulfoxide (DMSO)] is diluted 1:1000 in HBSS and is applied to the cells for 30 min at 37°. After washes with PBS, cells can be fixed with 3% paraformaldehyde diluted in PBS for 30 min and washed with PBS. Coverslips are mounted with glycergel (Dako) and examined by fluorescence microscopy (Axioplan, Zeiss, excitation filter: 340–380 nm, barrier filter: 430 nm). Because fixation weakens the staining signal, it can be omitted if not necessary in the course of the experiment.

Note. MDC staining is a convenient method to assay autophagy quickly; however, results should be confirmed by electron microscopy (Bursch et al., 2000) and/or LC3 labeling or processing (see later) because of the possible labeling of nonautophagic compartments.

Intracellular Localization of MAP-LC3 Protein

Mammalian MAP-LC3 is the homolog of yeast Atg8. This protein remains associated with the outer and inner membranes of autophagic vacuoles once their formation is completed (Kabeya et al., 2000) (Fig. 1). During autophagy, the MAP-LC3 protein is processed from an LC3-I (18 kDa) cytosolic form to a membrane-associated LC3-II (16 kDa) (Kabeya et al., 2000; Fig. 4). Based on these properties, LC3 can be used to monitor autophagy by microscopic methods (antibodies, GFP constructs) and biochemically by Western blotting, which allows the detection of LC3-II. Experiments can be done on cell homogenates and do not need fractionation prior to Western blotting experiments.

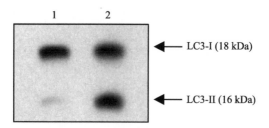

Fɪɢ. 4. Immunoblot analysis of MAP-LC3. Fifty micrograms of HT-29 cell lysates were separated on a 15% SDS–PAGE, transferred onto nitrocellulose, and incubated with a rabbit antisera specific for MAP-LC3. Lane 1, HT-29 cultured in complete medium. Lane 2, HT-29 cells cultured for 1 h in amino acid and serum-free medium in the presence of 50 μM vinblastine to inhibit the microtubule-dependent fusion of autophagic vacuoles with the lysosomal compartment. Molecular masses are indicated parentheses.

For Western blot experiments, cells are lysed in standard lysis buffer [50 mM Tris–HCl, pH 7.4, 150 mM NaCl, 1 mM EDTA, 100 μg/ml phenylmethylsulfonyl fluoride (PMSF), 1% Triton X-100] for 1 h at 4°. Fifty micrograms of proteins are submitted to SDS–PAGE and transferred onto nitrocellulose membrane. After blocking for 1 h in PBS, 5% milk, and 0.1% Tween 20, the membrane is incubated with a rabbit antibody directed against LC3 (kindly provided by T. Yoshimori), washed three times in PBS and 0.1% Tween 20, and finally incubated with the secondary antirabbit antibody conjugated to horseradish peroxydase (HRP). Bound antibodies are detected by enhanced chemiluminescence (ECL, Amersham Pharmacia Biotech).

It must be kept in mind that the vast majority of LC3 associated with autophagic vacuoles is degraded in late autophagic vacuoles/autolysosomes. In order to increase the signal, it is recommended to interrupt the autophagic proteolysis and/or the progression of the autophagic pathway by treating cells with leupeptin or vinblastine. Alternatively, analysis can be performed rapidly after the induction of autophagy (30 min).

The creation of a chimeric LC3 protein fused with green fluorescent protein (GFP) has been used successfully to detect autophagosomes in cell culture and *in vivo* (Mizushima *et al.*, 2001, 2004).

Role of GAIP in Autophagy and Regulation of GAIP Activity

RGS proteins are a large family of proteins identified by reference to a conserved domain of about 115 amino acids known as the RGS domain (Hollinger and Hepler, 2002). RGS are GTPase-activating proteins (GAP) for Gα. RGS-GAIP interacts with Gα$_{i3}$ protein (De Vries *et al.*, 1995) and accelerates its intrinsic GTPase activity (Berman *et al.*, 1996).

GAIP Expression in Intestinal Cells and Its Role during Autophagy

Northern blot and Western blot analyses have shown that the expression of GAIP decreased during the enterocyte-like differentiation of cultured human epithelial intestinal cells (Ogier-Denis *et al.*, 1997). Although the mechanism responsible for the differentiation-dependent expression of GAIP remains to be investigated, it is of note to observe that autophagy is also downregulated during the differentiation of cultured intestinal cells (HT-29, Caco2) (Houri *et al.*, 1995). In order to directly investigate the role of GAIP in HT-29 cells, we modulated its expression by transfecting its cDNA in the sense or antisense orientation (Ogier-Denis *et al.*, 1997). HT-29 cells were transfected with 5 μg of cDNA encoding GAIP using the Superfect transfection reagent (Qiagen). Under these conditions, we observed a stimulation of autophagic rates (measurement of long-lived degradation of protein and sequestration), whereas autophagy was reduced greatly in HT-29 cells transfected with the antisense GAIP CDNA (Table I).

In conclusion, by stimulating the GTPase activity of the $G\alpha_{i3}$ protein, GAIP favors the GDP-bound form of $G\alpha_{i3}$ protein that was shown to stimulate autophagy (Ogier-Denis *et al.*, 1996).

TABLE I
EFFECT OF GAIP AND AGS3 ON AUTOPHAGY IN HT-29 CELLS

Cell population	Autophagy	
	Rate[a]	Method used[b]
WT-GAIP[c]	↗	EM, P, S
GAIP antisense[c]	↘	P, S
S151A-GAIP[d]	→	P, S
AGS3[e]	↗	EM, P, F
AGS3-7TPR[e]	↘	P, S
AGS3-4GoLoco[e]	↘	P, S
AGS3-1GoLoco[e]	↘	P, S
AGS3-1GoLoco R624F[e]	→	P, S

[a] Rates of autophagy were compared to HT-29 cells transfected with an empty vector.
[b] EM, morphometric quantification by electron microscopy; F, fluorescence (MDC staining); P, proteolysis; S, sequestration of LDH.
[c] From Ogier-Denis *et al.* (1997).
[d] From Ogier-Denis *et al.* (2000).
[e] From Pattingre *et al.* (2003b).

Regulation of GAIP Activity by Phosphorylation

Several studies have also shown that the activity of RGS proteins can be modified by posttranslational modifications (phosphorylation and/or acylation) (for a review, see Hollinger and Hepler, 2002). HT-29 cells were radiolabeled with 0.5 mCi of ^{32}P-labeled orthophosphoric acid for 3 h at 37°. After lysis, following immunoprecipitation with polyclonal anti-GAIP (Ogier-Denis *et al.*, 1997) and SDS–PAGE of the product of immunoprecipitation and autoradiography, we showed that GAIP is a phosphoprotein. The subsequent radiolabeled GAIP was then submitted to acid hydrolysis and thin layer electrophoresis to show that only serine residues are phosphorylated (for details, see Ogier-Denis, 1997), and phosphoamino acid detection has shown that GAIP is phosphorylated on serine residues (Ogier-Denis *et al.*, 2000).

Interestingly, the phosphorylation of GAIP was increased in conditions known to stimulate autophagy (amino acid deprivation). In order to identify the kinases involved in GAIP phosphorylation, we analyzed the effect of both protein kinases inhibitors and *in vitro* phosphorylation of recombinant GAIP (produced as a His–tag fusion protein) by commercially available protein kinases. When it was required, the GAP activity of recombinant GAIP was analyzed toward a $G\alpha_{i3}$ recombinant protein (His-tagged form) under conditions where the rate of GDP dissociation is not rate limiting (Berman *et al.*, 1996). From the results obtained, we concluded that ERK1/2 is responsible for the phosphorylation of GAIP and for the stimulation of its GAP activity toward the $G\alpha_{i3}$ protein (Ogier-Denis *et al.*, 2000). Sequence analysis and site-directed mutagenesis revealed that the serine at position 151 in the RGS domain is the site of ERK1/2 phosphorylation (Ogier-Denis *et al.*, 2000). Transfection of cDNA encoding a S151A GAIP mutant in HT-29 cells and analysis of the GAP activity of this His–tag recombinant GAIP mutant confirmed that the replacement of serine 151 by an alanine residue abolished both ERK phosphorylation and stimulation of GAP activity of GAIP toward the $G\alpha_{i3}$ protein and consequently the stimulation of autophagy induced by the wild-type GAIP (see Table I). The ERK1/2-dependent phosphorylation of GAIP was confirmed by transfecting HT-29 cells with the Ras(G12V, T25S) mutant, which specifically activates the Raf-1/MEK/ERK1/2 pathway (Rodriguez-Viciana *et al.*, 1997). In addition, this mutant has a stimulatory effect on autophagy (Pattingre *et al.*, 2003a). However, amino acids still have an inhibitory effect on both GAIP phosphorylation and autophagy in Ras(G12V, T35S)-expressing cells. Further studies have shown that activation of the kinase Raf-1 is sensitive to amino acids (Fig. 1). The mechanism by which amino acids modulate Raf-1 activation is under investigation.

Role of AGS3 in the Regulation of Autophagy

The AGS3 acts as a guanine nucleotide dissociation inhibitor (GDI) toward G_{i3} proteins and stabilizes the GDP-bound form of the $G\alpha_{i3}$ protein independently of the $\beta\gamma$ dimer (Bernard *et al.*, 2001; De Vries *et al.*, 2000; Takesono *et al.*, 1999).

AGS3 is a bimodular protein containing 650 amino acids. Its N-terminal part contains seven tetratricopeptide (TPR) repeats responsible for its interaction with membranes (Pizzinat *et al.*, 2001), whereas its C-terminal part contains four G-protein regulatory (GPR or GoLoco) motifs that interact with $G\alpha_i$ proteins.

Northern blot and reverse transcriptase polymerase chain reaction experiments have shown that the mRNA encoding for the full-length form of AGS3 is expressed in different human intestinal cell lines (including HT-29 cells) (Pattingre *et al.*, 2003b). The expression vectors encoding AGS3 and its domains were kindly provided by Luc de Vries (De Vries *et al.*, 2000). In these experiments, we detected a 2-kb mRNA compatible with the expression of a truncated form of AGS3 as shown in cardiac tissues (Pizzinat *et al.*, 2001). After overexpression of AGS3 in HT-29 cells by transfection of 5 μg cDNA with the Superfect transfection reagent (Qiagen), we showed that full-length AGS3 stimulates autophagy (Pattingre *et al.*, 2003b).

The expression of cDNAs encoding either the N-terminal part of AGS3 containing the 7TPR domain (1–348) or variable numbers of C-terminals GoLoco motifs (4GoLoco, 424–650, or 1GoLoco, 591–650) has an inhibitory effect on autophagy (Pattingre *et al.*, 2003b). The inhibitory effect of GoLoco motifs is probably due to their binding to the GDP-bound form of the $G\alpha_{i3}$ protein because a mutant (AGS3-1GoLoco R624F) that does not bind the $G\alpha_{i3}$ protein (De Vries *et al.*, 2000) did not interfere with autophagy (Table I). How AGS3 stimulates autophagy needs to be investigated carefully. A simple explanation is that AGS3 would stabilize the GDP-bound form of AGS3. However, the N-terminal TPR domain is necessary for controlling autophagy, suggesting that this domain can recognize partners needed for autophagy (Pattingre *et al.*, 2003b). Whether these partners are cytosolic or membrane bound is not known. Interestingly, a large number of proteins involved in the control of autophagy only interact transiently with membranes (Ohsumi, 2001).

Conclusion

A major point that remains to be solved is how the $G\alpha_{i3}$ protein and its associated proteins (GAIP and AGS3) control autophagy. As the $G\alpha_{i3}$ protein and its partners (GAIP and AGS3) act prior to the formation of the autophagosome, we can hypothesize that they are involved in the control of membrane flux to the autophagic pathway. The delimiting membrane of the autophagosome has probably different origins (including lipids emanating from ER and Golgi/Trans Golgi Network membranes) (for a discussion, see Pattingre *et al.*, 2003b). A $G\alpha_{i3}$ protein-based mechanism could control membrane fluxes emanating from different vacuolar compartments to form a preautophogosomal structure. It is worth noting that when the $G\alpha_{i3}$-dependent transport of proteoglycans along the exocytic pathway in epithelial cells is stimulated (Stow *et al.*, 1991; Wylie *et al.*, 1999), the macroautophagic pathway is impaired severely (Ogier-Denis *et al.*, 1996, 1997), and *vice versa*. This suggests that the $G\alpha_{i3}$ protein and its cohort proteins can control the balance between the flow of membrane into the exocytic pathway and the macroautophagic pathway. Another interesting point to consider is that amino acids, physiological inhibitors of autophagy (van Sluijters *et al.*, 2000), are able to modulate the GAP ativity of GAIP and, consequently, the nucleotide state of the $G\alpha_{i3}$ protein by acting on the Raf-1 kinase (Fig. 1).

Acknowledgments

We are grateful to Luc de Vries (Pierre Fabre Research Institute, Castres, France) for providing us with the cDNAs encoding AGS3, GoLoco domains, 7TPR motifs, and the polyclonal anti-AGS3 used in this study and to Tamotsu Yoshimori (National Institute of Genetics, Shizuoka, Japan) for providing us with the anti LC3 antibody. Thanks are also due to Zsolt Tallóczy (Department of Medicine, Columbia University, New York, NY), for helpful comments on the manuscript. Work in P. Codogno's laboratory was supported by institutional funding from The Institut National de la Santé et de la Recherche Médicale (INSERM) and grants from the Association pour la Recherche sur le Cancer. SP and AP studies in PC's lab were supported by ARC fellowships.

References

Arico, S., Pattingre, S., Bauvy, C., Gane, P., Barbat, A., Codogno, P., and Ogier-Denis, E. (2002). Celecoxib induces apoptosis by inhibiting 3-phosphoinositide-dependent protein kinase-1 activity in the human colon cancer HT-29 cell line. *J. Biol. Chem.* **277,** 27613–27621.

Berman, D. M., Wilkie, T. M., and Gilman, A. G. (1996). GAIP and RGS4 are GTPase-activating proteins for the G(i) subfamily of G protein alpha subunits. *Cell* **86,** 445–452.

Bernard, M. L., Peterson, Y. K., Chung, P., Jourdan, J., and Lanier, S. M. (2001). Selective interaction of AGS3 with G-proteins and the influence of AGS3 on the activation state of G-proteins. *J. Biol. Chem.* **276,** 1585–1593.

Biederbick, A., Kern, H. F., and Elsasser, H. P. (1995). Monodansylcadaverine (MDC) is a specific *in vivo* marker for autophagic vacuoles. *Eur. J. Cell Biol.* **66,** 3–14.

Blommaart, E. F. C., Krause, U., Schellens, J. P. M., Vreeling-Sindelárová, H., and Meijer, A. J. (1997). The phosphatidylinositol 3-kinase inhibitors wortmannin and LY294002 inhibit autophagy in isolated rat hepatocytes. *Eur. J. Biochem.* **243,** 240–246.

Bursch, W., Hochegger, K., Torok, L., Marian, B., Ellinger, A., and Hermann, R. S. (2000). Autophagic and apoptotic types of programmed cell death exhibit different fates of cytoskeletal filaments. *J. Cell Sci.* **113,** 1189–1198.

Caro, L. H., Plomp, P. J., Wolvetang, E. J., Kerhof, C., and Meijer, A. J. (1988). 3-Methyladenine, an inhibitor of autophagy, has multiple effects on metabolism. *Eur. J. Biochem.* **175**, 325–329.

Codogno, P., and Meijer, A. J. (2003). Signaling pathways in mammalian autophagy. *In* "Autophagy" (D. J., Klionsky, ed.), pp. 26–48. Landes Bioscience, Georgetown, TX.

De Vries, L., Fischer, T., Tronchere, H., Brothers, G. M., Strockbine, B. *et al.* (2000). Activator of G protein signaling 3 is a guanine dissociation inhibitor for Galpha i subunits. *Proc. Natl. Acad. Sci. USA* **97**, 14364–14369.

De Vries, L., Mousli, M., Wurmser, A., and Farquhar, M. G. (1995). GAIP, a protein that specifically interacts with the trimeric G protein G alpha(i3), is a member of a protein family with a highly conserved core domain. *Proc. Natl. Acad. Sci. USA* **92**, 11916–11920.

Dunn, W. A., Jr. (1990a). Studies on the mechanism of autophagy: Formation of the autophagic vacuole. *J. Cell. Biol.* **110**, 1923–1933.

Dunn, W. A., Jr. (1990b). Studies on the mechanism of autophagy: Maturation of the autophagic vacuole. *J. Cell. Biol.* **110**, 1935–1945.

Fengsrud, M., Roos, N., Berg, T., Liou, W. L., Slot, J. W., and Seglen, P. O. (1995). Ultrastructural and immunocytochemical characterization of autophagic vacuoles in isolated hepatocytes: Effects of vinblastine and asparagine on vacuole distributions. *Exp. Cell. Res.* **221**, 504–519.

Hollinger, S., and Hepler, J. R. (2002). Cellular regulation of RGS proteins: Modulators and integrators of G protein signaling. *Pharmacol. Rev.* **54**, 527–559.

Houri, J. J., Ogier-Denis, E., De Stefanis, D., Bauvy, C., Baccino, F. M. *et al.* (1995). Differentiation-dependent autophagy controls the fate of newly synthesized N-linked glycoproteins in the colon adenocarcinoma HT-29 cell line. *Biochem. J.* **309**, 521–527.

Kabeya, Y., Mizushima, N., Ueno, T., Yamamoto, A., Kirisako, T. *et al.* (2000). LC3, a mammalian homologue of yeast Apg8p, is localized in autophagosome membranes after processing. *EMBO J.* **19**, 5720–5728.

Kadowaki, M., Venerando, R., Miotto, G., and Mortimore, G. E. (1994). De novo autophagic vacuole formation in hepatocytes permeabilized by *Staphylococcus aureus* alpha-toxin. Inhibition by nonhydrolyzable GTP analogs. *J. Biol. Chem.* **269**, 3703–3710.

Klionsky, D. J., Cregg, J. M., Dunn, W. A., Emr, S. D., Sakai, Y., Sandoval, I. V., Sibirny, A., Subramani, S., Thumm, M., Veenhuis, M., and Ohsumi, Y. (2003). A unified nomenclature for yeast autophagy-related genes. *Dev. Cell.* **5**, 539–545.

Klionsky, D. J., and Emr, S. D. (2000). Autophagy as a regulated pathway of cellular degradation. *Science* **290**, 1717–1721.

Klionsky, D. J., and Ohsumi, Y. (1999). Vacuolar import of proteins and organelles from the cytoplasm. *Annu. Rev. Cell. Dev. Biol.* **15**, 1–32.

Kopitz, J., Kisen, G. O., Gordon, P. B., Bohley, P., and Seglen, P. O. (1990). Nonselective autophagy of cytosolic enzymes by isolated rat hepatocytes. *J. Cell Biol.* **111**, 941–953.

Liang, X. H., Jackson, S., Seaman, M., Brown, K., Kempkes, B. *et al.* (1999). Induction of autophagy and inhibition of tumorigenesis by beclin 1. *Nature* **402**, 672–676.

Liou, W., Geuze, H. J., Geelen, M. J. H., and Slot, J. W. (1997). The autophagic and endocytic pathways converge at the nascent autophagic vacuoles. *J. Cell Biol.* **136**, 61–70.

Melendez, A., Talloczy, Z., Seaman, M., Eskelinen, E. L., Hall, D. H., and Levine, B. (2003). Autophagy genes are essential for dauer development and life-span extension in *C. elegans*. *Science* **301**, 1387–1391.

Mizushima, N. Y., Yamamoto, A., Hatano, M., Kobayashi, Y., Kabeya, Y. *et al.* (2001). Dissection of autophagosome formation using Apg5-deficient mouse embryonic stem cells. *J. Cell Biol.* **152**, 657–667.

Mizushima, N., Yamamoto, A., Matsui, M., Yoshimori, T., and Ohsumi, Y. (2004). *In vivo* analysis of autophagy in response to nutrient starvation using transgenic mice expressing a fluorescent autophagosome marker. *Mol. Biol. Cell.* **15,** 1101–1111.

Munafò, D. B., and Colombo, M. I. (2001). A novel assay to study autophagy: Regulation of autophagosome vacuole size by amino acid deprivation. *J. Cell Sci.* **114,** 3619–3629.

Nishino, I. (2003). Autophagic vacuolar myopathies. *Curr. Neurol. Neurosci. Rep.* **1,** 64–69.

Niemann, A., Baltes, J., and Elsässer, H. P. (2001). Fluorescence properties and staining behavior of monodansylpentane, a structural homologue of the lysosomotropic agent monodansylcadaverine. *J. Histochem. Cytochem.* **49,** 177–185.

Niemann, A., Takatsuki, A., and Elsässer, H. P. (2000). The lysosomotropic agent monodansylcadaverine also acts as a solvent polarity probe. *J. Histochem. Cytochem.* **48,** 251–258.

Nuoffer, C., and Balch, W. E. (1994). GTPases: Multifunctional molecular switches regulating vesicular traffic. *Annu. Rev. Biochem.* **63,** 949–990.

Ogier-Denis, E., Couvineau, A., Maoret, J. J., Houri, J. J., Bauvy, C. *et al.* (1995). A heterotrimeric Gi3-protein controls autophagic sequestration in the human colon cancer cell line HT-29. *J. Biol. Chem.* **270,** 13–16.

Ogier-Denis, E., Houri, J. J., Bauvy, C., and Codogno, P. (1996). Guanine nucleotide exchange on heterotrimeric Gi3 protein controls autophagic sequestration in HT-29 cells. *J. Biol. Chem.* **271,** 28593–28600.

Ogier-Denis, E., Pattingre, S., El Benna, J., and Codogno, P. (2000). Erk1/2-dependent phosphorylation of Galpha-interacting protein stimulates its GTPase accelerating activity and autophagy in human colon cancer cells. *J. Biol. Chem.* **275,** 39090–39095.

Ogier-Denis, E., Petiot, A., Bauvy, C., and Codogno, P. (1997). Control of the expression and activity of the Galpha-interacting protein (GAIP) in human intestinal cells. *J. Biol. Chem.* **272,** 24599–24603.

Ohsumi, Y. (2001). Molecular dissection of autophagy: Two ubiquitin-like systems. *Nature Rev. Mol. Cell. Biol.* **2,** 211–216.

Pattingre, S., Bauvy, C., and Codogno, P. (2003a). Amino acids interfere with the ERK1/2-dependent control of macroautophagy by controlling the activation of Raf-1 in human colon cancer HT-29 cells. *J. Biol. Chem.* **278,** 16667–16674.

Pattingre, S., De Vries, L., Bauvy, C., Chantret, I., Cluzeaud, F. *et al.* (2003b). The G-protein regulator AGS3 controls an early event during macroautophagy in human intestinal HT-29 cells. *J. Biol. Chem.* **278,** 20995–21002.

Peterson, Y. K., Bernad, M. L., Ma, H. Z., Hazard, S., Graber, S. G., and Lanier, S. M. (2000). Stabilization of the GDP-bound conformation of Gi alpha regulatory motif of AGS3. *J. Biol. Chem.* **275,** 33193–33196.

Petiot, A., Ogier-Denis, E., Bauvy, C., Cluzeaud, F., Vandewalle, A., and Codogno, P. (1999). Subcellular localization of the G(Alpha i3) protein and G alpha interacting protein, two proteins involved in the control of macroautophagy in human colon cancer HT-29 cells. *Biochem. J.* **337,** 289–295.

Petiot, A., Ogier-Denis, E., Blommaart, E. F., Meijer, A. J., and Codogno, P. (2000). Distinct classes of phosphatidylinositol 3′-kinases are involved in signaling pathways that control macroautophagy in HT-29 cells. *J. Biol. Chem.* **275,** 992–998.

Pizzinat, N., Takesono, A., and Lanier, S. M. (2001). Identification of a truncated form of the G-protein regulator AGS3 in heart that lacks the tetratricopeptide repeat domains. *J. Biol. Chem.* **276,** 16601–16620.

Punnonen, E. L., Autio, S., Kaija, H., and Reunanen, H. (1993). Autophagic vacuoles fuse with the prelysosomal compartment in cultured rat fibroblasts. *Eur. J. Cell. Biol.* **61,** 54–66.

Punnonen, E. L., Marjomaki, V. S., and Reunanen, H. (1994). 3-Methyladenine inhibits transport from late endosomes to lysosomes in cultured rat and mouse fibroblasts. *Eur. J. Cell. Biol.* **65**, 14–25.

Rabouille, C., Strous, G. J., Crapo, J. D., Geuze, H. J., and Slot, J. W. (1993). The differential degradation of two cytosolic proteins as a tool to monitor autophagy in hepatocytes by immunocytochemistry. *J. Cell Biol.* **120**, 897–908.

Rodriguez-Viciana, P., Warne, P. H., Khwaja, A., Marte, B. M., Pappin, D. *et al.* (1997). Role of phosphoinositide 3-OH kinase in cell transformation and control of the actin cytoskeleton by Ras. *Cell* **89**, 457–467.

Seglen, P. O., and Bohley, P. (1992). Autophagy and other vacuolar protein degradation mechanisms. *Experientia* **48**, 158–172.

Seglen, P. O., and Gordon, P. B. (1982). 3-Methyladenine: Specific inhibitor of autophagic/lysosomal protein degradation in isolated rat hepatocytes. *Proc. Natl. Acad. Sci. USA* **79**, 1889–1892.

Stow, J. L., de Almeida, J. B., Narula, N., Holtzman, E., Ercolani, L., and Ausiello, D. (1991). A heterotrimeric G protein, Galpha i-3, on Golgi membrane regulates the secretion of a heparan sulfate proteoglycan in LLC-PK1 epithelial cells. *J. Cell Biol.* **114**, 1113–1124.

Stromhaug, P. E., and Klionsky, D. J. (2001). Approaching the molecular mechanism of autophagy. *Traffic* **2**, 524–531.

Stromhaug, P. E., and Seglen, P. O. (1993). Evidence for acidity of prelysosomal autophagic/endocytic vacuoles (amphisomes). *Biochem. J.* **291**, 115–121.

Takesono, A., Cismowski, M. J., Ribas, C., Bernard, M., Chung, P. *et al.* (1999). Receptor-independent activators of heterotrimeric G-protein signaling pathways. *J. Biol. Chem.* **274**, 33202–33205.

Thumm, M. (2000). Structure and function of the yeast vacuole and its role in autophagy. *Microsc. Res. Technique* **51**, 563–572.

van Sluijters, D. A., Dubbelhuis, P. F., Blommaart, E. F. C., and Meijer, A. J. (2000). Amino-acid-dependent signal transduction. *Biochem. J.* **351**, 545–550.

Wylie, F., Heimann, K., Le, T. L., Brown, D., Rabnott, G., and Stow, J. L. (1999). GAIP, a Galphai-3-binding protein, is associated with Golgi-derived vesicles and protein trafficking. *Am. J. Physiol.* **276**, C497–C506.

[3] Analysis of RGSZ1 Protein Interaction with Gα$_i$ Subunits

By Yuren Wang and Kathleen H. Young

Abstract

RGSZ1 has been reported to interact with G-protein subunits of the Gα$_i$ family and function as a GTPase-accelerating protein on intrinsic Gα$_i$ GTPase activity. This article describes several experimental approaches and assays used to investigate the effect of RGSZ1 on Gα$_i$ subunits. The formats described here include physical and functional interaction assays by which the association of RGSZ1 with $G\alpha_i$ is explored both *in vitro* and *in vivo*. The methods analyzing physical interaction include pull-down and coimmunoprecipitation assays. We also apply yeast two-hybrid techniques to detect RGSZ1 protein interaction with $G\alpha$ subunits. Additionally, we developed several functional assay systems to identify the functional relationship between RGSZ1 and $G\alpha_i$, such as the single turnover GTPase assay, yeast pheromone response assay, mitogen-activated protein kinase assay, and serum response element reporter assay.

Introduction

RGSZ1 was identified independently by two groups as a GTPase-accelerating protein (GAP) for $G\alpha_z$. Wang *et al.* (1998) directly purified the RGSZ1 protein from bovine brain tissue lysate by virtue of its GAP activity toward $G\alpha_z$, whereas Glick *et al.* (1998) identified RGSZ1 using a $G\alpha_z$ Q205L mutant as "bait" in a yeast two-hybrid screen of a human brain library. The sequence of the human RGSZ1 protein is 83% identical to bovine RET-RGS1 and 56% identical to human GAIP (RGS19). Analysis of human genomic databases revealed that RGSZ1 is one of six different splice forms of the RGS20 gene product, including the RET-RGS1 isoform (Barker *et al.*, 2001). These RGS20 gene products differ in size and sequence of the N terminus and may have different tissue distributions and functions. RGSZ1 is expressed exclusively in brain, with the highest concentrations in the caudate nucleus and the temporal lobe (Glick *et al.*, 1998; Wang *et al.*, 1998), implying that RGSZ1 may play a role in regulating signals in these brain regions. RGSZ1 was reported to show high GAP activity toward $G\alpha_z$ in an *in vitro* GTPase assay, but no GAP activities toward $G\alpha_q$ or $G\alpha_s$ (Wang *et al.*, 1997, 1998). More recently, it was found that RGSZ1 also interacts with $G\alpha_i$ subunits, accelerates $G\alpha_i$ GTPase activity, and regulates $G\alpha_i$-mediated cell signaling (Wang *et al.*, 2002).

Approaches to analyze protein interactions generally include physical interaction assays and functional interaction assays. Most RGS proteins bind to specific conformational states of $G\alpha$ subunits to form a stable complex, although it is transient natively (Berman *et al.*, 1996). The GAP activity of RGS proteins depends largely on the binding affinity for $G\alpha$ subunits. For example, RGS4 binds to the transition state (AlF_4^- bound) of $G\alpha_{i1}$ subunits (Berman *et al.*, 1996), whereas RGS2 binds to both transition state and active (GTPγS-bound) forms of $G\alpha_q$ (Heximer *et al.*, 1997). However, neither RGS4 nor RGS2 binds to the GDP-bound form of $G\alpha$ subunits. Because the RGS protein and the $G\alpha$ subunit can form a stable complex under specific conditions, the interaction between RGS and $G\alpha_{i1}$ can be detected by *in vitro*-binding assays, such as the pull-down assay (Natochin *et al.*, 1997), coimmunoprecipitation assays (Hunt *et al.*, 1996), and the surface plasmon resonance biosensor assay (Kimple *et al.*, 2001; Popov *et al.*, 1997). More recently, several fluorescence-based techniques, such as fluorescence resonance energy transfer (Kimple *et al.*, 2001, 2003), have been explored for the analysis of protein–protein interactions. Technically, it is very critical to include AlF_4^- in the assay to investigate RGS function, as AlF_4^- facilitates the transition state of $G\alpha$ and thus enables stable binding to the RGS protein.

Functional assays are essential to confirm functional attributes of protein–protein interactions, as some interactions observed in *in vitro*-binding experiments may not translate to biological validity; additionally, some transient protein interactions may not be detected easily due to limitations of the system used. Functional assays reveal the effects of these interactions based on direct activity changes in the protein partners, or downstream readouts either *in vitro* or in whole cells. For the interactions between $G\alpha$ subunits and RGS proteins, assays readouts commonly use $G\alpha$ functional activities, such as GTP hydrolysis, GTPγS binding, cellular calcium mobilization, cAMP inhibition, or ion channel activities.

This article discusses several assays by which the interactions of RGSZ1 with $G\alpha_i$ subunits are explored *in vitro* and *in vivo*. The methods analyzing physical interaction include pull-down and coimmunoprecipitation assays. We also applied yeast two-hybrid techniques to detect the RGSZ1 protein interaction with $G\alpha$ subunits. Additionally, we developed several assays to identify the functional relationship between RGSZ1 and $G\alpha_i$, such as the single turnover GTPase, yeast pheromone response, mitogen-activated protein (MAP) kinase, and serum response element (SRE) reporter assays.

Pull-Down Assay for GST-RGSZ1 Protein and $G\alpha_i$ Subunits

The pull-down assay is based on the remarkable selectivity and affinity of a tagged protein of interest, bound via its tag to an affinity matrix, to "pull down" or precipitate another protein partner of interest from the solution milieu. There are different pull-down systems based on the various protein tags and the affinity matrixes. The glutathione *S*-transferase (GST) fusion protein pull-down approach is a common method. The GST protein serves as the "tag" when fused to the protein of interest, and glutathione agarose beads serve as the pull-down matrix. Another common approach uses the polyhistidine tag (6His) and a nickel affinity matrix. Both systems have been applied successfully in the studies of RGS protein interaction with G proteins (Druey *et al.*, 1996; Lan *et al.*, 1998; Natochin *et al.*, 1997). The pull-down assays can be done with purified proteins or cell lysates in which the target proteins are overexpressed. This section describes an experimental protocol to evaluate the interaction of purified GST-RGSZ1 and *in vitro*-transcribed/translated, ^{35}S-labeled $G\alpha_{i1}$ subunit in a glutathione resin pull-down assay. The purpose of using ^{35}S-labeled proteins is to compare quantitatively the RGSZ1 interactions with $G\alpha_{i1}$ and $G\alpha_z$ subunits without the use of antibodies that usually have different affinities to different $G\alpha$ subunits.

Expression and Purification of GST-RGSZ1 Fusion Protein from Escherichia coli

The GST-RGSZ1 fusion protein is purified from *E. coli* strain JM109 using the B-PER GST fusion protein purification kit from Pierce (Rockford, IL). A 50-ml overnight culture of bacteria expressing a full-length RGSZ1 open reading frame in the plasmid pGEX-4T-1 (Amersham Pharmacia, Piscataway, NJ) is diluted 1:20 into 1 liter of prewarmed LB medium (with 50 μg/ml ampicillin) and grown at 37° with vigorous shaking until an OD_{600} of 0.5–0.6 is reached (about 2–3 h). Isopropyl-β-D-thiogalactopyranoside (IPTG) is added to the culture to a final concentration of 1 mM to induce protein expression. The culture is incubated for an additional 3 h and cells are harvested by centrifugation at 6000g for 15 min at 4°. The cell pellet is resuspended in 10 ml of B-PER lysis buffer (provided in the kit) and shaken gently for 10 min at room temperature. The cell lysate is centrifuged at 27,000g for 30 min. The lysate supernatant is applied to the immobilized glutathione-Sepharose column (1 ml), and the sample is allowed to flow through the column by gravity. The column is washed twice with 10 ml washing buffer (provided in the kit), and the bound GST-RGSZ1 protein is eluted from the column by 3 ml glutathione solution (provided in the kit). The protein fractions are dialyzed into buffer containing 50 mM Tris, pH 7.5, 150 mM NaCl, 1 mM EDTA, and 5 mM dithiothreitol (DTT) and concentrated using a Centricon centrifugal filter device from Millipore (Bedford, MA). The purified protein is examined by sodium dodecyl sulfate–polyacrylamide gel electrophoresis (SDS–PAGE), and the protein concentration is measured using the Bio-Rad protein assay kit (Hercules, CA).

Generation of [35S]Methionine-Labeled Gα Proteins in a Cell-Free System

[35]S]$G\alpha_{i1}$ and ^{35}S]$G\alpha_z$ proteins are generated by an *in vitro* TNT T7-coupled transcription/translation system as described by the manufacturer (Promega, Madison, WI). This system is a reticulocyte lysate system with

modifications to simplify the process and reduce the time. This system combines the RNA polymerase, nucleotides, salts, and recombinant RNasin ribonuclease inhibitor with the reticulocyte lysate solution to form a single master mix. Previous studies have indicated that Gα subunits generated from this system are conformationally intact and functionally normal (Fan *et al.*, 2000; Journot *et al.*, 1989). For this purpose, 1 μg of pcDNA3 plasmid (Invitrogen, Carlsbad, CA) encoding Gα$_{i1}$ wild-type, Gα$_{i1}$Q204L, Gα$_{i1}$R178C, or Gα$_z$ wildtype, Gα$_z$Q205L, or Gα$_z$R179C open reading frames (Wang *et al.*, 2002) are each mixed with 1 μCi [^{35}S]methionine and 40 μl master mix for up to a total volume of 50 μl. The mixture is incubated at 30° for 60 min. The [^{35}S]methionine-labeled protein products are analyzed by autoradiography after SDS–PAGE and stored in a −80° freezer until use.

Pull-Down Experiments Using GST-RGSZ1 and ^{35}S-Labeled Gα Subunits

Five micrograms GST-RGSZ1 is incubated with glutathione-agarose beads (50 μl of 50% beads) and 2 μl of ^{35}S-labeled G-protein α subunits in a total volume of 500 μl binding buffer [50 mM HEPES, pH 8.0, 1 mM EDTA, 100 μM GDP, 1 mM DTT, 10 mM MgSO$_4$, 0.05% lubrol, and 0.05% bovine serum albumin (BSA)] in the presence of GTPγS (100 μM) or AlF$_4^-$ (10 mM NaF + 100 μM AlCl$_3$) for 4 h at 4° with gentle shaking. The beads are pelleted at 1000 rpm for 3 min and washed three times with 1 ml of ice-cold binding buffer and then are resuspended in 25 μl of 1× SDS–PAGE sample buffer. Gα proteins bound to RGSZ1 on the beads are then resolved with SDS–PAGE and visualized by autoradiography after a 2 to 3-day exposure. As shown in Fig. 1, RGSZ1 binds to Gα$_{i1}$ and Gα$_z$ only in the presence of AlF$_4^-$, but not in the presence of GTPγS. RGSZ1 does not bind to the constitutively active, GTPase-deficient Q-to-L or R-to-C mutants of Gα$_z$ or Gα$_{i1}$. This observation indicates that RGSZ1 binds directly to both Gα$_{i1}$ and Gα$_z$ in their transition states. Similar findings have been reported for other RGS proteins, such as RGS4 and GAIP (Berman *et al.*, 1996). Interestingly, we observe that RGSZ1 binds to Gα$_{i1}$ and Gα$_z$ with a similar activity in this *in vitro*-binding assay.

Coimmunoprecipitation of Gα$_{i1}$ and RGSZ1 in CHO Cell Lysates

Coimmunoprecipitation is another valuable approach to test for physical interactions between proteins of interest. By comparison to the pull-down assay, which usually needs purified proteins, the coimmunoprecipitation approach can be used to analyze protein interactions in cell

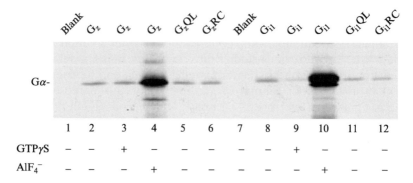

FIG. 1. Interaction of RGSZ1 with $G\alpha_{i1}$ and $G\alpha_z$ in a GST pull-down assay. Five micrograms of GST-RGSZ1 protein was incubated with 2 μl of each *in vitro*-translated, [^{35}S]methionine-labeled $G\alpha_z$, $G\alpha_z$ Q205L, $G\alpha_z$ R179C, or $G\alpha_{i1}$, $G\alpha_{i1}$ Q204L, $G\alpha_{i1}$ R178C proteins in the presence of GTPγS (100 μM) or AlF$_4^-$ (10 mM NaF + 100 μM AlCl$_3$) for 4 h at 4°. The $G\alpha_z$ and $G\alpha_{i1}$ subunits that bound to GST-RGSZ1 were pulled down by glutathione-agarose beads and subjected to SDS–polyacrylamide gel electrophoresis followed by autoradiography (2-day exposure). Reproduced with permission from Wang *et al.* (2002).

extracts in which the proteins of interest are overexpressed. The basic idea of the coimmunoprecipitation approach is to use a high-affinity antibody to a desired protein to precipitate the protein complex and thus associate the other protein partners with a solid-phase affinity matrix under nondenaturing conditions. The precipitate is then tested for the presence of secondary, specifically associated proteins of interest by immunoblotting or mass spectrometry. This approach can be used to isolate multiprotein complexes that presumably are present in the intact cell. This coimmunoprecipitation approach has been used for the investigation of RGS proteins (Hunt *et al.*, 1996). This section describes an application of this approach in the identification of the RGSZ1 interaction with the $G\alpha_{i1}$ protein.

Generation of an Antibody against RGSZ1

Rabbit polyclonal antiserum is generated against the extreme N-terminal 16 amino acids of human RGSZ1 (GenBank accession number AF074979). The peptide is synthesized with a C-terminal cysteine to allow cross-linking to keyhole limpet hemocyanin (KLH). Results from enzyme-linked immunosorbant assay (ELISA) and Western blotting analyses of the antiserum indicate high binding titer and specificity against human RGSZ1 protein, not to recombinant GAIP, RGS4 core domain, or RGS7 core domain protein. The RGSZ1 antibody is purified further by a protein-A IgG purification column from Pierce. The antibody is desalted by overnight dialysis against phosphate-buffered saline (PBS) buffer at $4°$ and concentrated using a Centricon centrifugal filter device from Millipore. The concentration of the antibody is determined by the Bio-Rad protein assay kit, and the titer of RGSZ1 antibody is determined in Western blot analysis.

Transfection of CHO Cells and Preparation of Whole Cell Extracts

CHO-K1 cells are maintained in Dulbecco's modified Eagle's medium (DMEM) containing 10% fetal bovine serum and are seeded in 100-mm-diameter dishes (3×10^6 cells/well) the day before transfection. Cells are transfected with RGSZ1 and $Gα_{i1}$ expression plasmids or with the empty pcDNA3(+) vector using LipofectAMINE PLUS reagent (Invitrogen). Cells are harvested 40–48 h after transfection in 1 ml of lysis buffer [20 mM HEPES, pH 7.5, 2 mM MgCl$_2$, 1 mM EDTA, 150 mM NaCl, 10 μg/ml aprotinin, 5 μg/ml leupeptin, and 1 mM phenylmethylsulfonyl fluoride (PMSF) with 2% Triton X-100] and are then incubated for 1 h with gentle shaking at room temperature. All subsequent procedures are carried out at $4°$. The unsolubilized fractions are removed by centrifugation at 100,000g for 1 h. The supernatant is collected and diluted with the lysis buffer without detergent to reach a final concentration of 1% Triton X-100.

Coimmunoprecipitation of $Gα_{i1}$ and RGSZ1 in Cell Extracts

One microgram of preimmune rabbit serum and 50 μl of 50% protein A-Sepharose beads (Amersham Pharmacia) are added to each 1 ml of cell supernatants. After incubation for 1 h at $4°$ with gentle shaking, a quick centrifugation at 1000g is carried out to remove any nonspecific-binding materials to the beads. Then 1 μg polyclonal antibody against RGSZ1 and 50 μl of 50% protein A-Sepharose beads are added to the supernatant with or without AlF$_4^-$ (10 mM NaF + 100 μM AlCl$_3$) and are incubated overnight at $4°$ with gentle shaking. Immunoprecipitates are collected by centrifugation at 1000g, washed three times with lysis buffer containing 1% Triton X-100, and resuspended in 1× sample buffer for resolution by SDS–PAGE. Following electrophoresis, proteins are transferred to a nitrocellulose membrane and probed with a monoclonal antibody against $Gα_{i1}$ (Upstate, Lake Placid, NY) using standard Western blotting procedures with ECL chemiluminescence (Amersham Pharmacia).

As shown in Fig. 2, $Gα_{i1}$ was only detected after immunoprecipitation with RGSZ1 antibody in the presence of AlF$_4^-$. The amount of $Gα_{i1}$ detected was increased when 1 μg purified recombinant RGSZ1 protein was added to the cell lysate to increase the availability of the RGSZ1 protein. These data further indicated that the transition state of $Gα_{i1}$ interacts specifically with RGSZ1. Notably, conditions for both cell extract

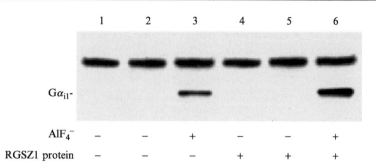

FIG. 2. Coimmunoprecipitation of RGSZ1 and $G\alpha_{i1}$ from the CHO-K1 cell lysate. CHO-K1 cells were cotransfected with the $G\alpha_{i1}$ plasmid (5 μg/well) and RGSZ1 plasmid (5 μg/well) in the following combinations: lanes 1 and 4, $G\alpha_{i1}$ and pcDNA3 and lanes 2, 3, 5, and 6, $G\alpha_{i1}$ and RGSZ1. Cell lysates were prepared in lysis buffer 48 h posttransfection, with and without AlF_4^- (10 mM NaF + 100 μM $AlCl_3$). One microgram of purified RGSZ1 protein was also added to some cell lysate fractions (lanes 4–6). The RGSZ1 protein was immunoprecipitated with a polyclonal antibody against RGSZ1 and protein A agarose beads. Beads were pelleted, washed, subjected to SDS–PAGE, and transferred to a membrane for Western blot analysis using a monoclonal antibody against $G\alpha_{i1}$. The upper bands are IgG heavy chains. Reproduced with permission from Wang *et al.* (2002).

preparation and coimmunoprecipitation needed to be optimized. The choice and the concentration of each detergent, as well as the amounts of salt, are essential for the success of the experiment. When no interaction is detected, it may be necessary to use less stringent conditions. For example, we use 2% Triton X-100 to prepare the cell extract because it was reported that RGSZ1 can only be solubilized from cell membranes in 2% Triton X-100 (Wang *et al.*, 1997). However, because 2% of Triton X-100 may be too stringent for the interaction of RGSZ1 with $G\alpha_i$, we dilute Triton X-100 to 1% in later steps of the experiment. In addition, a high background usually causes some problems for the interpretation of results. We found that the preclearing of cell lysates using preimmune rabbit serum and protein A-Sepharose beads was quite effective in reducing the background. Additionally, a high expression of desired proteins in cells is also helpful in increasing the signal-to-noise ratio and enhancing detection of the desired protein in the coimmunoprecipitation experiments.

Yeast Two-Hybrid Analysis of $G\alpha_{i1}$ and RGSZ1 Interaction

The yeast two-hybrid approach has been applied successfully in systems using $G\alpha$ subunits or RGS proteins as "bait" to identify their interacting partners by screening cDNA fusion libraries (De Vries and Farquhar, 2002; Glick *et al.*, 1998). One advantage of yeast two-hybrid screens is the ability to detect protein–protein interactions *in vivo*. This is of particular interest for some proteins that are difficult to purify or if the protein interactions are relatively weak under *in vitro* assay conditions. Several improved yeast two-hybrid systems have become available commercially. This section reports on the application of an improved yeast two-hybrid system (Nieuwenhuijsen *et al.*, 2003; Young *et al.*, 2004) to confirm the RGSZ1 interaction with $G\alpha_i$ subunits. The principle of this yeast two-hybrid system is to reestablish functionally the (singly) inactive DNA binding (BD) and the activation (AD) domains of Gal4p, resulting in an active transcription factor that drives expression of a functionally linked reporter gene (Fields and Song, 1989). We generated RGSZ1-AD and $G\alpha$ subunit-BD fusion plasmids that were used to transform yeast cells. Interaction of the protein partner pair functionally reconstitutes Gal4 transcription factor activity to activate a reporter gene. In this system, we use a firefly luciferase reporter gene with a *GAL4* promoter.

Generation of Plasmids

Human RGSZ1 was subcloned into the *NcoI/BamHI* sites of the pACT2 vector (Clontech, Palo Alto, CA) to generate a Gal4 DNA activation domain fusion protein. Human $G\alpha_{i1}$, $G\alpha_{i1}$Q204L mutant, $G\alpha_{i2}$, and $G\alpha_{i3}$ are subcloned individually into *EcoRI* and *PstI* sites of the pGBKT7 vector (Clonetech, Palo Alto, CA) to generate GAL4 DNA-binding domain fusion proteins. A Gal4 responsive firefly luciferase reporter gene is generated by cloning the firefly luciferase gene into the *BamHI* and *SalI* sites of the pEK1 vector (see Nieuwenhuijsen *et al.*, 2003). All constructs were confirmed by DNA sequencing.

Yeast Transformation

Yeast strains are generated by cotransformation of plasmids into yeast strain CY770 (Ozenberger and Young, 1995) using the lithium acetate method (Rose *et al.*, 1990). Briefly, the parent yeast strain is grown overnight in YPD medium at $37°$ on a rotor wheel. The culture is then diluted 50-fold with the same medium for continued growth for another 3–4 h. The yeast cells are then pelleted by centrifugation at 1500 rpm for 5 min at $4°$, washed with 10 ml sterile water, resuspended in 300–500 μl (~1:100 by volume) of 0.1 M lithium acetate, and incubated for 2 h with rotation at $30°$. One microgram of each plasmid as indicated in Fig. 3 is mixed with 1 μg of Gal4-luciferase reporter plasmid and 50 μg (~5 μl) of salmon testes DNA (carrier DNA) in 150 μl of suspended, lithium acetate-treated yeast cells and incubated for 30 min on a rotor wheel at $30°$. The samples are mixed with 800 μl of 0.1 M lithium acetate in 40% PEG and incubated for

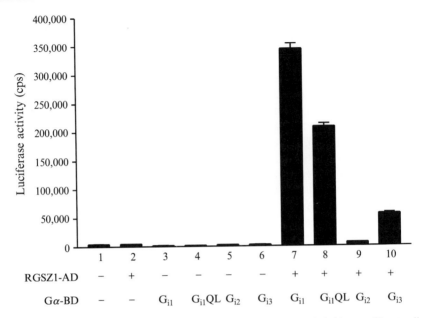

FIG. 3. Interaction of RGSZ1 with Gα_i subunits in the yeast two-hybrid assay. Yeast cells (CY770) were transformed with the following combinations of plasmids: lane 1, pACT2 and pGBKT7 empty control vectors; lane 2, RGSZ1-pACT2 and pGBKT7; lane 3, Gα_{i1}-pGBKT7 and pACT2; lane 4, Gα_{i1} Q204L-pGBKT7 and pACT2; lane 5, Gα_{i2}-pGBKT7 and pACT2; lane 6, Gα_{i3}-pGBKT7 and pACT2; lane 7, RGSZ1-pACT2 and Gα_{i1}-pGBKT7; lane 8, RGSZ1-pACT2 and Gα_{i1}Q204L-pGBKT7; lane 9, RGSZ1-pACT2 and Gα_{i2}-pGBKT7; and lane 10, RGSZ1-pACT2 and Gα_{i3}-pGBKT7. In all cases, a GAL4-luciferase reporter plasmid was cotransformed. The luciferase reporter gene activity of each strain was determined and presented as means ± SE of triplicate measurements from three independent experiments. Reproduced with permission from Wang *et al.* (2002).

1 h at 30° on a rotor. The transformations are then "heat shocked" for 10 min at 42°. Cells are pelleted by centrifugation at 1500 rpm for 5 min at room temperature. The cell pellets are resuspended in 150 μl of sterile water, plated evenly on a 10-cm agar plate (SC-trp-leu-ura), and incubated at 30° for 2–3 days. A "no DNA" control may be added to determine the overall transformation efficiency.

Yeast Two-Hybrid Assays Using a Luciferase Reporter

For each engineered yeast strain, 30 colonies are picked and mixed in liquid medium (SC-trp-leu-ura) and are grown overnight at 30° with continuous shaking. Yeast cultures are diluted to an OD_{600} of 0.15, aliquoted into a 96-well plate (100 μl/well), and grown for an additional 3 h at 30°.

Luciferase reporter gene activity is measured using a LucLite kit (Packard, Meriden, CT). One hundred microliters of luciferase substrate is added to each assay well of the 96-well plate and is incubated for 1 h at room temperature with shaking. Luminescence is determined using a single photon counting mode by a TopCount plate reader (Packard/Perkin Elmer, Meriden, CT). As shown in Fig. 3, transformants containing empty vectors alone, RGSZ1-AD alone, or Gα$_i$-BD alone had very low luciferase activity (~5000 counts), indicating a low basal activity of the luciferase reporter. By comparison, the yeast strain containing both RGSZ1-AD and Gα$_{i1}$-BD showed over ~70-fold increase (~350,000 counts) over the luciferase activity of controls, suggesting a strong interaction between RGSZ1 and Gα$_{i1}$ in yeast. RGSZ1 also interacted well with the Gα$_{i1}$Q204L mutant and Gα$_{i3}$ in this system, but interacted poorly with Gα$_{i2}$. According to previous reports, RGSZ1 also interacts with Gα$_o$ to a similar activity extent as with Gα$_z$ in a yeast two-hybrid experiment (Glick et al., 1998). Using the same assay, De Vries and co-workers (1995, 1996) showed that GAIP (RGS19), which is highly homologous to RGSZ1, interacts with all Gα$_i$ subfamily members (Gα$_{i3}$> Gα$_{i1}$, Gα$_o$> Gα$_z$ and Gα$_{i2}$), but not with other G proteins such as Gα$_s$, Gα$_q$, and Gα$_{13}$.

RGSZ1 Stimulates Gα$_{i1}$ GTPase Activity in a Single Turnover GTPase Assay

Most RGS proteins identified thus far function as GAPs for G-protein α subunits (Neubig and Siderovski, 2002; Ross and Wilkie, 2000). Thus assays to determine the effects of RGS proteins can be done by measuring the acceleration of GTP hydrolysis on Gα subunits. There are two commonly used assays to measure the GAP activity of RGS proteins. The first is the "single turnover" GTPase assay, which measures the rate of hydrolysis of Gα-bound GTP in a single enzymatic cycle. However, it is worth noting that the "single turnover" attribute refers to the hydrolysis of Gα-bound GTP. Actually, one molecule of RGS protein can catalyze hydrolysis by multiple Gα-GTP molecules during the assay. The single turnover GTPase assay is the simplest and most quantitative method used to measure the hydrolysis rate of Gα-GTP and the GAP activity of RGS proteins. The second method, the "steady-state" GTPase assay, measures overall hydrolysis rates of Gα-GTP in a membrane environment with GDP/GTP exchange promoted by agonist-occupied G-protein-coupled receptors. The steady-state GTPase assay is usually conducted using reconstituted lipid vesicles containing purified G proteins, receptors, and RGS proteins (Ross, 2002). The assay has been performed successfully using cell membrane fractions expressing a chimera of α_{2A}-adrenoreceptor and a pertussis toxin (PTX)-insensitive $G\alpha_{i1}$ mutant (C351V) (Cavalli *et al.*, 2000). The steady-state GTPase assay may provide additional information as to the dynamics of the overall $G\alpha$ GTPase cycle and the role of RGS proteins in accelerating GTP hydrolysis and is generally more sensitive than the single turnover GTPase assay in detecting the GAP activity of RGS proteins. For more information about these GTPase assays and their applications, please reference two excellent reviews published previously in this series (Krumins and Gilman, 2002; Ross, 2002). This section reports on application of the single turnover GTPase assay to examine the GAP activity of RGSZ1 on $G\alpha_i$.

Generation of RGSZ1-6His and $G\alpha_{i1}$-6His Proteins

Full-length recombinant RGSZ1-6His and $G\alpha_{i1}$-6His proteins are expressed and purified from *E. coli*. Human RGSZ1 and $G\alpha_{i1}$ genes are subcloned individually into the *NcoI* and *Bam*HI sites of the pQE60 vector (Qiagen, Valencia, CA). Competent JM109 cells are transformed using ~1 ng of plasmids using standard methods. Aliquots of transformants are plated on LB agar containing 100 μg/ml ampicillin and are incubated overnight at 37°. The JM109 strain is specifically chosen, as it harbors the *lacI* mutation and produces enough *lac* repressor to efficiently block transcription, which is ideal for storing and propagating pQE60 plasmids. Twenty milliliters of LB broth containing 100 μg/ml of ampicillin is inoculated with a single colony and grown at 37° overnight. On the second day, the 20-ml cultures are used to inoculate 1 liter of prewarmed LB broth, containing ampicillin, and are incubated at 37° with vigorous shaking until an OD_{600} of 0.6 is reached (~1 h). Protein expression is induced by adding IPTG (final concentration 1 mM), and incubation is continued for 4 h. Cells are collected by centrifugation at 4000g for 20 min and are then broken using a French press (model X2) in PBS buffer containing 20 mM imidazole. Triton X-100 is then added to the sample to a final concentration of 1%. The cell lysate is centrifuged at 30,000g for 10 min at 4° to remove cellular debris, and the supernatant is applied slowly to a prepacked, and lysate buffer-equilibrated, Ni-NTA column (1 ml size) in a fast-protein liquid chromatography (FPLC) apparatus. After washing twice with the lysis buffer, the protein is eluted with an imidazole gradient from 20 to 400 mM. The eluted protein fractions are purified further by direct application to a Hitrap Q anion-exchange column from Amersham Biotech and eluted with a 20–400 mM NaCl gradient. Protein samples are dialyzed and concentrated in a buffer containing 50 mM Tris, pH 8.0, 100 mM NaCl, and 10 mM DTT. The purity of the proteins is analyzed by SDS–PAGE and generally shows that both RGSZ1-6His and $G\alpha_{i1}$-6His proteins are over 90% pure (data not shown). GTPγS-binding analyses show that about 65% of the recombinant Gα$_{i1}$ protein is capable of binding GTPγS.

Preparation of [γ-³²P]GTP-Loaded Gα$_{i1}$

Gα$_{i1}$ protein (5 μM) is incubated with 5 μM [γ-³²P]GTP (Amersham Biotech) in a buffer containing 10 mM HEPES (pH 7.5), 5 mM EDTA, 2 μM GTP, 1 mM DTT, and 0.05% Lubrol with a total volume of 30 μl at 30° for 30 min. The [γ-³²P]GTP-loaded Gα$_{i1}$ is separated from unbound [γ-³²P]GTP by passing through a gel filtration spin column (Princeton Separations, NJ), centrifuged at 2000g for 3 min, and diluted to 500 nM with ice-cold assay buffer, described earlier, and then kept on ice.

Single Turnover GTPase Assay

The GTPase activity of the Gα$_{i1}$ protein is measured by monitoring free, inorganic, radiolabeled phosphate released from bound [γ-³²P]GTP, as described previously (Berman *et al.*, 1996). Briefly, Gα$_{i1}$ GTPase activity is initiated by adding 10 mM MgCl$_2$ and 100 μM GTP (final) to the [γ-³²P]GTP-loaded Gα$_{i1}$ samples prepared as described earlier with or without 100 nM of RGS protein. For a time course study, timed 50-μl aliquots of reaction sample are transferred to 500 μl of buffer (5% Norit-activated charcoal in 50 mM NaPO$_4$, pH 3.0) to stop the reaction. Following centrifugation (15,000g for 30 min in a tabletop microfuge), the supernatant containing free ³²Pi is collected and counted by liquid scintillation. All assays are performed at 0 or 15°.

As shown in Fig. 4, RGSZ1 accelerates the GTPase activity of Gα$_{i1}$ dramatically: 100 nM RGSZ1 maximally activated the GTP hydrolysis of 500 nM Gα$_{i1}$ in less than 1 min at 0°. The k_{obs} value is increased from 0.335 m^{-1} in the absence of RGSZ1 to 3.885 m^{-1} in the presence of RGSZ1 at 0°. By comparison, the basal rate of GTP hydrolysis was much faster at 15°. The k_{obs} value is increased from 0.652 m^{-1} without RGSZ1 to 3.985 m^{-1} with RGSZ1 at 15°. It is worth nothing that the fold stimulation by RGSZ1 of Gα$_{i1}$ GTPase activity is minimized by a longer reaction time due to the increasing basal GTPase activity. This is particularly noted at 15°.

The conditions for performing this GTPase assay with different Gα subunits require modification to obtain optimal signal-to-noise ratios. This is particularly important for G proteins such as Gα$_{i1}$ and Gα$_z$, which have different basal GTPase activities. The intrinsic GTPase activity of Gα$_z$ is approximately 200-fold slower than that of Gα$_{i1}$ (Casey *et al.*, 1990). Consequently, the GTPase assay for Gα$_z$ needs to be conducted at higher temperature, such as 15°, and with a longer reaction time (longer than 10 min). However, the Gα$_i$ GTPase assay needs to be conducted at 0–4°

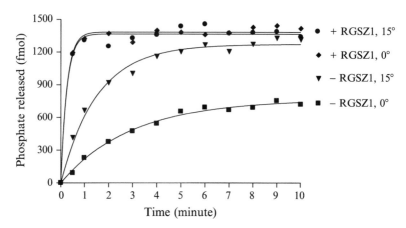

FIG. 4. RGSZ1 accelerates the GTP hydrolysis of $G\alpha_{i1}$ in GTPase assays. The RGSZ1 protein (100 nM) was incubated with 500 nM $G\alpha_{i1}$-[^{32}P]GTP at 0 or 15°. Reactions were stopped at the indicated times (0, 0.5, 1, 2, 3, 4, 5, 6, 7, 8, 9, and 10 min). All data represent one reaction from at least three independent experiments showing similar results. Reproduced with permission from Wang *et al.* (2002).

with shorter reaction times (less than 5 min) (Wang *et al.*, 1998). Data show that the effect of RGSZ1 in stimulating the GTPase activity of $G\alpha_{i1}$ may be mostly overshadowed by the high basal GTPase activity, if the assay temperature is too high or the assay time lasts too long.

RGSZ1 Attenuates Pheromone Signaling in a Yeast Pheromone Response Assay

Pheromone signaling in the yeast *Saccharomyces cerevisiae* is mediated through receptors of the seven transmembrane domain class and is coupled to downstream events via a G protein that controls the events of cell mating, such as cell cycle arrest and induction of gene transcription. The yeast G-protein α subunit Gpa1 is a $G\alpha_i$-like protein (Dietzel and Kurjan, 1987; Miyajima *et al.*, 1987). Previous studies have shown that mammalian RGS proteins, such as RGS1, RGS3T, RGS4, and RGS16, which interact well with mammalian $G\alpha_i$ proteins, can also complement the role of SST2 (a yeast RGS protein) in modulating Gpa1 and attenuating the pheromone response in yeast (Chen *et al.*, 1997; Druey *et al.*, 1996). There are, in general, two types of assays to evaluate RGS function in yeast employing the pheromone response pathway: one assay links the pathway to a β-galactosidase (or luciferase) reporter gene, which measures pheromone-dependent gene transcription, while the "halo assay" provides a measure of pheromone-induced yeast cell growth arrest. Both assays measure RGS protein activity by monitoring downstream events following yeast cell exposure to the α-factor pheromone. The β-galactosidase or luciferase assay is typically quick (1–2 h after α-factor treatment), whereas the "halo assay" usually takes 2–3 days after α-factor exposure. It is worth noting that, in some cases, the two assay formats may not correlate directly, as they measure different readouts and require different time frames. Additionally, the β-galactosidase and luciferase assays are generally more quantitative than the halo assay. However, the halo assay is still used most commonly because of its stability. For further discussions of these yeast assays, please consult Hoffman *et al.* (2002). This section describes a functional complementation study conducted to ascertain the interaction of RGSZ1 with Gpa1, as measured in a halo assay using the SST2 knockout strain.

Generation of Plasmids and Transformation of SST2 Knockout Yeast Strain

The cDNA of human RGSZ1, human RGS2, and SST2 are cloned into the *Bam*HI/*Cla*I sites of the p426TEF vector (ATCC, Manassas, VA) by standard molecular biology techniques. Yeast strains expressing hRGSZ1, hRGS2, or SST2 are generated using standard lithium chloride transformation (Rose *et al.*, 1990) of appropriate plasmids into yeast strain yKY113 [a yeast RGS protein (SST2) knockout strain described in Young *et al.* (2004)] and are cultured on plasmid retention medium plates (SC-leu).

Yeast Pheromone Response Assay

A single yeast colony from each transformation is selected, cultured using standard conditions in plasmid retention liquid medium (SC-leu), diluted to a density of 0.5 (OD_{600}) in 5 ml of medium, and applied on the surface of an agar plate (SC-leu) to produce a lawn of yeast cells. Plates are dried for 2 h at 30°, and then 5 μl of the α-factor pheromone is spotted on the surface of the agar plate in a series of dilutions (1, 10, 100, 1000 μM) and is incubated at 30° for 2–3 days. Treatment with α factor causes dose-dependent cell cycle arrest, as observed by the increase of the halo size in the empty vector control strain (Fig. 5). By comparison, α factor spotted at 100 μM induced a halo in the RGSZ1-transformed strain that was ~70% smaller than that observed for the vector control yeast strain (Fig. 5), suggesting that RGSZ1 partially blocks the pheromone response via modulation of Gpa1. Yeast cells expressing SST2 (a positive control) showed a stronger reversal of cell cycle arrest than that observed for yeast cell expressing RGSZ1, as measured in the halo assay. RGS2, which has lower GAP activity for $G\alpha_i$ than for $G\alpha_q$ (Chen *et al.*, 1997;

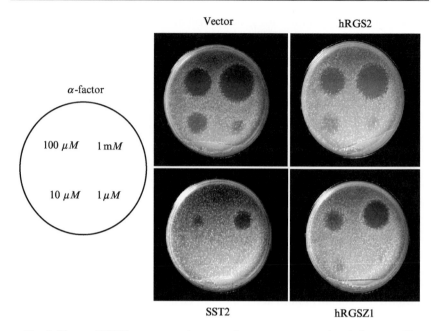

Fig. 5. Human RGSZ1 attenuates the yeast pheromone response in a halo assay. Yeast strain yKY113 (sst2Δ) was transformed with the empty p426-TEF vector or plasmids encoding human RGS2, human RGSZ1, or SST2. A halo bioassay was performed on each transformant strain with four doses of α factor (1 μM, 10 μM, 100 μM, and 1 mM). Reproduced with permission from Wang *et al.* (2002).

Heximer *et al.*, 1997, 1999), did not show a significant effect in modulating the yeast pheromone response.

RGSZ1 Blocks α_{2A}-Adrenergic Receptor-Mediated MAP Kinase Activity in PC12 Cells

The α_{2A}-adrenergic receptor (α_2-AR) couples preferentially to $G\alpha_i$-containing heterotrimers to transduce cellular signals (Chabre *et al.*, 1994). Activation of α_2-AR results in MAPK activation in various cell types by coupling to $G\alpha_i$. This activation is mediated by $G\beta\gamma$ subunits and occurs as a direct result of Ras activation (Alblas *et al.*, 1993). Previous studies have indicated that RGS4 attenuates MAPK activation by the interleukin 8 (IL-8) receptor and the D2 dopamine receptor, both of which couple to $G\alpha_i$ proteins (Druey *et al.*, 1996; Yan *et al.*, 1997). The role of RGSZ1 in modulating Gα$_i$ thus can be examined by monitoring the MAPK activity induced by the activation of endogenous α$_2$-AR in PC12 cells.

Generation of PC12 Stable Cell Lines

PC12 cells are maintained in DMEM containing 10% horse serum plus 5% fetal bovine serum and are seeded in 100-mm-diameter dishes (3 × 10^6 cells/well) the day before transfection. Cells are then transfected with the RGSZ1-pEGFP recombinant expression vector or pEGFP vector alone using Lipofectamine 2000 reagent (Invitrogen) following the manufacturer's recommendations. Cells are passaged by trypsinization into a 150-mm cell culture dish in fresh medium 24 h after transfection. On the following day, the medium is replaced with medium containing G418 (1 μg/ml). Clonal isolates are picked and cultured in DMEM containing 10% horse serum, 5% fetal bovine serum, and 200 μg/ml G418. The expression of either the eGFP-RGSZ1 fusion protein or the eGFP protein alone can be monitored with a fluorescence microscope. Western blot results show relatively identical expression levels of eGFP-RGSZ1 or eGFP in the PC12 stable cell lines (data not shown).

MAP Kinase Assay

p44/42 MAPK activity is measured using an MAP kinase assay kit from Cell Signaling Technology (Beverly, MA). Briefly, cells are starved overnight in serum-free medium and are then treated with the α$_2$-AR agonist 5-bromo-N-[4,5-dihydro-1H-imidazol-2-yl]-6-quinoxalinamine UK14304 (Sigma, St. Louis, MO) for 5 min. Thereafter, cells are immediately transferred on ice and washed once with ice-cold PBS. Cells are harvested using a cell lysis buffer from the kit plus 1 mM PMSF. Cells are sonicated four to five times for 5 s each on ice and are then centrifuged at 12,000g for 10 min at 4°. The supernatant is collected and diluted to 1 mg/ml of protein with the lysis buffer. Fifteen microliters of resuspended agarose immobilized with the phospho-p44/42 MAP kinase (Thr202/Tyr204) monoclonal antibody is added to each 200-μl cell lysate and incubated overnight at 4° with gentle shaking. The sample is centrifuged at 1000g for 1 min at 4°, and the pellet is washed twice with 500 μl of lysis buffer. The pellet is washed again twice with 500 μl of kinase buffer. The pellet is resuspended in 50 μl kinase buffer supplemented with 200 μM ATP and 2 μg ELK-1 fusion protein and is incubated for 30 min at 30°. Then, 25 μl of 3× SDS sample buffer is added to stop the reaction. The phospho-ELK1 is visualized by Western blotting using the phospho-ELK-1 antibody after SDS–PAGE. As shown in Fig. 6, treatment of the eGFP cell line with UK14304 causes a

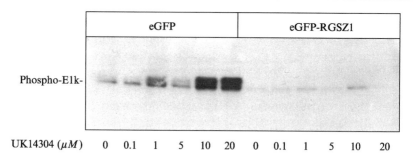

UK14304 (μM)   0   0.1   1   5   10   20   0   0.1   1   5   10   20

FIG. 6. RGSZ1 blocks MAPK activation by UK14304 in PC12 cells. PC12 cells stably expressing either eGFP or eGFP-RGSZ1 were cultured in serum-free DMEM overnight, and then UK14304 was added to the medium in the indicated final concentrations (0, 0.1, 1, 5, 10, and 20 μM) for 5 min. Cells were harvested, lysed, and the phospho-p44/42 MAPK content in cell lysates was immunoprecipitated using a monoclonal antibody immobilized on agarose beads. The activity of p44/42 MAPK was measured by incubating the complex with the Elk-1 fusion protein and ATP in the kinase reaction buffer for 30 min at 30°. The phosphorylation level of Elk-1 was analyzed by a Western blot using a phospho-Elk-1 antibody. Reproduced with permission from Wang et al. (2002).

dose-dependent activation of p42/44 MAPK (as shown by the increase in phospho-ELK levels). The activation of MAPK induced by 10 μM UK14304 reached the maximal level in 5 min. By comparison, the activation of p42/44 MAPK by UK14304 was blocked substantially in the eGFP-RGSZ1 cell line. Previous results showed that the activation of MAPK by UK14304 was also blocked by treatment with PTX (50 ng/ml) (not shown), suggesting that UK14304-induced MAPK activation was mediated by the PTX-sensitive $G\alpha_i$ protein, not by other endogenous PTX-insensitive $G\alpha$ subunits such as $G\alpha_z$, $G\alpha_q$, or $G\alpha_s$, which may also interact weakly with the α_{2A} receptor in PC12 cells (Chabre et al., 1994).

RGSZ1 Regulates the D2 Dopamine Receptor (D2R) Agonist-Induced SRE Pathway

The effect of RGSZ1 in attenuating $G\alpha_i$-coupled signal transduction in mammalian cells can be demonstrated further in the D2R pathway. The D2R is coupled predominantly to $G\alpha_i$ subunits. We have found previously that the activation of D2R by quinpirole, a D2R agonist, results in a substantial activation of the serum response element. This activation is $G\alpha_i/\beta\gamma$ dependent, as treatment with either PTX or the $G\beta\gamma$-sequestering C terminus of β-adrenergic receptor kinase (βARKct) blocks SRE activation. Thus, SRE transcriptional activity can be used as another functional readout for testing the regulation of $G\alpha_i$ by RGSZ1.

Generation of DNA Constructs and General Reagents

The human RGSZ1 is cloned into the eukaryotic expression vector pCR3.1 (Invitrogen) using standard molecular cloning techniques. Quinpirole is purchased from Sigma. The luciferase reporter gene assay system is purchased from Applied Biosystems (Foster City, CA). Cis-reporter plasmid pSRE-luc, which contains an SRE-dependent luciferase gene, is obtained from Stratagene (La Jolla, CA). The generation of CHO-K1 cells stably expressing the D2 dopamine receptor has been reported previously (Cockett *et al.*, 1997).

Serum Response Element Reporter Gene Assay

CHO-K1 cells stably expressing the D2R are maintained in 6-well plates in DMEM containing 10% fetal bovine serum, 5 μg/ml mycophenolic acid, 0.25 mg/ml xanthine, and HT supplement. Cells are transfected with cis-reporter plasmid pSRE-luc (Stratagene), which contains an SRE-dependent luciferase gene, and RGSZ1 expression plasmids. Cells are then starved in serum-free medium overnight and treated for 5 h with 10 μM quinpirole, an agonist of D2R. After washing twice with PBS, cell lysates are prepared by scraping the cells into 250 μl of lysis buffer containing 100 mM potassium phosphate, pH 7.8, 0.2% Triton X-100, and 0.5 mM DTT. The cell lysate is centrifuged for 2 min at 14,000g to remove cell debris, and the supernatant is used for the luciferase assay. Luciferase activity is measured using a reporter gene assay kit from Applied Biosystems. Measurements are performed with a luminometer with 0.1-s readings per well. Most often, the luciferase activities are normalized to β-galacosidase activities obtained by cotransfection of a *lacZ* expression plasmid. Figure 7 shows an example of these experiments: cotransfection of the RGSZ1 expression plasmid resulted in a 50% decrease of quinpirole-induced SRE activation. These data further indicate that RGSZ1 can play a role in the negative regulation of Gα$_i$-coupled signal transduction in cells.

Conclusion

We described several methods to analyze RGSZ1 interaction with Gα$_i$, including *in vitro*-binding assays and functional studies. Using GST-fusion protein pull-down assays, the GST-RGSZ1 protein was shown to bind the ^{35}S-labeled Gα$_{i1}$ protein in an AlF$_4^-$-dependent manner. The interaction between RGSZ1 and Gα$_i$ was confirmed further by coimmunoprecipitation studies and yeast two-hybrid experiments using a quantitative luciferase reporter gene. Extending these observations to functional studies, RGSZ1 accelerated the intrinsic GTPase activity of Gα$_{i1}$ in single turnover GTPase

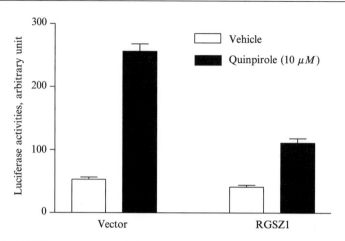

Fig. 7. RGSZ1 attenuates quinpirole-stimulated SRE activation. CHO cells stably expressing the D2 dopamine receptor were transfected with either empty vector or RGSZ1, and with pSRE-Luc reporter plasmid and pβGal reporter plasmid. Cells were then cultured in serum-free DMEM medium overnight and treated with 10 μM quinpirole for 5 h. Luciferase and β-galacosidase activities of cell lysates were measured with a luminometer. The activity of luciferase was normalized with βGal activity. Data shown are means ± SE of triplicate transfections in three independent experiments. Reproduced with permission from Wang et al. (2002).

assays. Human RGSZ1 inhibited the Gpa1-mediated yeast pheromone response when expressed in an SST2 yeast knockout strain. In PC12 cells, transfected RGSZ1 blocked MAPK activity induced by UK14304, an α_2-adrenergic receptor agonist. Furthermore, RGSZ1 attenuated D2 dopamine receptor agonist-induced SRE reporter gene activity in CHO cells. In summary, these data suggest that RGSZ1 serves as a GAP for $G\alpha_i$ and regulates $G\alpha_i$-mediated signaling, thus expanding the potential role of RGSZ1 in G-protein-mediated cellular activities.

References

Alblas, J., van Corven, E. J., Hordijk, P. L., Milligan, G., and Moolenaar, W. H. (1993). G_i-mediated activation of the p21^{ras}-mitogen-activated protein kinase pathway by α_2-adrenergic receptors expressed in fibroblasts. *J. Biol. Chem.* **268**, 22235–22238.

Barker, S. A., Wang, J., Sierra, D. A., and Ross, E. M. (2001). RGSZ1 and ret RGS: Two of several splice variants from the gene *RGS20*. *Genomics* **78**, 223–229.

Berman, D. M., Kozasa, T., and Gilman, A. G. (1996). The GTPase-activating protein RGS4 stabilizes the transition state for nucleotide hydrolysis. *J. Biol. Chem.* **271**, 27209–27212.

Casey, P. J., Fong, H. K., Simon, M. I., and Gilman, A. G. (1990). Gz, a guanine nucleotide-binding protein with unique biochemical properties. *J. Biol. Chem.* **265**, 2383–2390.

Cavalli, A., Druey, K. M., and Milligan, G. (2000). The regulator of G protein signaling RGS4 selectively enhances α_{2A}-adreoreceptor stimulation of the GTPase activity of $G_{o1}\alpha$ and $G_{i2}\alpha$. J. Biol. Chem. **275,** 23693–23699.

Chabre, O., Conklin, B. R., Brandon, S., Bourne, H. R., and Limbird, L. E. (1994). Coupling of the α_{2A}-adrenergic receptor to multiple G-proteins. J. Biol. Chem. **269,** 5730–5734.

Chen, C., Zheng, B., Han, J., and Lin, S. C. (1997). Characterization of a novel mammalian RGS protein that binds to Gα proteins and inhibits pheromone signaling in yeast. J. Biol. Chem. **272,** 8679–8685.

Cockett, M. I., Ochalski, R., Benwell, K., Franco, R., and Wardell-Swanson, J. (1997). Simultaneous expression of multi-subunit proteins in mammalian cells using a convenient set of mammalian cell expression vectors. Biotechniques **23,** 402–405.

De Vries, L., Elenko, E., Hubler, L., Jones, T. Z., and Farquhar, M. G. (1996). GAIP is membrane-anchored by palmitoylation and interacts with the activated (GTP-bound) form of $G\alpha_1$ subunits. Proc. Natl. Acad. Sci. USA **93,** 15203–15208.

De Vries, L., and Farquhar, M. G. (2002). Screening for interacting partners for Gαi3 and RGS-GAIP using the two-hybrid system. Methods Enzymol. **344,** 657–673.

De Vries, L., Mousli, M., Wurmser, A., and Farquhar, M. G. (1995). GAIP, a protein that specifically interacts with the trimeric G protein $G\alpha_{i3}$, is a member of a protein family with a highly conserved core domain. Proc. Natl. Acad. USA **92,** 11916–11920.

Dietzel, C., and Kurjan, J. (1987). The yeast SCG1 gene: A α-like protein implicated in the A- and α-factor response pathway. Cell **50,** 1001–1010.

Druey, K. M., Blumer, K. J., Kang, V. H., and Kehrl, J. H. (1996). Inhibition of G-protein-mediated MAP kinase activation by a new mammalian gene family. Nature **379,** 742–746.

Fan, X., Brass, L. F., Poncz, M., Spitz, F., Maire, P., and Manning, D. R. (2000). The α subunits of G_z and G_i interact with the eyes absent transcription confactor Eya2, preventing its interaction with the six class of homeodomain-containing proteins. J. Biol. Chem. **275,** 32129–32134.

Fields, S., and Song, O. (1989). A novel genetic system to detect protein-protein interactions. Nature **340,** 245–246.

Glick, J. L., Meigs, T. E., Miron, A., and Casey, P. J. (1998). RGSZ1, a G_z-selective regulator of G protein signaling whose action Is selective to the phosphorylation state of $G_z\alpha$. J. Biol. Chem. **273,** 26008–26013.

Heximer, S. P., Srinivasa, S. P., Bernstein, L. S., Bernard, J. L., Linder, M. E., Hepler, J. R., and Blumer, K. J. (1999). G protein selectivity is a determinant of RGS2 function. J. Biol. Chem. **274,** 34253–34259.

Heximer, S. P., Watson, N., Linder, M. E., Blumer, K. J., and Hepler, J. R. (1997). RGS2/G0S8 is a selective inhibitor of Gqα function. Proc. Natl. Acad. Sci. USA **94,** 14389–14393.

Hoffman, G. A., Garrison, T. R., and Dohlman, H. G. (2002). Analysis of RGS proteins in Saccharomyces cerevisiae. Methods Enzymol. **344,** 617–631.

Hunt, T. W., Fields, T. A., Casey, P. J., and Peralta, E. G. (1996). RGS10 is a selective activator of Gαi GTPase activity. Nature **383,** 175–177.

Journot, L., Bockaert, J., and Audigier, Y. (1989). Reconstitution of cyc-S49 membranes by in vitro translated Gsα membrane anchorage and functional implications. FEBS Lett. **251,** 230–236.

Kimple, R. J., Jones, M. B., Shutes, A., Yerxa, B. R., Siderovski, D. P., and Willard, F. S. (2003). Established and emerging fluorescence-based assays for G-protein function: Heterotrimer G-protein alpha subunits and regulator of G-protein signaling (RGS) proteins. Comb. Chem. High Throughput Screen **6,** 399–407.

Kimple, R. J., Vries, L. D., Tronchere, H., Behe, C. I., Morris, R. A., Farquhar, M. G., and Siderovski, D. P. (2001). RGS12 and RGS14 GoLoco motifs are $G_{\alpha i}$ interaction sites with guanine nucleotide dissociation inhibitor activity. *J. Biol. Chem.* **276**, 29275–29281.

Krumins, A. M., and Gilman, A. G. (2002). Assay of RGS protein activity *in vitro* using purified components. *Methods Enzymol.* **344**, 673–685.

Lan, K.-L., Sarvazyan, N. A., Taussig, R., Mackenzie, R. G., DiBello, P. R., Dohlman, H. G., and Neubig, R. R. (1998). A point mutation in Galpha o and Galpha i1 blocks interaction with regulator of G protein signaling proteins. *J. Biol. Chem.* **273**, 12794–12797.

Miyajima, I., Nakafuku, M., Nakayama, N., Brenner, C., Miyajima, A., Kaibuchi, K., Arai, K., Kaziro, Y., and Matsumoto, K. (1987). GPA1, a haploid-specific essential gene, encodes a yeast homolog mammalian protein which may be involved in mating factor signal transduction. *Cell* **50**, 1011–1019.

Natochin, M., Granovsky, A. E., and Artemyev, N. O. (1997). Regulation of transducin GTPase activity by human retinal RGS. *J. Biol. Chem.* **272**, 17444–17449.

Neubig, R. R., and Siderovski, D. P. (2002). Regulators of G-protein signalling as new central nervous system drug targets. *Nature Rev. Drug Disc.* **1**, 187–197.

Nieuwenhuijsen, B. W., Huang, Y., Wang, Y., Ramirez, F., Kalgaonkar, G., and Young, K. H. (2003). A dual luciferase multiplexed high-throughput screening platform for protein-protein interactions. *J. Biomol. Screen.* **6**, 676–684.

Ozenberger, B. A., and Young, K. H. (1995). Functional interaction of ligands and receptors of the hematopoietic superfamily in yeast. *Mol. Endocrinol.* **9**, 1321–1329.

Popov, S., Yu, K., Kozasa, T., and Wilkie, T. M. (1997). The regulators of G protein signaling (RGS) domains of RGS4, RGS10, and GAIP retain GTPase activating protein activity *in vitro*. *Proc. Natl. Acad. Sci. USA* **94**, 7216–7220.

Rose, M. D., Winston, F., and Heiter, P. (1990). Methods in Yeast Genetics – A Laboratory Course Manual. *Cold Spring Harbor Laboratory Press.*

Ross, E. M. (2002). Quantitative assays for GTPase-activating proteins. *Methods Enzymol.* **344**, 601–617.

Ross, E. M., and Wilkie, T. M. (2000). GTPase-activating proteins for heterotrimeric G proteins: Regulators of G protein signaling (RGS) and RGS-like proteins. *Annu. Rev. Biochem.* **69**, 795–827.

Wang, J., Ducret, A., Tu, Y., Kozasa, T., Aebersold, R., and Ross, E. M. (1998). RGSZ1, a Gz-selective RGS protein in brain: Structure, membrane association, regulation by $G_{\alpha z}$ phosphorylation, and relationship to a Gz GTPase-activating protein subfamily. *J. Biol. Chem.* **273**, 26014–26025.

Wang, J., Tu, Y., Woodson, J., Song, X., and Ross, E. M. (1997). A GTPase-activating protein for the G protein $G_{\alpha z}$: Identification, purification, and mechanism of action. *J. Biol. Chem.* **272**, 5732–5740.

Wang, Y., Ho, G., Zhang, J. J., Nieuwenhuijsen, B., Edris, W., Chanda, P. K., and Young, K. H. (2002). Regulator of G protein signaling Z1 (RGSZ1) interacts with $G_{\alpha i}$ subunits and regulates $G_{\alpha i}$-mediated cell signaling. *J. Biol. Chem.* **277**, 48325–48332.

Yan, Y., Chi, P. P., and Bourne, H. R. (1997). RGS4 inhibits G_q-mediated activation of mitogen-activated protein kinase and phosphoinositide synthesis. *J. Biol. Chem.* **272**, 11924–11927.

Young, K. H., Wang, Y., Bender, C., Ajit, S., Ramirez, F., Gilbert, A., and Nieuwenhuijsen, B. W. (2004). Yeast-based screening for inhibitors of regulators of G-protein signaling proteins. *Methods Enzymol.* **389**, 277–301.

[4] Analysis of the Regulation of Microtubule Dynamics by Interaction of RGSZ1 (RGS20) with the Neuronal Stathmin, SCG10

By ANDREW B. NIXON and PATRICK J. CASEY

Abstract

Regulators of G protein signaling (RGS proteins) are a diverse family of proteins that act to negatively regulate signaling by heterotrimeric G proteins; however, recent data have implied additional functions for RGS proteins. Previously, we employed the yeast two-hybrid system and identified the microtubule-destabilizing protein, superior cervical ganglia neural-specific 10 protein (SCG10), as a potential effector protein of RGSZ1. This article describes the expression and biochemical purification of both RGSZ1 and SCG10 and details the development of various *in vitro* assays to evaluate microtubule polymerization/depolymerization. Both turbidimetric and microscopy-based assays can be employed to study the impact that RGS proteins have on SCG10 function. The application of these *in vitro* assays may help identify a novel role for RGS proteins in regulating the cytoskeletal network.

Introduction

Regulators of G protein signaling (RGS) proteins comprise a class of signaling molecules that play major roles in mediating the GTP hydrolysis of heterotrimeric G proteins (Hepler, 1999; Ross and Wilkie, 2000). RGS proteins act as GTPase-accelerating proteins (GAPs) by hastening the hydrolysis of GTP to GDP by G protein α subunits. All identified RGS proteins contain a conserved region of approximately 120 residues referred to as the RGS domain; however, the family members differ greatly in size, sequence outside the RGS domain, cellular localization, and tissue distribution (Burchett, 2000; Ross and Wilkie, 2000). These features have led to the hypothesis that at least some RGS proteins can bind alternative target proteins in addition to Gα subunits. Data indicate that this can indeed occur, and a variety of motifs are present in RGS proteins that are important for a number of protein–protein interactions. For example, the R7 family of RGS proteins contains a DEP domain and a novel G protein γ-like (GGL) domain. The DEP domain is found in a variety of proteins and has been hypothesized to play a role in targeting DEP-containing proteins to G protein-coupled receptors (Ponting and Bork, 1996). GGL domains have been shown to be responsible for selective binding to Gβ5, a Gβ isoform quite divergent from Gβ isoforms 1–4 (Cabrera et al., 1998; Snow et al., 1998; Sondek and Siderovski, 2001). Additionally, RZ family members contain a cysteine string motif that is thought to play a role in membrane attachment (De Vries et al., 1996; Glick et al., 1998; Wang et al., 1998). These data suggest that portions outside the RGS domain play important roles in determining Gα specificity, membrane attachment, cellular localization, and other important functions.

Nixon et al. (2002) and Liu et al. (2002) have shown that RGS proteins can interact with a microtubule-destabilizing protein termed superior cervical ganglia neural-specific 10 protein (SCG10). SCG10 is a member of the stathmin family of proteins, all of which bind to tubulin to act as a sequestering agent, as well as to promote microtubule catastrophes (Schubart et al., 1989; Stein et al., 1988). SCG10 is a developmentally regulated, neuronal-specific protein, exhibiting high expression levels during periods of axonal growth and regeneration (Grenningloh et al., 2004). Furthermore, when compared to the ubiquitous stathmin, SCG10 has an additional N-terminal domain that is critical for membrane binding (Ozon et al., 1997). We initially identified SCG10 as a binding partner for RGSZ1 through yeast two-hybrid experiments and then sought to identify a functional consequence of this interaction. This article describes in detail the development of a variety of in vitro assays to investigate the effects of RGS proteins on microtubule stability through their abilities to interact with SCG10.

Protein Expression and Purification

One of the hurdles that needs to be overcome to evaluate biochemically the effect of RGS proteins on SCG10 function is the purification of SCG10 itself. Initial reports indicated that SCG10 was difficult to purify and even more difficult to obtain concentrated preparations of the protein (Antonsson et al., 1997). We were able to overcome this problem by expressing a glutathione-S-transferase (GST) fusion protein that could be recovered from insoluble inclusion bodies using urea denaturation.

The initial step in the process of producing recombinant SCG10 is to subclone the SCG10 gene into an appropriate GST-expression vector. We have had great success using many of the pGEX vectors (Amersham Pharmacia Biotech, Inc., Piscataway, NJ) for the expression of GST-tagged proteins. These vectors have been optimized for use in a variety of bacterial hosts, yield high-level bacterial expression, and are often engineered with a cleavage site for easy removal of the tag, if desired. Once subcloning is complete, the DNA should always be sequenced to validate proper insertion of the gene.

After sequence validation, plasmids are transformed into BL21DE3 (pLysS) cells (Invitrogen Corporation, CA), grown to an optical density (A_{600}) of 0.6 at 37°, and protein production induced with 0.5 mM isopropyl β-D-thiogalactopyranoside (IPTG) for 3.5 h at 37°. Cells (1 liter) are subjected to centrifugation at 5000g and are resuspended in 20 ml of 0.5 M sucrose, 50 mM Tris–HCl (pH 7.7), 7.5 mM KCl, 1 mM EDTA, 1 mM dithiothreitol (DTT), and a mixture of protease inhibitors referred to as "PI mix" [0.27 mM phenylmethylsulfonyl fluoride, 0.06 mM L-1-chloro-3-(4-tosylamido)-4-phenyl-2-butanone-HCl, 0.06 mM L-1-chloro-3-(4-tosylamido)-7-amino-2-heptanone-HCl]. The resuspended cells are lysed mechanically using a French press and the resulting extract is centrifuged at 20,000g for 30 min at 4°.

As mentioned earlier, full-length SCG10 fused to GST is insoluble when produced in bacteria. In order to solubilize GST-SCG10, the 20,000g pellet obtained after centrifugation of the lysate is resuspended in 8 M urea (5 ml urea/L cell pellet), incubated for 2 h at 4°, and next centrifuged at 30,000g for 30 min at 4° in order to obtain solubilized protein. The protein contained within the supernatant is then renatured by dialysis against 2 liters of 50 mM Tris–HCl (pH 7.7), 1 mM EDTA, and 1 mM DTT for 12 h at 4°. As is often the case with renaturation protocols, some protein will precipitate during the renaturation process. This precipitate is removed by centrifugation at 20,000g for 30 min at 4°, and the soluble GST-SCG10 is then purified by affinity chromatography. For the batch affinity chromatography purification of SCG10, glutathione-Sepharose beads (Amersham Pharmacia Biotech, Inc.) are added to the solubilized preparation of SCG10 (1–2 ml resin/L bacterial culture) and incubated for 1 h at 4°. The GST-SCG10-coupled beads are washed three times with 20 ml of 50 mM Tris–HCl (pH 7.7), 1 mM EDTA, and 1 mM DTT (buffer A), and the protein is eluted from beads in 1.5 ml of 50 mM Tris–HCl (pH 7.7), 1 mM EDTA, containing 50 mM glutathione. It is crucial to adjust the pH of this solution after glutathione addition, as the glutathione will dramatically lower the pH of the elution buffer. Further, by eluting GST-SCG10 from the beads in a small volume, no subsequent concentration step is needed. Eluted GST-SCG10 is then dialyzed directly into PEM buffer [80 mM PIPES (pH 6.9), 1 mM EGTA, and 1 mM MgCl$_2$]. This buffer is required for subsequent assays in which microtubule assembly/disassembly is measured; however, other buffers may be substituted for use in other assay systems.

RGSZ1 and an N-terminal truncation mutant (ΔRGSZ1, composed of residues 87–217) can be readily produced and purified as His-tagged proteins (Glick *et al.*, 1998). The cDNA encoding full-length RGSZ1 can be obtained from the Guthrie cDNA Resource Center (Sayre, PA), which has a continually growing repository of G protein-related cDNAs for purchase. The RGSZ1 cDNA can be subcloned into a pRSET expression vector (Invitrogen Corporation), and plasmids containing the RGSZ1 insert are transformed into BL21DE3(pLysS) cells (Invitrogen Corporation). We found a wide discrepancy in the expression of RGSZ1 among different bacterial colonies upon induction with IPTG. As a result, usually six to eight different colonies are grown individually at 30° to an A_{600} of 0.6. Following induction with 0.5 mM IPTG for 4 h at 37°, RGSZ1 expression is monitored by Coomassie Blue staining of sodium dodecyl sulfate–polyacrylamide gel electrophoresis (SDS–PAGE) gels. Only cells that produce RGSZ1 are harvested by centrifugation and resuspended in 20 mM Tris–HCl (pH 7.7), 10 mM β-mercaptoethanol, and PI mix. The resuspended cells (20 ml/L bacterial culture) are lysed mechanically using a French press, and the extract is diluted 1:1 with 20 mM Tris–HCl (pH 7.7), 10 mM β-mercaptoethanol, 300 mM NaCl, 0.2% lubrol, and PI mix. The diluted extract is centrifuged at 30,000g for 60 min at 4°, and soluble RGSZ1 is purified by affinity chromatography using Ni-NTA resin (Qiagen Inc., Valencia, CA). The supernatant is applied to a Ni-NTA column (0.75 ml resin/1 liter bacterial culture) preequilibrated in 20 mM Tris–HCl (pH 7.7), 10 mM β-mercaptoethanol, 150 mM NaCl, 0.1% lubrol, and PI mix. The column is next washed with 4 column volumes of 20 mM Tris–HCl (pH 7.7), 10 mM β-mercaptoethanol, 150 mM NaCl, 0.1% lubrol, 5 mM imidazole, and PI mix. A stepwise gradient of imidazole is initiated by using 3 column volumes each of 20 mM Tris–HCl (pH 7.7), 10 mM β-mercaptoethanol, 150 mM NaCl, and 0.1% lubrol containing increasing concentrations of imidazole (5, 10, 25, 50, 100, 150, 250, and 500 mM). Fractions are analyzed by SDS–PAGE; we find that RGSZ1 usually elutes around 150 mM imidazole. Fractions containing RGSZ1 are dialyzed against the desired buffer (i.e., PEM buffer), and aliquots are frozen rapidly in liquid nitrogen and stored at −80°.

Microtubule Assembly and Disassembly Assays:
Turbidimetric Assays

A well-characterized activity of SCG10 is its ability to impact microtubule assembly (Riederer *et al.*, 1997). One method for evaluating the assembly and disassembly of tubulin *in vitro* is a light-scattering assay (Gaskin *et al.*, 1974). This assay is based on the fact that assembled microtubules cause a solution to become turbid, while free tubulin does not. In assays such as these, the key factor is temperature. Tubulin can undergo a temperature-cycling phenomenon where at low temperatures (i.e., 4°), the tubulin remains unpolymerized; however, elevation of the temperature to 37° causes rapid polymerization of the free tubulin into microtubules (Fig. 1). This reversible cycling of tubulin from an unpolymerized state to an assembled state can be measured quantitatively using a spectrophotometer.

The first step in performing microtubule assembly assays is obtaining purified tubulin. There are commercial sources from which one can purchase tubulin (i.e., Cytoskeleton Inc., Denver, CO), but isolating tubulin from porcine brain is a fairly simple and straightforward procedure

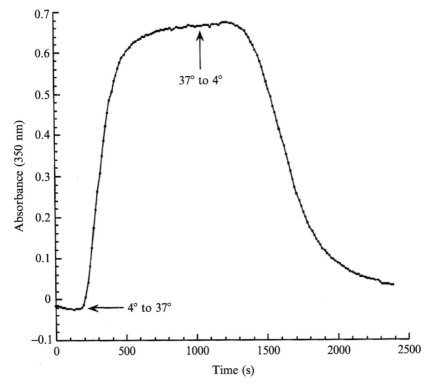

FIG. 1. Temperature cycling of microtubules induces polymerization and depolymerization *in vitro*. Mixed tubulin (tubulin with MAPs, 3.25 mg/ml) was incubated in PEM buffer with 1 mM GTP at 4° for 1 min. Polymerization was initiated by raising the temperature to 37°, and microtubule formation was allowed to proceed until reaching equilibrium. At this point, indicated by the arrow, the temperature was adjusted back to 4° to induce microtubule depolymerization. Data are from a single experiment that is representative of four separate experiments.

(Shelanski *et al.*, 1973; Williams and Lee, 1982) and is not discussed here. It must be noted that tubulin isolated from brain contains a variety of microtubule-associated proteins (MAP1, MAP2, tau, etc.) that aid in the stabilization of microtubules. While the presence of MAPs in these assembly/disassembly assays will lower the critical concentration of tubulin needed for assembly to occur, MAPs are not necessary components for this assay system. If desired, MAPs can be removed readily from tubulin preparations using phosphocellulose chromatography (see Williams and Lee, 1982).

For microtubule assembly assays, MAP-rich tubulin (3.25 mg/ml final concentration) in PEM buffer containing 1 mM GTP at 4° is mixed with SCG10, RGSZ1, or a combination of these proteins. Microtubule assembly is initiated by raising the temperature to 37°, and the absorbance at 350 nm is monitored over 15 min in a Hewlett Packard 8453 spectrophotometer. This particular spectrophotometer is equipped with a water-jacketed cuvette holder that allows for the maintenance of constant temperature. In order to change the temperature rapidly, it is essential to have two water baths to which the cuvette holder was attached, one at 37° and one at 4°. We have found that after obtaining a good baseline measurement, rapidly switching the water lines to the spectrophotometer at the start of each experiment gave excellent and reproducible results. Addition of SCG10 elicits a dose-dependent decrease in tubulin polymerization under the assay conditions (Fig. 2A). However, even at concentrations as high as 40 μM, RGSZ1 has no discernible effect on the ability of SCG10 to attenuate microtubule assembly (Fig. 2B). This was true whether RGSZ1 and SCG10 were preincubated at 4° for 1 h or added simultaneously to the tubulin-containing cuvette; in either case, no effect of RGSZ1 on SCG10-induced inhibition of microtubule assembly could be observed.

In addition to its ability to prevent microtubule assembly, SCG10 is also capable of initiating microtubule disassembly *in vitro*. Microtubule disassembly experiments are more difficult technically to perform, but generally follow the procedure described earlier. Tubulin is initially assembled in GTP-containing PEM buffer at 37°, added to the cuvette, and the temperature is maintained at 37°. The absorbance at 350 nm is monitored over 10 min at a constant temperature of 37°. Under these conditions, the addition of SCG10 promotes the disassembly of assembled microtubules in a concentration-dependent manner (Fig. 3A). Under these assay conditions, we observed complete disassembly of assembled microtubules upon inclusion of 15–20 μM SCG10. Using an intermediate dose of SCG10 (5 μM), preincubation of SCG10 with RGSZ1 blocks SCG10-induced microtubule disassembly in a dose-dependent manner (Fig. 3B). In these

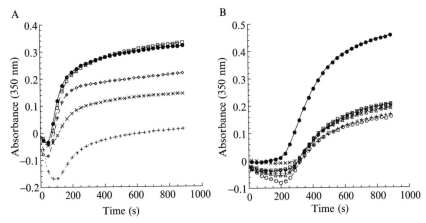

FIG. 2. Effect of SCG10 and RGSZ1 on microtubule assembly *in vitro*. (A) Effect of SCG10 on microtubule assembly. Mixed tubulin (tubulin with MAPs, 3.25 mg/ml) was incubated in PEM buffer with 1 mM GTP at 4° in the presence of 0 μM (●), 1 μM (□) 2.5 μM (◇), 5 μM (×), or 10 μM (+) GST-SCG10. Polymerization was induced by elevating the temperature to 37° and monitored as described. Data are from a single experiment that is representative of four separate experiments. (B) Effect of RGSZ1 on the ability of SCG10 to attenuate microtubule assembly. Mixed tubulin (tubulin with MAPs, 3.25 mg/ml) was incubated in PEM buffer with 1 mM GTP at 4° either alone (●) or in the presence of 5 μM SCG10 preincubated with 0 μM (□), 2.5 μM (◇), 5 μM (×), 10 μM (+), 20 μM (△), or 40 μM (○) RGSZ1. Polymerization was induced by elevating the temperature to 37° and monitored as described. Data are from a single experiment that is representative of four separate experiments. Reprinted with permission from Nixon *et al.* (2002).

experiments, the two proteins (SCG10 and RGSZ1) were preincubated at 4° for 1 h prior to addition to the cuvette; we have found that this is necessary to observe complete inhibition by RGSZ1 of SCG10-induced microtubule disassembly.

Microtubule Assembly and Disassembly Assays: Fluorescence Analysis

In addition to the indirect method of monitoring microtubule formation using light scattering, methods have been developed to visualize this process directly using fluorescence microscopy (Belmont and Mitchison, 1996). Using rhodamine-labeled tubulin in *in vitro* assembly/disassembly systems, microtubule formation can be viewed directly by fluorescence microscopy. These assays allow for the direct observation of microtubules and are a nice

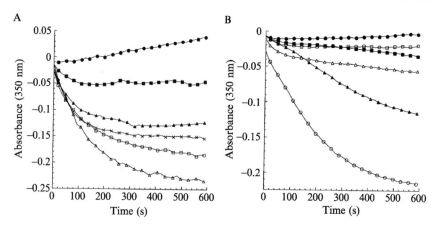

FIG. 3. Effect of SCG10 and RGSZ1 on microtubule disassembly *in vitro*. (A) Effect of SCG10 on microtubule disassembly. Microtubules were formed initially by incubating mixed tubulin (tubulin with MAPs, 3.25 mg/ml) in PEM buffer with 1 mM GTP at 37° for 30 min. SCG10 [0 μM (●), 1 μM (■), 2.5 μM (▲), 5 μM (×), 10 μM (□), or 15 μM (△)] was added to the microtubules, the spectrophotometer was zeroed, and the absorbance was measured as described. Data are from a single experiment that is representative of four separate experiments. (B) Effect of RGSZ1 on SCG10-induced microtubule disassembly. Microtubules were formed initially by incubating mixed tubulin (tubulin with MAPs, 3.25 mg/ml) in PEM buffer with 1 mM GTP at 37° for 30 min. Disassembly of microtubules alone (●) or microtubules treated with SCG10 (5 μM) that had been preincubated with 0 μM (○), 1 μM (▲), 2.5 μM (△), 5 μM (■), or 10 μM (□) RGSZ1 was monitored as described. Data are from a single experiment that is representative of four separate experiments. Reprinted with permission from Nixon *et al.* (2002).

complement to the turbidimetric assays. A key component for this assay is rhodamine-labeled tubulin. There are protocols that describe the synthesis of rhodamine-labeled tubulin (see Hyman *et al.*, 1991); however, very small quantities are actually needed to perform these assays. As a result, we obtained our rhodamine-labeled tubulin from Cytoskeleton Inc. The fluorescence-based assays can be set up to evaluate either assembly, by allowing the free tubulin to assemble under experimental conditions, or disassembly, by first forming microtubules and subsequently monitoring the loss of microtubules. For assembly experiments, MAP-free tubulin and rhodamine-labeled tubulin are premixed at a ratio of 4:1 (unlabeled:labeled) to a final concentration of 3.25 mg/ml in PEM buffer containing 2 mM GTP at 4° and GST-SCG10, GST-SCG10 plus RGSZ1, or buffer alone is added. These mixtures are then incubated at 37° for 10 min, and the assembly of microtubules is monitored. For disassembly experiments,

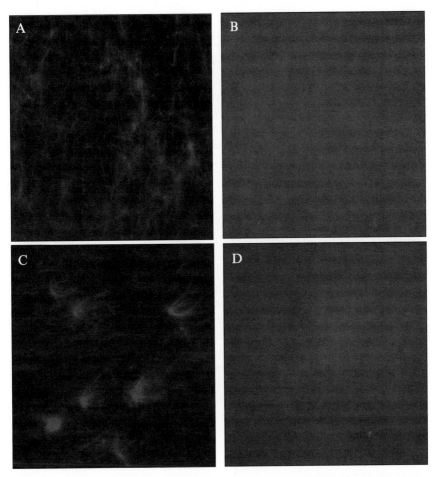

Fig. 4. Effect of RGSZ1 on SCG10-induced disassembly of rhodamine-labeled microtubules. Rhodamine-labeled microtubules (3.25 mg/ml) were formed by incubating MAP-free tubulin and rhodamine-labeled tubulin premixed at a ratio of 4:1 (unlabeled: labeled) to a final concentration of 3.25 mg/ml in PEM buffer containing 2 mM GTP. Microtubules were incubated with buffer (A), 5 μM GST-SCG10 (B), or the combination of 5 μM GST-SCG10 and 10 μM RGSZ1 (C) for 10 min at 37°. Microtubule disassembly was monitored by fluorescence microscopy using the same conditions for each image. Data are from a single experiment that is representative of four separate experiments. As a control (D), rhodamine-labeled tubulin (3.25 mg/ml) was incubated in PEM buffer at 4° without GTP for 30 min and viewed as just described. Reprinted with permission from Nixon *et al.* (2002).

MAP-free tubulin and rhodamine-labeled tubulin are premixed at a ratio of 4:1 (unlabeled:labeled) to a final concentration of 3.25 mg/ml in PEM buffer containing 2 mM GTP at 37° for 30 min to form microtubules. Following this incubation period, GST-SCG10, GST-SCG10 plus RGSZ1, or buffer alone is added to the mixture, which is then incubated at 37° for 10 min. In either the assembly or the disassembly protocols, reactions (10 μl) are diluted into 50 μl of 60% glycerol and mixed gently. A 4-μl aliquot of the glycerol suspension is then placed on a glass slide with a coverslip, and microtubule patterns are observed by fluorescence microscopy using a Nikon Eclipse TE300 inverted microscope (Nikon, Melville, NY).

Incubation of rhodamine-labeled tubulin with GTP at 37° induces the assembly of microtubules (Fig. 4A). For control purposes, incubation in the absence of GTP at 4° prevented microtubule assembly, which is indicated by a diffuse staining pattern (Fig. 4D). Addition of 5 μM SCG10 to the rhodamine-labeled microtubule mixture results in almost complete loss of the polymerized structure, which also coincides with an increase in the background fluorescence pattern, indicative of soluble tubulin (Fig. 4B). Preincubation of SCG10 with a twofold excess of RGSZ1 blocks the ability of SCG10 to induce microtubule disassembly (Fig. 4C). The inhibition of SCG10-induced disassembly in this assay system mirrors the observations made in the turbidimetric assays. The finding that RGSZ1 specifically blocks microtubule disassembly but not assembly is very intriguing and may allow for a biochemical separation of these two activities. However, additional studies are needed to clarify the mechanism by which RGSZ1 acts on the ability of SCG10 to promote microtubule disassembly.

Concluding Remarks

We described herein the production and purification of bacterially produced RGSZ1 and SCG10 as well as various *in vitro* assay systems to evaluate the impact of RGS proteins on SCG10-induced effects on microtubule structure. Reports indicate that at least two RGS proteins, RGSZ1 and RGS6, can interact with SCG10. These data provide the first evidence that RGS proteins can directly modulate cytoskeletal organization. Given the fact that SCG10 is a neuronal-specific protein, the ability to study these effects *in vitro* offers researchers a more amenable system to evaluate the impact of RGS proteins on SCG10 function. However, there is a clear need for an appropriate experimental system to understand further the biological process(es) associated with this interaction.

Acknowledgments

We thank Gabriele Grenningloh for providing various reagents and valuable discussions throughout the project. This work was supported by National Institutes of Health Grant RO1 GM55717 (to PJC) and National Research Service Award Fellowship F32 GM19663 (to ABN).

References

Antonsson, B., Lutjens, R., Di Paolo, G., Kassel, D., Allet, B., Bernard, A., Catsicas, S., and Grenningloh, G. (1997). Purification, characterization, and *in vitro* phosphorylation of the neuron-specific membrane-associated protein SCG10. *Protein Expr. Purif.* **9,** 363–371.

Belmont, L. D., and Mitchison, T. J. (1996). Identification of a protein that interacts with tubulin dimers and increases the catastrophe rate of microtubules. *Cell* **84,** 623–631.

Burchett, S. A. (2000). Regulators of G protein signaling: A bestiary of modular protein binding domains. *J. Neurochem.* **75,** 1335–1351.

Cabrera, J. L., de Freitas, F., Satpaev, D. K., and Slepak, V. Z. (1998). Identification of the Gbeta5-RGS7 protein complex in the retina. *Biochem. Biophys. Res. Commun.* **249,** 898–902.

De Vries, L., Elenko, E., Hubler, L., Jones, T. L., and Farquhar, M. G. (1996). GAIP is membrane-anchored by palmitoylation and interacts with the activated (GTP-bound) form of G alpha i subunits. *Proc. Natl. Acad. Sci. USA* **93,** 15203–15208.

Gaskin, F., Cantor, C. R., and Shelanski, M. L. (1974). Turbidimetric studies of the *in vitro* assembly and disassembly of porcine neurotubules. *J. Mol. Biol.* **89,** 737–755.

Glick, J. L., Meigs, T. E., Miron, A., and Casey, P. J. (1998). RGSZ1, a Gz-selective regulator of G protein signaling whose action is sensitive to the phosphorylation state of Gzalpha. *J. Biol. Chem.* **273,** 26008–26013.

Grenningloh, G., Soehrman, S., Bondallaz, P., Ruchti, E., and Cadas, H. (2004). Role of the microtubule destabilizing proteins SCG10 and stathmin in neuronal growth. *J. Neurobiol.* **58,** 60–69.

Hepler, J. R. (1999). Emerging roles for RGS proteins in cell signalling. *Trends Pharmacol. Sci.* **20,** 376–382.

Hyman, A., Drechsel, D., Kellogg, D., Salser, S., Sawin, K., Steffen, P., Wordeman, L., and Mitchison, T. (1991). Preparation of modified tubulins. *Methods Enzymol.* **196,** 478–485.

Liu, Z., Chatterjee, T. K., and Fisher, R. A. (2002). RGS6 interacts with SCG10 and promotes neuronal differentiation. *J. Biol. Chem.* **277,** 37832–37889.

Nixon, A. B., Grenningloh, G., and Casey, P. J. (2002). The interaction of RGSZ1 with SCG10 attenuates the ability of SCG10 to promote microtubule disassembly. *J. Biol. Chem.* **277,** 18127–18133.

Ozon, S., Maucuer, A., and Sobel, A. (1997). The stathmin family: Molecular and biological characterization of novel mammalian proteins expressed in the nervous system. *Eur. J. Biochem.* **248,** 794–806.

Ponting, C. P., and Bork, P. (1996). Pleckstrin's repeat performance: A novel domain in G-protein signaling? *Trends Biochem. Sci.* **21,** 245–246.

Riederer, B. M., Pellier, V., Antonsson, B., Di Paolo, G., Stimpson, S. A., Lutjens, R., Catsicas, S., and Grenningloh, G. (1997). Regulation of microtubule dynamics by the neuronal growth-associated protein SCG10. *Proc. Natl. Acad. Sci. USA* **94,** 741–745.

Ross, E. M., and Wilkie, T. M. (2000). GTPase-activating proteins for heterotrimeric G proteins: Regulators of G protein signaling (RGS) and RGS-like proteins. *Annu. Rev. Biochem.* **69,** 795–827.

Schubart, U. K., Banerjee, M. D., and Eng, J. (1989). Homology between the cDNAs encoding phosphoprotein p19 and SCG10 reveals a novel mammalian gene family preferentially expressed in developing brain. *DNA* **8,** 389–398.

Shelanski, M. L., Gaskin, F., and Cantor, C. R. (1973). Microtubule assembly in the absence of added nucleotides. *Proc. Natl. Acad. Sci. USA* **70,** 765–768.

Snow, B. E., Krumins, A. M., Brothers, G. M., Lee, S. F., Wall, M. A., Chung, S., Mangion, J., Arya, S., Gilman, A. G., and Siderovski, D. P. (1998). A G protein gamma subunit-like domain shared between RGS11 and other RGS proteins specifies binding to Gbeta5 subunits. *Proc. Natl. Acad. Sci. USA* **95,** 13307–13312.

Sondek, J., and Siderovski, D. P. (2001). Ggamma-like (GGL) domains: New frontiers in G-protein signaling and beta-propeller scaffolding. *Biochem. Pharmacol.* **61,** 1329–1337.

Stein, R., Orit, S., and Anderson, D. J. (1988). The induction of a neural-specific gene, SCG10, by nerve growth factor in PC12 cells is transcriptional, protein synthesis dependent, and glucocorticoid inhibitable. *Dev. Biol.* **127,** 316–325.

Wang, J., Ducret, A., Tu, Y., Kozasa, T., Aebersold, R., and Ross, E. M. (1998). RGSZ1, a Gz-selective RGS protein in brain: Structure, membrane association, regulation by Galphaz phosphorylation, and relationship to a Gz gtpase-activating protein subfamily. *J. Biol. Chem.* **273,** 26014–26025.

Williams, R. C., and Lee, J. C. (1982). Preparation of tubulin from brain. *Methods Enzymol.* **85,** 376–385.

Subsection B

R4 Subfamily

[5] RGS2-Mediated Regulation of Gqα

By Scott P. Heximer

Abstract

Regulators of G-protein signaling (RGS) proteins are GTPase-activating proteins (GAPs) that attenuate signaling by heterotrimeric G proteins. In RGS2, three unique residues within the G-protein-binding (RGS) domain have been shown to direct its selective inhibition of Gqα function. The function of RGS2 as a regulator of Gq is also dependent on regulatory sequences in its amino-terminal domain that direct its localization to the plasma membrane. This work details various approaches that have been used to characterize the relative contribution of the RGS and regulatory domains in RGS2 to its function as a regulator of Gq signaling. Specifically, assays describing (i) the identification of α subunit binding partners for RGS2 (ii) the characterization of RGS2-mediated inhibition of Gq-dependent phosphatidylinositol signaling in tissue culture models, and (iii) the measurement of Gq-dependent calcium responses in vascular smooth muscle cells from RGS2-deficient mice are presented. Results from these studies have been used to demonstrate the high relative potency of RGS2 for the regulation of Gq signaling at the biochemical, cellular, and physiologic level.

Introduction

The regulator of G-protein signaling (RGS) protein family is composed of >30 members, several of which are expressed in a given cell or tissue type. Since many of the RGS isoforms can regulate several different types of heterotrimeric G-protein α subunits, it is an ongoing challenge to identify unique biologic functions for each RGS family member. All RGS proteins contain a characteristic ~110 amino acid domain (RGS box) that mediates their Gα GTPase-activating protein (GAP) function (Berman *et al.*, 1996; Watson *et al.*, 1996). Some RGS box-containing proteins are large multidomain molecules that possess the potential to interact with a diverse set of signaling cascades. Other RGS protein family members are much smaller proteins, consisting of an RGS box flanked by short amino- and carboxyl-terminal extensions. Therefore, these "small" RGS proteins are likely to function primarily as modulators of G-protein function. We hypothesized that G-protein selectivity is one of several mechanisms that determine the biological functions of small RGS proteins.

From the crystal structure of RGS4 bound to AlF$_4$-activated Giα (Tesmer et al., 1997), it is known that the RGS box folds as a bundle of nine α helices. Alignment of the primary amino acid sequences for several of the RGS box-containing proteins suggests that they also have similar tertiary structures. However, the putative G-protein-binding surface in RGS2 displays a number of unique amino acid features relative to the other small RGS proteins (Fig. 1). We predicted that these differences might indicate the unique ability of RGS2 to selectively bind to one or more classes of Gα subunits. Indeed, studies to date have shown that RGS2 is a highly potent and relatively selective inhibitor of Gqα function (Heximer et al., 1997b, 1999) and that this activity may be important for the normal control of vascular function (Heximer et al., 2003; Tang et al., 2003). Although RGS2 has also been reported to inhibit other G-protein signaling pathways by direct (Giα) (Ingi et al., 1998) and indirect (inhibition of adenylyl cyclase)(Salim et al., 2003; Sinnarajah et al., 2001) mechanisms,

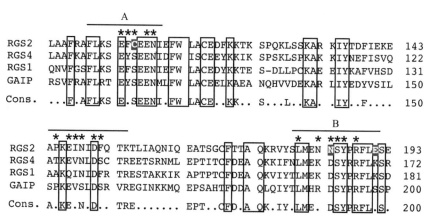

FIG. 1. Potential G-protein selectivity determinants of the RGS2 core domain. The RGS core domain (RGS box) sequences of RGS1, RGS2, RGS4, and RGS-GAIP (RGS19) were aligned by the CLUSTAL method. Amino acid residues that contribute to the Gα interaction surface of RGS4 (Tesmer et al., 1997) are indicated (asterisks). Highly conserved residues in the RGS core domain are boxed. The putative Gα interaction surface of RGS2 is indicated (overlines); amino acid residues within two of these regions (A and B) that are unique to RGS2 and proved to be functionally important are highlighted (gray). Adapted from Heximer et al. (1999), with permission.

the potent inhibition of Gqα function remains among its most widely appreciated biologic functions. Accordingly, the work in this article summarizes the various experimental approaches used to characterize the function of RGS2 as an inhibitor of Gqα signaling.

In most cases, the design of these techniques was specific to the task of studying RGS2 function; however, the challenges we faced during their development were similar to those encountered by most of the researchers in the RGS protein field. We first needed to develop an assay that would allow us to screen rapidly and easily for potential α-subunit-binding partners for RGS2. The result was the development of a nickel affinity column "pull-down style" assay that could be used either to identify new Gα subunit-binding partners from complex membrane fractions or to characterize the affinity of RGS/Gα interactions using purified proteins. While this technique was not expected to provide direct information regarding the potency of RGS2 as a GAP for the various Gα, we hypothesized that it would be able to predict the subset of Gα subunits to which RGS2 could bind. Thus, we used this method to demonstrate the relative biochemical RGS2 for binding to activated Gqα (Heximer et al., 1997b).

The development of assays to study RGS protein function by its over-expression in living cells provided a different set of challenges. We chose to develop protocols in which phosphoinositide hydrolysis was used to measure RGS inhibition of Gqα activity. We felt it was very important to ensure that Gqα activation occurred only in cells that contained the RGS construct to be analyzed. This condition was satisfied in our studies by measuring phosphoinositide hydrolysis following (i) endogenous muscarinic receptor-mediated activation of Gqα in HEK293 cells stably transfected with RGS proteins or (ii) contransfection of activated Gqα(R183C) with the appropriate RGS construct. We further recognized the need to correlate the expression levels of an RGS protein in transfected cells with function to determine its relative efficacy as a G-protein regulator. This was accomplished by designing RGS protein expression constructs that were identically epitope tagged with either a triple myc epitope or a green fluorescent protein and could be used in mammalian transfection assays. Although these transfection assays did not provide the same flexibility with respect to kinetic analyses or controlling RGS2 protein levels that were afforded by electrophysiology approaches used in other laboratories (Xu et al., 1999), they ensured that RGS2 was expressed under conditions where it could be properly post-translationally modified. These considerations guided the development of two new quantitative techniques for the determination of relative RGS protein function in transfected cells, which were required to demonstrate the high relative potency of RGS2 compared to RGS4 for the inhibition of receptor-mediated Gqα signaling (Heximer et al., 1999).

We further predicted that the *rgs2* knockout mouse (Oliveira-dos-Santos *et al.*, 2000) would be a powerful experimental tool to study the function of RGS2 as a regulator of receptor-mediated Gqα signaling in living cells. We are grateful to Drs. David Siderovski and Josef Penninger, who generated this mouse line together at the Amgen Institute in Toronto, Ontario, and kindly sent it to us for our studies. We used these mice to study the role of RGS2 as a regulator of Gqα-mediated intracellular calcium signaling in primary vascular smooth muscle cells (VSMCs) in culture. In these experiments, intracellular calcium levels in Fura-2-loaded VSMCs were analyzed to study Gqα-mediated signaling following vasoconstrictor stimulation. This analysis allowed us to compare both the intensity and the kinetic regulation of the intracellular calcium wave response between wild-type and RGS2-deficient VSMCs and has led to a new appreciation of RGS2-mediated Gqα regulation during vasoconstrictor-mediated calcium signaling and blood pressure control (Heximer *et al.*, 2003).

Biochemical Characterization of G-Protein-Binding Partners for RGS2

As described previously (Krumins and Gilman, 2002; Ross, 2002), single turnover GTPase assays using purified G-protein α subunits are the benchmark means to assess RGS function as GTPase-activating proteins. However, single turnover assays require purification of the various α subunits to be tested, thus making pan-screens for RGS/Gα binding specificity a fairly labor-intensive experiment for most laboratories. Furthermore, it is very difficult to perform GAP assays with some purified wild-type Gα subunits, in particular Gqα, due to its low affinity for GTP. While alternative approaches such as single turnover assays using the Gqα(R183C) mutant or steady-state GTPase assays using Gqα coreconstituted with receptors in lipid vesicles (Chidiac *et al.*, 2002) have proven effective, these techniques are highly specialized methods unavailable to most laboratories. Thus, we set out to design a set of experimental alternatives that would allow most investigators to study the specificity of RGS/Gα interactions *in vitro*. These protocols took advantage of the fact that AlF_4^--bound α subunits mimic the transition state conformation of the α subunit during the GTP hydrolysis reaction. In this conformation, the α subunit has been shown to have a very high affinity for its respective RGS protein partners. Furthermore, RGS/G-protein complexes are resistant to 1% cholate, a detergent that can be used to extract G proteins from biological membranes. Thus, by adding purified histidine-tagged RGS2 to bovine brain membrane preparations in which the G proteins had been activated with $GDP + AlF_4^-$, cholate-extracted RGS2/G-protein complexes could be isolated by nickel affinity chromatography and analyzed by western blotting. This strategy was used to identify G-protein α subunit-binding partners for RGS2 and RGS4 in the context of complex biological membrane preparation. To further characterize RGS/G-protein interactions, similar "pull-down" assays can also be performed using purified RGS proteins and G-protein α subunits.

Production and Purification of Recombinant (His)₁₀-RGS2 Protein

Production and Purification of Recombinant $(His)_{10}$-RGS2 Protein

The RGS2 coding sequence is polymerase chain reaction amplified from a full-length human cDNA template (Siderovski *et al.*, 1994) and cloned into pET19b (Novagen) to generate an amino-terminal $(His)_{10}$-tagged RGS2 bacterial expression construct. Decahistidine-tagged RGS2 protein is expressed in *Escherichia coli* [BL21(DE3)] cells and purified under nondenaturing conditions to yield a soluble protein with an apparent molecular mass of 30 kDa on 12% SDS–PAGE gels (Heximer *et al.*, 1997a). Freshly transformed BL21(DE3) colonies containing the $(His)_{10}$-RGS2 expression plasmid are grown at 30° in LB medium containing 100 μg/ml ampicillin. When cultures reach $OD_{600} = 0.6$, $(His)_{10}$-RGS2 expression is induced by the addition of 0.4 mM isopropyl β-D-thiogalactoside (IPTG) for 3 h. Bacterial cell pellets are collected and lysed by repeated sonication in a nickel affinity column-binding buffer (20 mM Tris–HCl, pH 7.9, 0.5 M NaCl, 5 mM imidazole) containing 0.1% NP-40, 1 mM phenylmethylsulfonyl fluoride, 2 μg/ml leupeptin, and 20 μg/ml aprotinin. Soluble proteins are bound in batchwise fashion to 5 ml of Ni^{2+}-charged Ni^{2+}-NTA agarose (Qiagen). Contaminating bacterial proteins are removed by successive washes (2 × 20 ml) of the nickel resin with the column-binding buffer and column wash buffer (20 mM Tris–HCl, pH 7.9, 0.5 M NaCl, 60 mM imidazole). $(His)_{10}$-RGS2 is eluted in nickel affinity column elution buffer (20 mM Tris–HCl, pH 7.9, 0.5 M NaCl, 0.5 M imidazole) and dialyzed into cryo-storage buffer [50 mM HEPES, pH 8, 0.5 M NaCl, 1 mM EDTA, 3 mM dithiotheitol (DTT), 10% glycerol] prior to storage at −80°. Protein prepared in this manner is stable for at least 4 years.

Technical Notes

Technical Notes

We have found that $(His)_{10}$-RGS2 protein purified according to these methods remains soluble at concentrations up to 20 mg/ml in aqueous buffers (pH 7.0–8.5), provided the solutions contain >350 mM NaCl. In biochemical assay buffers with salt concentrations below 350 mM, $(His)_{10}$-RGS2 will remain soluble at concentrations up to 0.125 mg/ml. Under low salt conditions, higher concentrations of $(His)_{10}$-RGS2 are tolerated if phospholipid vesicles (containing >20% phosphatidylserine) are present. Storage of $(His)_{10}$-RGS2 under reducing conditions (1–3 mM DTT) is also necessary to prevent the formation of inter- and intramolecular disulfide bonds.

Additional Reagents

RGS4 expression constructs were generated in previous studies (Srinivasa *et al.*, 1998). Wild-type and inactive mutant forms (E87A, N88A double mutant; N128A single mutant) of RGS4 were produced and purified essentially as described earlier. Baculoviruses encoding untagged Gqα, Gβ, and his-tagged Gγ subunits and methods for the expression and purification of untagged Gqα from Sf9 cells were as described (Kozasa and Gilman, 1995). Recombinant Goα (provided by Dr. M. Linder, Washington University, St. Louis) was purified from *E. coli* as described previously (Linder and Gilman, 1991).

Identification of G-Protein α Subunit-Binding Partners for RGS2 in Bovine Brain Membranes

Bovine brain membranes (0.5 mg protein) in buffer A (20 mM HEPES, pH 8.0, 500 mM NaCl, 3 mM DTT, 6 mM MgCl$_2$) containing 100 μM GDP or 100 μM GDP, 30 μM AlCl$_3$, and 10 mM NaF are, incubated 30 min at 5° with his-tagged RGS proteins (10 μg). Membranes are solubilized with 1% cholate, and detergent-soluble extracts obtained after centrifugation at 100,000g are added to Ni^{2+}-NTA agarose beads equilibrated with buffer A containing 20 mM imidazole, 0.1% $C_{12}E_{10}$, 10 μM GDP or 10 μM GDP, 30 μM AlCl$_3$, and 10 mM NaF. Bound proteins are eluted with 500 mM imidazole, resolved by SDS–PAGE, transferred to nitrocellulose membranes, and detected by Ponceau S staining (Fig. 2A) and immunoblotting (Fig. 2B) with antisera directed against different α subunits.

Results and Interpretations. Because brain membrane fractions are known to contain readily detectable levels of most isoforms of G-protein α subunits, cholate-extracted RGS–Gα complexes from AlF$_4^-$-activated membranes provided a comprehensive screen of the potential G-protein-binding partners for RGS2 and RGS4. In this assay, one of the most abundant G-protein α subunits in brain membrane preparations, Goα, was found to be associated very strongly with RGS4 but not RGS2 (Fig. 2A and B). Conversely, both RGS2 and RGS4 were able to bind to activated Gqα. RGS4 isoforms containing mutations along the RGS/G-protein-binding interface did not associate with Goα or Gqα. No complexes were detected between RGS2 or RGS4 and Gsα or G12/13α (data not shown). These data provided the first evidence to suggest that

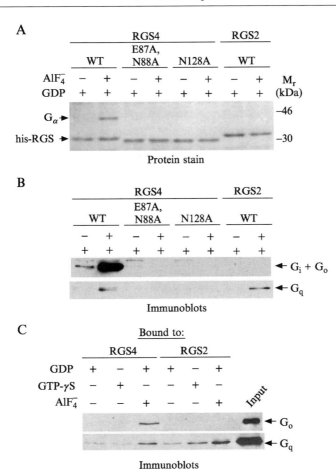

FIG. 2. Binding of RGS proteins and Gα subunits. (A and B) Binding of RGS2 and RGS4 to G-protein α subunits in membrane fractions. The indicated wild-type or mutant forms of his-tagged RGS4 or wild-type RGS2 were incubated with bovine brain membranes treated with GDP or GDP and AlF$_4^-$. Complexes containing RGS proteins were purified from detergent extracts using Ni^{2+}-NTA chromatography. (A) Polypeptides bound to RGS proteins were resolved by SDS–PAGE, transferred to nitrocellulose blots, and detected by staining with Ponceau S. The position where α subunits of the Gi family migrate is indicated. Mutant forms of RGS4 are smaller because they lack the first 12 amino acids of the protein, which is dispensable for GAP activity. (B) Identification of Gα subunits bound by RGS proteins. Nitrocellulose blots as in A were probed with antisera specific for the indicated G-protein α subunits (and G12/13 and Gsα; data not shown) and HRP-coupled secondary antibodies. Enhanced chemiluminescence detection was used. (C) Interaction of RGS proteins with purified G-protein α subunits in their inactive (GDP), active (GTPγS), and transition state (GDP + AlF$_4^-$) conformations. Binding of the indicated G-protein α subunits and his-tagged RGS proteins was detected by isolating RGS–G protein complexes on Ni^{2+}-NTA

recombinant RGS2 displays biochemical selectivity for binding to Gqα and pointed to the utility of this screening method for the identification of G-protein-binding partners for RGS proteins.

Characterization of RGS/Gα Binding Selectivity Using Purified Proteins

In assays using purified proteins, G-protein α subunits (110 ng) are incubated (20 min at 22° for Goα, and 30 min at 30° for Gqα) in HEDL (buffer with 1 mM GDP, GTPγS, or with 1 mM GDP, 30 μM AlCl$_3$, and 10 mM NaF). These Gα subunits then are incubated with his-tagged RGS4 or RGS2 (330 ng) in 40 μl buffer B (50 mM Na-HEPES, pH 8.0, 1 mM MgCl$_2$, 20 mM imidazole, 0.025% C$_{12}$E$_{10}$, 10 mM 2-mercaptoethanol, 10% glycerol) containing the appropriate guanine nucleotides (1 mM) with or without 30 μM AlCl$_3$, and 10 mM NaF. The RGS–Gα complexes are isolated and detected as described previously.

Results and Interpretations. The relative specificities with which RGS2 and RGS4 bind Gqα and Goα were determined by performing experiments using purified proteins (Fig. 2C). To determine whether RGS–Gα binding was affected by the activation state of Gα subunits, we performed *in vitro*-binding experiments under conditions where recombinant Goα and Gqα were in their inactive (GDP), active (GTP-γS), and transition state (GDP + AlF$_4^-$) conformations. Whereas RGS4 bound to Goα in its transition state (but not to inactive, GDP-bound or active GTP-γS-bound Goα), RGS2 did not bind detectably to Goα, regardless of the activation state of the G protein. In contrast, both RGS2 and RGS4 bound Gqα in its transition state; binding of RGS2 appeared to be more efficient. RGS2 and RGS4 also bound activated (GTP-γS-bound) Gqα; again, binding of RGS2 was apparently more efficient. Results of these experiments indicated that, in contrast to RGS4, RGS2 binds relatively selectively to Gqα. The advantage of this type of experiment is that it is a rapid and efficient alternative to the membrane pull-down experiments. We anticipate a wide number of uses for this assay as we begin to understand how RGS protein activity is modulated in living cells. In particular, we have found these assays to be especially useful in our ongoing studies to analyze the effect of specific point mutations and posttranslational modifications on the formation of RGS2/Gqα complexes.

---

beads and subjecting the eluted proteins to immunoblotting experiments. In experiments using Goα, 10 and 30%, respectively, of the input and eluted samples were analyzed. In those using Gqα, 10% of the input and eluted samples were analyzed. Adapted from Heximer *et al.* (1997b). Copyright National Academy of Sciences, USA.

Using Phosphoinositide Hydrolysis Assays to Measure
RGS2 Function in Living Cells

Rationale

A well-defined consequence of Gqα-coupled receptor activation is the activation of phospholipase Cβ to result in the production of second messenger signaling molecules, including [Ins(1,4,5)P$_3$]. Previous articles in this series have detailed how phosphoinositide hydrolysis can been used in biochemical assays to demonstrate the potential of RGS proteins to inhibit Gqα-dependent activation of PLC-β (Chidiac *et al.*, 2002; Hains *et al.*, 2004). Work in this article describes how the production of inositol phosphates by PLC-β may also be used to study RGS2 function as a Gqα inhibitor in live cells.

*RGS2-Mediated Inhibition of Gqα Signaling by Endogenous
Muscarinic Receptors in Stably Transfected HEK293 Cells*

This experimental strategy takes advantage of the fact that HEK293 cells express endogenous Gqα-coupled muscarinic receptors. Thus, assuming that the clonality of muscarinic receptor-mediated signaling is preserved in this cell line, then every cell should possess the same potential to activate inositol phosphate signaling in response to carbachol stimulation. By using these cells to generate a series of RGS protein stable transfectants with varying expression levels, it is possible to correlate the relative Gqα inhibitory function of an RGS protein with its level of expression. This approach was developed in our laboratory as a means of demonstrating the biological potency of RGS proteins as inhibitors of receptor-mediated Gqα signaling in live cells.

Generation of HEK293 Cell Lines That Stably Express Triple-myc Epitope-Tagged RGS2 and RGS4 Proteins. Low-passage HEK293 cells (7×10^6 cells in 10-cm plates) are transfected (see later) with 5 μg of mammalian expression constructs that direct the expression of RGS2 or RGS4 that had been tagged at their carboxyl termini with three tandem copies of the c-myc epitope. Stable RGS–(myc)$_3$-expressing clones are selected for growth in 0.5 mg/ml geneticin (G418). Subclones are screened for RGS expression levels by Western blotting and for clonality by immunofluorescence staining using the mouse 9E10 monoclonal antibody(Covance Research Products, Denver, PA). To determine the relative expression levels of RGS proteins in stably transfected cell lines, cells from trypsinized plates are counted, pelleted, and lysed (2×10^7 cells/ml) in Laemmli sample buffer. Following sonication and boiling, samples are resolved by SDS–PAGE. Protein expression levels are determined by immunoblotting using the 9E10 antibody, ECL (Amersham), and densitometry.

Technical Note. The inclusion of tandem (triple-myc) epitopes was very helpful for detecting RGS2 expression in our stable cell lines, as for reasons that are not well understood, RGS2 protein expression levels in our stable cell lines were typically much lower than those observed for RGS4 lines.

Measurement of Carbachol-Induced Inositol Phosphate Production in HEK293 Cell Lines That Stably Express RGS2 or RGS4 Proteins. Measurement of inositol phosphate production is measured essentially as described by Venkatakrishnan and Exton (1996). Briefly, 0.5×10^6 cells/well in a 12-well dish are labeled in complete Dulbecco's modified Eagle's medium (DMEM, without inositol) containing 4 μCi/ml [^{3}H]-inositol (DuPont-New England Nuclear) for 17–24 h. Cells are washed once with DMEM (without bicarbonate) containing 25 mM HEPES, pH 8.0, and incubated for 10 min in 2 ml of the same medium containing 5 mM LiCl. Inositol phosphate production is initiated by the addition of 40 μl of a 1 mM carbachol stock to each well. Incubations are continued for exactly 45 min before stopping them by aspiration of medium and immediate addition of 750 μl ice-cold 20 mM formic acid. Following incubation for 30 min at 4°, contents of the entire well are collected and spun at 15,000g for 15 min in a microcentrifuge. The supernatant fraction (700 μl) is mixed with 214 μl of 0.7 M NH$_4$OH in a new microfuge tube to neutralize the samples before proceeding to the ion-exchange chromatography steps.

Isolation of Inositol Phosphate Fraction Using Ion-Exchange Chromatography. For each sample to be measured, a 3-ml Dowex resin (AG 1-X8, 200–400 mesh, formate form) column is prepared by hydration with 10 ml water, charging with 5 ml 2 M ammonium formate in 0.1 M formic acid, and washing with another 10 ml of water. Three 20-ml liquid scintillation counting vials are prepared for each sample to be measured. With columns over the first set of collection vials, the entire sample is added to each column and is washed into a collection tube with 4 ml water. This sample represents the total inositol-containing lipid fraction. Columns are placed over a second set of collection tubes and washed with 4 ml 40 mM ammonium formate in 0.1 M formic acid. Inositol phosphates are eluted into a third set of collection vials using 5 ml 1.2 M ammonium formate, and 0.5 ml of each sample from total lipid and inositol phosphate-containing fractions is added to 10 ml of Ready-Safe scintillant. Inositol phosphate levels are expressed as the fraction of the total soluble inositol-labeled material for each sample.

Technical Notes. We found that the highest level of reproducibility for this assay was achieved when similar numbers of cells were plated in each well prior to the [^{3}H]-inositol-labeling step. It is also important to ensure that neutralization is complete before proceeding to the chromatography step. The residual phenol red in the formic acid cell lysate mixture can be used conveniently as a pH indicator to ensure that the mixture has been neutralized by the ammonium hydroxide (sample turns light pink).

Results. To characterize the phosphoinositide hydrolysis assay in our HEK293 cells, we showed that carbachol treatment resulted in a four-fold increase in inositol phosphate production relative to untreated controls (data not shown). Because this response was insensitive to pertussis toxin, we assumed that PLCβ activation was mediated through Gqα-coupled m1 or m3 muscarinic acetylcholine receptors in these cells. Two stable HEK293 lines expressing similar levels of RGS2-myc (RGS2-1 and RGS2-2) and four lines expressing various levels of RGS4-myc (RGS4-1 through −4) were obtained by limited dilution cloning. We compared carbachol-induced inositol phosphate production in the RGS2-1 line with that in the four cell lines expressing various levels of RGS4-myc (Fig. 3). In RGS2-expressing cells (RGS2-1), carbachol-induced inositol phosphate production was nearly completely inhibited compared to the control population. In the case of RGS4-expressing cell lines, the extent of inhibition of inositol phosphate production could be correlated directly with RGS4-myc expression levels. Importantly, however, inositol phosphate production was

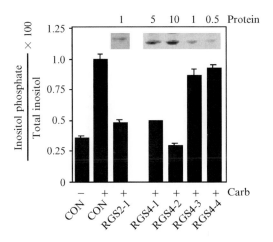

FIG. 3. Inhibition of agonist-stimulated production of inositol trisphosphate in HEK293 cells expressing RGS2 or RGS4. Carbachol-induced (Carb) inositol phosphate production by control cells (Con), RGS2-myc expressing cells (RGS2–1), and RGS4-myc expressing cells (RGS4–1 through RGS4-4) after 45 min with agonist treatment. The relative levels of RGS2-myc and RGS4-myc protein (*top*) were determined by immunoblotting (*inset*) and densitometry. Inositol phosphate values represent the average of triplicate samples and were expressed as the percentage of soluble inositol phosphates relative to total soluble inositol-labeled material; SEM are indicated. These data are representative of three independent experiments. Adapted from Heximer *et al.* (1999), with permission.

inhibited to a similar extent only when RGS4-myc expression was several-fold higher than that of RGS2-myc (compare RGS2-1 with RGS4-1 and RGS4-2). Furthermore, expression of RGS4-myc at a level similar to that of RGS2-myc had little inhibitory effect (compare RGS2-1 with RGS4-2 and RGS4-3). The results of these assays suggest that RGS2 is approximately five-fold more potent than RGS4 as an inhibitor of Gqα-mediated phosphoinositide signaling in HEK293 cells.

RGS2-Mediated Inhibition of Constitutively Active GqR183C in Mammalian Cells

In this line of experiments, cotransfection of RGS2 with GqR183C is used to study the relative function of various RGS2 constructs. The utility of recombinant GqR183C as a tool in the biochemical characterization of RGS proteins as GAPs for Gqα has been described in detail elsewhere (Chidiac et al., 2002). Here, we take advantage of the fact that GqR183C transfection into HEK293 cells results in the activation of phosphoinositide hydrolysis pathways in the absence of RGS2 coexpression. Accordingly, the relative Gq inhibitory function of wild-type and mutant forms of RGS2 may be tested by their simultaneous introduction into HEK293 cells with GqR183C using cotransfection assays. Results using this assay have shown that the amino terminus of RGS2 is essential for its effective inhibition of Gq-mediated signaling (Fig. 4).

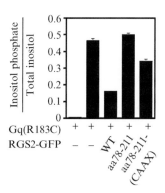

FIG. 4. Ability of RGS2 core and plasma membrane-targeted core domains to inhibit Gq-mediated phosphoinositide hydrolysis. HEK293 cells were cotransfected with control or constitutively active Gqα, the indicated GFP control, or RGS2 constructs and a *Renilla* luciferase control. Triplicate wells containing transfected cells were labeled with [³H]inositol and 10 m*M* LiCl. Steady-state inositol phosphate levels were determined. Values represent the average of triplicate samples normalized for luciferase activity and were expressed as the percentages of soluble relative to total soluble inositol-containing material; SEM are indicated by *error bars*. Adapted from Heximer et al. (2001), with permission.

Reagents. The CMV promoter on plasmid pEGFP-C1 (Clontech, Palo Alto, CA) is used to express various RGS2-GFP constructs. In each case, RGS2 is inserted ahead of GFP at the *Nhe*I and *Age*I sites in this vector. The Ki-Ras CAAX-box sequence (KDGKKKKKKSKTKCVIM) is added to the carboxyl-terminal end of RGS2-GFP to generate plasma membrane-targeted RGS2 constructs. The constitutively active Gqα (R183C) construct in pCIS was a kind gift from Dr. John Hepler (Emory University, Atlanta, GA). The thymidine kinase promoter-driven *Renilla* luciferase reporter (pRL-TK) used as a transfection control in transient assays was a kind gift from Dr. Murphy (Washington University, St. Louis, MO). Polyclonal GFP antibody was the kind gift of Dr. Silver (Dana Farber Cancer Institute, Boston, MA).

Cells (7×10^6 in 10-cm plates) are washed in serum-free medium and incubated with 6.4 ml Opti-MEM (Gibco/Life Technologies) containing a mixture of the following DNA constructs: 1 μg pRL-TK; 1.5 μg Gq(R183C); and 3 μg RGS2-GFP or empty vector control (5.5 μg total plasmid DNA/well) and 10 μl lipofectamine. Following a 5-h incubation, media are removed and replaced with DMEM supplemented with 10% (v/v) fetal calf serum (FCS), 2 mM glutamine, 10 μg/ml streptomycin, and 100 units/ml penicillin. After a 17-h recovery, cells from each transfection are trypsinized and replated in 6-well plates (3 wells $\times$ 0.5 $\times$ 10^6 cells/well) for phosphoinositide hydrolysis assays and in luciferase-based quantitation of transfection efficiency. The remaining cells are replated in 10-cm dishes for Western blot characterization of Gαq and RGS2-GFP expression. For inositol hydrolysis assays, at 22 h post-transfection, culture media are replaced with labeling medium (see earlier discussion) containing 10 mM LiCl, and cells are incubated overnight. Phosphoinositide hydrolysis is determined as described earlier, and values are normalized to *Renilla* luciferase levels to account for differences in transient transfection efficiency. For determination of *Renilla* luciferase levels, cells at 22–25 h post-transfection are washed twice in ice-cold phosphate-buffered saline (PBS) and harvested in 250 μl hypotonic lysis buffer (10 mM KH$_2$PO$_4$, pH 7.5, 1 mM EDTA) and spun at 12,000g to remove cellular debris. Duplicate 50-μl samples of the supernatant are measured for *Renilla* luciferase activity in a Optocomp II luminometer (MGM Instruments, Hamden, CT). The *Renilla* luciferase substrate is diluted 1/25,000 in 10 mM KH$_2$PO$_4$, pH 7.5, 0.5 M NaCl, and 1 mM EDTA for the assays. Levels of RGS and G-protein expression are monitored on western blots using ECL detection (Amersham, Arlington Heights, IL).

Results and Interpretations. Previous data from studies of RGS2 function in yeast suggested that the amino terminus [RGS2(aa1-78)] was required for proper localization to the plasma membrane and function

(Heximer et al., 1999). Notably, the activity lost after deletion of this domain was restored by targeting RGS2 to the plasma membrane using a carboxyl-terminal CAAX box domain from Ras2. We wondered whether inhibition of Gqα signaling pathways in mammalian cells might require similar functional properties of the amino-terminal domain in RGS2. We therefore used our cotransfection/phosphoinositide hydrolysis assay to analyze the function of these RGS2 constructs in HEK293 cells. Similar to the results obtained in yeast, deletion of the N terminus from RGS2-GFP markedly reduced its ability to inhibit Gq(R183C)-mediated signaling in HEK293 cells (Fig. 4). Appending the polybasic region and prenylation signal of Ki-Ras to the C terminus of the RGS2 core domain (RGS2(aa78-211)-GFP-CAAX) was able to partially restore its inhibitory function (Fig. 4), suggesting that plasma membrane localization is an important determinant of RGS2 function as a Gqα inhibitor. These studies highlight the utility of RGS2/GqR183C cotransfection assays combined with phosphoinositide hydrolysis to study functional determinants during the RGS2-mediated inhibition of Gqα signaling.

Studying Gqα-Mediated Calcium Signaling in Primary Vascular
    Smooth Muscle Cells from RGS2 Knockout Mice

Rationale

Because RGS2 is predicted to act as a GAP for Gqα in the wide number of cell types in which it is expressed, we examined the RGS2 knockout mouse for defects in Gqα signaling. Previous studies had shown that *Rgs2* mRNA levels were increased in rat vascular smooth muscle cells treated with angiotensin II, suggesting a possible role for this gene in the modulation of vasoconstrictor signaling (Grant et al., 2000). We predicted that the loss of RGS2 might lead to either (i) prolonged Gqα activity following agonist-dependent stimulation or (ii) increased sensitivity to Gqα-coupled agonists. Because previous work showed that RGS2-deficient mice were both hypertensive and showed increased responsiveness to acute stimulation by a Gqα-coupled α1-adrenergic agonist (Heximer et al., 2003), we set out to study Gqα-dependent signaling kinetics in vascular smooth muscle cells. We reasoned that the primary effector of Gqα-mediated signaling in VSMCs was PLC-β and that a primary effect of the loss of RGS2 would be increased IP3 production and augmented calcium responses following GPCR activation. The following experiment describes the protocols used to compare intracellular calcium signaling responses in wild-type and RGS2-deficient VSMCs.

Materials

RGS2-deficient mice were kindly sent to us by Drs. Siderovski and Penninger (Amgen Institute, Toronto, Ontario, Canada).

Measurement of Agonist-Induced Changes in Intracellular Calcium in Primary Aortic VSMC Cultures

Vascular smooth muscle cells from explanted thoracic aortic tissue are prepared as follows. Cells are dissociated by gentle agitation for 30 min at 37° in DMEM digestion buffer containing elastase (0.03%, w/v) and collagenase (0.07% w/v), sieved through 50-μm nylon mesh, washed, and seeded at low density onto polylysine-coated 13-mm #1 glass coverslips. Cells in culture are maintained in DMEM/F12 medium containing 20% FCS and 2 mM glutamine with penicillin/streptomycin. Immunostaining with anti-smooth muscle α actin antibody (1A4; Sigma) indicates that cell preparations are >80% VSMCs. Cells are washed and incubated in 4 μM Fura-2/AM in Cal solution (11 mM glucose, 130 mM NaCl, 4.8 mM KCl, 1.2 mM MgCl$_2$, 17 mM HEPES, 1 mM CaCl$_2$ pH 7.3) for 30 min at 37°, washed once, and incubated in Cal solution without Fura-2 for 10 min at 37° to allow hydrolysis of the acetoxymethyl ester. Coverslips are placed in an RC-26G large open-bath recording chamber (Warner Instrument Corp., Hamden, CT) that is mounted on a Nikon Eclipse E600 FN microscope equipped with a Nikon fluor 40×/0.80 water immersion lens. Excitation light (340/380 nm) is provided by a Till Polychrome IV monochronomator (TiLL Photonics, Inc., Martinsried, Germany) in conjunction with a 495-nm dichroic mirror and a 525 ± 20-nm emission filter (Chroma Technology Corp, Brattleboro, VT). Fluorescence ratio imaging is performed with TiLLvisION 4.0 imaging software (TiLL), and images are acquired with a PCO Sensicam cooled CCD camera. Image pairs are taken every 2 s, and background-corrected ratio images (340/380) are analyzed with individual cells defined as regions of interest (ROI). Data are presented as relative ratio values over time for each ROI.

Results and Interpretations

We examined vasoconstrictor signaling in aortic smooth muscle cells cultured from wild-type and $Rgs2^{-/-}$ mice. The cellular responses to rATP stimulation of Gqα-coupled P2Y receptors were assessed using Fura-2 imaging to follow relative changes in intracellular calcium levels (Fig. 5). Cells from $Rgs2^{-/-}$ mice were approximately three-fold more sensitive than cells from wild-type mice (Fig. 5A), consistent with a role for RGS2 in the normal attenuation of calcium responses in VSMCs. At the highest doses

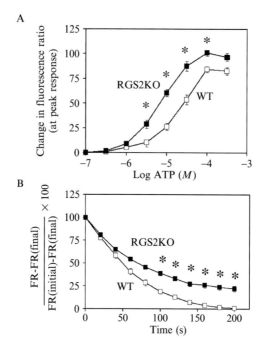

FIG. 5. Changes in intracellular calcium levels in response to vasoconstrictors in primary aortic vascular smooth muscle cells derived from wild-type and $Rgs2^{-/-}$ mice. (A) Dose–response relationships for P2Y receptor-mediated increases in intracellular calcium levels. Cells were treated with the indicated doses of ATP, an agonist for P2Y receptors. Changes in intracellular calcium levels were measured as changes in fluorescence ratio [FR ± SEM = (emission at 510 nm upon excitation at 340 nm)/(emission at 510 nm upon excitation at 380 nm)] in Fura 2-loaded cells isolated from wild-type (WT; n = 15–22 individual cells for each agonist dose) and $Rgs2^{-/-}$ (RGS2KO; n = 15–22 individual cells for each agonist dose) mice. (B) Rates of intracellular calcium level decay [100% × (FR-FR (final)/FR (initial)-FR (final)) ± SEM] after a maximal response had been elicited by treating cells from wild-type (WT; n = 39 individual cells) and $Rgs2^{-/-}$ (RGS2KO; n = 46 individual cells) mice with 100 μM ATP. Pairwise comparisons of the FR values of wild-type and $Rgs2^{-/-}$ (RGS2KO; n = 46 individual cells) mice with 100 μM ATP. Pairwise comparisons of the FR values of wild-type and $Rgs2^{-/-}$ cells were made at similar agonist concentrations in A or times after maximal FR response in B. Statistically different values are indicated (*p < 0.001). Reproduced from Heximer et al. (2003), with permission.

of ATP, peak calcium responses were greater in cells from $Rgs2^{-/-}$ compared to wild-type mice, indicating that both the potency and the efficacy of ATP were also greater. Fura-2 imaging of relative intracellular calcium changes was also used to examine the kinetics of signal termination following the peak response elicited by challenge with a maximal dose of ATP.

As indicated by the kinetics with which intracellular Ca^{2+} levels declined to baseline levels after reaching a peak response, the signal termination kinetics of aortic smooth muscle cells from RGS2-deficient mice were markedly slower (Fig. 5B), consistent with a role for RGS2 in the normal termination of Gqα-coupled vasoconstrictor responses in VSMCs. It is important to recognize that the molecular mechanisms leading to abnormal calcium signaling in RGS2-deficient cells remain to be determined. If, as expected, the loss of RGS2 results in prolonged activity of Gqα in stimulated VSMCs, there are a number of downstream effectors of Gqα that might ultimately impact Ca^{2+} signaling and homeostasis. Clearly, PLC-β is a primary candidate (see rationale); however, other putative Gqα effectors, including PKC and tyrosine kinases, may also play a role in determining the calcium response through IP_3-independent mechanisms.

Summary

RGS2 is unique among the subfamily of small RGS proteins due to its biochemical selectivity for activated Gqα versus Giα. The approaches detailed in this article have been used to measure RGS2-mediated inhibition of Gqα under a variety of experimental conditions. We anticipate that these protocols will be useful in the future discovery of novel regulatory motifs in RGS2 and other small RGS proteins that inhibit Gqα signaling.

Acknowledgment

I am grateful to Dr. K. Blumer for his valuable discussions and critical review of this work.

References

Berman, D. M., Wilkie, T. M., and Gilman, A. G. (1996). GAIP and RGS4 are GTPase-activating proteins for the Gi subfamily of G protein alpha subunits. *Cell* **86,** 445–452.

Chidiac, P., Gadd, M. E., and Hepler, J. R. (2002). Measuring RGS protein interactions with Gq alpha. *Methods Enzymol.* **344,** 686–702.

Grant, S. L., Lassegue, B., Griendling, K. K., Ushio-Fukai, M., Lyons, P. R., and Alexander, R. W. (2000). Specific regulation of RGS2 messenger RNA by angiotensin II in cultured vascular smooth muscle cells. *Mol. Pharmacol.* **57**(3), 460–467.

Hains, M. D., Siderovski, D. P., and Harden, T. K. (2004). Application of RGS box proteins to evaluate G-protein selectivity in receptor-promoted signaling. *Methods Enzymol.* **389,** 71–88.

Heximer, S. P., Cristillo, A. D., and Forsdyke, D. R. (1997a). Comparison of mRNA expression of two regulators of G-protein signaling, RGS1/BL34/1R20 and RGS2/G0S8, in cultured human blood mononuclear cells. *DNA Cell Biol.* **16**(5), 589–598.

Heximer, S. P., Knutsen, R. H., Sun, X., Kaltenbronn, K. M., Rhee, M. H., Peng, N., Oliveira-dos-Santos, A., Penninger, J. M., Muslin, A. J., Steinberg, T. H., Wyss, J. M., Mecham,

R. P., and Blumer, K. J. (2003). Hypertension and prolonged vasoconstrictor signaling in RGS2-deficient mice. *J. Clin. Invest.* **111**(4), 445–452.

Heximer, S. P., Srinivasa, S. P., Bernstein, L. S., Bernard, J. L., Linder, M. E., Hepler, J. R., and Blumer, K. J. (1999). G protein selectivity is a determinant of RGS2 function. *J. Biol. Chem.* **274**(48), 34253–34259.

Heximer, S. P., Watson, N., Linder, M. E., Blumer, K. J., and Hepler, J. R. (1997b). RGS2/G0S8 is a selective inhibitor of Gqalpha function. *Proc. Natl. Acad. Sci. USA* **94**(26), 14389–14393.

Ingi, T., Krumins, A. M., Chidiac, P., Brothers, G. M., Chung, S., Snow, B. E., Barnes, C. A., Lanahan, A. A., Siderovski, D. P., Ross, E. M., Gilman, A. G., and Worley, P. F. (1998). Dynamic regulation of RGS2 suggests a novel mechanism in G-protein signaling and neuronal plasticity. *J. Neurosci.* **18**(18), 7178–7188.

Kozasa, T., and Gilman, A. G. (1995). Purification of recombinant G proteins from Sf9 cells by hexahistidine tagging of associated subunits: Characterization of alpha 12 and inhibition of adenylyl cyclase by alpha z. *J. Biol. Chem.* **270**(4), 1734–1741.

Krumins, A. M., and Gilman, A. G. (2002). Assay of RGS protein activity *in vitro* using purified components. *Methods Enzymol.* **344**, 673–685.

Linder, M. E., and Gilman, A. G. (1991). Purification of recombinant Gi alpha and Go alpha proteins from *Escherichia coli*. *Methods Enzymol.* **195**, 202–215.

Oliveira-dos-Santos, A. J., Matsumoto, G., Snow, B. E., Bai, D., Houston, F. P., Whishaw, I. Q., Mariathasan, S., Sasaki, T., Wakeham, A., Ohashi, P. S., Roder, J. C., Barnes, C. A., Siderovski, D. P., and Penninger, J. M. (2000). Regulation of T cell activation, anxiety, and male aggression by RGS2. *Proc. Natl. Acad. Sci. USA* **97**, 12272–12277.

Ross, E. M. (2002). Quantitative assays for GTPase-activating proteins. *Methods Enzymol.* **344**, 601–617.

Salim, S., Sinnarajah, S., Kehrl, J. H., and Dessauer, C. W. (2003). Identification of RGS2 and type V adenylyl cyclase interaction sites. *J. Biol. Chem.* **278**(18), 15842–15849.

Siderovski, D. P., Heximer, S. P., and Forsdyke, D. R. (1994). A human gene encoding a putative basic helix-loop-helix phosphoprotein whose mRNA increases rapidly in cycloheximide-treated blood mononuclear cells. *DNA Cell Biol.* **13**(2), 125–147.

Sinnarajah, S., Dessauer, C. W., Srikumar, D., Chen, J., Yuen, J., Yilma, S., Dennis, J. C., Morrison, E. E., Vodyanoy, V., and Kehrl, J. H. (2001). RGS2 regulates signal transduction in olfactory neurons by attenuating activation of adenylyl cyclase III. *Nature* **409**(6823), 1051–1055.

Srinivasa, S. P., Watson, N., Overton, M. C., and Blumer, K. J. (1998). Mechanism of RGS4, a GTPase-activating protein for G protein alpha subunits. *J. Biol. Chem.* **273**(3), 1529–1533.

Tang, M., Wang, G., Lu, P., Karas, R. H., Aronovitz, M., Heximer, S. P., Kaltenbronn, K. M., Blumer, K. J., Siderovski, D. P., Zhu, Y., and Mendelsohn, M. E. (2003). Regulator of G-protein signaling-2 mediates vascular smooth muscle relaxation and blood pressure. *Nature Med.* **9**(12), 1506–1512.

Tesmer, J. J., Berman, D. M., Gilman, A. G., and Sprang, S. R. (1997). Structure of RGS4 bound to A1F4–activated G(i alpha 1): Stabilization of the transition state for GTP hydrolysis. *Cell* **89**(2), 251–261.

Venkatakrishnan, G., and Exton, J. H. (1996). Identification of determinants in the alpha-subunit of Gq required for phospholipase C activation. *J. Biol. Chem.* **271**(9), 5066–5072.

Watson, N., Linder, M. E., Druey, K. M., Kehrl, J. H., and Blumer, K. J. (1996). RGS family members: GTPase-activating proteins for heterotrimeric G-protein alpha-subunits. *Nature* **383**, 172–175.

Xu, X., Zeng, W., Popov, S., Berman, D. M., Davignon, I., Yu, K., Yowe, D., Offermanns, S., Muallem, S., and Wilkie, T. M. (1999). RGS proteins determine signaling specificity of Gq-coupled receptors. *J. Biol. Chem.* **274**(6), 3549–3556.

[6] Analysis of the Interaction between RGS2 and Adenylyl Cyclase

By SAMINA SALIM and CARMEN W. DESSAUER

Abstract

Regulators of G-protein signaling (RGS) are a family of highly diverse, multifunctional signaling proteins that enhance the intrinsic GTPase rate of certain heterotrimeric G-protein α subunits. New findings indicate that RGS proteins act not only as dedicated G-protein inhibitors, but rather as tightly regulated modulators of many aspects of G-protein signaling. Like other RGS proteins, RGS2 lacks GTPase-activating protein activity for Gsα; however, it directly inhibits the activity of several adenylyl cyclase (AC) isoforms. This article discusses methods, including AC binding assays, cAMP accumulation assays, *in vitro* AC activity assays, and gel filtration, used to identify the interaction site of RGS2 and type V adenylyl cyclase.

Overview

Regulators of G-protein signaling (RGS) proteins are a relatively new family of proteins that contain a signature RGS domain. So far, more than 30 mammalian family members have been identified and classified into seven subfamilies based on sequence identity and functional similarities (De Vries *et al.*, 2000; Hollinger *et al.*, 2002; Ross *et al.*, 2000). Although many RGS proteins are relatively simple, others are more complex and contain multiple domains and motifs that regulate their actions and/or enable them to interact with protein-binding partners with diverse cellular roles (Hepler, 1999; Siderovski *et al.*, 1999). Early evidence suggested that RGS proteins acted primarily as negative regulators of G-protein signaling, serving as GTPase-activating proteins (GAPs) for heterotrimeric G proteins. More recently, broader cellular functions of RGS proteins have been established, and these proteins are now recognized as central participants in receptor signaling and cell physiology. One such novel role of RGS proteins is the inhibitory effects on various adenylyl cyclase (AC) isoforms (Salim *et al.*, 2003; Sinnarajah *et al.*, 2001).

Mammalian AC consists of a short N terminus and a repeating unit of six transmembrane spans and a cytoplasmic domain (Smit *et al.*, 1998). The two cytoplasmic domains (C_1 and C_2) can be expressed individually in

Escherichia coli and, upon mixing, retain most of the properties of the native enzyme (Dessauer *et al.*, 1997, 1998; Whisnant *et al.*, 1996; Yan *et al.*, 1996). The nine isoforms of membrane-bound AC are regulated differentially. RGS2 can inhibit types III, V, and VI AC (Roy *et al.*, 2003; Salim *et al.*, 2003; Sinnarajah *et al.*, 2001). Type III AC is the predominant isoform in olfactory neurons, and injection of RGS2 blocking antibodies increases cAMP in these cells (Sinnarajah *et al.*, 2001). For type V AC, inhibition by RGS2 is due to a direct interaction between the N terminus of RGS2 and the C_1 domain of AC (Salim *et al.*, 2003). The N-terminal 38 amino acids of RGS2 are unique among RGS family members and are not present in closely related RGS proteins such as RGS4. This article describes the biochemical protocols used to analyze the interaction between RGS2 and the C_1 domain of type V AC.

Construction of Expression Vectors

RGS2 and AC Binding Assays

For RGS2 pull-down assays, C_1 and C_2 cytoplasmic domains of type V AC containing hexa-histidine or GST tags are required. Hexa-histidine-tagged RGS4 and C_1 (VC$_1$(670)H$_6$) and C_2 (H$_6$VC$_2$) AC domains have been described previously (Berman *et al.*, 1996; Dessauer *et al.*, 1998; Sunahara *et al.*, 1997). The plasmid encoding GST-VC$_1$ is created by subcloning VC$_1$(670)H$_6$ (Sunahara *et al.*, 1997) using the restriction enzymes *Nco*I and *Hind*III into a modified form of pGEX-cs (Wang *et al.*, 1997) containing a *Hind*III site in the 3' portion of the polylinker. Two different bacterial expression vectors for RGS2 can be utilized. To create the first, the polymerase chain reaction (PCR) fragment from RGS2 is inserted in-frame with the hexa-histidine tag in pET15b (Sinnarajah *et al.*, 2001). The second construct inserts an RGS2 fragment flanked by *Bsp*HI and *Bam*HI sites in-frame with the hexa-histidine tag in *Nco*I/*Bam*HI-digested H$_6$pQE60 [a modified form of Qiagen's pQE60 (Lee *et al.*, 1994)]. This creates an N-terminal extension of only eight residues (MH$_6$A).

RGS2 and AC cAMP Accumulation Assays

Mammalian expression vectors for RGS2, adenylyl cyclase, and constitutively active Gα_s are needed for transient expression assays measuring cAMP in HEK293 cells as described later. The expression vector for Gα_s-Q227L was provided by Dr. J. Silvio Gutkind (National Institutes of Health) and contains a C-terminal HA tag. Full-length type V AC and the E411A and L472A mutant clones are subcloned from the pVL1392 vector (Dessauer *et al.*, 1998; Taussig *et al.*, 1993) into the mammalian expression vector pcDNA3 using the restriction sites *Bam*HI and *Xba*I.

A number of truncations and mutations of RGS2 have been used to identify an AC-binding site (Fig. 1). All of the RGS2 truncations and mutations are created using the wild-type RGS2 sequence containing a triple HA tag at the C-terminal residue in the vector pCDNA3 [HA tag sequence: TACCCATACGATGTTCCAGATTACGCTTACCCATA-CGATGTTCCAGATTACGCTTACCCATACGATGTTCCAGATTA-CGCTTGA (Beadling *et al.*, 1999; Sinnarajah *et al.*, 2001)]. RGS2 N-terminal truncations are created using PCR by deletion of DNA encoding amino acid residues 1–19 (ΔNT1), 1–39 (ΔNT2), and 1–62 (ΔNT3) (see Fig. 1A). All of the truncations contain an N-terminal initiating methionine. The RGS2-ΔBox is created by the deletion of DNA encoding amino acid residues 72–211 employing the following forward and reverse primers, 5′-GCCACCATGCAAAGTGCTATG-3′ and 5′-GTTGCGGCCGCTA-CTTGATGAAAGCTTGCTGTTG-3′. RGS2-ΔC is constructed by deleting the last 12 residues at the C terminus of RGS2. Deletion of the first 19 amino acids of RGS2 is sufficient to destroy binding to and inhibition of AC. Deletion of the RGS box or C-terminal residues has no effect on RGS2 interactions with AC.

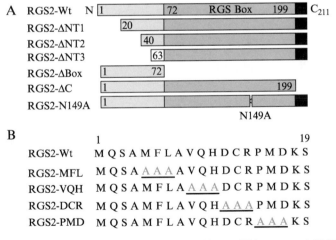

Fig. 1. (A) Schematic representation of RGS2 wild-type (Wt), truncated (RGS2-ΔNT1, ΔNT2, ΔNT3, ΔBox, ΔC), and mutant (RGS2-N149A) RGS2 constructs used to identify regions of RGS2 necessary for binding and inhibition of AC. (B) Series of mutations constructed within the N-terminal 19 amino acids of RGS2 required for AC binding and inhibition. Groups of three amino acids (MFL, VQH, DCR, PMD) are mutated to alanine.

Mutations within RGS2 are generated using a QuickChange mutagenesis kit (Stratagene). Asparagine at position 149 of RGS2 is mutated to alanine using the oligonucleotides 5′-GCTCCAAAAGAGATAGCCA-TAGATTTTCAAACC-3′ and 5′-GGTTTGAAAATCTATGGCTATC-TCTTTTGGAGC-3′. Four mutants are generated in the first 19 amino acids at the N terminus of RGS2, namely RGS2-MFL, RGS2-VQH, RGS2-DCR, and RGS2-PMD (see Fig. 1B). Residues MFL, VQH, DCR, and PMD are mutated to alanine using the following oligonucleotides: for RGS2-MFL, 5′-ATGCAAAGTGCTGCGGCCGCGGCTGTTCAACA-C-3′ and 5′-CGTGTTGAACAGCCGCGGCCGCAGCACTTTGCAT-3′; for RGS2-VQH, 5′-GCTATGTTCTTGGCTGCTGCTGCTGACTGCA-GACCCATG-3′ and 5′-CATGGGTCTGCAGTCAGCAGCAGCAGC-CAAGAACATAGC-3′; for RGS2-DCR, 5′-GGCTGTTCAACACGC CGCCGCACCCATGGACAAGAGCG-3′ and 5′-CGCTCTTGTCCAT GGGTGCGGCGGCGTGTTGAACAGCC-3′; and for RGS2-PMD, 5′-ACACGACTGCAGAGCCGCGGCCAAGAGCGCAGGC-3′   and 5′-GCCTGCGCTCTTGGCCGCGGCTCTGCAGTCGTGT-3′. The veracity of all of the DNA constructs is verified by nucleotide sequencing. A schematic showing the expected RGS proteins expressed from these constructs is shown (Fig. 1). The mutant RGS2-VQH displayed no inhibition of AC, whereas all other mutations were similar to wild-type RGS2. Only two deletions of RGS2 are tagged at their N termini, ΔBox (72–211), and ΔC (199–211), which contain three copies of the HA epitope. Wild-type and N-terminal HA-tagged RGS2 can be obtained from the Guthrie cDNA Resource Center.

Adenylyl Cyclase-Binding Assays

Two different types of pull-down assays can be used to detect interactions between RGS proteins and adenylyl cyclase. Due to the low levels of AC expression, binding assays using the full-length AC protein are not feasible, although weak interactions between ACIII and RGS2 have been detected using olfactory epithelial membranes (data not shown). Due to these limitations, both assays described herein make use of the cytoplasmic domains of AC. The first assay uses HA-tagged RGS proteins obtained from adenoviral-infected (Ia) or transiently transfected (Ib) HEK293 cells. The cellular extracts are incubated with purified hexa-histidine-tagged C_1 or C_2 domains from type V AC. Use of RGS adenoviruses has the advantage of easy scale-up while transient transfection of HEK293 cells allows rapid screening of a number of RGS mutants without the need for creating individual adenoviruses for each.

The second assay relies completely on purified proteins using the GST-tagged C_1 domain and hexa-histidine-tagged RGS2 and RGS4. Because all proteins used in this second assay are purified, any interactions observed must be direct and not mediated by an accessory protein. Both assays show strong interactions between the C_1 domain of AC and RGS2, but not RGS4 (Salim et al., 2003).

Materials for Pull-Down Assay I

Recombinant adenoviruses[1] encoding RGS2 and RGS4 (method Ia)
FuGENE 6 transfection Reagent (Roche Molecular Biochemicals) and plasmid DNAs encoding tagged RGS proteins (method Ib)
Ni-NTA agarose resin
$VC_1(670)$-H_6[2]
H_6-VC_2[2]
Anti-HA antibody (12CA5; Roche Molecular Biochemicals)

Solutions

Lysis buffer A: 50 mM Tris, pH 7.4, 5 mM MgCl$_2$, 1 mM EGTA, 250 mM sucrose, and protease inhibitors
Phosphate-buffered saline (PBS)
Lysis buffer B: 50 mM Tris, pH 7.5, 4 mM EDTA, and protease inhibitors
Wash buffer A: 20 mM HEPES, pH 8.0, 2 mM MgCl$_2$, 200 mM NaCl, 1 mM β-mercaptoethanol, and 0.1% $C_{12}E_{10}$
Wash buffer B: 20 mM HEPES, pH 8.0, 2 mM MgCl$_2$, 200 mM NaCl, 1 mM β-mercaptoethanol
Protease inhibitors: phenylmethylsulfonyl fluoride (0.1 mM), Nα-p-tosyl-L-lysine chloromethyl ketone (0.1 mM), and N-tosyl-L-phenyl alanine chloromethyl ketone (0.1 mM)

---

[1]The preparation of recombinant adenoviruses is described in Graham et al. (1991) and McGrory et al. (1988). Adenoviruses for RGS2 and RGS4 were generated using the adenoviral shuttle vector pACsk2.cmv and the viral DNA pJM17 (Salim et al., 2003). Newer methods allow for bacterial recombination and speed up viral production greatly (He et al., 1998).
[2]Type V AC domains are expressed with an N-terminal (VC_2) or C-terminal (VC_1) hexa-histidine tag and are purified as described previously (Dessauer et al., 1998).

Preparation of Cell Lysates from Adenoviral-Infected HEK293 Cells (Method Ia)

HEK293 cells are maintained in Dulbecco's modified Eagle's medium (DMEM) supplemented with 10% (v/v) fetal bovine serum, 1 mM glutamine, 50 μg/ml streptomycin, and 50 units/ml penicillin at 37° with 5% CO_2.

1. HEK293 cells (~70% confluent) are infected with recombinant adenoviruses encoding triple HA-tagged RGS2 or RGS4 at a multiplicity of infection of 2 in 10-cm dishes.

2. Cells 24 h postinfection are rinsed with 5 ml cold PBS per dish, scraped in 2.5 ml PBS, and transferred to a centrifuge tube.

3. The plate is rinsed with another 3 ml PBS and transferred to the same centrifuge tube.

4. The cells are pelleted at 2000g (Beckman J6-HC) at 4° for 5 min, resuspended in 1 ml cold lysis buffer A, and incubated on ice for 10 min.

5. Cells are lysed using a 1-ml Dounce homogenizer (50 strokes on ice), taking care not to froth the suspension.

6. Cytosolic proteins are obtained by an ultracentrifuge (Beckman TL100) at 100,000g at 4° for 20 min.

7. The supernatant is collected and tested for RGS2HA and RGS4HA protein expression by SDS–PAGE and immunoblotting with the anti-HA antibody. High overexpression from adenoviral expression leads to a doublet of bands on SDS–PAGE, which is most likely the result of proteolytic clipping of one or two copies of the triple HA tag present at the C terminus of RGS2 and RGS4. This is not observed with lower expression of RGS2 and RGS4 in transient transfections.

8. When appropriate expression of RGS proteins is thus confirmed, the pull-down assays are conducted with the prepared HEK293 cell lysates as described next.

Preparation of Cell Lysates from Transiently Transfected HEK293 Cells (Method Ib)

1. HEK293 cells are plated in 10-cm dishes 1 day before transient transfection to obtain a cell density of 60–70% at the time of transfection. Two micrograms of triple HA-tagged RGS plasmid DNA is transfected using the FuGENE6 transfection reagent at a DNA:FuGENE ratio of 1:3. Serum free-DMEM (100 μl) is first added to an Eppendorf tube followed by 6 μl of FuGENE. Great care is taken not to dispense any FuGENE on the walls of the tube as it hinders efficient DNA uptake. Serum-free DMEM and FuGENE are mixed by tapping the tube gently and are

incubated for 10 min at room temperature. DNA (2 μg/10-cm dish) is then added carefully at the center of the tube and incubated for 20 min at room temperature. Add the complex to cells and incubate at 37° with 5% CO_2.

2. Cells at 36 h posttransfection are washed once with 1 ml ice-cold PBS and harvested in 300 μl of hypotonic lysis buffer B (50 mM Tris, pH 7.5, 4 mM EDTA plus protease inhibitors) followed by Dounce homogenization and centrifugation to remove cellular debris as described earlier.

3. The lysates thus obtained are checked for RGS2 protein expression as detailed previously and are used in the following pull-down assay.

Method for Pull-Down Assay I

1. To prepare the Ni-NTA resin, swirl the resin bottle and remove 1 ml suspension to an Eppendorf tube. Allow the resin to settle at 4° or touch spin in a microcentrifuge and wash the resin twice with 1 ml wash buffer B. Approximately 500 μl packed resin is obtained per 1-ml suspension.

2. Lysates prepared from HEK293 cells expressing RGS proteins (50 μg) are incubated with 15 μg of $VC_1(670)H_6$ or H_6VC_2 for 30 min at 4°. Ni-NTA-agarose resin (100 μl packed) is then added and incubated on a rotating mixer at 4° for 2 h.

3. The complex is pelleted by centrifugation at 16,000g (microfuge) at 4° for 1 min.

4. The supernatant is removed but saved for future analysis on SDS–PAGE if desired.

5. The resin is washed twice with 200–500 μl of wash buffer A.

6. Laemmli buffer (20 μl) is added, and the complex is boiled at 100° for 5 min. The resin is pelleted by centrifugation, and the supernatant is loaded on a 13% SDS–PAGE gel and probed with anti-HA antibody at a dilution of 1:400.

Materials for Pull-Down Assay II

Escherichia coli purified hexa-histidine tagged RGS2 and RGS4[3]
Glutathione resin
GST-VC_1
GST
Ni-NTA horseradish peroxidase antibody (Qiagen)

---

[3]Hexa-histidine tagged RGS proteins are expressed in E. coli and purified under nondenaturing conditions using Ni-NTA-agarose resin (Berman et al., 1996; Salim et al., 2003).

Preparation of GST-VC₁

Solutions

Lysis buffer: 50 mM Tris, pH 7.7, 1 mM EDTA, 2 mM dithiothreitol (DTT), 120 mM NaCl, and protease inhibitors

Elution buffer: 20 mM HEPES, pH 8.0, 1 mM EDTA, 2 mM DTT, 100 mM NaCl, 5% glycerol, 12 mM glutathione, and protease inhibitors

GST-VC₁ is purified using glutathione affinity resin. BL21(DE3) cells containing the GST-VC₁ expression plasmid are grown pre- and postinduction at 30° in T7 medium (2% tryptone, 1% yeast extract, 0.5% NaCl, 0.2% glycerol, and 50 mM KH_2PO_4, pH 7.2). Synthesis of GST-VC₁ is induced with 0.5 mM isopropyl β-D-thiogalactoside (OD_{600} = 1.0). Cells are harvested after 4 h and resuspended in lysis buffer. Lysozyme (0.2 mg/ml) is added for 30 min at 4° followed by incubation with 5 mM $MgCl_2$ and DNase (0.01 mg/ml) for 30 min. Cellular debris is removed by centrifugation (Beckman Ti45) at 100,000g for 30 min. The supernatant is applied to a pre-equilibrated glutathione column (1 ml of packed resin). The column is washed with lysis buffer containing 400 mM NaCl (25 column volumes) followed by 5 column volumes of lysis buffer without NaCl. The protein is eluted with buffer containing 12 mM glutathione. The protein is concentrated and dialyzed overnight in elution buffer lacking glutathione.

Method for Pull-Down Assay II

Glutathione agarose resin is allowed to swell for 1 h at room temperature at a concentration of 5 mg/ml distilled water. When the resin has settled, aspirate water from the resin, add 1 ml water, and pellet the resin for 1 min at 14,000 rpm on a microfuge. Wash the resin twice with 50 μl of wash buffer A and continue with the pull-down assay.

1. GST or GST-VC₁ (15 μg) is first bound to 100 μl of packed glutathione resin by incubation on a rotating mixer at 4° for 2 h.
2. Pellet the resin by centrifugation for 1 min at 16,000g (microfuge) at 4°.
3. Remove the supernatant and wash the resin with 500 μl wash buffer (20 mM HEPES, pH 8.0, 2 mM $MgCl_2$, 200 mM NaCl, 1 mM β-mercaptoethanol, and 0.1% $C_{12}E_{10}$).
4. Incubate the resin with 2 μM purified hexa-histidine-tagged RGS2 or RGS4 in a total volume of 500 μl of wash buffer for 2–4 h at 4° on a rotating mixer.
5. Wash the resin twice with 500 μl wash buffer.

6. Add Laemmli buffer and boil for 5 min. Subject the supernatant to SDS–PAGE (13% gel) and analyze by immunoblotting using the Ni-NTA horseradish peroxidase antibody at a 1:500 dilution.

In Vivo cAMP Detection Assay

To examine the regulation of AC by RGS2 in cells, cAMP accumulation is measured in HEK293 cells cotransfected with type V adenylyl cyclase and RGS2 in the presence of either a GTPase-deficient, constitutively active form of $G\alpha_s$, $G\alpha_s$-Q227L, or β_2-adrenergic receptor. Coexpression of $G\alpha_s$-Q227L and type V AC increases cAMP accumulation 3-fold. Additional expression of RGS2 reduces $G\alpha_s$-Q227L-stimulated cAMP accumulation to nearly basal levels (Fig. 2A). Similar results are obtained using the β_2-adrenergic receptor (Fig. 2B). Because HEK293 cells have an endogenous β_2-adrenergic receptor [approximately 7–20 fmol/mg protein (Friedman et al., 2002)], a concentration of 100 nM isoproterenol is chosen carefully so as to activate largely the overexpressed receptor population. This concentration of drug produces a substantially greater increase in cAMP in β_2-adrenergic receptor-transfected cells as opposed to vector

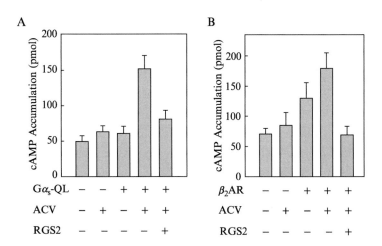

Fig. 2. Inhibition of cAMP accumulation by RGS2 in the presence of $G\alpha_s$-Q227L (A) or β_2-adrenergic receptor (B). Production of cAMP is measured in HEK293 cells transfected as indicated with type V AC, $G\alpha_s$-Q227L (A) or β_2-adrenergic receptor (B) and wild-type RGS2. Data are expressed as the means ± SE from three different experiments, each performed in duplicate. HEK293 cells transfected with β_2AR or control constructs are stimulated with 100 nM isoproterenol for 15 min prior to IBMX treatment, lysis, and cAMP detection.

controls. Expression of β_2-adrenergic receptor and type V AC results in a 2.5-fold increase in cAMP, whereas additional expression of RGS2 results in inhibition to basal levels. As shown elsewhere, the inhibitory effect of RGS2 was also reported in HEK293 cells on another cardiac isoform, type VI AC (Roy et al., 2003). The only major difference in this method from that outlined later is the detection of intracellular cAMP by labeling with [^{3}H]adenine (Salomon et al., 1974). Under these conditions, cAMP increased 2-fold in cells transfected with $G\alpha_s$ alone and approximately 10-fold in the presence of $G\alpha_s$ and type VI AC. RGS2 reduced intracellular cAMP by 51% (Roy et al., 2003). This assay allows for the rapid screening of RGS2 truncations and mutations outlined in Fig. 1.

Reagents

FuGENE 6 transfection reagent and plasmid DNAs
cAMP enzyme immunoassay kit (Assay Designs, Inc., Ann Arbor, MI)
3-Isobutyl-1-methylxanthine (IBMX; 1 mM final)
Isoproterenol (100 nM final)
Lysis buffer: 50 mM Tris–HCl (pH 7.5), 4 mM EDTA, and protease inhibitors

Method for Detecting cAMP Accumulation of Transiently Transfected HEK293 Cells

1. For transient expression of proteins, subconfluent HEK293 cells are plated on 6-well plates 1 day before transfection. Cells are transfected at 60–70% confluence.

2. The FuGENE 6 reagent is diluted in serum-free DMEM for a final DNA:FuGENE ratio of 1:3. Serum-free DMEM (100 μl) is added first to an Eppendorf tube followed by 5.1 μl of FuGENE per well, taking utmost precaution not to dispense any FuGENE on the walls of the tube.

3. The serum-free DMEM and FuGENE 6 reagent are mixed by tapping the tube gently and are incubated for 10 min at room temperature. Constructs expressing either β_2-adrenergic receptor (0.5 μg/well) or $G\alpha_s$-Q227L (0.5 μg/well) and RGS2 (1.0 μg/well) and type V AC (0.2 μg/well) are added carefully to the center of the tube containing the FuGENE mix and are incubated for 20 min at room temperature. The vector pcDNA3 (Invitrogen) is used to normalize all of the DNA concentrations to 1.7 μg/well.

4. The complex is added to cells and incubated at 37° with 5% CO_2.

5. After 36 h of transfection, cells are starved overnight ($\sim$12 h) in DMEM containing 1% fetal bovine serum. It is necessary to add the low serum medium to the side of the well, taking care not to damage the cells. At this point the cells can lift off the plate easily if treated roughly.

6. After overnight starvation, the medium is again aspirated and serum-free medium is added for 2 h.

7. To measure cAMP accumulation in cells transfected with Gα_s-Q227L or corresponding control DNA constructs, the cells are treated with 1 mM IBMX for exactly 15 min to inhibit cellular cAMP phosphodiesterases. Medium devoid of serum is aspirated, and serum-free medium containing 1 mM IBMX is added. An increment of 10 or 20 s is usually maintained between the aspiration of the serum-free medium and the addition of medium containing IBMX to keep all timed incubations constant between wells.

8. After exactly 15 min, medium containing IBMX is aspirated and 250 μl hypotonic lysis buffer is added and cells are placed on ice for 15 min.

9. Cells are harvested with a cell scraper while still on ice.

10. A portion of the cell lysate (50 μl) is used for Western blotting to monitor protein expression, and the remaining lysate is boiled for 10 min at 100° and centrifuged at 16,000g (microfuge) for 10 min to remove cellular debris (Rall *et al.*, 1957).

11. The supernatant thus obtained is used to measure total cAMP accumulation by cAMP enzyme immunoassay detection as per the manufacturer's protocol.

12. For cells transfected with β_2-adrenergic receptor and corresponding controls, cells are treated with isoproterenol (100 nM) and IBMX (1 mM) for 15 min before subsequent harvesting, lysis, and cAMP detection.

13. Data are represented as the means $\pm$ SE, and statistical significance is determined by the Student's t test ($p < 0.05$ is considered significant).

Western Blotting

Cell lysates are boiled for 5 min in Laemmli sample buffer and bath sonicated for 5–10 min to shear DNA. The proteins are resolved by SDS–PAGE and transferred to nitrocellulose. Protein expression levels are determined by immunoblotting using the anti-HA antibody (1:400). Immunoreactive protein bands are detected by a horseradish peroxidase-conjugated secondary antibody and enhanced chemiluminescence.

Technical Considerations

The ratio of DNA concentrations mentioned earlier may need to be altered to observe significant inhibition of cAMP accumulation by RGS2, depending on the source of HEK293 cells. For example, it was necessary to transfect HEK293 cells obtained from a source other than the one used in the reported study (Salim *et al.*, 2003) with different concentrations of DNA in order to see comparable inhibition in cAMP accumulation by RGS2. In this system, optimal DNA concentrations consisted of Gsα-Q227L (0.25 μg/well), type V AC (0.025 μg/well), and RGS2 (2.0 μg/well), keeping the DNA:Fugene ratio constant as stated previously. A dose–response of RGS2 inhibition is strongly recommended. The shift in the dose–response curve could be attributed to the quality of the DNA used and/or the sheer variable nature of immortal cell lines.

Subcellular Distribution of RGS2 Proteins

Subcellular distribution of RGS proteins is still insufficiently documented. However, in order to understand many of the cellular functions they may regulate, their precise intracellular localization is essential. Efforts to detect native RGS proteins effectively are hampered by the lack of effective antibodies as experimental tools. Limited reports have addressed this question to assess intracellular distribution using the overexpression of recombinant proteins. Many of these have used subcellular fractionation methods and have found RGS proteins associated with the membrane fraction or distributed between the membrane and cytosolic fractions (reviewed in De Vries *et al.*, 2000). However, several reports of nuclear localization of RGS2 and other RGS proteins raise the possibility that RGS proteins may contribute to more cellular functions than currently envisioned (Bowman *et al.*, 1998; Dulin *et al.*, 2000; Heximer *et al.*, 2001; Song *et al.*, 2001). It is necessary to examine the distribution of any truncation or mutation of RGS2 that is utilized, as previous reports have shown dramatic changes in RGS2 subcellular localization upon truncation of the protein (Heximer *et al.*, 2001). In accordance with earlier evidence, we have found HA-tagged RGS2 to be distributed equally between the membrane and the cytosol of HEK293 cells upon subcellular fractionation (Salim *et al.*, 2003). However, nuclear localization of RGS2 may have been overlooked due to the limitations of the technique used. In fact, overexpression of RGS2 using adenoviruses leads to significant nuclear localization, as well as plasma membrane and cytoplasmic localization as measured by fractionation methods that maintain intact nuclei. Deletion of the N-terminal 19 amino acids of RGS2 or mutation of residues within this region (see Fig. 1) does not alter the relative proportion of RGS2 protein found in the membrane or cytosol.

In Vitro Assay of Adenylyl Cyclase Activity

To examine direct inhibition of AC enzymatic activity, *in vitro* AC assays are employed. Two different systems can be utilized: assay of membranes prepared from Sf9 (*Spodoptera frugiperda* ovary) cells overexpressing an individual isoform of AC or assay of reconstituted purified cytoplasmic domains of AC. Detailed reviews of the basic adenylyl cyclase assay are available (Dessauer, 2002; Johnson *et al.*, 1991) and thus are not emphasized here; rather features of the assay unique to RGS2 inhibition are discussed.

AC Assay of C_1 and C_2 Cytoplasmic Domains

As mentioned earlier, the C_1 and C_2 domains of AC are expressed and purified individually from *E. coli*. In order to reconstitute activity, a limiting concentration of the C_1 domain from type V AC [$VC_1(670)H_6$; ~60 nM] is mixed with the C_2 domain of type V AC (H_6VC_2; 0.5 μM final concentration) in a buffer containing 50 mM HEPES, pH 8.0, 2 mM MgCl$_2$, 1 mM EDTA, 2 mM DTT, and 0.5 mg/ml bovine serum albumin (BSA). BSA is critical for stabilizing the diluted C_1 domain and RGS2, and one domain of AC must be maintained at a concentration of 0.5 μM or above in order to drive the interaction between C_1 and C_2. RGS proteins or control buffer is added to the cytoplasmic domains and preincubated for 20 min on ice. The volume of RGS protein added to this reaction should be limited due to the high salt content of their storage buffer. In general, the diluted concentrations are sufficiently low, such that precipitation in lower salts is no longer an issue for RGS2. GTPγS-Gα_s (400 nM) is added next, and the reaction is immediately started upon the addition of the prewarmed AC reaction mix containing 0.5 mM [α-^{32}P]ATP and 5 mM MgCl$_2$ in a final volume of 50 μl. The reactions are terminated after 10 min at 30°, and the products are separated by sequential chromatography on Dowex-50 and Al$_2$O$_3$.

Assay of Membrane-Bound AC

The inhibition of AC activity by RGS2 using the cytoplasmic domains is extremely reliable and straightforward. The *in vitro* assay of membrane-bound AC can be more problematic. Membranes from Sf9 cells expressing type V AC are prepared as described by Dessauer *et al.* (2002). The typical assay involves preincubation of membranes (~8 μg) and RGS2 (50 nM to 1 μM) for 30 min at room temperature. GTPγS-Gα_s (100 nM) is added and the reaction is immediately started upon the addition of the pre-warmed AC reaction mix containing 0.5 mM [α-^{32}P]ATP and 5 mM MgCl$_2$ in a final volume of 50 μl. The reactions are stopped after 5 min at 37°. The problem with this assay is due to the unstable nature of RGS2 in low salts and the hydrophobic nature of the membranes. Higher salt concentrations tend to stabilize RGS2, but also serve to inhibit AC. Detergents will prevent RGS2 aggregation and precipitation (as described for gel filtration later), but also tend to inhibit the enzymatic activity of AC. Therefore, a careful balance must be maintained to stabilize both RGS2 and AC in order to observe inhibition. In general, the cAMP accumulation assays and the *in vitro* AC assay of the cytoplasmic domains described earlier have proven more reliable.

Gel Filtration Profiles of RGS2

Due to the limitations of AC expression, biochemical studies have relied on the use of the cytoplasmic domains of type V AC. We originally wanted to use gel filtration to examine the potential interactions of RGS2 with the individual or combined domains of AC in order to map the binding site of RGS2. To our surprise, most of RGS2 alone runs as a dimer and elutes in the molecular mass range of 50,000 Da in the presence of 200 mM NaCl and 5 mM CHAPS (Fig. 3). We had hoped to get a shift in the molecular weight of RGS2 upon the addition of C_1, C_2, and activated Gα_s. However, it is very difficult to distinguish RGS2 dimers from potential RGS2–C_1 interactions. A small fraction of RGS2 also runs as a dimer on SDS–PAGE. This is observed with the H_6-tagged purified protein from *E. coli* and the HA-tagged RGS2 protein from transfected or adenoviral-infected HEK293 cells. This can be prevented to a large extent with sufficient amounts of fresh DTT. To add to the complexity, RGS2 does not elute from the Superdex 200 column at physiological salt concentrations in the absence of any detergent. Either high salt or detergent is necessary to prevent RGS2 from adhering to the column and thus smearing across a wide range of fractions. The presence of a carrier protein, such as BSA, increases the yield of protein that elutes from the column (Fig. 3A vs Fig. 3C). RGS2 will run as a monomer if 500 mM NaCl and 3 mM DTT are included (Heximer, 1997). These caveats have limited the usefulness of this technique for analyzing RGS2 protein–protein interactions.

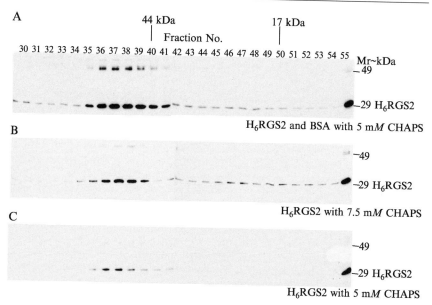

FIG. 3. Gel filtration profiles of RGS2. H_6RGS2 (50 μg) was subjected to gel filtration on an FPLC superdex 200 column in the presence of 50 μg BSA and 5 mM CHAPS (A) or in the absence of BSA and 7.5 mM CHAPS (B) or 5 mM CHAPS (C). Fractions (0.2 ml) were collected and run on SDS–PAGE (13%) and were probed with the anti-RGS2 antibody. The elution positions of gel filtration standards are indicated.

Reagents

Superdex 200 HR 10/30 (Amersham Biosciences)
Gel filtration Standards (Bio-Rad #151–1901)
H_6RGS2 (50 μg)
Anti-RGS2 antiserum generated in rabbits against a C-terminal peptide (PQITEPHAT) coupled to keyhole limpet hemocyanin
Gel filtration buffer: 20 mM HEPES, 5 mM MgCl2, 1 mM EDTA, 5 mM CHAPS, 200 mM NaCl, and 2 mM DTT

References

Beadling, C., Druey, K. M., Richter, G., Kehrl, J. H., and Smith, K. A. (1999). Regulators of G protein signaling exhibit distinct patterns of gene expression and target G protein specificity in human lymphocytes. *J. Immunol.* **162,** 2677–2682.
Berman, D. M., Wilkie, T. M., and Gilman, A. G. (1996). GAIP and RGS4 are GTPase-activating proteins for the G_i subfamily of G protein α subunits. *Cell* **86,** 445–452.

Bowman, E. P., Campbell, J. J., Druey, K. M., Scheschonka, A., Kehrl, J. H., and Butcher, E. C. (1998). Regulation of chemotactic and proadhesive responses to chemoattractant receptors by RGS (regulator of G-protein signaling) family members. *J. Biol. Chem.* **273,** 28040–28048.

Dessauer, C. W. (2002). Kinetic analysis of the action of P-site analogs. *Methods Enzymol.* **345,** 112–126.

Dessauer, C. W., Chen-Goodspeed, M., and Chen, J. (2002). Mechanism of Galpha i-mediated inhibition of type V adenylyl cyclase. *J. Biol. Chem.* **277,** 28823–28829.

Dessauer, C. W., and Gilman, A. G. (1997). The catalytic mechanism of mammalian adenylyl cyclase: Equilibrium binding and kinetic analysis of P-site inhibition. *J. Biol. Chem.* **272,** 27787–27795.

Dessauer, C. W., Tesmer, J. J. G., Sprang, S. R., and Gilman, A. G. (1998). Identification of a $G_{i\alpha}$ binding site on type V adenylyl cyclase. *J. Biol. Chem.* **273,** 25831–25839.

De Vries, L., Zheng, B., Fischer, T., Elenko, E., and Farquhar, M. G. (2000). The regulator of G protein signaling family. *Annu. Rev. Pharmacol. Toxicol.* **40,** 235–271.

Dulin, N. O., Pratt, P., Tiruppathi, C., Niu, J., Voyno-Yasenetskaya, T., and Dunn, M. J. (2000). Regulator of G protein signaling RGS3T is localized to the nucleus and induces apoptosis. *J. Biol. Chem.* **275,** 21317–21323.

Friedman, J., Babu, B., and Clark, R. B. (2002). Beta(2)-adrenergic receptor lacking the cyclic AMP-dependent protein kinase consensus sites fully activates extracellular signal-regulated kinase 1/2 in human embryonic kidney 293 cells: Lack of evidence for G(s)/G(i) switching. *Mol. Pharmacol.* **62,** 1094–1102.

Graham, F. L., and Prevec, L. (1991). Manipulation of adenovirus vectors. *In* "Methods in Molecular Biology" (E. J. Murray, ed.), Vol. 7, pp. 109–128. Humana Press, Clifton, NJ.

He, T. C., Zhou, S., da Costa, L. T., Yu, J., Kinzler, K. W., and Vogelstein, B. (1998). A simplified system for generating recombinant adenoviruses. *Proc. Natl. Acad. Sci. USA* **95,** 2509–2514.

Hepler, J. R. (1999). Emerging roles for RGS proteins in cell signalling. *Trends Pharmacol. Sci.* **20,** 376–382.

Heximer, S. P. (1997). "Studies on Two Putative G0/G1 Switch Genes in Human T-Lymphocytes, RGS2/G0S8 and G0S24" Ph.D. thesis dissertation, Queen's University.

Heximer, S. P., Lim, H., Bernard, J. L., and Blumer, K. J. (2001). Mechanisms governing subcellular localization and function of human RGS2. *J. Biol. Chem.* **276,** 14195–14203.

Hollinger, S., and Hepler, J. R. (2002). Cellular regulation of RGS proteins: Modulators and integrators of G protein signaling. *Pharmacol. Rev.* **54,** 527–559.

Johnson, R. A., and Salomon, Y. (1991). Assay of adenylyl cyclase catalytic activity. *Methods Enzymol.* **195,** 3–21.

Lee, E., Linder, M. E., and Gilman, A. G. (1994). Expression of G protein α subunits in *Escherichia coli. Methods Enzymol.* **237,** 146–164.

McGrory, W. J., Bautista, D. S., and Graham, F. L. (1988). A simple technique for the rescue of early region I mutations into infectious human adenovirus type 5. *Virology* **163,** 614–617.

Rall, T. W., Sutherland, E. W., and Berthet, J. (1957). The relationship of epinephrine and glucagon to liver phosphorylase. *J. Biol. Chem.* **224,** 463–475.

Ross, E. M., and Wilkie, T. M. (2000). GTPase-activating proteins for heterotrimeric G proteins: Regulators of G protein signaling (RGS) and RGS-like proteins. *Annu. Rev. Biochem.* **69,** 795–827.

Roy, A. A., Lemberg, K. E., and Chidiac, P. (2003). Recruitment of RGS2 and RGS4 to the plasma membrane by G proteins and receptors reflects functional interactions. *Mol. Pharmacol.* **64,** 587–593.

Salim, S., Sinnarajah, S., Kehrl, J. H., and Dessauer, C. W. (2003). Identification of RGS2 and type V adenylyl cyclase interaction sites. *J. Biol. Chem.* **278,** 15842–15849.

Salomon, Y., Londos, C., and Rodbell, M. (1974). A highly sensitive adenylate cyclase assay. *Anal. Biochem.* **58,** 541–548.

Siderovski, D. P., Strockbine, B., and Behe, C. I. (1999). Whither goest the RGS proteins? *Crit. Rev. Biochem. Mol. Biol.* **34,** 215–251.

Sinnarajah, S., Dessauer, C. W., Srikumar, D., Chen, J., Yuen, J., Yilma, S., Dennis, J. C., Morrison, E. E., Vodyanoy, V., and Kehrl, J. H. (2001). RGS2 regulates signal transduction in olfactory neurons by attenuating activation of adenylyl cyclase III. *Nature* **409,** 1051–1055.

Smit, M. J., and Iyengar, R. (1998). Mammalian adenylyl cyclases. *Adv. Second Messenger Phosphoprotein Res.* **32,** 1–21.

Song, L., Zmijewski, J. W., and Jope, R. S. (2001). RGS2: Regulation of expression and nuclear localization. *Biochem. Biophys. Res. Commun.* **283,** 102–106.

Sunahara, R. K., Dessauer, C. W., Whisnant, R. E., Kleuss, C., and Gilman, A. G. (1997). Interaction of G$_{s\alpha}$ with the cytosolic domains of mammalian adenylyl cyclase. *J. Biol. Chem.* **272,** 22265–22271.

Taussig, R., Iñiguez-Lluhi, J., and Gilman, A. G. (1993). Inhibition of adenylyl cyclase by G$_{i\alpha}$. *Science* **261,** 218–221.

Wang, C. R., Esser, L., Smagula, C. S., Sudhof, T. C., and Deisenhofer, J. (1997). Identification, expression, and crystallization of the protease-resistant conserved domain of synapsin I. *Protein Sci.* **6,** 2264–2267.

Whisnant, R. E., Gilman, A. G., and Dessauer, C. W. (1996). Interaction of the two cytosolic domains of mammalian adenylyl cyclase. *Proc. Natl. Acad. Sci. USA* **93,** 6621–6625.

Yan, S. Z., Hahn, D., Huang, Z. H., and Tang, W.-J. (1996). Two cytoplasmic domains of mammalian adenylyl cyclase form a G$_{s\alpha}$- and forskolin-activated enzyme *in vitro*. *J. Biol. Chem.* **271,** 10941–10945.

[7] Assays of RGS3 Activation and Modulation

By Patrizia Tosetti *and* Kathleen Dunlap

Abstract

Regulator of G-protein signaling (RGS) proteins accelerate the intrinsic GTPase activity of Gα subunits, terminating G-protein signaling. Although the structure and the downstream effectors of RGS proteins have been investigated extensively, the mechanisms underlying their activation are still largely unexplored. Recent investigations of RGS proteins in intact cells are starting to shed some light on this issue. In particular, the retrovirus-mediated overexpression of individual RGS3 isoforms showed that activation of RGS3s requires Ca^{2+} binding to an EF-like domain, and such Ca^{2+} binding promotes the rapid desensitization of G$_o$-mediated inhibition of N-type Ca^{2+} (N) channels. This article provides detailed information on the use and power of retrovirus-mediated overexpression of RGS proteins in dorsal root ganglion neurons to isolate and identify the mechanisms underlying the initiation of RGS3-dependent inhibitory activity.

Introduction

The intrinsic GTPase activity of $G\alpha$ subunits controls cellular signaling by heterotrimeric G proteins by promoting reassociation of the inactive $G\alpha\beta\gamma$ heterotrimer. In recent years, regulator of G-protein signaling (RGS) proteins have been identified as key regulator of this process. Members of this large gene family share a well-conserved "RGS domain" that interacts directly with GTP-bound $G\alpha$ subunits and terminates signaling by either accelerating the rate of intrinsic GTP hydrolysis of $G\alpha$ subunits (Berman et al., 1996) or binding to $G\alpha$ subunits and physically blocking their interaction with downstream signaling molecules (Melliti et al., 2000).

Although much is known about the structure and the downstream effectors of RGS proteins, whether RGS proteins themselves are under cellular control and the mechanisms underlying their activation have been little explored. Recent investigations of RGS proteins in intact cells are starting to shed some light on this issue (Shin et al., 2003; Tosetti et al., 2003). Protocols in this article describe how we have used dorsal root ganglion (DRG) neurons from embryonic chick to investigate the mechanisms underlying the initiation of RGS3-dependent signaling.

RGS3 is an RGS family member that opposes cellular signaling via $G_{i/o}$ and $G_{q/11}$ protein-coupled receptors (Druey et al., 1996; Scheschonka et al., 2000). The RGS3 gene encodes several splice variants characterized by a common C terminus, including the RGS domain, and a divergent N terminus of variable length and domain composition (Kehrl et al., 2002). Chick DRG neurons express at least four RGS3 variants, ranging in size from 165 (with no N-terminal domain) to 799 amino acids (with several N-terminal motifs, including a Ca^{2+}-binding EF-like domain, PDZ and PH domains) (Fig. 1). Two intermediate-length RGS3 variants (RGS3s and RGS3ss), when overexpressed in DRG neurons, promote the termination of G-protein–Ca^{2+} channel signaling. The actions of RGS3s and RGS3ss are initiated by calcium influx through the Ca^{2+} channels themselves (Tosetti et al., 2003).

Ca^{2+} differentially activates the two splice variants. RGS3s is activated predominantly by direct Ca^{2+} binding to an EF–like domain, triggering a rapid termination of G-protein-dependent inhibition of Ca^{2+} channels. In contrast, RGS3ss lacks an EF–like domain, and its activation appears to be predominantly mediated indirectly via calmodulin, producing a slower termination of G-protein signaling. Thus, activity-evoked Ca^{2+} entry in

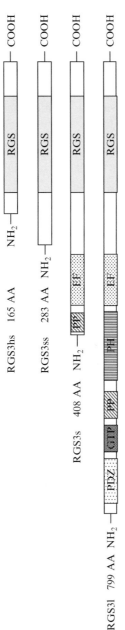

FIG. 1. Schematic structure of RGS3 splice variants identified in chick DRG neurons. Motif analysis identified the following domains: RGS, RGS domain, the region responsible for binding Gα and accelerating its GTPase activity; EF, EF-like domain, a Ca²⁺-binding motif; PP, proline-rich region; GTP, P-loop, a putative ATP/GTP-binding region; PH, pleckstrin homology, a protein–protein interaction domain; PDZ, PSD95/Discs large/ZO-1 homology region, another protein–protein interaction domain.

sensory neurons appears to provide differential negative feedback control of G-protein signaling via splice isoforms of RGS3. The distinct time courses for the termination of G-protein-dependent inhibition of the Ca^{2+} channels *in vivo* may ensure faithful transmission of information from the most active sensory inputs.

This article describes in detail the methodologies used (1) to clone the chick RGS3 variants, (2) to isolate variant-specific effects by retroviral-mediated overexpression of the variant of interest, and (3) to assess the Ca^{2+} dependence of RGS3 activation.

Identification of RGS3 Splice Variants in Chick DRG Neurons

The strategy used to identify RGS3 mRNAs expressed in chick DRG neurons consisted of a combination of reverse transcriptase-polymerase chain reaction (RT-PCR) and library screening. At first, a partial chick RGS3 fragment was amplified by RT-PCR using degenerate primers directed toward regions of the RGS domain highly conserved among species. The newly identified cDNA was then labeled radioactively and used as a probe to screen a chick embryonic DRG cDNA library in order to obtain the full-length sequences of the four chick RGS3 splice variants. The variant identity was confirmed by similarity searches, and protein sequences were subject to standard database motif analysis.

This cloning strategy is recommended when all published genes related to the gene of interest show strong divergence in the N and C termini or when only poorly related genes are available for primer design.

Experimental Results

Chick DRG Neurons Express at Least Four Distinct RGS3 Isoforms. RNA from chick DRG neurons was used as a template for RT-PCR with degenerate primers derived from regions of the RGS domain that are highly conserved among RGS proteins (Koelle and Horvitz, 1996). Along with a number of other RGS sequences, we obtained 13 independently cloned PCR products that encoded a 240-bp cDNA with 90% identity to human RGS3. The PCR fragment was then labeled radioactively and used to screen an embryonic chick DRG cDNA library in order to obtain the full-length RGS3 sequence.

Several positive clones were isolated from the screening and sequenced. All of them shared a common C-terminal RGS domain to which the probe hybridized. The N-terminal sequences of these clones, however, diverged significantly. Overall, we identified four distinct full-length isoforms of chick RGS3, varying in predicted lengths from 165 to 799 amino acids

(Fig. 1). We called them (in order of increasing length) RGS3hs, 3ss, 3s, and 3l (GenBank accession numbers: AY124773–7). Sequence analysis of the variants using the InterProScan tool yielded interesting results. The two shortest variants, 3hs and 3ss, have very simple structures dominated by the RGS domain. The third variant, 3s, consists of 408 amino acids and its N terminus carries a Ca^{2+}-binding domain (similar to an EF hand, thus termed EF-like domain) as well as a proline-rich (PP) region. The longest variant, RGS3l, contains an additional ATP/GTP-binding region (P loop) as well as PH and PDZ domains (Fig. 1).

Protocols

Total RNA Extraction from Chick DRGs. To avoid RNA degradation by exogenous ribonucleases, gloves and RNase-free material should be used throughout the extraction process. Total RNA from DRG neurons from 11- or 12-day-old chicken embryos (Charles River SPAFAS, North Franklin, CT) is extracted in a single step using an acidic phenol/guanidinium thiocyanate solution (STAT-60 reagent, Tel-Test Inc., Friendswood, TX). For a detailed description of DRG dissection from embryonic chick, see Anantharam and Diversé-Pierluissi (2002). DRGs from four chick embryos are collected in an RNase-free Eppendorf tube and 1 ml of STAT-60 is added. The mixture is thoroughly homogenized in a motor-driven, tight-fitting, glass–Teflon homogenizer and is then left at room temperature for 5 min to allow the complete dissociation of nucleoprotein complexes. Guanidinium thiocyanate, a powerful protein denaturant, inactivates all endogenous ribonucleases released during tissue homogenization. After the addition of 200 μl of CH$_3$Cl, the mixture is shaken vigorously for 15 s, left at room temperature for 3 min, and then centrifuged for 15 min at 4°. Under the acidic conditions dictated by STAT-60 (pH 4), RNA remains in the upper aqueous phase, whereas proteins and DNA localize at the interface between the two phases and in the lower organic phase, and cellular debris are insoluble. The upper phase is transferred to a new, RNase-free tube, paying careful attention to avoid the interface, and 0.5 ml of isopropanol is added to the tube. After leaving the mixture at room temperature for 10 min to allow RNA precipitation, the process is brought to completion by centrifuging the tube at 13,000g for 15 min at 4°. The supernatant is discarded, and the pellet is washed with 1 ml of 75% EtOH to eliminate residual contaminating guanidinium, vortexed, and centrifuged at 7500g for 5 min at 4°. After pouring off the supernatant, the pellet is partially air dried for 10–15 min, avoiding complete dehydration, which would decrease pellet solubility greatly. The RNA is finally dissolved in RNase-free, diethylpyrocarbonate-treated water, and its concentration and purity are determined by spectrophotometric analysis.

RT-PCR. Freshly prepared RNA (0.5–1 μg) is used as a template for reverse transcription using Moloney murine leukemia–virus reverse transcriptase (2.5 U/μl), 4 m*M* dNTP mix, and a 1:1 mix of random–hexamer and oligo (dT) (2.5 μ*M*) as primers. The reaction mixture is heated at 42° for 15 min for first strand synthesis, and the resulting cDNA mix is used as a template for PCR. Degenerate primers are designed from highly conserved sequences of the RGS domain (Koelle and Horvitz, 1996). After the addition of 1 U of *Taq* DNA polymerase and 1 μ*M* each of degenerate sense (5'-GAG AAC AT(G/T) (C/T)T(C/A) TTC TGG-3') and antisense (5'-TA(G/T) GAG TCC C(G/T)G TGC AT-3') primers, PCR is performed using the following cycling protocol: 94° for 1' (1 cycle), 94° for 30 s, 45° for 1 min, 72° for 1 min (40 cycles), and 72° for 5 min (1 cycle). Reaction products of the expected size (240 bp) are purified by gel extraction using the QIAquick gel extraction kit (Qiagen, Valencia, CA), subcloned into pCR2.1 (Invitrogen, Carlsbad, CA), mapped by restriction digestion, and sequenced.

cDNA Library Construction and Screening. Approximately 20 μg of poly(A)$^+$ RNA is purified from 1.5 mg of total RNA (from ~150 chick embryos) using the poly(A) tract mRNA purification system (Promega Co., Madison, WI). The mRNA is sent to Stratagene (La Jolla, CA) for the construction of an embryonic chick DRG cDNA library (in λZAP II phage vector). This cDNA library contains 1.7×10^6 primary inserts, with an average length of 1.55 kb, size selected at 600 bp.

For DNA screening, 1.1×10^5 phage units are added to 600 μl of XL1-Blue host bacteria at $OD_{600} = 0.5$ and are grown for 8–12 h on 150-mm NYZ plates at 37°. Nine plates are used to screen about one million phage particles per round. Plates are then chilled for 2 h at 4°, and plaques are transferred to Nytran membranes (Schleicher & Schuell, Inc., Keene, NH) in duplicate. Phage DNA is denatured for 2 min in 0.5 *M* NaOH, 1.5 *M* NaCl, neutralized for 5 min in 0.5 *M* Tris–HCl, 1.5 *M* NaCl, pH 8.0, and rinsed for 30 s in 0.2 m*M* Tris–HCl (pH 7.5). Membranes are air dried at room temperature for 1 h and are then UV cross-linked to immobilize the phage DNA. Cross-linked membranes are prehybridized for 2 h at 42° in 50% deionized formamide, 0.8 *M* NaCl, 20 m*M* PIPES (pH 6.5), 0.5% sodium dodecyl sulfate (SDS), and 100 μg/ml denatured salmon sperm DNA. During membrane prehybridization, a 240-bp cDNA probe, corresponding to the PCR-amplified fragment of the chick RGS3 RGS domain (see earlier discussion), is labeled with $[\alpha-^{32}P]$ dCTP using a random priming kit (Amersham Pharmacia Biotech, Piscataway, NJ) and added to fresh prehybridization solution. Membranes are transferred to the probe-containing solution and incubated overnight at 42°. After two 45-min washes at 58° in 0.1 × SSC/0.1% SDS, the membranes are dried for 1 h on Whatman 3MM paper, wrapped in plastic, and sandwiched with Kodak X-ray film between intensifying screens in an autoradiography cassette. The film is exposed overnight at −80°. Viral plaques positive in both original and duplicate membranes are identified following film development and alignment. Double-positive plaques are isolated, amplified, and subjected to a second round of screening. Individual plaques positive after the second screening are subjected to *in vitro* excision using the ExAssist/SOLR system (Stratagene). Each phage gave rise to a phagemid (pBluescript) containing an individual cDNA clone. The presence of a cDNA insert in the phagemid was confirmed by restriction digestion and sequence analysis.

Sequence Analysis. The PCR products and cDNA clones were translated into amino acid sequences using the Translate tool at the Expasy web site (http://us.expasy.org/tools/dna.html). The web-based program provided the user with six possible amino acid sequences, corresponding to the translation of all putative cDNA reading frames in both the 5'-3' and the 3'-5' direction. The correct reading frame could usually be identified by the absence of or a low number of stop codons. The identity of the cDNA was then determined by means of a similarity search against a nonredundant protein database using the BLAST algorithm at the National Center for Biotechnology Information (http://www.ncbi.nlm.nih.gov/blast/index.shtml). To predict the function of an unknown protein, it has become increasingly useful to compare the protein sequence with databases of known protein/functional domains. The translated protein sequences were, thus, scanned for functional motifs using InterProScan (European Bioinformatics Institute, http://www.ebi.ac.uk/interpro/scan.html). This web-based tool accesses many commonly used signature databases and allows a nonredundant characterization of a given protein domain or functional site.

Overexpression of RGS3 Splice Variants by *In Ovo*
  Retroviral Infection

RGS protein action is influenced by the type and availability of interacting Gα subunits, their intracellular localization, and the local environment in general (Chidiac and Roy, 2003; Ross and Wilkie, 2000). Studying RGS proteins in native environments is the strategy of choice to obtain a faithful picture of their physiological roles. In order to characterize chick RGS3 proteins in their tissue of origin (the DRG neuron), a method was optimized that permitted us to isolate the effects of each variant from the background by individually overexpressing each of the four RGS3 iso-forms. Overexpression was achieved by subcloning each RGS3 cDNA into a replication-competent retrovirus and injecting the virus into the neural tubes of early stage chick embryos *in ovo*. The infected DRG neurons were then tested for the ability of the overexpressed RGS3 variant to alter G-protein-dependent inhibition of N-type Ca^{2+} (N) channels produced through the activation of native $GABA_B$ receptors. The biochemical path-ways underlying the GABA-induced inhibition of N current in these neu-rons have been well characterized (Diversé-Pierluissi and Dunlap, 1993; Diversé-Pierluissi *et al.*, 1995, 1997). GABA, acting through G_o-coupled $GABA_B$ receptors, strongly inhibits these channels (up to 80%) by two distinct cascades, each giving rise to a separate form of inhibition: $G_o\beta\gamma$ binds directly to the channel to produce voltage-dependent inhibition (Diversé-Pierluissi *et al.*, 1995), whereas $G_o\alpha$ activates a protein tyrosine kinase pathway to produce voltage-independent inhibition (Diversé-Pierluissi *et al.*, 1997). Thus, DRG neurons represent an excellent native preparation in which to study the molecular mechanisms and complexities of G-protein modulators.

Experimental Results

Structural differences among the RGS3 variants and the abundance of protein–protein interaction domains in the longest isoform raised the in-teresting possibility that individual variants may exhibit unique character-istics and functions. To test this, we developed a method in which the RGS3 isoforms could be individually overexpressed to high levels in DRG neurons and tested for consequences in a standard electrophysiologi-cal assay to measure G-protein–Ca^{2+} channel signaling. As a first step to understanding the role and differential function of RGS3 proteins, we studied the activity of three isoforms: RGS3ss, characterized by a short sequence containing only the RGS domain; RGS3s, which contains two additional N-terminal domains, including a putative Ca^{2+}-binding EF-like motif; and RGS3l, the longest variant that possesses six putative protein–protein interaction domains in addition to the C-terminal RGS domain. Three different RGS3-containing viruses and a sham virus (to be used as control) were produced and overexpressed in chick DRG neurons.

Neural tube injections into stage 9 embryos are technically challenging and only rarely yielded 100% survival rates. The percentage of embryos surviving to E12 (the age of the neurons used in the electrophysiological assays) varied considerably, depending on the biological activity of the injected virus. Between 75 and 90% of embryos reached the E12 stage when injected with virus lacking RGS3 cDNA. Injection of certain RGS3 constructs (e.g., RGS3ss) led to a 50–60% survival rate, whereas others (e.g., RGS3s) showed much greater lethality (~5–10% survival). RGS3s-expressing embryos surviving to day 12 were somewhat smaller but otherwise indistinguishable from embryos exposed to virus alone. We found that of the three isoforms tested, RGS3s exhibited by far the most profound alterations in G-protein signaling (see next section) (Tosetti *et al.*, 2003, 2004).

Overexpression of the variant of interest was verified by immunocytochemical detection of the HA epitope engineered at the C-terminus of each construct. Because endogenous RGS proteins did not contain the tag, the assay detected exclusively the isoforms introduced by viral injection. The percentage of infected DRG cells did not vary significantly among viruses or egg batches. A slight caudal-to-rostral difference in expression could be detected, with the more rostral DRGs showing higher infection rates. DRGs from lumbar, thoracic, and cervical regions used in the electrophysiological assays exhibited consistent infection rates above 90% (Fig. 2).

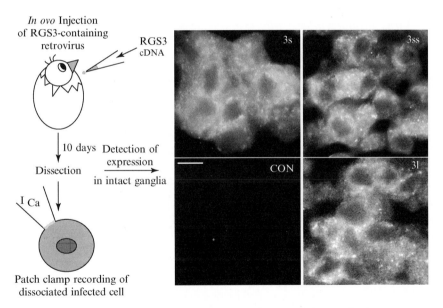

FIG. 2. *In ovo* injection of chick embryos with RGS3-containing, replication-competent retroviruses (*left, top*) led to >90% of infected DRG neurons (panels at right). There is no apparent difference in protein expression among viruses. (*Right*) Frozen sections of intact dorsal root ganglia from 12-day-old embryos, 10.5 days following infection with either empty retroviral vector (CON) or retroviral vectors carrying HA-tagged RGS3 variants (3ss, 3s, 3l). Calibration bar: 20 μm.

Protocols

Production of Replication-Competent Retroviruses. A 5′ *Xba*I restriction site (TCT AGA) followed by a translation/transcription accessory sequence (CCA CGT TGG CCA GGC GGT AGC TGG GAC GTG CAG CCG ACC ACC) and a 3′ HA epitope (TAC GAC GTG CCC GAC TAC GCC) followed by a stop codon (TAG) and a *Hin*dIII restriction site (AAG CTT) are engineered in-frame into RGS3 variant cDNAs by PCR, using primers carrying the appropriate sequences. The PCR products are then digested and subcloned in frame into the *Xba*I/*Hin*dIII sites of the adaptor plasmid Cla12 NCO. The sequences of interest are then excised from the adaptor plasmid by *Cla*I digestion and are subcloned nondirectionally into the *Cla*I site of RCASBP retroviral variants of the B subgroup (both plasmids were a generous gift of Dr. Donna Fekete, Purdue University, West Lafayette, IN). All ligations are carried out using the rapid DNA ligation kit from Boehringer Mannheim (Indianapolis, IN), according to the manufacturer's instructions. The restriction enzymes are from New England Biolabs (Beverly, MA). Properly oriented constructs are identified by PCR and sequencing.

Virus Production and Infection. Avian-specific, replication-competent retroviruses are generated by standard $Ca_3(PO_4)_2$ transfection (Morgan and Fekete, 1996) of chick fibroblasts with the RGS3-containing RCASB genome. Viral particles are released in the culture medium 24–48 h posttransfection. The medium from five 150-mm dishes is collected and concentrated to the optimal titer for injection (10^9 infectious units/μl) by centrifuging it for 3 h in a Beckman SW 28 rotor at 4° and aspirating the supernatant to a final volume of 200 μl (Morgan and Fekete, 1996). The viral pellet is then resuspended by gentle pipetting, divided into 20-ml aliquots, and stored at −80° or lower. The viral titering procedure is carried out as described by Morgan and Fekete (1996). When stored in liquid N_2, such viruses remain viable for at least 1 year. To infect proliferating DRG neuron precursors in the neural crest, the engineered retrovirus is microinjected into the neural tubes of stage 9 white leghorn chick embryos (Spafas, MA) *in ovo*. After fertilization, eggs are maintained in a nonturning, forced air-draft, humidified egg incubator at 37° for 36–38 h. Using fine scissors, a circular opening is cut into the blunt end of the eggshell, being careful not to disturb the embryo or damage the yolk. The yolk protective membrane is removed with fine forceps, and the embryo is exposed. The egg is then visualized under a dissecting microscope (at 10× magnification), and 1–3 μl of viral stock is injected into the neural tube via a glass pipette driven by mechanical micromanipulators. The viral solution is pressure injected via a 50-ml syringe connected to the glass pipette by a 1.2-mm inner diameter polypropylene tube. The entire apparatus is air tight and filled with mineral oil to allow hydraulic transfer of pressure to the pipette. The viral solution is then front loaded into the glass pipette by suction, being careful not to generate air bubbles. The water-based viral solution does not mix with the oil. The addition of bromphenol blue to the injection solution allows visual monitoring. After injection, the eggshells are closed with cellophane tape, and embryos are returned to the incubator to resume growth at 37° until their use at day 11–12.

Immunocytochemistry. Intact DRGs, dissected from wild-type or infected 11- to 12-day-old chick embryos, are fixed in 2% freshly prepared HCHO for 30 min, cryoprotected in 30% sucrose for 30 min, embedded into OTC4 (Electron Microscopy Sciences, Fort Washington, PA), and flash frozen in liquid N_2. Frozen sections (7 μm) are cut using a cryostat, transferred to glass slides, and air dried. For immunocytochemical detection of infected cells, sections are first incubated 30 min with 0.25% gelatin [in phosphate-buffered saline (PBS)] to block nonspecific binding, followed by 1 h with rabbit anti-HA primary antiserum (1:500 dilution, Jackson Labs, Bar Harbor, ME), a 10-min wash in PBS, 1 h with a biotinylated goat antirabbit secondary antibody (1:150 dilution, Jackson Labs), another 10-min PBS wash, and 40 min with FITC-conjugated streptavidin (1:1000, Jackson Labs). Sections are rinsed in PBS, fixed with cold methanol, mounted (Vectashield, Vector Labs), and analyzed on a Zeiss Axioskop.

Calcium Dependence of RGS3 Activity

When overexpressed in chick DRG neurons, chick RGS proteins accelerate the termination of G-protein signaling (Diversé-Pierluissi *et al.*, 1999; Tosetti *et al.*, 2002, 2003). In particular, the retrovirus-mediated overexpression of individual RGS isoforms showed that chick RGS3 proteins differentially desensitize G_o-mediated inhibition of N channels (Tosetti *et al.*, 2003). Because motif analysis identified an EF-like domain in RGS3s (which in other proteins is known to confer Ca^{2+} sensitivity), we tested for Ca^{2+} dependence of RGS3 function in DRG neurons. Ca^{2+} currents were recorded electrophysiologically, and their G-protein-dependent inhibition was evoked by activating native G_o-coupled GABA$_B$ receptors in control cells or in cells overexpressing one of the RGS3 isoforms. We found that RGS3s dramatically accelerated the termination of G_o-dependent inhibition of Ca^{2+} current, whereas RGS3ss-expressing neurons showed faster termination than control cells but were significantly slower than RGS3s-expressing neurons. Interestingly, the acceleration produced by both variants was dependent on Ca^{2+} influx through the same channels that are subject to inhibition by G_o, although the mechanisms underlying the

Ca^{2+}-dependent effects differed for the two isoforms. The rapid RGS3s-dependent termination of G_o activity is due to direct interaction of Ca^{2+} with the EF-like domain (as discussed here), whereas the slower RGS3ss-dependent effect is likely to be mediated by Ca^{2+}–calmodulin (CaM) (Tosetti *et al.*, 2003). Curiously, RGS3l, which contains both of the putative Ca^{2+}-sensitive domains present on the other RGS3 isoforms, nevertheless exhibits no Ca^{2+}-dependent effects on G-protein channel signaling (Tosetti *et al.*, 2004), suggesting the interesting possibility that the protein–protein interaction domains that are unique to RGS3l target this protein to regions of the neurons lacking G_o–Ca^{2+} channel complexes.

Experimental Results

This section describes in detail the four distinct approaches used to study the Ca^{2+} dependence of RGS3 activation: (1) Substitution of Ca^{2+} ions with Ba^{2+} ions during electrophysiological recordings, (2) construction of a mutant lacking the putative Ca^{2+}-binding EF-like domain, (3) measuring direct Ca^{2+}–RGS3 interactions via a gel shift assay, and (4) pharmacological block of CaM-mediated processes.

Substitution of Ca^{2+} Ions with Ba^{2+} Ions during Electrophysiological Recordings. Ba^{2+}, like Ca^{2+}, is permeant through voltage-gated Ca^{2+} channels, yet its ability to activate Ca^{2+}-dependent proteins is typically much weaker than that of Ca^{2+} (Falke *et al.*, 1994). For this reason, we compared the time courses of G_o-dependent modulation under conditions in which Ca^{2+} or Ba^{2+} was the primary charge carrier. The rapid, high-affinity Ca^{2+} chelator, BAPTA, was included at 5 mM in the recording pipette solution to prevent bulk changes in intracellular Ca^{2+} concentration, thereby confining Ca^{2+}-dependent activation to intracellular microdomains near sites of Ca^{2+} entry (Naraghi and Neher, 1997).

With Ca^{2+} as the charge carrier, saturating concentrations of GABA (100 μM) inhibit N channels by an average $37 \pm 2\%$ in control cells. Prolonged applications of neurotransmitter are accompanied by a slow desensitization process, with the inhibition waning at a rate of $\sim 10\%$/min. Overexpression of RGS3s produced a reduction in maximal GABA-induced inhibition and a dramatic >10-fold acceleration in the rate of desensitization (Fig. 3A) (Tosetti *et al.*, 2003). GABA-induced responses of cells expressing RGS3ss also desensitize more rapidly than control cells, but the rates were much slower than those observed in RGS3s-expressing cells (Fig. 3A).

When Ba^{2+} was substituted for Ca^{2+} in the extracellular recording solution, the maximal GABA-induced inhibition of Ba^{2+} current through N channels was no different from that observed for Ca^{2+} currents. However,

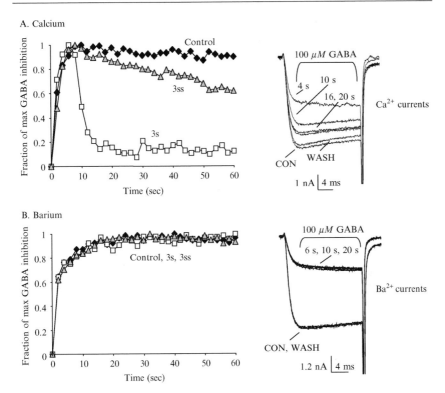

FIG. 3. RGS3s-induced termination of GABA_B receptor response required the presence of Ca²⁺. (*Left*) Time courses of GABA-induced inhibition of Ca²⁺ (A) and Ba²⁺ (B) currents in individual control (◆), RGS3ss- (△), and RGS3s-expressing (□) neurons. Representative currents shown at right before (CON), during, and after (WASH) exposure to 100 μM GABA for the times indicated. Both RGS3-dependent, fast and slow termination of GABAergic inhibition are eliminated when Ca²⁺ is replaced by Ba²⁺.

the attenuation in maximal inhibition and the increased rate of desensitization observed in neurons expressing RGS3s and RGS3ss were no longer observed (Fig. 3B). These results indicate that Ca²⁺ influx through N channels is required for RGS3 activity.

Mutagenesis Approach Used to Identify the Domain on RGS3s Responsible for Fast, Ca²⁺-Dependent Desensitization. The presence of the EF-like domain in RGS3s but not RGS3ss prompted us to test whether the fast desensitization rates in RGS3s-expressing cells result from Ca²⁺–EF-like motif interactions. This was done first through mutagenesis of RGS3s. As described in the following section, results were also confirmed using a biochemical assay–Ca²⁺ mobility gel shift. Site-directed cDNA mutagenesis is a strategy used commonly to identify the residues

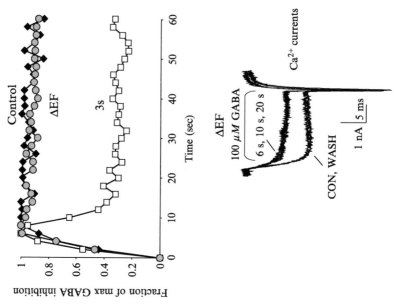

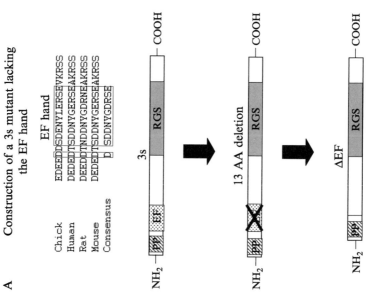

A   Construction of a 3s mutant lacking
                    the EF hand

essential for protein function. We applied this method to identify the region of RGS3s underlying fast Ca^{2+}-dependent desensitization by deleting 39 bp of the in-frame coding sequence, effectively eliminating the predicted 12 residue EF-like domain in RGS3s. This mutation (ΔEF) eliminated the ability of RGS3s to attenuate GABA-mediated inhibition and accelerate its desensitization rate (Fig. 4). Eliminating Ca^{2+}-dependent activity did not interfere, however, with other functional properties of the recombinant protein, indicating that the deletion did not result in global alterations in protein structure. In fact, the ΔEF mutant was still able to induce the rightward shift in the voltage dependence of N channel activation that characterizes wild-type RGS3s (Tosetti and Dunlap, personal observations; Tosetti et al., 2003). These results indicate that the EF-like domain plays a fundamental role in RGS3s activation by Ca^{2+} ions.

Measuring Direct Ca^{2+} Binding via a Gel Shift Assay. The ability of Ca^{2+} to bind directly to an EF-like domain can be assessed with a simple electrophoretic assay. Although the precise mechanism has not been established, Ca^{2+} is known to accelerate the electrophoretic mobility of EF hand-containing proteins, migrating through polyacrylamide gels, presumably by altering the charge and/or the conformation of the protein (Grab et al., 1979; Kameshita and Fujisawa, 1997). EF hands are stable structures capable of maintaining the ability to bind Ca^{2+} even under mildly denaturing conditions (Grab et al., 1979). Thus, comparing the electrophoretic mobility of a protein in the presence and absence of Ca^{2+} or Ba^{2+} allows an assessment of the ability of the protein to bind divalent cations directly.

Although the gel shift assay *per se* is technically straightforward, preparation of the recombinant protein samples may require some special attention. This is particularly true for insoluble proteins (such as chick RGS3). If this is the case, the protein purification procedure will require strong denaturing conditions and partial renaturation prior to electrophoresis (see protocols given later) to allow the protein to retain at least some of its native conformation. Renaturation decreased protein yield (approximately 60% of RGS3ss, 40% of ΔEF, and 40% of RGS3s were recovered from the column following refolding) but was essential to the success of the gel shift assay.

FIG. 4. The N-terminal EF-like domain is required for RGS3s-mediated effects. (A, *top*). Alignment of the putative EF-like domain in RGS3 from a number of species; (*bottom*) schematic illustration of the EF-like domain deletion (ΔEF). (B) Timecourse (top) of GABA-induced inhibition of Ca^{2+} currents in a control cell (♦) and in individual cells expressing either wild-type RGS3s (□) or ΔEF (○). (Bottom) Representative, superimposed Ca^{2+} currents in a ΔEF-expressing cell exposed to GABA for the times indicated. Fast desensitization is eliminated in cells expressing the EF-like domain deletion mutant of RGS3s.

Ca^{2+} added to the sample buffer significantly accelerated the mobility of recombinant RGS3s, but not RGS3ss or ΔEF (Fig. 5), indicating that Ca^{2+} interacts directly with RGS3s. Boiling RGS3s eliminated all effects of Ca^{2+} (not shown). Importantly, no shift in mobility was observed in the presence of Ba^{2+} (not shown), consistent with the inability of Ba^{2+} to accelerate desensitization of the GABA response in RGS3s-expressing neurons.

Pharmacological Block of Calmodulin-Mediated Processes. Desensitization rates in RGS3ss-expressing cells were still faster than in control cells (although significantly slower than in RGS3s-expressing neurons), despite the absence of the EF-like domain in RGS3ss (Fig. 3A). Interestingly, the RGS3ss-mediated effect was also Ca^{2+} dependent (Fig. 3B). These results suggest a second mechanism for calcium-dependent RGS activation independent of the EF-like domain. The Ca^{2+}/CaM complex has been shown to modulate GAP activity through binding to a domain in the RGS box (Popov *et al.*, 2000). As RGS3ss contains this putative CaM-binding site, we tested for an involvement of CaM in the calcium-dependent activation of RGS3ss by employing the CaM inhibitor, calmidazolium. This CaM inhibitor is lipophilic and can permeate cell membranes easily

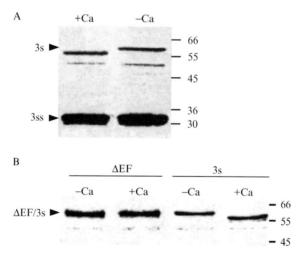

Fig. 5. Ca^{2+} interacts directly with the EF-like domain in RGS3s. Coomassie Blue-stained gels of recombinant RGS3s and RGS3ss proteins (A) or RGS3s and ΔEF proteins (B) separated electrophoretically in sample buffer containing either 2 m*M* Ca^{2+} (+Ca) or 2 m*M* EGTA (with no added calcium, −Ca), as marked. Size markers (in kDa) are indicated at right; arrows at left indicate expected molecular weights of the three proteins. The electrophoretic mobilities of the two isoforms lacking the EF-like domain, 3ss and ΔEF, are not affected by the presence of Ca^{2+}.

(Burgoyne and Norman, 1984; Peracchia, 1984; Wolberg and Zimmerman, 1984). It is, therefore, sufficient to add calmidazolium to the extracellular solution for 15–45 min to allow intracellular access of the inhibitor and block CaM-mediated processes. In cells expressing RGS3ss, calmidazolium completely blocked the slow component of calcium-dependent desensitization (Tosetti *et al.*, 2003). Conversely, the CaM inhibitor had no effect on the rapid, calcium-dependent desensitization in cells expressing wild-type RGS3s (Tosetti *et al.*, 2003). These data are consistent with slow termination of signaling being mediated by calcium binding to CaM, whereas fast desensitization is produced by the direct binding of calcium to the EF-like domain in RGS3s. These results demonstrate the power of combining overexpression of recombinant proteins in a native cell environment with sensitive single cell electrophysiological assays to probe the mechanisms of RGS function and activation.

Protocols

Preparation of and Recording from Chick DRG Neurons. Dorsal root ganglia are dissected from 11- or 12-day-old chick embryos (Spafas, MA) and incubated for 10 min at 37° in Ca^{2+}-, Mg^{2+}-free saline containing 0.1% collagenase A (Boehringer Mannheim, Indianapolis, IN). For a detailed description of DRG dissection from chick embryo, see Anantharam and Diversé-Pierluissi (2002). After removing the collagenase-containing solution, ganglia are washed twice and resuspended in ~1 ml of culture medium (DMEM, supplemented with 10% heat-inactivated horse serum, 5% chicken embryo extract, 50 U/ml penicillin, 50 mg/ml streptomycin, 1 mM glutamine and 0.1% nerve growth factor), and are mechanically dissociated by trituration through a small-bore, fire-polished Pasteur pipette. A small drop of the cell suspension (~100 μl) is placed in the center of a 35-mm tissue culture dish and incubated at 37° to allow the cells to attach to the substrate. After ~1 h of incubation, the plated dishes are filled with culture medium.

Four to 48 h after plating, cultured DRG neurons are employed for electrophysiological recordings of Ca^{2+} currents using standard tight-seal, whole cell methods (Hamill *et al.*, 1981). Cells are visualized on the stage of an inverted microscope, and the cell culture medium is replaced with (in mM) 133 NaCl, 1 CaCl₂, 0.8 MgCl₂, 25 HEPES, 12.5 NaOH, 5 glucose, 10 tetraethylammonium Cl, 3×10^{-4} tetrodotoxin, and 10^{-2} bicuculline, pH 7.4. When studying the divalent cation dependence of desensitization, 1 mM CaCl₂ is replaced by 750 μM BaCl₂ in the extracellular solution. Lowering the BaCl₂ concentration allows us to maintain approximately constant current amplitudes, as N-type Ca^{2+} channels are more permeable to

Ba^{2+} than to Ca^{2+}. All reagents were from Sigma Chemical Company (St. Louis, MO).

Recording pipettes are fabricated from microhematocrit tubing (Fisher) and filled with an internal recording solution containing (in mM) 150 CsCl, 5 BAPTA, 5 MgATP, and 10 HEPES, pH 7.4. Pipette resistances vary between 0.8 and 1.5 MΩ when filled with recording solution. Currents are recorded using a List Biologic (Campbell, CA) EPC-7 patch-clamp amplifier, filtered at 3 kHz, digitized at 10 kHz using an ITC16 A/D interface (Instrutech Corp., Great Neck, NY), and stored on a Macintosh G3 PowerPC running Pulse software (HEKA Electronik, Germany). Capacitive transients and series resistances are compensated with the EPC-7 circuitry, and the leak (passive) currents are subtracted with a standard P/4 protocol (Hamill *et al.*, 1981). All recordings are performed at room temperature.

Ca^{2+} and Ba^{2+} Current Analysis. Igor (Wavemetrics, Lake Oswego, OR) and Excel (Microsoft) software are employed for data analysis. Ca^{2+} and Ba^{2+} currents are evoked by a 20-ms test pulse to 0 mV in the presence or absence of GABA. All current amplitudes are measured at the time that control currents peaked (T_p), and modulated currents are normalized to control and expressed as a percentage. The difference between control and modulated current amplitude at T_p constitutes the total GABA-induced inhibition. Ca^{2+} currents are corrected for rundown by measuring Ca^{2+} currents as a function of time in control cells in the absence of neurotransmitter. Analysis is confined to data from cells that exhibit a rundown of less than 1%/min. Ba^{2+} currents showed no rundown over a recording period of ~10 min.

Solutions and Drug Delivery. Drug solutions are prepared from stock solutions immediately before the experiment; they are delivered rapidly (<1 s) via a pressurized system of multiple converging quartz tubes (140 μm internal diameter) positioned in the vicinity of the cell of interest. For all experiments, GABA is diluted into the extracellular solution at a working concentration of 100 μM. For experiments requiring the block of CaM-mediated activities, 10 μM calmidazolium is added to the DRG culture medium 15–45 min prior to electrophysiological recordings to allow intracellular access of the inhibitor (Burgoyne and Norman, 1984; Peracchia, 1984; Wolberg and Zimmerman, 1984). The culture medium is then replaced with regular external recording solution, and electrophysiological experiments are carried out as usual.

Statistics. All results are expressed as means ± standard errors of the mean. Statistical differences are analyzed using either paired or unpaired Student's *t* tests, as appropriate, and differences are considered significant when $p < 0.05$.

Generation of a Mutant Virus Lacking the RGS3s Ca²⁺-Binding Domain. A mutant RGS3s cDNA lacking the EF-like domain ("ΔEF") is generated by mutating the RGS3s/RCASB construct using the QuikChange site-directed mutagenesis kit (Stratagene) according to the manufacturer's instructions. The following primer set is used: 5' GAT GAG GAT GAA GAG AAG CGC AGC AGC ATG 3' (sense) and 5' CAT GCT GCT GCG CTT CTC TTC ATC CTC ATC 3' (antisense). The presence of the deletion of interest is confirmed by sequencing. The mutated construct is then used to transfect chick fibroblasts to generate the corresponding retrovirus (see earlier discussion).

Recombinant Proteins. Escherichia coli, strain BL21(DE3), are transformed with C-terminal His₆-tagged RGS3s, RGS3ss, and ΔEF subcloned into pET11a (Novagen). Cells are grown in LB medium with ampicillin (100 μg/ml) at 37° to an OD_{600} = 0.6–0.8 and are then induced with isopropyl-1-thio-β-D-galactopyranoside (0.5 mM) for 4 h. Cells are concentrated by centrifugation and lysed by freezing and sonication in urea buffer (8 M urea, 20 mM Tris–HCl, pH 8, 100 mM NaCl). Because all three proteins are insoluble, the purification procedure is carried out under strong denaturing conditions, i.e., in the presence of 8 M urea, to allow the dissolution of insoluble bodies and maximal recovery of the proteins. The lysate is centrifuged (12,000g for 30 min), and the supernatant fraction (containing the denatured proteins) is loaded onto a minicolumn containing 50 μl of Ni-NTA affinity resin (Qiagen, Hilden, Germany). Denatured proteins are adsorbed onto the column and refolded using buffers of decreasing urea concentration (6, 4, 2, and 0.5 M). To minimize sample precipitation and loss, washes are applied very gently by letting the liquid slide along the column walls and accumulate on top of the resin. All eluates from the washing steps are collected and examined with SDS–PAGE to monitor the progressive elution of unbound proteins. A final wash with buffer A (50 mM HEPES, pH 8, 20 mM β-mercaptoethanol, 100 mM NaCl, 0.1 mM phenylmethylsulfhonyl fluoride) removes the residual urea. The refolded proteins are eluted from the column with 250 μl of buffer A and either 250 or 150 mM imidazole (for RGS3ss or RGS3s, respectively), concentrated to a volume of 100 μl, checked for purity using SDS–PAGE, and then stored at −80°. The amount of protein precipitated in the column following refolding is then estimated by first treating the resin with 1 ml of buffer A containing 1 M imidazole to elute all residual soluble proteins. A second wash with 1 ml of 8 M urea buffer dissolves all precipitated proteins, which are eluted with 250 μl of urea buffer containing 1 M imidazole. The eluate is collected, concentrated to a volume of 100 μl, and quantified using SDS–PAGE. Under these experimental conditions, approximately 60% of refolded RGS3ss, 40% of refolded ΔEF, and 40% of refolded RGS3s could be recovered.

Calcium-Induced Mobility Shift Assay. A mobility shift assay using classical denaturing polyacrylamide gel electrophoresis (SDS–PAGE) is used to test for direct interactions between Ca^{2+} and RGS3s, RGS3ss, and ΔEF (Grab *et al.*, 1979; Rosel *et al.*, 2000). Ten microliters of each of the three recombinant proteins is diluted 1:1 into Laemmli sample buffer modified to contain a final concentration of 2 mM Ca^{2+} or 2 mM EGTA (with no added Ca^{2+}). Samples are not heat denatured before loading on gels, with the exception of the fully inactivated negative controls that are incubated for 20 min at 100° in a water bath. Unheated samples are kept on ice and loaded as rapidly as possible to avoid protease degradation. Proteins are separated electrophoretically using standard 12.5% Tris–glycine SDS–PAGE. After electrophoresis, gels are stained for 30–60 min with 0.006% Coomassie Blue G250, 10% acetic acid, destained for 2–3 h with 10% acetic acid, 15% MeOH, and photographed.

Acknowledgments

This work was supported by National Institutes of Health Grant NS16483 (KD) and Human Frontier Science Program Long- and Short-Term Fellowships (PT).

References

Anantharam, A., and Diversé-Pierluissi, M. A. (2002). Biochemical approaches to study interaction of calcium channels with RGS12 in primary neuronal cultures. *Methods Enzymol.* **345**, 60–70.

Berman, D. M., Wilkie, T. M., and Gilman, A. G. (1996). GAIP and RGS4 are GTPase-activating proteins for the G_i subfamily of G protein alpha subunits. *Cell* **86**, 445–452.

Burgoyne, R. D., and Norman, K. M. (1984). Effect of calmidazolium and phorbol ester on catecholamine secretion from adrenal chromaffin cells. *Biochim. Biophys. Acta* **805**, 37–43.

Chidiac, P., and Roy, A. (2003). Activity, regulation, and intracellular localization of RGS proteins. *Receptors Channels* **9**, 135–147.

Diversé-Pierluissi, M., and Dunlap, K. (1993). Distinct, convergent second messenger pathways modulate neuronal calcium currents. *Neuron* **10**, 753–760.

Diversé-Pierluissi, M., Goldsmith, P. K., and Dunlap, K. (1995). Transmitter-mediated inhibition of N-type calcium channels in sensory neurons involves multiple GTP-binding proteins and subunits. *Neuron* **14**, 191–200.

Diversé-Pierluissi, M., Remmers, A. E., Neubig, R. R., and Dunlap, K. (1997). Novel form of crosstalk between G protein and tyrosine kinase pathways. *Proc. Natl. Acad. Sci. USA* **94**, 5417–5421.

Diversé-Pierluissi, M. A., Fischer, T., Jordan, J. D., Schiff, M., Ortiz, D. F., Farquhar, M. G., and De Vries, L. (1999). Regulators of G protein signaling proteins as determinants of the rate of desensitization of presynaptic calcium channels. *J. Biol. Chem.* **274**, 14490–14494.

Druey, K. M., Blumer, K. J., Kang, V. H., and Kehrl, J. H. (1996). Inhibition of G-protein-mediated MAP kinase activation by a new mammalian gene family. *Nature* **379**, 742–746.

Falke, J., Drake, S., Hazard, A., and Peersen, O. (1994). Molecular tuning of ion binding to calcium signaling proteins. *Q. Rev. Biophys.* **27**, 219–290.

Grab, D. J., Berzins, K., Cohen, R. S., and Siekevitz, P. (1979). Presence of calmodulin in postsynaptic densities isolated from canine cerebral cortex. *J. Biol. Chem.* **254,** 8690–8696.

Hamill, O. P., Marty, A., Neher, E., Sakmann, B., and Sigworth, F. J. (1981). Improved patch-clamp techniques for high-resolution current recording from cells and cell-free membrane patches. *Pflüg. Arch.* **391,** 85–100.

Kameshita, I., and Fujisawa, H. (1997). Detection of calcium binding proteins by two-dimensional sodium dodecyl sulfate-polyacrylamide gel electrophoresis. *Anal. Biochem.* **249,** 252–255.

Kehrl, J., Srikumar, D., Harrison, K., Wilson, G., and Shi, C. (2002). Additional 5′ exons in the RGS3 locus generate multiple mRNA transcripts, one of which accounts for the origin of human PDZ-RGS3. *Genomics* **79,** 860–868.

Koelle, M. R., and Horvitz, H. R. (1996). EGL-10 regulates G protein signaling in the C. Elegans nervous system and shares a conserved domain with many mammalian proteins. *Cell* **84,** 115–125.

Melliti, K., Meza, U., and Adams, B. (2000). Muscarinic stimulation of alpha1E Ca channels is selectively blocked by the effector antagonist function of RGS2 and phospholipase C-beta1. *J. Neurosci.* **20,** 7167–7173.

Morgan, B. A., and Fekete, D. M. (1996). Manipulating gene expression with replication-competent retroviruses. *Methods Cell Biol.* **51,** 185–218.

Naraghi, M., and Neher, E. (1997). Linearized buffered Ca²⁺ diffusion in microdomains and its implications for calculation of [Ca²⁺] at the mouth of a calcium channel. *J. Neurosci.* **17,** 6961–6973.

Peracchia, C. (1984). Communicating junctions and calmodulin: Inhibition of electrical uncoupling in Xenopus embryo by calmidazoliam. *J. Membr. Biol.* **81,** 49–58.

Popov, S. G., Krishna, U. M., Falck, J. R., and Wilkie, T. M. (2000). Ca²⁺/calmodulin reverses phosphatidylinositol 3,4, 5-trisphosphate-dependent inhibition of regulators of G protein-signaling GTPase-activating protein activity. *J. Biol. Chem.* **275,** 18962–18968.

Rosel, D., Puta, F., Blahuskova, A., Smykal, P., and Folk, P. (2000). Molecular characterization of a calmodulin-like dictyostelium protein CalB. *FEBS Lett.* **473,** 323–327.

Ross, E. M., and Wilkie, T. M. (2000). GTPase-activating proteins for heterotrimeric G proteins: Regulators of G protein signaling (RGS) and RGS-like proteins. *Annu. Rev. Biochem.* **69,** 795–827.

Scheschonka, A., Dessauer, C. W., Sinnarajah, S., Chidiac, P., Shi, C. S., and Kehrl, J. H. (2000). RGS3 is a GTPase-activating protein for $G_i\alpha$ and $G_q\alpha$ and a potent inhibitor of signaling by GTPase-deficient forms of $G_q\alpha$ and $G_{11}\alpha$. *Mol. Pharmacol.* **58,** 719–728.

Shin, D. M., Dehoff, M., Luo, X., Kang, S. H., Tu, J., Nayak, S. K., Ross, E. M., Worley, P. F., and Muallem, S. (2003). Homer 2 tunes G protein-coupled receptors stimulus intensity by regulating RGS proteins and PLCβ GAP activities. *J. Cell Biol.* **162,** 293–303.

Tosetti, P., Pathak, N., Covic, L., Koliopulos, A., Jacob, M. H., and Dunlap, K. (2004). RGS3 splice variants differentially modulate N-type calcium channels in chick sensory neurons. In preparation.

Tosetti, P., Pathak, N., Jacob, M. H., and Dunlap, K. (2003). RGS3 mediates a calcium-dependent termination of G protein signaling in sensory neurons. *Proc. Natl. Acad. Sci. USA* **100,** 7337–7342.

Tosetti, P., Turner, T., Lu, Q., and Dunlap, K. (2002). Unique isoform of Galpha-interacting protein (RGS-GAIP) selectively discriminates between two Go-mediated pathways that inhibit Ca²⁺ channels. *J. Biol. Chem.* **277,** 46001–46009.

Wolberg, G., and Zimmerman, T. P. (1984). Effects of calmodulin antagonists on immune mouse lymphocytes. *Mol. Pharmacol.* **26,** 286–292.

[8] Analysis of PDZ-RGS3 Function in Ephrin-B Reverse Signaling

By QIANG LU, EDNA E. SUN, and JOHN G. FLANAGAN

Abstract

Transmembrane B ephrins and their Eph receptors signal bidirectionally. However, the molecular and cellular mechanisms of the "reverse signaling" by B ephrins are not fully understood. The identification of several B ephrin cytoplasmic domain-interacting proteins has begun to shed light on the signal transduction mechanisms and the cellular effects of B ephrin reverse signaling. This article describes the use of a Transwell migration assay system to characterize the function of a PDZ domain and a regulator of G-protein signaling (RGS) domain-containing protein, PDZ-RGS3, in mediating B ephrin reverse signaling in cultured neurons. The Transwell system discussed here can provide an effective cell-based assay for elucidating RGS protein-mediated signaling interactions in neurons.

Introduction

The Eph receptor tyrosine kinases and their ephrin ligands are critically important in diverse neural functions, including axon guidance, neuron migration, tissue compartmentation, and modulation of synaptic organization (Cowan and Henkemeyer, 2002; Flanagan and Vanderhaeghen, 1998; Holmberg and Frisen, 2002; Knoll and Drescher, 2002; Kullander and Klein, 2002; Murai and Pasquale, 2002; Thompson, 2003; Wilkinson, 2001). All known ephrins are membrane associated and are subgrouped into two classes: glycosyl-phosphatidyl-inositol (GPI)-anchored ephrin-A and transmembrane ephrin-B. The structure of ephrins allows Eph/ephrin family molecules to mediate bidirectional signaling, with a forward signal through the receptor and a reverse signal through the ligand. A current focus of study is to understand the mechanistic basis of Eph receptors and the ephrins in relation to their specific biological actions. The identification of several ephrin-B cytoplasmic domain-associating proteins (Bruckner *et al.*, 1999; Cowan and Henkemeyer, 2001; Lin *et al.*, 1999; Lu *et al.*, 2001; Torres *et al.*, 1998) has started to provide insights on the signaling mechanisms of the reverse signaling by ephrins. This chapter discusses the structure and function of a regulator of G-protein signaling (RGS)

METHODS IN ENZYMOLOGY, VOL. 390

protein, PDZ-RGS3, in ephrin-B reverse signaling. Acting as a GTPase-accelerating protein (GAP), PDZ-RGS3 links ephrin-B reverse signaling to selective inhibition of G-protein-coupled chemoattraction (Lu *et al.*, 2001).

Identification of PDZ-RGS3

B ephrins have the same domain structures: an extracellular domain, a single transmembrane segment, and a cytoplasmic domain. The intracellular portions of B ephrins are approximately only 90 amino acids in length, but are highly conserved among all three known B ephrins. These strong homologies in sequence are indicative of domain motifs for protein–protein interactions. Indeed, the carboxyl-terminal sequence YYKV of B ephrins fits well into the description of a PDZ (PSD95/Dlg/ZO-1)-binding motif (Songyang *et al.*, 1997). We reasoned that the identification of target-interacting proteins of these domain motifs would be an important first step in dissecting ephrin-B reverse signaling. We therefore screened a yeast two-hybrid mouse cDNA library using the entire cytoplasmic domain of ephrin-B2 as the bait. This approach led to the identification of several cDNAs for candidate ephrin-B cytoplasmic domain-interacting proteins. Subsequent studies focused on one of the cDNAs, encoding a previously unidentified 930 amino acid protein sequence. This protein sequence contains two motifs, a PDZ domain at the N terminus, which mediates the interaction with ephrin-B, and an RGS domain at the C terminus.

Our initial screen of the two-hybrid library identified a fragment of the cDNA containing the translation initiation site and the PDZ domain. The remaining cDNA sequence, including that coding for the RGS domain, was assembled by 3′ RACE. Database searching revealed strong homology of the C-terminal half of the mouse sequence to full-length human RGS3 (Druey *et al.*, 1996), a shorter sequence (519 amino acids) identified previously (Fig. 1A). Hence we named the protein encoded by the mouse cDNA PDZ-RGS3. The N-terminal half of PDZ-RGS3 was later found highly homologous to C2PA (Linares *et al.*, 2000), reported to be a nuclear protein identified during mouse spermatogenesis (Fig. 1A).

Mouse genome sequences reveal that PDZ-RGS3, RGS3, C2PA, or C2PA-RGS3 may be differentially spliced variants of the same gene that is located on chromosome 4. The exon–intron organization of the gene is illustrated in Fig. 1B. The PDZ-RGS3 sequences contain 17 exons (P1–P17). The N-terminal PDZ domain-containing sequence and the RGS3 sequence are encoded by exons 1 to 10 and exons 11 to 17, respectively. The C2PA sequences have 20 exons (C1–C20, Fig. 1B), some of which are shared with the PDZ-RGS3 sequences.

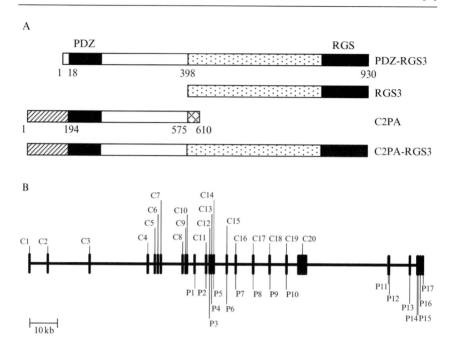

FIG. 1. Domain structure and gene structure of PDZ-RGS3 and related molecules. PDZ-RGS3, RGS3, C2PA, and C2PA-RGS3 appear to be differentially spliced variants of the same gene. Note that the N- and C-terminal sequence of C2PA diverges from PDZ-RGS3. (A) Numbers indicate the amino acid positions where PDZ-RGS3 and C2PA diverge. (B) Exons of PDZ-RGS3 and C2PA are labeled as P1–P17 and C1–C20, respectively. PDZ-RGS3 and C2PA share several exons from P2/C11 to P10/C19. The PDZ domains of PDZ-RGS3 and C2PA are encoded by exons P2/C11, P3/C12, and P4/C13.

Functional Studies of PDZ-RGS3 in Ephrin-B Reverse Signaling

Our biochemical characterization of PDZ-RGS3 demonstrated that PDZ-RGS3 and B ephrins bind to each other and have overlapping expression in mouse tissues (Lu *et al.*, 2001), suggesting that PDZ-RGS3 is a biologically relevant interacting protein for ephrin-B. The activity of RGS domains to downregulate G-protein signaling (Berman *et al.*, 1996; Hunt *et al.*, 1996; Watson *et al.*, 1996) led us to a molecular model for ephrin-B reverse signaling (Fig. 2) in which ephrin-B can negatively regulate G-protein signaling through PDZ-RGS3.

To study the function of PDZ-RGS3 in ephrin-B reverse signaling, we established a Transwell migration assay using mouse cerebellar granule neurons. This assay was based on the endogenous expression of ephrin-B2, PDZ-RGS3, and chemokine receptor CXCR4 in granule cells and the

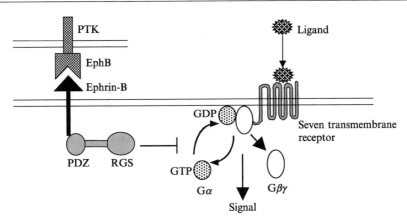

FIG. 2. A molecular mechanism of ephrin-B reverse signaling. Binding of B ephrins and their EphB receptors results in bidirectional signaling, with a forward signal perhaps through the protein tyrosine kinase (PTK) domain of EphB and a reverse signal through the cytoplasmic domain of ephrin-B. Heterotrimeric G-protein signaling is activated by ligands that act through seven transmembrane receptors, such as the chemoattractant SDF-1 and its receptor CXCR4. PDZ-RGS3 binds the C-terminal sequence of B ephrins through its PDZ domain and inhibits heterotrimeric G-protein signaling through the GAP activity of its RGS domain. These interactions provide a link between ephrin-B reverse signaling and G-protein-coupled signaling. Adapted from Lu *et al.* (2001).

ability of granule cells to migrate toward a chemoattractant SDF-1, a ligand of the G-protein-coupled receptor CXCR4. This section discusses the assay system in detail.

Preparation of Culture Flasks and Transwells

We start the assay by preparing poly-D-lysine (PDL; Sigma P7280)-coated tissue culture flasks (25 cm²) and laminin (BD Biosciences #354232)-coated Transwells. To coat the flasks, add 2 ml of PDL (20 µg/ml) into the flask and incubate at 37° for 4–5 h or until use. We use Costar Transwells (5 µm in pore size) and coat membranes on both sides with laminin. Laminin on the upper side of the membrane helps granule cells adhere, a necessary step for initiating migration. Laminin on the lower side membrane helps retain granule cells on the membrane after they cross to the outer side, which facilitates the detection of migrated cells (see "Detection of Cell Migration"). We determined that a laminin concentration of 20 µg/ml was sufficient to provide enough adhesion to the membrane but not so much to render the granule cells immobile. To coat Transwell membranes, add 600 µl of laminin (20 µg/ml in phosphate-buffered saline, PBS) into each well. This volume is just enough to cover the entire membrane of the Transwell inserts in order to save the amount of laminin used. The membrane and laminin are incubated at 37° for 4–5 h or until use. Before use, wash the flasks and the Transwells twice with PBS.

Isolation of Granule Cells

Brains are removed from postnatal day 8–9 (P8–P9) mice and cerebella are collected in ice-cold Hank's balanced salt solution (HBSS, with calcium and magnesium but without phenol red) in a 60-mm petri dish. We routinely add penicillin (100 units/ml)/streptomycin (100 μg/ml) into HBSS to minimize the possibility of bacterial contamination in the final Transwell cultures. Remove meninges from the cerebella as much as possible with fine-tipped forceps in HBSS. Then cut each cerebellum into seven to eight smaller pieces with the tips of forcep. Transfer all pieces of cerebella into a 15-ml tube using a blue-tip cut off to give it a wider opening. After removing the HBSS from the cerebella pieces, add 5 ml of PBS (without calcium and magnesium), 2.5 ml 2.5% trypsin (GIBCO) and 250 μg DNase I (Worthington). Incubate the cerebella at 37° for 10 min, occasionally mixing the tube by swirling every 2–3 min. At the end of the incubation, add 1 ml fetal bovine serum and 250 μg DNase I and incubate for an additional 5 min at 37°, occasionally mixing the tube. Let the pieces of cerebella settle in the tube and remove the solution. Add to the cerebella 8 ml of neural basal media (GIBCO) plus B27 supplement (GIBCO) and 100 μg DNase I. Triturate with a 5-ml pipette five to six times or until pieces of cerebella disappear. Let the cell suspension stand at room temperature for 5 min to allow any large chunks of tissue to settle. The dissociated cells are now ready for the Percoll step gradient purification as described by Hatten (1985).

Percoll Step Gradient Purification

Add 26 ml of 20× PBS into 500 ml of Percoll (Sigma P1644) to make the Percoll solution isotonic. Dilute the Percoll solution using neural basal-B27 medium to make two working concentrations: 60 and 35%. These two stock solutions can be stored in the refrigerator for several weeks. For six to eight cerebella, we use four 15-ml tubes for the purification step. To prepare the step gradient, first add 2 ml of 35% Percoll into each 15-ml tube with a 5-ml pipette. Then, using a Pasteur pipette, add 2 ml of 60% Percoll into the bottom of each tube (below the 35% Percoll) and carefully withdraw the pipette at the completion of loading without disrupting the upper Percoll layer. Load about 2 ml granule cell suspension over the 35% Percoll layer. Centrifuge at 3500 rpm for 30 min at room temperature and allow the rotor to reach a full stop with the breaks off. At the end of this centrifugation, granule cells are located at the interface between the two Percoll gradients.

To collect granule cells, first remove the top gradient of the Percoll. Then use a Pasteur pipette to transfer and pool granule cells into a clean 15-ml tube. To pellet the cells, add 4 volumes of neural basal-B27 medium to dilute the Percoll in the granule cell preparation and centrifuge at 1000 rpm for 5 min. Remove the medium, resuspend the cell pellet in 5 ml of neural basal-B27 medium, and transfer into a PDL-coated flask for incubation.

Recovery of Granule Cells

We found that purified granule cells need a recovery period of 2 h or more in our migration assay, perhaps for restoration of ephrin-B2 expression after trypsin cleavage. We suggest such a recovery period when studying a receptor molecule that may be susceptible to trypsin degradation. Purified granule cells are incubated at 37° on a PDL-coated culture flask for 2 h. PDL helps maintain the health of granule cells during this incubation. After recovery, granule cells are shaken off the plate by holding the flask in one hand and hitting against the other hand and are then collected for the Transwell assay.

For migration assays on infected granule cells, we used the Sindbis viral expression system (Invitrogen) to introduce a dominant-negative PDZ-RGS3 mutant protein (dnPDZ-RGS3, made by replacing the RGS domain with EGFP) into granule cells. We modified the recovery period to the following: immediately after the 2-h regular recovery period, EGFP or dnPDZ-RGS3 virus, with a similar titer, is added. The cells are infected for 1 h at room temperature on an orbital shaker and then for another hour at 37°.

Transwell Migration Assay

Recovered granule cells are placed in a 15-ml tube and centrifuged at 1000 rpm for 5 min. Cells are resuspended in 5 ml neural basal-B27 medium for counting and are then diluted to a concentration of 10^6 cells/ml. We would generally obtain more than 10^7 granule cells from six to eight P8 cerebella.

To set up the Transwell system, first set the inserts aside into the wells that will not be used. Add to the experimental wells 600 μl/well neural basal-B27 medium containing the appropriate amount of chemoattractant and/or modulators. Load 100 μl of purified granule cells, at a concentration of 10^6 cells/ml, into the bottom of each insert using a 200-μl pipette tip. Once cells are distributed into all the inserts, carefully transfer the inserts into the wells containing medium and avoid creating air bubbles at the interface between the surface of the medium and the lower side of the insert. Incubate the cells at 37° for 16 h. For virally infected granule cells, this incubation is shortened to 6 h to ensure cell survival during Sindbis expression of the protein.

Detection of Cell Migration

At completion of the Transwell assay, remove media and add 1 ml of methanol to each well to fix the cells at room temperature for 5 min. Remove methanol and let the membrane air dry for 5 min. Add to each well 1 ml of Giemsa (Sigma GS-500, dilute 10-fold in water) to stain cells for 1 h. Remove the staining solution and wash two to three times with water. After the final wash, use a cotton swab to wipe off cells that adhered to the upper side of the membrane. Cells that have migrated and attached to the lower side are counted. For each membrane, we count cells in four central fields with a 16× objective. The average of the four fields gives the final number for migrated cells under different conditions. We routinely test each condition in duplicate or, in some cases, in triplicate per experiment. For assays with virally infected granule cells, the Transwell membrane is fixed with 4% paraformaldehyde in PBS for 10 min, washed once with PBS, and slide mounted with Fluoromount-G (Fisher NC9749960). Migrated cells carrying EGFP or EGFP-fusion proteins are counted under a fluorescence microscope.

PDZ-RGS3 Mediates a Selective Inhibition of G-Protein Signaling by B Ephrins

Using the Transwell migration assay system, we first established a chemoattractant effect of chemokine SDF-1 on purified granule cells (Fig. 3A; Lu *et al.*, 2001). Activation of ephrin-B reverse signaling by the dimeric EphB2-Fc fusion protein inhibited the chemoattractant effect of SDF-1, whereas control Fc protein had no detectable effect. EphB2-Fc protein did not inhibit granule cell migration toward another chemoattractant, BDNF (Fig. 3B), a ligand for TrkB receptor tyrosine kinase. This demonstrated that ephrin-B reverse signaling could selectively inhibit G-protein-coupled signaling. PDZ-RGS3 appeared to mediate inhibition of the chemoattractant effect of SDF-1 by EphB2-Fc because a dominant-negative mutant of the PDZ-RGS3 protein blocked inhibition (Fig. 3C).

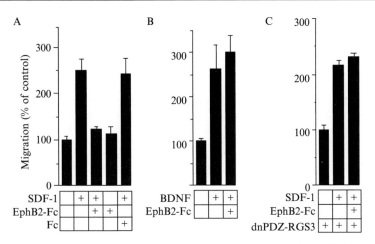

FIG. 3. Regulation of cerebellar granule cell chemotaxis. Granule cells purified from P8 or P9 mouse cerebella were placed in the upper chamber of a Transwell apparatus, and cells migrating to the lower side of the membrane were counted. Error bars show SEM. (A) Granule cells showed chemotaxis to SDF-1 placed in the lower chamber. This chemotaxis was inhibited by EphB2-Fc placed in the upper chamber. EphB2-Fc by itself, placed in the upper chamber, had no detectable effect. Control Fc did not block the effect of SDF-1. (B) Granule cells also showed chemotaxis to BDNF placed in the lower chamber. EphB2-Fc did not inhibit chemotaxis to BDNF. (C) dnPDZ-RGS3, a dominant-negative truncated form of PDZ-RGS3 where the RGS domain is replaced by an EGFP fluorescent marker, was introduced into purified cerebellar granule cells using a Sindbis viral vector. Fluorescently labeled granule cells expressing dnPDZ-RGS3 still showed chemotaxis to SDF-1, but the inhibitory effect of EphB2-Fc on chemotaxis was now blocked. Adapted from Lu *et al.* (2001).

Concluding Remarks

RGS proteins constitute a large family of negative regulators for G-protein signaling. An emerging feature of these molecules is that many RGS proteins contain additional domains that could potentially allow control of G proteins by other signaling pathways. PDZ-RGS3 is one such example. Identification of PDZ-RGS3 led us to study the regulatory role of ephrin-B reverse signaling on G-protein-mediated cellular guidance in neurons. In our study, we utilized the Transwell migration assay system, which has been widely used for hematopoietic cells but not yet applied in cultured neurons. Because G-protein-coupled signaling is closely linked to the chemotactic property of a cell and migration is a prominent feature of neurons, the Transwell assay discussed here could provide an effective cell-based assay system for elucidating RGS protein-mediated signaling inter-actions, as well as other pathways that regulate guidance and migration in neurons.

Acknowledgment

We thank Xiuyun Wang for constructing Fig. 1B.

References

Berman, D. M., Wilkie, T. M., and Gilman, A. G. (1996). GAIP and RGS4 are GTPase-activating proteins for the Gi subfamily of G protein alpha subunits. *Cell* **86,** 445–452.

Bruckner, K., Pablo Labrador, J., Scheiffele, P., Herb, A., Seeburg, P. H., and Klein, R. (1999). EphrinB ligands recruit GRIP family PDZ adaptor proteins into raft membrane microdomains. *Neuron* **22,** 511–524.

Cowan, C. A., and Henkemeyer, M. (2001). The SH2/SH3 adaptor Grb4 transduces B-ephrin reverse signals. *Nature* **413,** 174–179.

Cowan, C. A., and Henkemeyer, M. (2002). Ephrins in reverse, park and drive. *Trends Cell Biol.* **12,** 339–346.

Druey, K. M., Blumer, K. J., Kang, V. H., and Kehrl, J. H. (1996). Inhibition of G-protein-mediated MAP kinase activation by a new mammalian gene family. *Nature* **379,** 742–746.

Flanagan, J. G., and Vanderhaeghen, P. (1998). The ephrins and Eph receptors in neural development. *Annu. Rev. Neurosci.* **21,** 309–345.

Hatten, M. E. (1985). Neuronal regulation of astroglial morphology and proliferation *in vitro*. *J. Cell Biol.* **100,** 384–396.

Holmberg, J., and Frisen, J. (2002). Ephrins are not only unattractive. *Trends Neurosci.* **25,** 239–243.

Hunt, T. W., Fields, T. A., Casey, P. J., and Peralta, E. G. (1996). RGS10 is a selective activator of G alpha i GTPase activity. *Nature* **383,** 175–177.

Knoll, B., and Drescher, U. (2002). Ephrin-As as receptors in topographic projections. *Trends Neurosci.* **25,** 145–149.

Kullander, K., and Klein, R. (2002). Mechanisms and functions of Eph and ephrin signalling. *Nature Rev. Mol. Cell Biol.* **3,** 475–486.

Lin, D., Gish, G. D., Songyang, Z., and Pawson, T. (1999). The carboxyl terminus of B class ephrins constitutes a PDZ domain binding motif. *J. Biol. Chem.* **274,** 3726–3733.

Linares, J. L., Wendling, C., Tomasetto, C., and Rio, M. C. (2000). C2PA, a new protein expressed during mouse spermatogenesis. *FEBS Lett.* **480,** 249–254.

Lu, Q., Sun, E. E., Klein, R. S., and Flanagan, J. G. (2001). Ephrin-B reverse signaling is mediated by a novel PDZ-RGS protein and selectively inhibits G protein-coupled chemoattraction. *Cell* **105,** 69–79.

Murai, K. K., and Pasquale, E. B. (2002). Can Eph receptors stimulate the mind? *Neuron* **33,** 159–162.

Songyang, Z., Fanning, A. S., Fu, C., Xu, J., Marfatia, S. M., Chishti, A. H., Crompton, A., Chan, A. C., Anderson, J. M., and Cantley, L. C. (1997). Recognition of unique carboxyl-terminal motifs by distinct PDZ domains. *Science* **275,** 73–77.

Thompson, S. M. (2003). Ephrins keep dendritic spines in shape. *Nature Neurosci.* **6,** 103–104.

Torres, R., Firestein, B. L., Dong, H., Staudinger, J., Olson, E. N., Huganir, R. L., Bredt, D. S., Gale, N. W., and Yancopoulos, G. D. (1998). PDZ proteins bind, cluster, and synaptically colocalize with Eph receptors and their ephrin ligands. *Neuron* **21,** 1453–1463.

Watson, N., Linder, M. E., Druey, K. M., Kehrl, J. H., and Blumer, K. J. (1996). RGS family members: GTPase-activating proteins for heterotrimeric G-protein alpha-subunits. *Nature* **383,** 172–175.

Wilkinson, D. G. (2001). Multiple roles of EPH receptors and ephrins in neural development. *Nature Rev. Neurosci.* **2,** 155–164.

[9] Biochemical and Electrophysiological Analyses of RGS8 Function

By OSAMU SAITOH and KUBO YOSHIHIRO

Abstract

RGS8 was identified as a regulator of G-protein signaling (RGS) protein induced in neuronally differentiated P19 cells. This article presents methods used to estimate the binding activity and selectivity between RGS8 and $G\alpha$ subunits. It describes three kinds of *in vitro*-binding experiments using RGS8 proteins generated by three different techniques: recombinant protein from *Escherichia coli, in vitro*-translated protein, and protein expressed in cDNA-transfected cultured cells. It also presents methods of the functional analysis of RGS protein using the *Xenopus* oocyte expression system. Electrophysiological procedures, which were used to examine the effects of RGS8 on G_i- and G_q-mediated responses, are described.

Introduction

Numerous extracellular signals are translated into intracellular signaling cascades through seven transmembrane-spanning receptors that activate heterotrimeric G proteins. Such receptor activation stimulates nucleotide exchange and dissociation of the G protein, releasing $G\alpha$ in its GTP-bound state from the $G\beta\gamma$ complex. The released subunits stimulate cellular effectors, including enzymes and ion channels. Although the regulatory mechanisms used to adjust the strength and duration of signaling for an appropriate response have not been fully understood, studies have discovered a new family of regulators of G-protein signaling (RGS) (Berman and Gilman, 1998; Dohlman and Thorner, 1997; Druey *et al.*, 1996; Koelle and Horvitz, 1996). RGS proteins comprise a large family of more than 30 members which modulate heterotrimeric G-protein signaling. They share a homologous domain, the RGS domain, which is flanked by diverse N and C termini. Biochemical studies demonstrated that RGS members function as GTPase-accelerating proteins (GAP) for the α subunits of heterotrimeric G proteins by using their RGS domains (Berman *et al.*, 1996; Hunt *et al.*, 1996; Watson *et al*, 1996). Using a culture system of P19 cells, we previously isolated the cDNA of RGS8 as an RGS protein induced in neuronally differentiated P19 cells (Saitoh *et al.*, 1997). RGS8

METHODS IN ENZYMOLOGY, VOL. 390

is a small RGS protein composed of the short N terminus and the RGS domain and belongs to the B/R4 subfamily (Hollinger and Hepler, 2002). *In vivo*, RGS8 is specifically expressed in Purkinje cells of the cerebellum (Saitoh *et al.*, 1999). We have studied biochemical and physiological properties of RGS8, and this article presents methods for these studies. As biochemical properties, we focused on the binding activity between RGS8 and $G\alpha$ subunits. For physiological analysis, we described the electrophysiological recording using *Xenopus* oocytes.

Biochemical Analysis of RGS8 Function

$G\alpha$-Binding Assay

Overview. RGS proteins bind directly and preferentially to active GTP-bound forms of $G\alpha$ subunits. Then they act as GAPs to greatly accelerate the rate of $G\alpha$-GTP hydrolysis. The selectivity of RGS proteins for $G\alpha$ targets contributes mainly to which and how G-protein pathways are modulated by RGS proteins. Most RGS proteins interact with the $G\alpha_i$ and $G\alpha_q$ families (Hollinger and Hepler, 2002). However, RGS2 is reported to be a selective GAP for $G\alpha_q$ (Heximer *et al.*, 1997), the RGS domains of p115RhoGEF (Kozasa *et al.*, 1998), LARG (Fukuhara *et al.*, 2000), and PDZ-RhoGEF (Fukuhara *et al.*, 1999) interact specifically with the $G\alpha_{12}$ family, and RGS-PX1 is thought to be selective for $G\alpha_s$ (Zheng *et al.*, 2001). Therefore, when a new RGS protein is identified, it is important as a first step to determine which $G\alpha$ subunits are recognized by the new RGS protein. Because the binding affinity between RGS proteins and active forms of $G\alpha$ subunits is rather strong, both proteins form stable complexes *in vitro* and their complexes can be isolated from the protein solution with coprecipitation procedures, such as immunoprecipitation. Analysis of the complex reveals selectivity of RGS proteins for $G\alpha$ targets. After cloning of RGS8 cDNA, to perform the coprecipitation experiment, we generated the protein of RGS8 using three kinds of experimental procedures: recombinant protein from *Escherichia coli*, *in vitro*-translated protein, and protein expressed in cDNA-transfected culture cells.

$G\alpha$-Binding Assay of RGS8 Using Recombinant Protein. We first describe the $G\alpha$-binding assay using the recombinant RGS8 protein. Because no information concerning $G\alpha$ selectivity is available when characterizing a new RGS protein, as a source of G proteins, the brain membrane fraction is useful instead of collecting purified G proteins. Brain membranes are highly rich in several subtypes of G proteins, although $G\alpha_o$ is present most abundantly in the brain. We determined previously the $G\alpha$-binding selectivity of RGS8 using His-tagged RGS8 (His-RGS8) and brain membranes

(Saitoh *et al.*, 1997). Briefly, the His-RGS8 protein is purified from recombinant *E.coli*, the isolated protein is incubated with brain membranes, and the formed complex containing His-RGS8 is recovered with Ni^{2+}-NTA beads. Examination of the presence and subtypes of $G\alpha$ subunits in the isolated complex with sodium dedecyl sulfate–polyacrylamide gel electrophoresis (SDS–PAGE) and western blotting reveals the properties and specificity of $G\alpha$ binding of RGS8.

EXPRESSION VECTOR AND *E. COLI* STRAIN. The coding sequence of RGS8 is amplified using the corresponding primers with the appropriate restriction enzyme sites and is cloned into the pQE30 vector (Qiagen, Hilden, Germany). This expression vector contains 6xHis tag 5′ to the polylinker, and the His-tagged protein expressed in *E.coli* can be purified using Ni^{2+}-NTA agarose (Qiagen). As a host strain, we use M15 containing pREP4, which confers kanamycin resistance and constitutively expresses the *lac* repressor (M15[pREP4]).

PURIFICATION OF HIS-RGS8 PROTEIN FROM *E. COLI*. Before performing a large-scale preparation, several different conditions should be examined to determine the suitable extraction method for *E. coli* expressing His-RGS8.

A single colony of M15[pREP4] containing His-RGS8 cDNA is selected and grown in the presence of 25 μg/ml kanamycin and 100 μg/ml ampicillin overnight. The overnight culture (1 ml) is used to seed 20 ml of prewarmed medium (with antibiotics) and is grown at 37° with vigorous shaking until an OD_{600} of 0.6 is reached. A 1-ml sample is taken and then expression of the His-RGS8 is induced by adding isopropyl-thio-β-D-galactoside (IPTG) to 1 mM. The culture is incubated for an additional 4 h and is divided into several microcentrifuge tubes (1 ml per 1.5-ml tube). Cells are harvested by centrifugation for 1 min at 18,400g and the supernatant is discarded. Cells are resuspended in 100 μl of five different lysis buffers: (1) 50 mM sodium phosphate buffer, pH 7, 0.3 M NaCl; (2) 0.5% Triton X-100 in buffer 1; (3) 0.5% Tween 20 in buffer 1; (4) 1 M NaCl in buffer 1; and (5) 1 mg/ml lysozyme in buffer 1.

Cells are lysed by sonication for 30 s at maximum intensity and are centrifuged at 18,400g for 10 min. Each supernatant is transferred to fresh microcentrifuge tubes, and 20 μl of a 50% slurry of Ni^{2+}-NTA agarose is added to each tube. The extract is mixed gently with Ni^{2+}-NTA agarose for 20 min at 4° and centrifuged for 1 min at 2000g. The supernatant is discarded, and the resin pellet is washed twice with 100 μl of each extraction buffer by resuspension and centrifugation. Then, to elute the His-tagged protein, 30 μl of 1 M imidazole–HCl (pH 7.0) is added and mixed well by vortexing. The mixture is centrifuged for 5 min at 8200g, and the supernatant is examined by SDS–PAGE. As shown in Fig. 1, condition 4 is better for the purification of His-RGS8 from *E. coli*. For other RGS proteins,

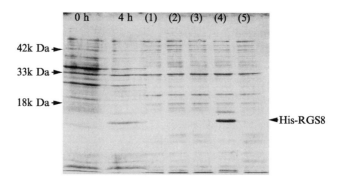

FIG. 1. Examination of extraction conditions of His-RGS8 from *E. coli*. M15[pREP4] *E. coli* containing RGS8-pQE30 was grown and treated further with 1 m*M* IPTG for 4 h. Direct SDS extracts of *E. coli* before and after IPTG treatment are shown as 0 and 4 h. After IPTG treatment, cells were extracted with 50 m*M* sodium phosphate buffer, pH 7.0, 0.3 *M* NaCl (A, lane 1), 0.5% Triton X-100 in A (lane 2), 0.5% Tween 20 in A (lane 3), 1 *M* NaCl in A (lane 4), 1 mg/ml lysozyme in A (lane 5), and their extracts were examined with SDS–PAGE and Coomassie Brilliant blue (CBB) staining. The position of His-RGS8 was determined by immunoblotting using an anti-His tag antibody (Qiagen)(data not shown).

different conditions may be suitable for purification. Based on the pilot experiment described earlier, we developed the following procedures for the large-scale purification of His-RGS8.

The *E. coli* culture induced by 1 m*M* IPTG (500 ml) is prepared as described earlier. The culture is divided into ten 50-ml tubes (Falcon 352070). Cells are harvested by centrifugation for 15 min at 3500*g*, and supernatants are discarded. Two milliliters of ice-cold lysis buffer (50 m*M* sodium phosphate buffer, pH 7, 1.3 *M* NaCl, 10% glycerol, 40 m*M* imidazole–HCl, pH 7.0, 14 m*M* β-mercaptoethanol, 0.1 m*M* phenylmethylsulfonyl fluoride) is added to each tube. Cell precipitates are resuspended and sonicated for 30 s at maximum intensity on ice. Cell lysates are centrifuged for 15 min at 9800*g*, and supernatants are collected in a Falcon 50-ml tube. To this cell extract, 6 ml of 50% slurry of Ni^{2+}-NTA agarose pre-equilibrated with lysis buffer is added and mixed with a rotator for 30 min min at 4°. The mixture is centrifuged for 5 min at 900*g* to precipitate Ni^{2+}-NTA resin, and the supernatant is discarded. The Ni^{2+}-NTA resin is then washed with ice-cold lysis buffer five times. The washed resin is transferred into a small chromatography column (371–1550, Bio-Rad). The His-tagged protein is eluted stepwise with 2 ml of 0.1, 0.2, 0.3, 0.4, and 0.5 *M* of imidazole–HCl (pH 7.0). The pattern of protein elution is analyzed with SDS–PAGE. A typical example of purification using this procedure is illustrated in Fig. 2. Fractions eluted with 0.2–0.4 *M* imidazole buffer are

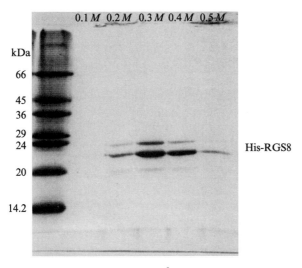

Fig. 2. Elution pattern of His-RGS8 from Ni^{2+}-NTA agarose. M15[pREP4] containing RGS8-pQE30 was extracted with lysis buffer (composition is indicated in the text), and the extract was incubated with Ni^{2+}-NTA agarose. After washing with lysis buffer, proteins were eluted from the agarose resin with 0.1–0.5 M imidazole–HCl, pH 7.0. The eluted proteins were analyzed with SDS–PAGE and stained with CBB.

combined, gel filtrated to remove imidazole buffer using PD-10 (Amersham Biosciences, NJ), equilibrated with 50 mM HEPES, pH 8.0, 5 mM EDTA, and 2 mM dithiothreitol, and concentrated to 1–2 mg/ml with Centricon 10 (Millipore Corporation, Bedford, MA). The resultant protein sample is used as the His-RGS8 for the $G\alpha$-binding assay.

MEMBRANE PREPARATION OF RAT BRAIN. Rat whole brains are minced and placed in ice-cold 20 mM Tris–acetate, pH 7.2, containing a protease inhibitor mixture. They are homogenized with a polytron and centrifuged at 1000g for 5 min at 4°. The supernatant is centrifuged again at 38,000g for 20 min at 4°, and the supernatant is discarded. The resultant pellet is washed by resuspension in 10 volumes of 20 mM Tris–acetate, pH 7.2, containing a protease inhibitor mixture and by centrifugation at 38,000g for 20 min. This washing is repeated three times, and the final pellet is resuspended in the same buffer as the brain membrane suspension.

BINDING ASSAY OF HIS-RGS8 TO $G\alpha$ PRESENT IN BRAIN MEMBRANES. This binding assay was initially demonstrated by Watson *et al.* (1996). They showed that various RGS family members interact specifically with $G\alpha$ subunits present in membranes. His-tagged RGS1, RGS4, and GAIP were incubated with bovine brain membrane fractions treated with GDP or

$GDP + AlF_4^-$. In membranes treated with GDP, G proteins are present as heterotrimers. Treatment with $GDP + AlF_4^-$ induces the transition state and dissociation of $G\alpha$ from $G\beta\gamma$. After detergent extraction, they examined complexes containing RGS proteins purified by Ni^{2+}-NTA chromatography. Applying this procedure, we studied the binding properties of RGS8 to $G\alpha$ using His-RGS8 and rat brain membranes, which were prepared as described earlier.

His-RGS8 (10 μg) and the rat brain membrane fraction (0.5 mg) are incubated for 30 min at 4° in 200–400 μl of binding buffer [20 mM HEPES, pH 8.0, 0.38 M NaCl, 3 mM dithiothreitol, 6 mM MgCl$_2$, 10 μM GDP, 30 μM AlF$_4^-$ (30 μM AlCl$_3$, 10 mM NaF), 40 mM imidazole–HCl, pH 7.0]. Sodium cholate (10%, w/v in H$_2$O) is added to 1% of the final concentration, and membranes are solubilized with gentle shaking for 1 h at 4°. After centrifugation at 44,000g for 20 min, the supernatant is transferred to a fresh microcentrifuge tube. To each tube containing the supernatant, 30 μl of 50% slurry of Ni^{2+}-NTA agarose pre-equilibrated with binding buffer is added. After gentle mixing for 30 min at 4°, the resin is precipitated by centrifugation at 2000g for 5 min. The resin precipitates are washed five times with washing buffer [20 mM HEPES, pH 8.0, 3 mM dithiothreitol, 0.1% polyoxyethylene 10-lauryl ether (C$_{12}$E$_{10}$), 1 μM GDP, 30 μM AlF$_4^-$, 40 mM imidazole–HCl, pH7.0]. The final pellet is treated with 30 μl of SDS sample buffer. After centrifugation at 18,400g for 5 min, the supernatant is investigated for the presence and subtypes of $G\alpha$ subunits by SDS–PAGE and western blotting. A typical result using this procedure is shown in Fig. 3. His-RGS8 co-precipitated with the 40-kDa protein from brain membranes. This 40-kDa protein was recognized with antibodies to $G\alpha_{i3}$ and $G\alpha_o$. Results demonstrate that RGS8 binds preferentially to $G\alpha_o$ and $G\alpha_{i3}$ (Saitoh et al., 1997).

In Vitro-*Binding Assay Using Biotinylated RGS8 Protein.* Instead of purifying recombinant protein from *E. coli* as a way to have proteins encoded by the gene of interest, the method of *in vitro* transcription/ translation of cDNA can be used. To examine $G\alpha$ binding of RGS8, we used the products of this method and $G\alpha_o$ purified from brain (Saitoh et al., 2001). The biotinylated RGS8 protein was generated by *in vitro* transcription/translation and incubated with purified $G\alpha_o$. Immunoprecipitation with the anti-$G\alpha_o$ antibody from the incubated mixture was performed, and co-immunoprecipitation of the biotinylated RGS8 protein was examined with streptavidin-horseradish peroxidase (HRP).

GENERATION OF BIOTINYLATED RGS8 PROTEIN. RGS8 cDNA is cloned into the pCXN2 expression vector (Niwa et al., 1991) with the polylinker containing T3 and T7 promoters. Although we simply need T3, T7, or SP6 promoters in vectors to synthesize sense RNA, we observed an enhanced

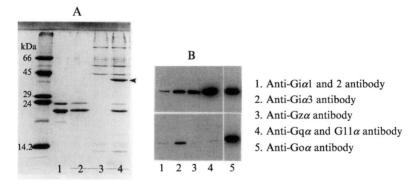

FIG. 3. Interaction of His-RGS8 with Gα in brain membranes. His-tagged RGS8 was incubated with rat brain membranes treated with GDP and AlF$_4^-$. After extraction with detergent, complexes containing His-tagged RGS8 were purified by Ni^{2+}-NTA-agarose. (A) Proteins bound to RGS8 were resolved by SDS–PAGE and detected by staining with CBB. Lane 1, purified His-tagged RGS8; lane 2, recovered His-tagged RGS8 from buffer alone; lane 3, proteins recovered in the absence of His-tagged RGS8; and lane 4, proteins recovered in the presence of His-tagged RGS8. The 40-kDa protein recovered with His-tagged RGS8 is indicated by an arrowhead. (B) Identification of Gα subunit bound to RGS8. Rat brain membrane proteins (*top*) and proteins bound to His-tagged RGS8 (*bottom*) were analyzed by immunoblotting with antibodies to various Gα subunits. Reprinted with permission from Saitoh *et al.* (1997).

translation of proteins using DNA containing a poly(A)$^+$ sequence downstream of the cDNA. Such vectors are available from Promega (Madison, WI). The resultant plasmid DNA is prepared by the standard alkaline lysate method. After final ethanol precipitation, the DNA pellet should be dissolved in RNase-free H$_2$O. Although linear DNA templates, such as those generated by polymerase chain reaction or by restriction enzyme digestion, can be used, circular plasmid DNA gives the best translation results. As directed by the manufacturer's instructions (TNT-coupled reticulocyte lysate system, Promega), the biotinylated RGS8 protein is generated. The following reaction components are assembled: TNT rabbit reticulocyte lysate, reaction buffer, T3 RNA polymerase, amino acids mixture, RNase inhibitor, circular plasmid DNA (1 μg), and Transcend biotin-lysyl-tRNA. The translation reaction is then incubated at 30° for 90 min. Using this reaction system, biotinylated lysine residues are incorporated into nascent proteins during translation, eliminating the need for labeling with radioactive amino acids. This biotinylated protein is stable for more than a year and can be visualized easily using chemiluminescent detection.

CO-IMMUNOPRECIPITATION OF BIOTINYLATED RGS8 PROTEIN. The biotinylated RGS8 protein is incubated with the purified $G\alpha_o$ protein. G proteins are purified from the bovine cerebral cortex according to the method of Sternweis and Robishaw (1984), and $G\alpha_o$ is purified as described by Asano et al. (1988). The incubation buffer for binding contains 20 mM HEPES, pH 8.0, 0.1 M NaCl, 1 mM dithiothreitol, 6 mM MgCl$_2$, 10 μM GDP, 30 μM AlF$_4^-$ (30 μM AlCl$_3$, 10 mM NaF), and 0.1% polyoxyethylene 10-lauryl ether (C$_{12}$E$_{10}$). Protein G-Sepharose is pre-equilibrated with the incubation buffer. Out of 50 μl in vitro transcription/translation product, 10–20 μl is incubated with 2–4 μg of the $G\alpha_o$. One binding assay is performed in 50–100 μl of the incubation buffer. After incubation for 4 h at 4°, the reaction mixture is incubated with 30 μl of 50% (v/v) protein G-Sepharose for 1 h at 4° and is precleared by centrifugation at 2000g for 5 min. The supernatant is incubated with the anti-bovine $G\alpha_o$ antibody (MBL, Nagoya, Japan) at 4° for 2 h and is then incubated further with 30 μl of 50% (v/v) protein G-Sepharose at 4° for 2 h. After centrifugation at 2000g for 5 min, protein G beads are washed four times in the incubation buffer, suspended in SDS sample buffer, and boiled for 5 min. The resultant protein samples are separated by SDS–PAGE and transferred to a nitrocellulose membrane. Coimmunoprecipitation of biotinylated RGS8 is examined with streptavidin-HRP (Promega). The transferred membrane is blocked by incubation in Tris-buffered saline (TBS) containing 0.5% Tween 20 (TBST) for 1 h. The membrane is incubated further in 1:5000 diluted streptavidin-HRP in TBST. After a 60-min incubation, the membrane is washed three times with TBST and then twice with TBS. Signals on the washed membrane are detected with an ECL system (Amersham Biosciences, Piscataway, NJ). Immunoprecipitation of $G\alpha_o$ is confirmed by Western blotting using the anti-$G\alpha_o$ antibody (Santa Cruz Biotechnology, Santa Cruz, CA). A typical result of this method is shown in Fig. 4. We generated biotinylated proteins of the RGS8 and ΔNRGS8 mutant lacking N-terminal 35 amino acids outside of the RGS domain. Both biotinylated proteins were coimmunoprecipitated with $G\alpha_o$ using the anti-$G\alpha_o$ antibody. Results indicate that ΔNRGS8 retains G-protein-binding activity and that the N terminus of RGS8 is not involved directly in $G\alpha$ binding (Saitoh et al., 2001).

Gα Binding of RGS8 Estimated with Coimmunoprecipitation of Proteins Expressed by cDNA Transfection in Tissue Culture Cells. To examine the binding activity of wild-type and mutant RGS8 to the $G\alpha$ protein, we performed coimmunoprecipitation experiments using HEK293T cells cotransfected with cDNAs of RGS8 constructs and $G\alpha_o$ (Masuho et al., 2004). Because previous protein purification is not required, this procedure might be useful for screening binding-defective mutants of

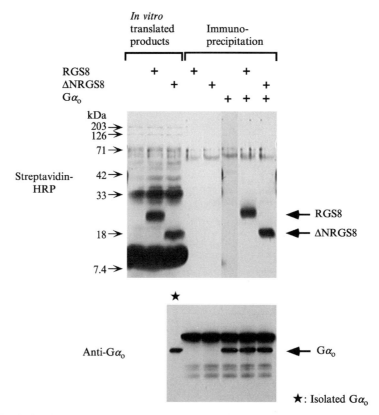

FIG. 4. Gα binding of biotinylated proteins of RGS8 and ΔNRGS8. Biotinylated proteins of RGS8 and ΔNRGS8 were generated by *in vitro* transcription/translation and mixed with purified bovine Gα_o. After incubation for 4 h at 4°, the complex containing Gα_o was immunoprecipitated with the anti-bovine Gα_o antibody. Precipitated proteins were separated by a SDS–polyacrylamide gel and transferred to nitrocellulose membranes. Coimmunoprecipitation of the biotinylated protein was examined with streptavidin-horseradish peroxidase (*top*). The right three lanes contained immunoprecipitated Gα_o as confirmed by western blotting using the anti-Gα_o antibody (*bottom*). In the lane indicated with a star, isolated Gα_o was applied. Reprinted with permission from Saitoh *et al.* (2001).

other known RGS proteins. To facilitate immunoprecipitation, the RGS8 or RGS8 mutant is expressed as a chimeric protein with green fluorescent protein (GFP) at the C terminus. With such a GFP construct, the constitutively active form of Gα_o (GαoQ205L) is coexpressed in HEK293T cells. As a transfection procedure, transient transfection using Effectene (Qiagen, Hilden, Germany) is performed. At 48 h after transfection, cells are scraped with a cell scraper (Falcon 3085) from a 10-mm culture dish and are precipitated by centrifugation at 6300g for 5 min. Five hundred microliters of Tx buffer (50 mM Tris–HCl, pH 7.5, 300 mM NaCl, 1% Triton X-100, 0.1 mM phenylmethylsulfonyl fluoride) is added to the cell precipitates, and the cell lysate is prepared by sonication (20 s at maximum intensity) in Tx buffer. Protein G-Sepharose is pre-equilibrated with Tx buffer. After clarifying by centrifugation at 18,400g for 10 min, 30 μl of 50% (v/v) protein G-Sepharose is added to the supernatant, and the mixture is incubated at 4° for 30 min. The mixture is precleared by centrifugation at 2000g for 5 min, and the supernatant is incubated with 4 μg of anti-GFP antibody (Roche, Indianapolis, IN) at 4° for 1 h followed by incubation with 50 μl of 50% (v/v) protein G-Sepharose at 4° for 1 h. After centrifugation at 2000g for 5 min, resin precipitates are washed five times with 500 μl of Tx buffer. The final pellet is treated with 40 μl of SDS sample buffer, and the SDS extract is examined for co-precipitation of GαoQ205L with western blotting using the anti-Gα_o antibody (Santa Cruz Biotechnology). Immunoprecipitation of various RGS8-GFP chimeras is confirmed by western blotting using the anti-GFP antibody (as an example of results, see Masuho *et al.*, 2004).

Electrophysiological Analysis of RGS Function

Rationale of Electrophysiological Analysis

RGS proteins were initially identified as GAPs or as negative regulators of G-protein signaling using a yeast growth arrest assay system (e.g., Druey *et al.*, 1996). These analyses focus on the effect of RGS proteins on the cumulative aspects of G-protein signaling.

We previously identified a novel neuron-specific RGS, RGS8 (Saitoh *et al.*, 1997). To know the functional significance of RGS proteins in neurons, it is critically important to clarify the kinetic aspect of the modulation of G-protein signaling by RGS proteins. We, therefore, adopted an electrophysiological analysis. We expressed G$_i$-coupled receptors and G-protein-coupled inwardly rectifying K$^+$ (GIRK) channels (Kubo *et al.*, 1993) in *Xenopus* oocytes and evaluated the turning-on and -off kinetics of the GIRK current on application and removal of the receptor ligand. We observed unexpectedly that not only the turning-off but also the turning-on kinetics was accelerated by RGS8.

In terms of the turning-off kinetics, it can also be measured biochemically by estimating the GAP activity, as has been done by many groups (Berman *et al.*, 1996; Hunt *et al.*, 1996; Watson *et al.*, 1996). In contrast, the turning-on kinetics (e.g., GDP/GTP exchange speed) in the presence of activated metabotropic receptor is too fast to be analyzed biochemically even at low temperatures. From this point of view, electrophysiological analysis using the GIRK channel as a direct G-protein effector is unique and powerful. We believe that this is the reason why we could obtain an unexpected interesting finding. Doupnik *et al.* (1997) also carried out a similar analysis on the RGS4 protein using a mammalian culture cell as an expression system and reached the same conclusion.

This Section describes details of the functional analysis of RGS protein using the *Xenopus* oocyte expression system.

G_i-Mediated Responses by Xenopus Oocyte Expression System

Overview. As RGS8 was demonstrated to bind to $G\alpha_{i/o}$ biochemically, it is necessary to use an effector involved in this pathway. GIRK1/4 and GIRK1/2 heterotetramers are known to be major GIRK channels in the heart and the brain, respectively. Both of them are known to be activated by direct interaction with the $G\beta\gamma$ subunit released from the $G_{i/o}$ protein. Thus, activation of G_i-coupled receptors such as the m2 muscarinic or D2 dopamine receptor is known to cause an increase in GIRK channel activity.

For quantitative analysis and for kinetics analysis of G-protein function, it is critical to avoid amplification and time lag. It is not ideal to use effectors activated by second messenger system and downstream enzymes, including IP_3, Ca^{2+}, cAMP, and protein kinase A and C. As the GIRK channel is a direct effector of G protein and as no second messenger systems or enzymes are involved for activation, it is advantageous to use it as an effector for the analysis of RGS8. We can trust that the activity of the GIRK channel reflects G-protein activity directly.

We coexpressed the G_i-coupled receptor and GIRK1/2 channel with or without RGS8 in *Xenopus* oocytes. We did not especially induce expression of the G protein, as it is expressed endogenously in oocytes.

cRNA Synthesis. As we inject cRNA of each clone to oocytes to induce expression, the first step is to prepare cRNA. The details have been described in detail elsewhere (Kubo and Murata, 2001). Here we describe only the outline and points where attention should be paid. We linearize the plasmid by digesting at a unique restriction site at the $3'$ end of the insert and transcribe cRNA using an appropriate RNA polymerase. Both T7 and T3 are satisfactory, but the SP6 polymerase gives lower yield. Therefore, we subclone the insert so that we can use T7 and T3 polymerase to transcribe the sense cRNA. We mostly use pBluescript (Stratagene, La Jolla, CA) as a standard vector, but in cases where a low expression level is anticipated, the use of pGEMHE (Tytgat *et al.*, 1994) is recommended. It has $5'$ and $3'$ noncoding sequences of *Xenopus* β-globin gene before and after the multicloning site. Attachment of $5'$ and $3'$ noncoding sequences of the *Xenopus* β-globin gene to the coding region of the subcloned cDNA enhances the translation efficiency of cRNA to the protein. Therefore, the protein expression level is higher than when using other vectors. The final cRNA product should not be dissolved in diethyl pyrocarbonate (DEPC)-treated water, but in human injection grade water (Otsuka Pharm., Tokyo, Japan), as even a trace amount of DEPC is harmful to oocytes. The cRNA should be divided to small aliquots and kept at $-80°$ until use.

Preparation of Oocytes. The maintenance of frogs and how to isolate oocytes from frogs have been already described in detail elsewhere (Kubo and Murata, 2001). It is important to note that the enzymatic activity of collagenase, which makes oocytes detach from each other, and toxicity to oocytes differ significantly depending on lots. Therefore, we suggest testing some lots before purchasing in a large scale.

Injection of cRNA. We use the nanoinjector of Drummond Co. Ltd. The most critical point is the tip size of the injection pipettes. It goes without saying that thinner pipettes are better for minimizing damage to the oocytes. We prepare very thin pipettes from capillary glass using a programmable puller (Sutter Instrument, Novato, CA). It is recommended to make short tapered pipettes by pulling by multiple stages. We then set the pipette onto the injector and break the tip little by little so that we can manage to fill up the cRNA solution. Once the cRNA solution can be filled up from the tip, it can be easily injected to the oocytes. By doing this, we can keep the tip size and damage to the oocytes to a minimum. To carry out injection efficiently, we move the injection pipette only in one dimension, not in three dimensions. Instead we move the dish in which oocytes are set by the other hand so that the targeted oocyte locates in the center. Oocytes injected with cRNA are kept in a 6-cm sterile plastic dish at $17°$, and we change the dish and solution to a fresh one everyday to avoid bacteria growth. Oocytes suspected of being unhealthy are discarded everyday.

Two Electrode Voltage-Clamp Recording. We record under a two electrode voltage clamp using the oocyte clamp OC-725C (Warner Co. Ltd.) amplifier. It is important to monitor the actual voltage recording to confirm that the voltage clamping condition is satisfactory. To achieve a good voltage clamp, the most important factor is the resistance of the pipettes, especially the current-passing one. We use pipettes with a resistance of 0.1–0.3 $M\Omega$ when filled with 3 M KCl. In the case of GIRK current, which shows inward rectification, we record the inward current by holding constitutively at -120 mV in 90 mM K$^+$ (E_K = approximately 0 mV), with the DC boost circuit turned on to increase the stability of voltage clamp. Although this circuit decelerates the voltage-clamp speed upon step pulse application, there is no demerit to turn on this circuit when we record by holding at a certain membrane potential without applying step pulses.

There is a basal GIRK current even before application of the ligand. To discriminate GIRK current from simple leak current, we apply a short depolarizing pulse for estimation of the leak outward current level. If the leak outward current is negligible when compared to the inward current, the inward current can be interpreted to be mostly GIRK current. We discard data with a significant level of leak current.

Application and Washout of the Ligand. By applying the ligand that activates the coexpressed receptor, the GIRK current amplitude increases and upon wash-out it decreases. For the kinetic analysis of these steps, the speed of application and washout is critically important. We use a recording chamber of 160 μl volume, and apply 40 μl of five times concentrated stock of ligand solution to the chamber so that the flow is not directed at the oocytes and then mix immediately by pipetting. To wash out agonist in the bath, we use standard rapid perfusion and vacuum suction. The exchange rate of solutions in the bath was determined by changing the K^+ concentration using the same methods. The τ_{in} value ranged from 299 to 533 ms, and τ_{out} ranged from 840 to 1132 ms; this speed is sufficient for the accurate analysis of the off rate. The fast on rate in the presence of RGS8 is estimated to be slightly slower than the real value, but this error does not affect the conclusion of the study.

Analysis. We analyze data off line. We fit the increasing phase of GIRK current upon ligand application and decreasing phase upon removal with single exponential functions, and obtain the time constants τ_{on} and τ_{off}. We then calculate the mean and standard deviation (SD) of these values in the absence and presence of RGS8 and examine the difference statistically by the Student's t test. Using this analysis, we concluded previously that RGS8 accelerates not only the turning-off step but also the turning-on step of G-protein signaling (Saitoh *et al.*, 1997) (Fig. 5).

Implication. Our results on the turning-off step go well with the previous results of biochemical studies that RGS functions as a GAP. Our results not only confirmed the previous finding, but also demonstrated that regulation of the effector function by this mechanism actually occurs *in vivo*. In contrast, as to the acceleration of the turning-on step, no clear biochemical data explain our observations at a molecular level. The most simple explanation is that RGS8 functions as a GDP/GTP exchange factor or its accelerator. Another possibility is that RGS proteins facilitate the release of $G\beta\gamma$ subunits by binding to $G\alpha$ and by expelling $G\beta\gamma$. As the turning-on process in the presence of an activated receptor is too fast to be analyzed biochemically, the answer is still unclear.

Other Expression Systems. Doupnik *et al.* (1997) reported similar electrophysiological results on RGS1, RGS3, and RGS4. A clear difference between their study and ours is that they used a mammalian culture cell as

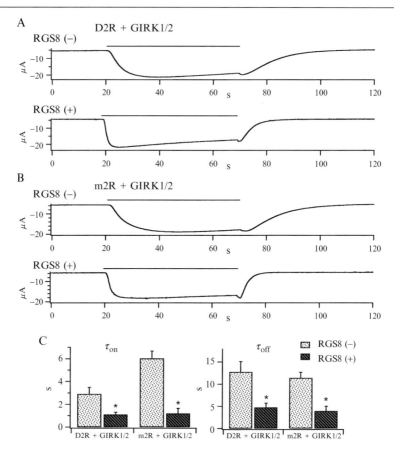

FIG. 5. Effects of RGS8 on the turning-on and -off kinetics of the G-protein-coupled inwardly rectifying K$^+$ channel (GIRK1/2) current upon stimulation of the receptors. (A) The GIRK1/2 and D2 dopamine receptor without (upper trace) or with (lower trace) RGS8 were co-expressed in *Xenopus* oocytes. Current traces at the holding potential of −120 mV are shown. By applying 50 nM dopamine, at the time indicated by the bars, an increase in the current amplitude on top of the basal level was observed. The kinetics of the turning on and off are obviously faster when RGS8 is coexpressed. (B) Instead of the D2 dopamine receptor, the m2 muscarinic receptor was coexpressed. Acetylcholine (Ach, 4 M) was applied at the time indicated by the bars. (C) Comparison of the τ_{on} and τ_{off} of GIRK1/2 current upon stimulation of the D2 or m2 receptor in the absence or presence of RGS8. From the traces in A and B, the increasing and decreasing phases were fitted by a single exponential function, and the time constants (τ_{on} and τ_{off}) were obtained. The mean and standard deviation of τ_{on} ($n = 14$) were D2 receptor, RGS(−): 2905 ± 576 ms, D2 receptor, RGS(+): 1103 ± 206 ms, m2 receptor, RGS(−): 6021 ± 651 ms, and m2 receptor, RGS(+): 1188 ± 444 ms. Those of τ_{off} ($n = 14$) were D2 receptor, RGS(−): 12761 ± 2360 ms, D2 receptor, RGS(+): 4714 ± 998 ms, m2 receptor, RGS(−): 11361 ± 1304 ms, and m2 receptor, RGS (+): 3909 ± 1 087 ms. Asterisks indicate data judged to be significantly different from the control values (p value < 0.05) by the Student's unpaired t test. Reprinted with permission from Saitoh *et al.* (1997).

their expression system. A merit to the use of mammalian cells is that it is relatively easy to apply and wash out ligands very rapidly using a piezo-motor-driven application pipette. An obvious merit using oocytes is the feasibility of recording. We can examine various combinations of the expressing clones. Another merit is that it is easy to adjust the expression level by changing the amount of injected cRNA.

A demerit to the use of oocytes is a variation of data depending on the batch. The τ_{on} and τ_{off} values in the absence of RGS proteins show significant variation from batch to batch. Therefore, it is necessary to carry out experiments of control and RGS group always in parallel. We always compare data obtained from the same batch of oocytes. The reason of the variation is not clear, but it is possible that the amount of G protein or endogenous RGS protein differs depending on the batch.

G_q-Mediated Responses

Overview. The N-terminal domain of RGS4 was reported to confer receptor-selective inhibition of G_q signaling (Zeng *et al.*, 1998). There is significant sequence conservation in the N-terminal region of RGS4 and RGS8. Therefore, we also examined the effect of RGS8 on G_q-coupled responses. We also analyzed the effect of the splice variant of RGS8, RGS8S (Fig. 6).

As players downstream of G_q are expressed endogenously in *Xenopus* oocytes (i.e., phospholipase Cβ, IP$_3$ and diacylglycerol, Ca^{2+} stores, as well as Ca^{2+} activated Cl$^-$ channels), we can monitor the G_q-coupled responses as an increase in the current amplitude of the Ca^{2+}- Cl$^-$ channel. The only molecule necessary is a G_q-coupled metabotropic receptor such as m1 or m3 muscarinic and substance P receptor. This is the reason why *Xenopus* oocytes have been used often for the expression cloning of metabotropic receptors, including substance P, 5-HT1c, and the metabotropic glutamate receptor.

There is also a demerit in monitoring G_q-coupled responses by the increase in the Ca^{2+}-activated Cl$^-$ current. It involves mobilization of the second messenger system, and amplification and/or a time lag obviously exists to some extent. Therefore, this assay system is not suitable in discussing the turning-on and turning-off kinetics in detail. Also, the amplitude of responses does not necessarily reflect the extent of the activated G protein, as some amplification factors are involved between the G protein and the Ca^{2+}-Cl$^-$ channel. There is obviously no linearity. With these points in mind, we studied the effect of RGS8 and RGS8S on G_q-coupled responses.

Electrophysiological Recording. The Ca^{2+}-activated Cl$^-$ current has been recorded very often in frog ringer solution (Kubo *et al.*, 1998) as an

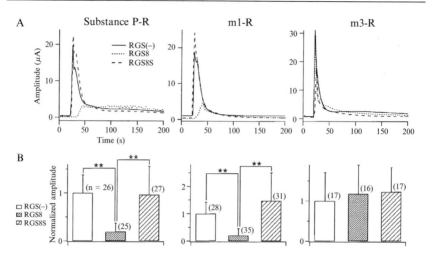

FIG. 6. Effects of RGS8 and RGS8S on $G\alpha_q$-coupled responses. The SP receptor (*left*), m1 muscarinic receptor (*middle*), or m3 muscarinic receptor (*right*) was expressed alone, with RGS8, or with RGS8S. Responses were monitored as an increase in the current amplitude of the Ca^{2+}-Cl^- current. (A) Typical examples of responses to the application of 10 nM SP (*left*) and 10 μM ACh (*middle* and *right*). (B) Comparison of the peak amplitudes of the responses. To normalize variation in the expression level among batches of oocytes, each recording was normalized by dividing the average of the responses of RGS (−) oocytes of the same batch. Normalized data obtained from three (SP-R, m3-R) or five (m1-R) batches were pooled, and the mean and standard deviation of each group were plotted. The n values are indicated in the graph. Differences judged to be significant ($p < 0.01$) by the Student's unpaired t test are marked by double asterisks. Reprinted with permission from Saitoh *et al.* (2002).

inward current by holding at about −80 mV. However, as the Ca^{2+}-activated Cl^- current shows strong outward rectification, we monitor the Ca^{2+}-Cl^- current by applying depolarizing step pulses of 200 ms every 2 s from the holding potential of −80 mV. In the course of repeated application of step pulses, we apply the receptor ligand by pipetting as described earlier. The sensitivity of detection is much higher than recording at the hyperpolarized potential, especially when the increase in Ca_i^{2+} is small. Because the responses decline even in the continuous presence of the ligand, probably by the decline of Ca_i^{2+}, we do not analyze the turning-off step. We record in the continuous presence of the ligand for 90 s.

Analysis. We measure the peak outward current amplitude induced by each step pulse and plot the time course of the change of the peak amplitude, which reflects the time course of the responses. In terms of the turning-on step, as the latency of the response cannot be interpreted in a

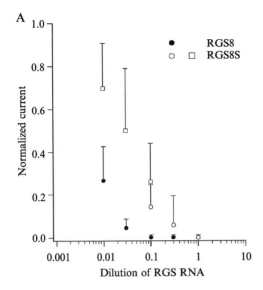

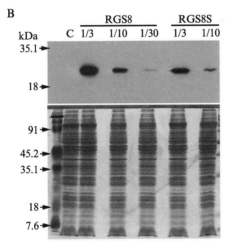

FIG. 7. Inhibition of m1 receptor-coupled response by various doses of RGS8 and RGS8S subcloned into a high expression vector and comparison of the amount of the expressed proteins. (A) RGS8 and RGS8S were subcloned into the pGEMHE vector. It yields cRNA with 5'- and 3'-untranslated regions of the *Xenopus* β-globin gene, which expresses at a very high level in *Xenopus* oocytes. The amplitudes of Ca^{2+}-Cl^- current upon application of 10 μM ACh were normalized by the mean of the amplitudes in the absence of RGS proteins and were plotted against the dilution of RGS cRNA. Dilution ×1 means full concentration, approximately 1 $\mu g/\mu l$. Filled circles indicate RGS8, and empty circles indicate the results of RGS8S in the same set of experiments. Empty squares represent data of RGS8S injected into the same batch of oocytes on the following day. Bars indicate SD ($n = 10$ or 13).

straightforward manner, we do not analyze it. We evaluate the peak amplitude of the response, calculate the mean and SD, and then compare data in the presence and absence of RGS statistically by the Student's t test.

Observation. We obtained the following unexpected results. (a) RGS8, which does not show clear binding to $G\alpha_q$, suppressed the responses mediated by G_q-coupled receptors. (b) The suppression was receptor-type specific. Responses mediated by m1 muscarinic and substance P receptors were clearly suppressed, whereas responses by m3 muscarinic receptor were not. (c) The N terminus region of RGS8 distinct from the RGS domain is critical for suppression because RGS8S, a splice variant of RGS8 at the N terminus, did not show clear suppression. Taken together, we speculated that RGS8 inhibits $G\alpha_q$–mediated responses in a receptor-type specific manner and that the N terminus is involved in this suppression (Saitoh *et al.*, 2002).

Quantitative Aspect

When we analyzed the function of RGS8, we compared the properties of two groups, with and without RGS8. This analysis is straightforward. If there is a certain difference, it reflects the effect of RGS8 (Fig. 7).

In contrast, when we compared the effects of RGS8 and RGS8S, the story is not so simple. Actually we misunderstood at first that RGS8S has little effect in suppressing G_q-mediated responses by injecting the same amount of cRNA with RGS8. However, we later noticed that RGS8S also shows inhibitory effects when nondiluted concentrated cRNA was used for injection. As RGS8S is a splice variant at the N terminus, the sequence around the first methionine differs, resulting in the difference of translation efficiency. Therefore, it is necessary to take the protein expression level into account. We compared the amplitude of responses and the protein expression level in parallel. By injecting cRNA of various dilutions, we could obtain RGS8 and RGS8S groups of similar expression levels and compared the response amplitudes. We concluded that the apparent difference of RGS8 and RGS8S groups is not only due to the difference of the

(B) Oocytes injected with RGS8 or RGS8S cRNA (from part A) were frozen after electrophysiological analysis, and the whole protein was extracted. The amount of RGS proteins was compared by Western blotting using the RGS8 antibody (*top*). Lanes correspond to noninjected control (C), 1/3, 1/10, and 1/30 dilutions of RGS8, and 1/3 and 1/10 dilutions of RGS8S, from left to right. The intensities of the bands quantified by Densitograph (ATTO, Japan) are 3841, 1345, 96, 2048, and 442 from left to right. The staining pattern with CBB of the whole protein extracts is also shown (*bottom*). Reprinted with permission from Saitoh *et al.* (2002).

expression level, but is also due to a qualitative difference (Saitoh *et al.*, 2002).

Conclusion

This article described detailed methods of Gα binding and electrophysiological recording for RGS8. These procedures are probably applicable to the characterization of other RGS proteins.

Acknowledgments

We thank Dr. T. Asano for helpful discussions. This work was supported by research grants from the Ministry of Education, Culture, Sports, Science, and Technology of Japan (to O.S. and Y.K.). O.S. and Y.K. are also supported by Core Research for Evolutional Science and Technology of the Japan Science and Technology Corporation.

References

Asano, T., Kamiya, N., Morishita, R., and Kato, K. (1988). Immunoassay for the βγ subunits of GTP-binding proteins and their regional distribution in bovine brain. *J. Biochem.* **103,** 950–953.

Berman, D. M., Wilkie, T. M., and Gilman, A. G. (1996). GAIP and RGS4 are GTPase-activating proteins for the Gi subfamily of G protein alpha subunits. *Cell* **86,** 445–452.

Berman, D. M., and Gilman, A. G. (1998). Mammalian RGS proteins: Barbarians at the gate. *J. Biol. Chem.* **273,** 1269–1272.

Dohlman, H. G., and Thorner, J. (1997). RGS proteins and signaling by heterotrimeric G proteins. *J. Biol. Chem.* **272,** 3871–3874.

Doupnik, C. A., Davidson, N., Lester, H. A., and Kofuji, P. (1997). RGS proteins reconstitute the rapid gating kinetics of Gbetagamma-activated inwardly rectifying K$^+$ channels. *Proc. Natl. Acad. Sci. USA* **94,** 10461–10466.

Druey, K. M., Blumer, K. J., Kang, V. H., and Kehrl, J. H. (1996). Inhibition of G-protein-mediated MAP kinase activation by a new mammalian gene family. *Nature* **379,** 742–746.

Fukuhara, S., Chikumi, H., and Gutkind, J. S. (2000). Leukemia-associated Rho guanine nucleotide exchange factor (LARG) links heterotrimeric G proteins of the G12 family to Rho. *FEBS Lett.* **485,** 183–188.

Fukuhara, S., Murga, C., Zohar, M., Igishi, T., and Gutkind, J. S. (1999). A novel PDZ domain containing guanine nucleotide exchange factor links heterotrimeric G proteins to Rho. *J. Biol. Chem.* **274,** 5868–5879.

Heximer, S. P., Watson, N., Linder, M. E., Blumer, K. J., and Hepler, J. R. (1997). RGS2/G0S8 is a selective inhibitor of Gαq function. *Proc. Natl. Acad. Sci. USA* **94,** 14389–14393.

Hollinger, S., and Hepler, J. R. (2002). Cellular regulation of RGS proteins: Modulators and integrators of G protein signaling. *Pharm. Rev.* **54,** 527–559.

Hunt, T. W., Fields, T. A., Casey, P. J., and Peralta, E. G. (1996). RGS10 is a selective activator of G alpha i GTPase activity. *Nature* **383,** 175–177.

Koelle, M. R., and Horvitz, H. R. (1996). EGL-10 regulates G protein signaling in the *C. elegans* nervous system and shares a conserved domain with many mammalian proteins. *Cell* **84**, 115–125.

Kozasa, T., Jiang, X., Hart, M. J., Sternweis, P. M., Singer, W. D., Gilman, A. G., Bollag, G., and Sternweis, P. C. (1998). p115RhoGEF, a GTPase activating protein for $G\alpha12$ and $G\alpha13$. *Science* **280**, 2109–2111.

Kubo, Y., Miyashita, T., and Murata, Y. (1998). Structural basis for a Ca^{2+}-sensing function of the metabotropic glutamate receptors. *Science* **279**, 1722–1725.

Kubo, Y., and Murata, Y. (2001). Control of rectification and permeation by two distinct sites after the second transmembrane region in Kir2.1 K^+ channel. *J. Physiol.* **531**, 645–660.

Kubo, Y., Reuveny, E., Slesinger, P. A., Jan, Y. N., and Jan, L. Y. (1993). Primary structure and functional expression of a rat G-protein-coupled muscarinic potassium channel. *Nature* **364**, 802–806.

Masuho, I., Itoh, M., Itoh, H., and Saitoh, O. (2004). The mechanism of membrane-translocation of regulator of G-protein signaling (RGS) 8 induced by $G\alpha$ expression. *J. Neurochem.* **88**, 161–168.

Niwa, H., Yamamura, K., and Miyazaki, J. (1991). Efficient selection for high-expression transfectants with a novel eukaryotic vector. *Gene* **108**, 193–200.

Saitoh, O., Kubo, Y., Miyatani, Y., Asano, T., and Nakata, H. (1997). RGS8 accelerates G-protein-mediated modulation of K^+ currents. *Nature* **390**, 525–529.

Saitoh, O., Kubo, Y., Odagiri, M., Ichikawa, M., Yamagata, K., and Sekine, T. (1999). RGS7 and RGS8 differentially accelerate G protein-mediated modulation of K^+ currents. *J. Biol. Chem.* **274**, 9899–9904.

Saitoh, O., Masuho, I., Terakawa, I., Nomoto, S., Asano, T., and Kubo, Y. (2001). Regulator of G protein signaling 8 (RGS8) requires its NH_2 terminus for subcellular localization and acute desensitization of G protein-gated K^+ channels. *J. Biol. Chem.* **276**, 5052–5058.

Saitoh, O., Murata, Y., Odagiri, M., Itoh, M., Itoh, H., Misaka, T., and Kubo, Y. (2002). Alternative splicing of RGS8 gene determines inhibitory function of receptor-type-specific Gq signaling. *Proc. Natl. Acad. Sci. USA* **99**, 10138–10143.

Sternweis, P. C., and Robishaw, J. D. (1984). Isolation of two proteins with high affinity for guanine nucleotides from membranes of bovine brain. *J. Biol. Chem.* **259**, 13806–13813.

Tytgat, J., Vereecke, J., and Carmeliet, E. (1994). Reversal of rectification and alteration of selectivity and pharmacology in a mammalian Kv1.1 potassium channel by deletion of domains S1 to S4. *J. Physiol.* **481**, 7–13.

Watson, N., Linder, M. E., Druey, K. M., Kehrl, J. H., and Blumer, K. J. (1996). RGS family members: GTPase-activating proteins for heterotrimeric G-protein alpha-subunits. *Nature* **383**, 172–175.

Zeng, W., Xu, X., Popov, S., Mukhopadhyay, S., Chidiac, P., Swistok, J., Danho, W., Yagaloff, K. A., Fisher, S. L., Ross, E. M., Muallem, S., and Wilkie, T. M. (1998). The N-terminal domain of RGS4 confers receptor-selective inhibition of G protein signaling. *J. Biol. Chem.* **273**, 34687–34690.

Zheng, B., Ma, Y.-C., Ostrom, R. S., Lavoie, C., Gill, G. N., Insel, P. A., Huang, X.-Y., and Farquhar, M. G. (2001). RGS-PX1, a GAP for $G\alpha s$ and sorting nexin in vesicular trafficking. *Science* **294**, 1939–1942.

Subsection C
R7 Subfamily

[10] Biochemical Purification and Functional Analysis of Complexes between the G-Protein Subunit Gβ5 and RGS Proteins

By D. Scott Witherow and Vladlen Z. Slepak

Abstract

Regulator of G-protein signaling (RGS) proteins of the R7 subfamily (RGS6, 7, 9, and 11) contain a unique Gγ-like (GGL) domain that enables their association with the G-protein β subunit Gβ5. The existence of these complexes was demonstrated by their purification from native tissues as well as by reconstitution *in vitro*. According to pulse–chase analysis, Gβ5 and RGS7 monomers undergo rapid proteolytic degradation in cells, whereas the dimer is stable. Studies of the functional role of Gβ5–RGS dimers using GTPase activity, ion channel, and calcium mobilization assays showed that, similarly to other RGS proteins, they can negatively regulate G-protein-mediated signal transduction. Protein–protein interactions involving the Gβ5–RGS7 complex can be studied in cells using fluorescence resonance energy transfer utilizing Gβ5, RGS, and Gα subunits fused to the cyan and yellow versions of green fluorescent protein.

Introduction

Upon their agonist-mediated activation, seven transmembrane spanning receptors act as guanine nucleotide exchange factors (GEF) for heterotrimeric ($\alpha\beta\gamma$) G proteins. The exchange of GDP for GTP on the G protein leads to dissociation of the Gα subunit from the Gβγ dimer, and these two separate entities can modulate cellular effectors. This active state lasts until the GTP is hydrolyzed and the Gα and Gβγ reassociate. Two major mechanisms control the duration of the active state of the G-protein pathway. One is receptor phosphorylation by G-protein receptor kinases (GRKs) and the subsequent binding of arrestin proteins, which uncouples the receptor from the G protein and initiates downregulation of the receptor (Sibley and Lefkowitz, 1985). The second mechanism of quenching G-protein signaling is by the acceleration of GTP hydrolysis on the Gα subunits by GTPase-accelerating proteins (GAPs). Some G-protein effectors, such as PLCβ1 (Berstein *et al.*, 1992; Chidiac and Ross, 1999), cyclic GMP PDE (Arshavsky and Bownds, 1992), and adenylyl cyclase type V (Scholich *et al.*, 1999), have been shown to accelerate the GTPase activity of their cognate Gα subunits. Another class of GAPs is the large group of diverse proteins known as regulator of G-protein signaling (RGS) proteins (reviewed in Berman and Gilman, 1998; Hepler, 2003; Siderovski *et al.*, 1999). A characteristic domain that is found in all RGS proteins (the RGS domain) binds to Gα subunits and catalyzes GTP hydrolysis (Berman *et al.*, 1996a,b; De Vries *et al.*, 1995). Many RGS proteins contain additional structural domains and perform other functions in addition to serving as a GAP. This article focuses on a subfamily of RGS proteins consisting of RGS6, 7, 9, and 11 (R7 family) (Sondek and Siderovski, 2001; Witherow and Slepak, 2003). These RGS proteins have additional DEP (named for its homology to the *Drosophila* protein disheveled, the *C. elegans* protein EGL-10, and pleckstrin) and GGL (G protein gamma-like) domains and are associated with the G-protein β subunit Gβ5.

Gβ5 sets itself apart from the other four Gβs in amino acid sequence, expression pattern, and subcellular localization (Watson *et al.*, 1994). Gβ subunits 1–4 share ~90% homology to each other on the amino acid level; Gβ5, however, is only ~50% identical to the rest of the family. Gβ5 also has a photoreceptor-specific splice variant (Gβ5L), which has an additional 42 amino acids on its N terminus (Watson *et al.*, 1996). Furthermore, whereas Gβ subunits 1–4 are expressed ubiquitously, Gβ5 is predominantly neurospecific. While other Gβ subunits are membrane associated via prenylated Gγ subunits (Higgins and Casey, 1996; Muntz *et al.*, 1992), Gβ5 is distributed almost equally between the cytosol and crude membrane extracts in native tissues (Rose *et al.*, 2000; Watson *et al.*, 1996; Witherow *et al.*, 2000; Zhang *et al.*, 2000). There have also been reports depicting the nuclear localization of Gβ5 (Rojkova *et al.*, 2003; Zhang *et al.*, 2001). These unique properties of Gβ5 suggest that it might be acting differently than "conventional" Gβγ subunits, eliciting further studies of this protein.

Purification of the Gβ5–RGS7 Complex from the Retina

The original purification protocol was developed for the bovine retina, where nearly 100% of Gβ5 (short form) was found to be soluble (Cabrera *et al.*, 1998; Watson *et al.*, 1996). Cytosolic retinal extracts are prepared by the established method used to isolate transducin (Bigay and Chabre, 1994). One hundred dark-adapted bovine retinas are resuspended in 200 ml of an isotonic buffer containing 45% sucrose. After 3 min of vigorous shaking, the sample is centrifuged at 5000g for 10 min. The resulting supernatant fluid is diluted 1:1 with buffer to dilute the sucrose and is subjected to centrifugation at 7500g for 10 min. The supernatant fluid is then cleared once more by centrifugation at 40,000g for 60 min to remove any residual membranes that drastically reduce the binding capacity of the chromatography columns. This final supernatant fluid is used for the chromatographic steps discussed later.

Because Gβ5 is not a highly abundant protein (less than 0.1% of total protein), several consecutive steps of purification are required to achieve homogeneity. The only available method for detection of Gβ5 in the fractions is the use of Western blot with the anti-Gβ5 antibody. An initial screen of the types of chromatography media appropriate for Gβ5 purification revealed that Gβ5 bound not only to anion exchangers, like other Gβ subunits, but also to the cation exchanger SP-Sepharose (Cabrera et al., 1998). This, in itself, was an early indication that the physicochemical properties of the Gβ5 complex, specifically the apparent positive charge required to interact with a cation exchange column, was distinct from traditional Gβγ dimers. Incorporation of the cation exchange step led to the development of the original protocol used for the purification of Gβ5 complexes from the soluble fraction of the retina (Cabrera et al., 1998), as well as for Gβ5–RGS complexes from the rod outer segments (Makino et al., 1999) and brain (Witherow et al., 2000). Because monomeric Gβ5 does not bind to SP-Sepharose, whereas monomeric RGS7 and the Gβ5-RGS7 dimer both do bind (Levay et al., 1999), the ability to bind to cation exchangers can be recommended as a simple test of Gβ5 association with an R7 family RGS protein.

As most proteins, including Gβ1 and other G protein subunits, are negatively charged at neutral pH and are not retained by the cation exchanger SP-Sepharose, this chromatography gave the highest degree of purification (nearly 300-fold) and logically would have been a natural choice for the first step in the purification protocol. However, after this initial purification step, the Gβ5 complex became unstable in solution and could not be visualized after any subsequent chromatographic step. The addition of 0.5% sodium cholate to the lysate stabilized the protein throughout purification. Higher concentrations of detergent prevented Gβ5 from binding to SP-Sepharose. The presence of cholate also prevented it from binding to the anion-exchange Q-Sepharose (Cabrera and Slepak, unpublished observations). To circumvent this problem, the protocol described in detail by Cabrera et al. (1998) first uses fractionation over Q-Sepharose in the absence of detergent. Following elution with a linear NaCl gradient, an approximate 30-fold purification of Gβ5 is obtained, as a majority of the proteins elute at a salt concentration lower than Gβ5

(200 mM). It is also important to note that it is during this stage that Gβ5 is separated from Gβ1, the most prevalent Gβ subunit in the retinal extract, as Gβ1 elutes at a slightly higher salt concentration. Following fractionation over the Q-Sepharose column, the Gβ5-containing fractions are pooled, diluted to reduce the salt concentration, supplemented with sodium cholate to reach a final concentration of 0.5%, and loaded onto an SP-Sepharose column. This stage results in approximately a 50-fold further purification of Gβ5. Any residual Gβ1 is completely removed during this step, as Gβ1-containing complexes do not bind to SP-Sepharose. The final step in the purification process is a gel filtration step using a Superdex 200 column. A final purification of about 10-fold is achieved during this last step; the resulting Gβ5 complex elutes with an apparent molecular mass of 130 kDa. Silver staining of the Gβ5-containing fractions shows two bands, one corresponding to Gβ5 at 39 kDa and the other band at 55 kDa, which upon sequencing was determined to be RGS7. No Gγ subunit is detected in the fractions containing the Gβ5–RGS7 complex (Witherow et al., 2000), and the reasoning for this was not completely understood until the discovery of the GGL domain in a subfamily of RGS proteins (RGS6, 7, 9, and 11) (Levay et al., 1999; Snow et al., 1998).

Similar studies showing comigration of RGS9 and Gβ5L extracted from photoreceptor rod outer segments have also been performed (Makino et al., 1999). Rod outer segment fractions (containing 3 mg rhodopsin) are solubilized in 500 μl of 2% lauryl sucrose in either 20 mM HEPES-KOH (pH 6.0) with 2 mM MgCl$_2$ (for purification using the FPLC cation-exchange column, Mono S) or 50 mM Tris–HCl (pH 7.8) with 2 mM MgCl$_2$ (for purification using the anion-exchange column, Mono Q). The columns are equilibrated with the corresponding buffers containing 0.5% lauryl sucrose and are eluted using a 0–1 M gradient of NaCl at a flow rate of 1 ml/min. The Gβ5L–RGS9 complex was not purified to homogeneity, but Western blots of the chromatography fractions using anti-RGS9 and anti-Gβ5 antibodies showed that these proteins could not be found apart from each other following either Mono S or Mono Q chromatographies. Thus, in both the soluble retinal extract and the photoreceptor rod outer segments, Gβ5 is dimerized to a GGL-containing RGS protein.

Purification of Gβ5–RGS Dimers from the Brain

Following the initial finding that Gβ5 binds to RGS7 (and Gβ5L to RGS9) in the retina, the next question to be answered was whether Gβ5 is bound to RGS proteins in other tissues as well. Because Gβ5 was reported to be expressed in the central nervous system (Watson et al., 1994, 1996) and the localization pattern of the R7 family also suggested neurospecificity

(Gold *et al.*, 1997), brain was the most logical tissue to use for this study. For isolation of Gβ5 from brain, Witherow *et al.* (2000) used the following method. One rat brain (~2.5 g) is homogenized in 10 ml of buffer containing 20 mM Tris–HCl, pH 7.5, 1 mM EDTA, 2 mM β-mercaptoethanol, and 50 mM NaCl (TEBS). The total brain homogenate is centrifuged at 50,000g for 20 min at 4° and the resulting supernatant fluid is used as the cytosolic extract for purification. The supernatant fluid is applied directly to a 5-ml Q-Sepharose anion-exchange column that has been pre-equilibrated with TEBS. The flow rate for the column is 400 μl/min. After loading the sample, the column is washed with 5 column volumes of TEBS before being eluted with 25 ml of a 50–500 mM NaCl linear gradient. Fractions (800 μl) are collected and, together with the unbound material, are analyzed by Western blot using Gβ5 and RGS7 antibodies. Fractions containing Gβ5 and RGS7 are pooled, diluted to 50 mM NaCl with TEB buffer (without NaCl), and concentrated using an Amicon concentration cell with a 10-kDa cutoff filter to approximately 1 ml. The concentrated material is then loaded onto a 2-ml SP-Sepharose cation-exchange column that is pre-equilibrated with 50 mM TEBS + 0.5% sodium cholate at a flow rate of 50 μl/min. Following loading of the sample, the column is washed with at least 5 column volumes of TEBS + 0.5% sodium cholate and eluted in TEB with a 20-ml linear gradient of NaCl from 50 to 500 mM. The input, the unbound material, and the collected fractions (500 μl) are analyzed by Western blot using Gβ5 and RGS7 antibodies. This purification was not done to homogeneity as in the retina, and only the anion- and cation-exchange columns were used to show that Gβ5–RGS7 comigrated through these chromatographic stages and were completely separated from both Gβ1 and Gγ subunits.

Because Gγ subunits are prenylated and are found associated almost exclusively with the plasma membrane (Higgins and Casey, 1996), if Gβ5Gγ dimers existed, they would likely be found in brain membrane fractions. To isolate the membrane-bound Gβ5 complex, rat brain membranes are prepared from a homogenate centrifugation, as described by Witherow *et al.* (2000). The pellet remaining following cytosolic removal is washed twice in ice-cold TEBS buffer and is then resuspended in TEBS buffer containing 1% sodium cholate or 1% Genapol C-100. The suspension is incubated at 4° with constant mixing for 2 h before centrifugation at 100,000g for 1 h. The two different detergents allow for the direct comparison of the effect of an ionic (cholate) and nonionic (Genapol C-100) surfactant on the elution profile and properties of the Gβ5 dimer. The supernatant is diluted 1:2 using TEBS buffer and is loaded onto a 5-ml Q-Sepharose column pre-equilibrated with TEBS + 0.5% sodium cholate or Genapol C-100 at a flow rate of 400 μl/min. After washing with the starting buffer (5 column volumes), proteins are eluted in TEB + 0.5% sodium cholate or Genapol C-100 with a linear gradient of NaCl from 50 to 500 mM in a total volume of 25 ml. Fractions (800 μl) containing Gβ5 and RGS7, as detected by Western blot, are pooled, diluted four-fold in TEB buffer + 0.5% detergent (without NaCl) to ~30 mM NaCl, and concentrated using an Amicon cell. The concentrated sample is then applied to a 2-ml SP-Sepharose column pre-equilibrated with TEBS + 0.5% detergent using a flow rate of 50 μl/min. Following washing the column with at least 5 column volumes of TEBS + 0.5% detergent, the proteins are eluted in TEB + 0.5% detergent with a 10-ml NaCl gradient from 50 to 500 mM and 500-μl fractions are collected. Unbound material and fractions are analyzed by Western blot.

Unlike what was found previously in the retinal extracts, the Gβ5–RGS7 dimer bound to and eluted from the Q-Sepharose column, despite the presence of sodium cholate, which was used to extract the membranes. This allowed for the same consecutive steps of chromatography described earlier to be applied to the membrane extracts. Furthermore, the use of two different detergents did not seem to have an effect on the behavior or composition of the native Gβ5 dimer, as opposed to the recombinant Gβ5–Gγ2 dimer, which dissociates in the presence of sodium cholate or CHAPS (Jones and Garrison, 1999).

To achieve purification to homogeneity of the Gβ5 complex from the membrane extract fraction, the cholate extract of one bovine brain (used instead of rat brain because of its larger size) is carried through the initial steps of Q- and SP-Sepharose by scaling up the chromatography conditions proportional to the amount of starting material (Brandt and Slepak, unpublished observations). Following SP-Sepharose, the fractions are pooled and diluted to 50 mM NaCl and concentrated like in Cabrera *et al.* (1998). However, upon loading the sample on a gel filtration column, the sample could not be recovered, and under several different conditions, final purification of the dimer by gel filtration was not achieved. In contrast, using hydroxylapatite chromatography as a final step, the Gβ5–RGS complex was purified to show only two bands using silver and Coomassie staining. Western blot analysis was used to confirm that the bands correspond to Gβ5 and RGS7.

Immunoaffinity Purification of Gβ5–RGS Dimers

A panel of antibodies to RGS7 was obtained by immunizing rabbits with peptides corresponding to unique sequences in the N and C terminus, as well as a sequence in the RGS domain that is conserved between RGS7 and RGS9. The resulting serum was then affinity purified using an antipeptide column and was tested for its ability to immunoprecipitate the Gβ5–RGS7 dimer (Witherow and Slepak, unpublished observations). Of the panel of antibodies, only one that was raised against a C-terminal sequence (CGKTLTSKRLTSKLVQS) functioned to immunoprecipitate the dimer (Witherow *et al.*, 2000). In brain cytosol and membrane extracts prepared using a variety of detergents (0.1–1% sodium cholate, 0.1–1% Genapol C-100, 0.1–1% Triton X-100, and 60 mM *n*-octyl β-D-glucopyranoside), immunoprecipitation with the anti-RGS7 antibody coprecipitates Gβ5. The RGS7 antibody is premixed with protein A-Sepharose beads (20 μl of a 1:1 slurry) for 1 h at 4°. The beads are allowed to settle for 5 min, the supernatant is removed, and the brain extract (50 μg of either cytosolic or membrane extracts) is incubated with the protein A–antibody beads overnight with constant mixing at 4°. After allowing the beads to sediment, the unbound fraction is saved, and following three washes with TEBS buffer, the beads are eluted with 2× SDS–PAGE loading buffer in the same volume as the unbound material to enable a quantitative comparison of the amount of Gβ5 that coimmunoprecipitated with RGS7. In most experiments, 80–90% of the Gβ5 was pulled down using the anti-RGS7 antibody. The remaining Gβ5 was assumed to be bound to other GGL-RGS proteins, i.e., RGS6, RGS9, and RGS11. This was confirmed using a reciprocal approach by Zhang and Simonds (2000). The anti-Gβ5 antibody is attached covalently to protein A beads using the Immunopure Protein A IgG orientation kit (Pierce Biotechnology, Rockford, IL) and is used to purify the Genapol C-100 extract from mouse brain. Following washes, bound proteins are eluted with a Gβ5 immunogen peptide, and the eluate is analyzed by silver staining. Two prominent bands are seen by silver staining: a band at 39 kDa corresponding to Gβ5 and a band at approximately 55 kDa, which was microsequenced and shown to contain a mixture of RGS6 and RGS7 proteins. Again, no Gγ subunits were purified along with Gβ5. The presence of RGS6 and RGS7 proteins in the complex with Gβ5 was not surprising, as both contain the requisite GGL domain to bind Gβ5. It is possible that smaller amounts of the other GGL RGS proteins (RGS9 and 11) are also present, but are not expressed at levels detected by silver staining and mass spectrometry.

It is interesting to note that in these immunoprecipitation experiments (and in conventional chromatographies as well) there were no other proteins pulled down along with the Gβ5–RGS dimer. Attempts to detect other signaling components associated with the dimer (such as Gα subunits) using both chromatography and immunoprecipitation methods have been unsuccessful. Despite attempting these experiments in both the presence and the absence of aluminum tetrafluoride (AlF$_4^-$), which when bound to Gα proteins causes them to mimic their active state and heightens their affinity to RGS proteins (Berman *et al.*, 1996a; Tesmer *et al.*, 1997), no evidence of a Gα/Gβ5/RGS heterotrimer could be observed (Posner *et al.*, 1999; Snow *et al.*, 1998; Witherow *et al.*, 2000). However, it is known that Gβ5–RGS dimers can act as GTPase-activating proteins toward Gα subunits (Posner *et al.*, 1999; Snow *et al.*, 1998). Thus, the methods described earlier appear to be inadequate to detect the interaction between Gβ5–RGS dimers and Gα proteins, perhaps because they are transient in nature.

Pulse–Chase Analysis of the Stability of Gβ5 and RGS in Transfected Cells

Initial observations made upon studying the function of the Gβ5–RGS dimer in transfected cells were consistent in that levels of either Gβ5 and RGS6 (Snow *et al.*, 1999) or RGS7 (Witherow *et al.*, 2000) were about 10-fold lower when expressed alone than when coexpressed. This phenomenon was reminiscent of G$\beta\gamma$ dimers (Simonds *et al.*, 1991). To see if the increased expression upon coexpression was due to a proteolytic mechanism, as is the case with G$\beta\gamma$ dimers, pulse–chase-labeling experiments were performed to monitor the degradation of the proteins over time (Witherow *et al.*, 2000). One day after transfection with the appropriate plasmid(s), COS7 cells are labeled metabolically with a pulse of 200 μCi of [^{35}S]methionine and [^{35}S]cysteine (NEN Life Sciences)/60-mm dish for 1 h in methionine- and cysteine-free Dulbecco's modified Eagle's medium (DMEM) containing 10% dialyzed fetal bovine serum. After washing the cells twice using phosphate-buffered saline (PBS), the cells are chased with serum-free DMEM at 37°. The cells are harvested at the indicated times by scraping into 500 μl of ice-cold PBS containing 10 mM EDTA and 10 mM phenylmethylsulfonyl fluoride. Following freeze–thawing, the cells are centrifuged for 30 min at 16,000g at 4°. The supernatants are then precleared by incubation with 15 μl of washed protein A-Sepharose beads for 20 min at 4°. Following preclearing, the supernatant is immunoprecipitated using protein A-Sepharose coupled with the affinity-purified anti-RGS7 antibody or anti-Gβ5 antibody (1 μg IgG/20 μl protein A) at 4°. After mixing for 1 h, beads are washed twice with PBS and are then eluted using 2× SDS–PAGE loading buffer. The [^{35}S]methionine/cysteine-labeled proteins are resolved by 12% SDS–PAGE, transferred to nitrocellulose, visualized by autoradiography, and quantified. The results were conclusive in showing that the Gβ5 and RGS7 monomers were degraded rapidly ($t_{1/2} \cong 1.5$ h), whereas the Gβ5-RGS7 dimer had a half-life of over 24 h. This result implies that formation of a heterodimer is necessary for protection from proteolytic degradation in cells.

Fluorescence Resonance Energy Transfer (FRET) Studies of the Gβ5–RGS7 Dimer

While a Gβ5–RGS7–Gα complex could not be seen in immunoprecipitation experiments, other cellular functional data, such as GAP assays, suggested that an interaction exists between the dimer and Gα proteins. Fluorescence resonance energy transfer was used to further study the interaction between the Gβ5–RGS7 dimer and Gα subunits in intact cells (Witherow et al., 2003).

Briefly, the FRET pair of CFP (cyan fluorescent protein; excitation wavelength – 433 nm, emission wavelength – 480 nm) and YFP (yellow fluorescent protein; excitation wavelength – 488 nm, emission wavelength – 528 nm) is used in which the energy emitted from CFP following excitation can be transferred to YFP and lead to YFP fluorescence. The YFP and CFP fusion proteins of both Gβ5 and RGS7 were tested for their functionality and were proven to still be able to dimerize and inhibit calcium mobilization in cells, suggesting that the fluorescent tag does not affect their function drastically (Witherow et al., 2003).

FRET Spectroscopy

To measure FRET accurately using fluorescence spectroscopy, each experiment requires four samples: (1) untransfected cells, (2) cells transfected with only the YFP fusion protein, (3) cells transfected with only the CFP fusion protein, and (4) cells transfected with both YFP and CFP fusion proteins. Between 24 and 48 h after transfection in 10-cm plates, cells are washed twice with PBS, scraped from the plate using 1 ml of PBS, and stored on ice. Cells are transferred to a quartz cuvette and initially diluted to 2 ml, and fluorescence spectra of the samples are then recorded using a fluorescence spectrophotometer. Each of the four samples is excited at 433 nm (CFP excitation) and scanned from 450 to 550 nm. To enable accurate subtraction of the curves, the level of CFP in cells containing CFP only and cells containing both CFP and YFP is adjusted by dilution until the CFP peak at ~480 nm is identical between the samples. If the CFP levels between the two samples are drastically different (>1.2-fold), the experiment should be aborted because the dilution would significantly alter the number of cells in each sample. This would prevent the subsequent subtractions and corrections to be interpreted properly due to changes in background fluorescence. Following acquisition of CFP emission spectra, both the YFP only and the CFP + YFP samples are scanned from 500 to 550 nm upon excitation at 488 nm (YFP excitation).

Mathematical manipulation of spectral data is done by the following operations.

1. Subtract the CFP trace (excitation at 433 nm) of the untransfected cells from the CFP alone and the CFP + YFP traces. This corrects for autofluorescence of cells upon CFP excitation.

2. Subtract the resulting CFP alone trace from the CFP + YFP trace. This provides a subtracted trace for the CFP + YFP pair depicting fluorescence at 528 nm due to background excitation of YFP and/or energy transfer from CFP to YFP.

3. Normalize the YFP alone and the CFP + YFP trace obtained at excitation at 488 nm (YFP excitation) by determining the constant which is needed to multiply the CFP + YFP trace to cause it to overlap the YFP alone trace.

4. Multiply the corrected CFP + YFP trace (at 433 nm) by the normalization constant and compare to the YFP alone trace. FRET is seen as an increase of fluorescence in the CFP + YFP trace over the background excitation of YFP at 528 nm.

The FRET method just described was initially tested using RGS7YFP and $G\beta5$CFP and, as expected, showed strong FRET between the two proteins. Using $G\beta1$CFP or CFP alone as the FRET donor, no signficant energy transfer to RGS7YFP was observed. To be able to determine the degree of interaction between RGS7 and $G\alpha_q$ and $G\alpha_o$, the $G\alpha$CFP constructs were made so that the CFP fluorophore was engineered at the same site in the coding sequence of both proteins. For each $G\alpha$ subunit, two constructs were made with CFP placed on either the N or the C terminus. Thus, assuming that both $G\alpha$ subunits bind to RGS7 at the same binding site, this enables the CFP tethered to the $G\alpha$ proteins to be the same distance from the YFP of RGS7 when interacting. Because of this, the intensity of the FRET signal can be used as an indicator of the strength of interaction between the two $G\alpha$ subunits. When the interaction between RGS7YFP and CFP constructs of both $G\alpha_o$ and $G\alpha_q$ was tested, a definite interaction between RGS7 and $G\alpha_q$ was observed. However, there was little if any energy transfer between $G\alpha_o$ and RGS7. Thus, according to FRET spectroscopy, RGS7 interacts with $G\alpha_q$ much stronger than $G\alpha_o$ (Witherow *et al.*, 2003).

FRET Microscopy

All imaging is performed using a fluorescence imaging workstation consisting of a Nikon Eclipse TE2000-U microscope equipped with a 60× oil objective lens, a cooled charge-coupled device CoolSnap HQ camera (Photometrics), a z-step motor, dual filter wheels, and a Xenon 175-W light source, all controlled by MetaMorph 5.0r2 software (Universal Imaging Corporation). Filter sets used were YFP (excitation, 500/20 nm; emission,

535/30 nm) and CFP (excitation, 436/20 nm; emission, 480/40 nm), along with an 86004BS dichroic mirror (all from Chroma Technology Corporation).

To assay for FRET using the method of acceptor photobleaching (Miyawaki and Tsien, 2000; Siegel et al., 2000), transfected cells of low-to-moderate expression of both YFP and CFP fusion proteins are selected. Because GFP (and its variants) is known to dimerize at higher concentrations (Zacharias et al., 2002), controls are performed with proteins known not to interact and also with CFP and YFP alone to test fluorescence levels at which nonspecific interactions do not occur. Once this level is determined, cells well beneath the threshold for nonspecific interactions are chosen and imaged with the CFP filter using an exposure time to produce significant, but nonsaturating fluorescence of each representative pixel within the cell. For higher expressing cells, 1×1 binning is used; for lower expressing cells, binning is increased to 2×2. Next, the cell is imaged for YFP, again making sure to obtain a nonsaturating level of fluorescence. Following YFP imaging, the field is then photobleached using constant YFP excitation light until the YFP signal is undetectable using the identical conditions as used to visualize YFP before bleaching (generally 3–4 min). Immediately following photobleaching, a CFP image of the same field is captured using identical parameters used to obtain the CFP image before photobleaching.

Following acquisition of the images, data are interpreted by comparing the CFP levels before and after YFP photobleaching. FRET is identified as an increase in CFP fluorescence following photobleaching due to the inability of the photobleached YFP to accept energy from CFP emission. Simply, the CFP image obtained prior to photobleaching is subtracted from the CFP image following photobleaching. Because the amount of energy transferred from CFP to YFP represents a small portion of the overall CFP energy, the change in intensity is quite subtle. Thus, for representing the subtracted image, the scale on which the subtracted image is displayed has to be adjusted to 20% of the original scale. Thus, if the original CFP image is plotted on a scale of 0–3000 units of fluorescence, the subtracted image is plotted on a scale of 0–600 to allow for amplification and visualization of the signal.

Upon transfection of CHO cells, distinct FRET between RGS7 and $G\alpha_q$ was observed (Witherow et al., 2003). FRET was seen using the same $G\alpha_q$ constructs as used in the spectroscopy experiments. However, the $G\alpha_q$–CFP construct was localized throughout the cell and not primarily on the membrane as is assumed for $G\alpha$ subunits due to their palmitoylation. The fact that the CFP moiety affected the localization of the protein was not completely surprising due to its large size. To ascertain whether the interaction with RGS7 would still occur at the plasma membrane, a $G\alpha_q$–CFP construct that had the CFP engineered in the $\alpha B/\alpha C$ loop of the helical domain of $G\alpha_q$ was used (Hughes *et al.*, 2001). Using this construct, it was shown that the RGS7–$G\alpha_q$ interaction indeed occurs on the membrane. However, no FRET could be observed between RGS7 and $G\alpha_o$ using the photobleaching method.

The combination of two FRET methods (spectroscopy and microscopy) is beneficial for many reasons. One important reason is that spectroscopy allows FRET to be visualized in a group of cells, i.e., cells that are untransfected, transfected with only YFP or CFP, and cells expressing different levels of each fluorophore are measured together. This allows for an unbiased representation of the total FRET signal following the appropriate corrections. At the same time, high background and nonspecific effects could be observed because of aggregation and other artifacts of the overexpression of proteins in transfected cells. Microscopy can be used to limit the study to cells that fit certain parameters, such as healthy looking ones expressing a predefined level of each fluorophore. However, while this approach has advantages, it does lead to a bias toward a particular cell type and does not allow the cells to be analyzed as a group. Thus, combination of the two methods can be recommended for the correct FRET analysis.

Conclusions

Protein purification demonstrated that the interaction between $G\beta 5$ and R7 family RGS proteins occurs *in vivo*. Pulse–chase experiments showed that unassociated $G\beta 5$ and RGS7 proteins underwent rapid proteolytic degradation in cells, likely providing the mechanism maintaining the 1:1 stoichiometry of $G\beta 5$ and RGS subunits in cells. Although $G\beta 5$–RGS7 has been shown to have GAP activity and regulate G-protein-mediated signaling, pull-down assays failed to detect their binding interaction with $G\alpha$ subunits. In contrast, interaction between $G\beta 5$–RGS7 and $G\alpha_q$ has been shown using FRET in a cell-based assay, indicating that this approach is applicable to future studies of $G\beta 5$–RGS interactions with $G\alpha$ subunits.

References

Arshavsky, V., and Bownds, M. D. (1992). Regulation of deactivation of photoreceptor G protein by its target enzyme and cGMP. *Nature* **357**, 416–417.
Berman, D. M., and Gilman, A. G. (1998). Mammalian RGS proteins: Barbarians at the gate. *J. Biol. Chem.* **273**, 1269–1272.
Berman, D. M., Kozasa, T., and Gilman, A. G. (1996a). The GTPase-activating protein RGS4 stabilizes the transition state for nucleotide hydrolysis. *J. Biol. Chem.* **271**, 27209–27212.

Berman, D. M., Wilkie, T. M., and Gilman, A. G. (1996b). GAIP and RGS4 are GTPase-activating proteins for the Gi subfamily of G protein alpha subunits. *Cell* **86,** 445–452.

Berstein, G., Blank, J. L., Jhon, D. Y., Exton, J. H., Rhee, S. G., and Ross, E. M. (1992). Phospholipase C-beta 1 is a GTPase-activating protein for Gq/11, its physiologic regulator. *Cell* **70,** 411–418.

Bigay, J., and Chabre, M. (1994). Purification of transducin. *Methods Enzymol.* **237,** 139–146.

Cabrera, J. L., de Freitas, F., Satpaev, D. K., and Slepak, V. Z. (1998). Identification of the Gbeta5-RGS7 protein complex in the retina. *Biochem. Biophys. Res. Commun.* **249,** 898–902.

Chidiac, P., and Ross, E. M. (1999). Phospholipase C-beta1 directly accelerates GTP hydrolysis by Galphaq and acceleration is inhibited by Gbeta gamma subunits. *J. Biol. Chem.* **274,** 19639–19643.

De Vries, L., Mousli, M., Wurmser, A., and Farquhar, M. G. (1995). GAIP, a protein that specifically interacts with the trimeric G protein G alpha i3, is a member of a protein family with a highly conserved core domain. *Proc. Natl. Acad. Sci. USA* **92,** 11916–11920.

Gold, S. J., Ni, Y. G., Dohlman, H. G., and Nestler, E. J. (1997). Regulators of G-protein signaling (RGS) proteins: Region-specific expression of nine subtypes in rat brain. *J. Neurosci.* **17,** 8024–8037.

Hepler, J. R. (2003). RGS protein and G protein interactions: A little help from their friends. *Mol. Pharmacol.* **64,** 547–549.

Higgins, J. B., and Casey, P. J. (1996). The role of prenylation in G-protein assembly and function. *Cell Signal.* **8,** 433–437.

Hughes, T. E., Zhang, H., Logothetis, D. E., and Berlot, C. H. (2001). Visualization of a functional Galpha q-green fluorescent protein fusion in living cells: Association with the plasma membrane is disrupted by mutational activation and by elimination of palmitoylation sites, but not by activation mediated by receptors or AlF$_4$. *J. Biol. Chem.* **276,** 4227–4235.

Jones, M. B., and Garrison, J. C. (1999). Instability of the G-protein beta5 subunit in detergent. *Anal. Biochem.* **268,** 126–133.

Levay, K., Cabrera, J. L., Satpaev, D. K., and Slepak, V. Z. (1999). Gbeta5 prevents the RGS7-Galphao interaction through binding to a distinct Ggamma-like domain found in RGS7 and other RGS proteins. *Proc. Natl. Acad. Sci. USA* **96,** 2503–2507.

Makino, E. R., Handy, J. W., Li, T., and Arshavsky, V. Y. (1999). The GTPase activating factor for transducin in rod photoreceptors is the complex between RGS9 and type 5 G protein beta subunit. *Proc. Natl. Acad. Sci. USA* **96,** 1947–1952.

Miyawaki, A., and Tsien, R. Y. (2000). Monitoring protein conformations and interactions by fluorescence resonance energy transfer between mutants of green fluorescent protein. *Methods Enzymol.* **327,** 472–500.

Muntz, K. H., Sternweis, P. C., Gilman, A. G., and Mumby, S. M. (1992). Influence of gamma subunit prenylation on association of guanine nucleotide-binding regulatory proteins with membranes. *Mol. Biol. Cell* **3,** 49–61.

Posner, B. A., Gilman, A. G., and Harris, B. A. (1999). Regulators of G protein signaling 6 and 7: Purification of complexes with Gbeta5 and assessment of their effects on G protein-mediated signaling pathways. *J. Biol. Chem.* **274,** 31087–31093.

Rojkova, A. M., Woodard, G. E., Huang, T. C., Combs, C. A., Zhang, J. H., and Simonds, W. F. (2003). Ggamma subunit-selective G protein beta 5 mutant defines regulators of G protein signaling protein binding requirement for nuclear localization. *J. Biol. Chem.* **278,** 12507–12512.

Rose, J. J., Taylor, J. B., Shi, J., Cockett, M. I., Jones, P. G., and Hepler, J. R. (2000). RGS7 is palmitoylated and exists as biochemically distinct forms. *J. Neurochem.* **75,** 2103–2112.

Scholich, K., Mullenix, J. B., Wittpoth, C., Poppleton, H. M., Pierre, S. C., Lindorfer, M. A., Garrison, J. C., and Patel, T. B. (1999). Facilitation of signal onset and termination by adenylyl cyclase. *Science* **283**, 1328–1331.

Sibley, D. R., and Lefkowitz, R. J. (1985). Molecular mechanisms of receptor desensitization using the beta-adrenergic receptor-coupled adenylate cyclase system as a model. *Nature* **317**, 124–129.

Siderovski, D. P., Strockbine, B., and Behe, C. I. (1999). Whither goest the RGS proteins? *Crit. Rev. Biochem. Mol. Biol.* **34**, 215–251.

Siegel, R. M., Chan, F. K., Zacharias, D. A., Swofford, R., Holmes, K. L., Tsien, R. Y., and Lenardo, M. J. (2000). Measurement of molecular interactions in living cells by fluorescence resonance energy transfer between variants of the green fluorescent protein. *Sci. STKE* **2000**, PL1.

Simonds, W. F., Butrynski, J. E., Gautam, N., Unson, C. G., and Spiegel, A. M. (1991). G-protein beta gamma dimers: Membrane targeting requires subunit coexpression and intact gamma C-A-A-X domain. *J. Biol. Chem.* **266**, 5363–5366.

Snow, B. E., Betts, L., Mangion, J., Sondek, J., and Siderovski, D. P. (1999). Fidelity of G protein beta-subunit association by the G protein gamma-subunit-like domains of RGS6, RGS7, and RGS11. *Proc. Natl. Acad. Sci. USA* **96**, 6489–6494.

Snow, B. E., Krumins, A. M., Brothers, G. M., Lee, S. F., Wall, M. A., Chung, S., Mangion, J., Arya, S., Gilman, A. G., and Siderovski, D. P. (1998). A G protein gamma subunit-like domain shared between RGS11 and other RGS proteins specifies binding to Gbeta5 subunits. *Proc. Natl. Acad. Sci. USA* **95**, 13307–13312.

Sondek, J., and Siderovski, D. P. (2001). Ggamma-like (GGL) domains: New frontiers in G-protein signaling and beta-propeller scaffolding. *Biochem. Pharmacol.* **61**, 1329–1337.

Tesmer, J. J., Berman, D. M., Gilman, A. G., and Sprang, S. R. (1997). Structure of RGS4 bound to AlF4–activated G(i alpha1): Stabilization of the transition state for GTP hydrolysis. *Cell* **89**, 251–261.

Watson, A. J., Aragay, A. M., Slepak, V. Z., and Simon, M. I. (1996). A novel form of the G protein beta subunit Gbeta5 is specifically expressed in the vertebrate retina. *J. Biol. Chem.* **271**, 28154–28160.

Watson, A. J., Katz, A., and Simon, M. I. (1994). A fifth member of the mammalian G-protein beta-subunit family: Expression in brain and activation of the beta 2 isotype of phospholipase C. *J. Biol. Chem.* **269**, 22150–22156.

Witherow, D. S., and Slepak, V. Z. (2003). A novel kind of G protein heterodimer: The Gbeta5-RGS complex. *Receptors Channels* **9**, 205–212.

Witherow, D. S., Tovey, S. C., Wang, Q., Willars, G. B., and Slepak, V. Z. (2003). G beta 5.RGS7 inhibits G alpha q-mediated signaling via a direct protein-protein interaction. *J. Biol. Chem.* **278**, 21307–21313.

Witherow, D. S., Wang, Q., Levay, K., Cabrera, J. L., Chen, J., Willars, G. B., and Slepak, V. Z. (2000). Complexes of the G protein subunit Gbeta 5 with the regulators of G protein signaling RGS7 and RGS9: Characterization in native tissues and in transfected cells. *J. Biol. Chem.* **275**, 24872–24880.

Zacharias, D. A., Violin, J. D., Newton, A. C., and Tsien, R. Y. (2002). Partitioning of lipid-modified monomeric GFPs into membrane microdomains of live cells. *Science* **296**, 913–914.

Zhang, J. H., Barr, V. A., Mo, Y., Rojkova, A. M., Liu, S., and Simonds, W. F. (2001). Nuclear localization of G protein beta 5 and regulator of G protein signaling 7 in neurons and brain. *J. Biol. Chem.* **276**, 10284–10289.

Zhang, J. H., Lai, Z., and Simonds, W. F. (2000). Differential expression of the G protein beta(5) gene: Analysis of mouse brain, peripheral tissues, and cultured cell lines. *J. Neurochem.* **75**, 393–403.

[11]   Purification and *In Vitro* Functional Analysis of R7 Subfamily RGS Proteins in Complex with Gβ5

By SHELLEY B. HOOKS and T. KENDALL HARDEN

Abstract

The study of purified regulator of G-protein signaling (RGS) proteins in steady-state GTPase assays using reconstituted proteoliposomes is a powerful approach to characterizing the RGS protein-mediated acceleration of intrinsic Gα subunit GTPase activity in the context of various G-protein and G-protein-coupled receptor (GPCR) combinations. This approach has been applied successfully to the R7 subfamily of RGS proteins, RGS6, -7, -9, and -11, which form heterodimers with Gβ5 subunits via the G-protein γ-like domain of R7 proteins. This article describes the purification of heterodimers from Sf9 insect cells following the expression of recombinant R7 protein and histidine-tagged Gβ5 using affinity and ion-exchange chromatography. The ability of the heterodimers to accelerate the intrinsic GTPase activity of Gα subunits was assessed in steady-state GTPase assays performed on proteoliposomes consisting of phospholipids, purified G proteins, and purified GPCRs.

Introduction

R7-subfamily regulator of G-protein signaling (RGS) proteins are unique multidomain proteins, containing a conserved RGS box, a DEP (**d**isheveled, **E**GL-10, **p**leckstrin) domain implicated in recruitment to the plasma membrane (Martemyanov *et al.*, 2003), and a novel G-protein γ-like (GGL) domain homologous to the Gγ subunit of heterotrimeric G proteins. The GGL domain, found in the mammalian proteins RGS6, 7, 9, and 11, confers high-affinity binding to Gβ5 subunits, but not Gβ1–4 (Snow *et al.*, 1998, 1999). Heterodimeric association with Gβ5 is necessary for the stability and biological activity of R7 proteins (Chen *et al.*, 2003; Zhang *et al.*, 2000). *In vitro* measurements of single turnover (Posner *et al.*, 1999; Snow *et al.*, 1998) and steady-state (Hooks *et al.*, 2003) GTPase activity of purified Gα subunits suggest that R7 family members are Gα$_{i/o}$ selective GTPase-activating proteins (GAPs). While the physiologic functions of RGS6, 7, and 11 are unknown, a clear role for Gβ5/RGS9 dimers has been described in regulation of the phototransduction response. The heterodimer containing retinal-specific splice variants of RGS9 and Gβ5 (Gβ5L/RGS9-1) is the physiologic GAP for Gαt and therefore is intimately involved in recovery of the signaling response to light exposure in the retina (Makino *et al.*, 1999). Analogous R7 proteins in *Caenorhabditis elegans*, EAT-16 and EGL-10, form heterodimeric complexes with GPB-2, which is the *C. elegans* ortholog of Gβ5, and subserve important roles in a neuronal pathway controlling nematode motility and egg laying (Patikoglou and Koelle, 2002).

The biochemical activities and selectivities of RGS proteins have been difficult to establish *in vivo*; however, analysis of the activities of purified proteins overcomes the complexities of intact cell systems. For example, GAP activity of purified RGS proteins may be measured in *in vitro* steady-state GTPase assays with phospholipid vesicles reconstituted with G-protein-coupled receptors (GPCR) and heterotrimeric G proteins (Parker *et al.*, 1991). This article describes in detail methods for the purification of recombinant R7 family RGS proteins as dimers with their obligate binding partner Gβ5 using the baculovirus expression system in insect cells. It also describes methods to assay the effects of Gβ5/R7 dimers on the steady-state GTPase activity of purified G proteins reconstituted with GPCRs in proteoliposomes.

Protein Purification

Several RGS proteins have been purified utilizing the affinity chromatography of epitope-tagged recombinant proteins (e.g., Snow *et al.*, 1998). This approach was applied to the purification of Gβ5/R7 heterodimers (Fig. 1). Gβ5 containing a hexa histidine tag at the amino terminus was coexpressed with each of the mammalian R7 family RGS proteins in insect cells, and the resulting soluble heterodimer was purified by a combination of affinity (via nickel-agarose) and ion-exchange (via S- or Q-Sepharose) chromatography. Yield was approximately 1 mg of Gβ5/R7 dimer per 4 liters of Sf9 cell culture, except for Gβ5/RGS11, the purification of which yielded approximately 250 μg per 4 liters of Sf9 cell culture.

Virus Preparation

Baculoviruses for Gβ5, RGS6, RGS7, RGS9, and RGS11 are prepared in pFastBac plasmids (GibcoBRL) and amplified according to the manufacturer's instructions and as described previously (Posner *et al.*, 1999; Siderovski *et al.*, 2002; Snow *et al.*, 1998). The titer of working stocks of the virus is approximately 7×10^7 pfu/ml, and these stocks are stored at 4° in the dark and used within 6 months. P2 stocks are stored up to a year at 4° for reamplification of working stocks and also are maintained in aliquots at −80° for long-term storage (Siderovski *et al.*, 2002).

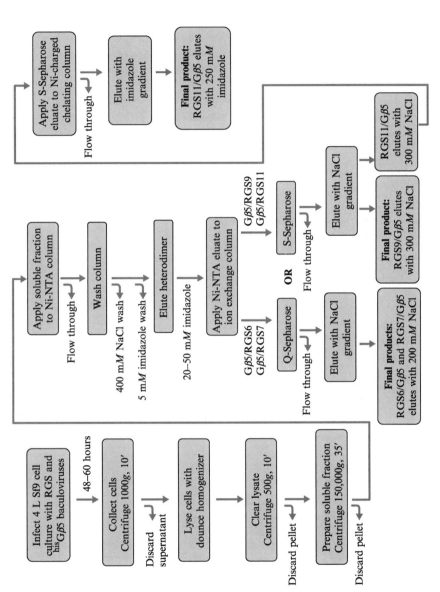

FIG. 1. Schematic of the methods used to purify Gβ5/R7 heterodimers.

Immunodetection

R7 RGS and Gβ5 proteins are tracked through the purification process by immunoblots with specific antibodies (Fig. 2A). Rabbit polyclonal antibodies raised against R7 family members and Gβ5 detect the Sf9 cell-expressed proteins under standard western blotting conditions using goat antirabbit HRP-conjugated secondary antibodies (Pierce) and enhanced chemiluminescent substrate (Super Signal, Pierce) [RGS antisera provided by A. Krumins and A. Gilman, University of Texas Southwestern Medical Center; Gβ5 antiserum provided by W. Simonds, NIH (antisera designation and appropriate dilution for immunoblotting: RGS6: U-1895, 1:500;

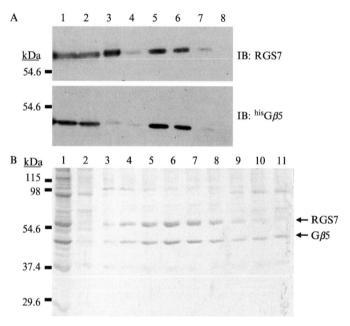

FIG. 2. Gβ5/RGS7 purification. (A) The presence of Gβ5 and RGS7 proteins during Ni-NTA purification of the heterodimer is determined by Western blotting. SDS–PAGE-resolved proteins were transferred to nitrocellulose and immunoblotted with a hexa-histidine monoclonal antibody for the detection of hisGβ5 and with the U-1480 polyclonal antibody for the detection of RGS7. Lane 1, cleared lysate; lane 2, soluble/cytosolic fraction (Ni-NTA column load); lane 3, particulate/membrane fraction; lane 4, Ni-NTA column flow through; lane 5, 20 mM imidazole eluate; lane 6, 40 mM imidazole eluate; lane 7, 120 mM imidazole eluate; and lane 8, 250 mM imidazole eluate. (B) The load and eluting fractions of the Q-Sepharose purification of Gβ5/RGS7 are shown on a Coomassie-stained gel. Lane 1, column load (Ni-NTA, 40 mM imidazole elution); lane 2, column flowthrough; and lanes 3–11, fractions 21–29.

RGS7: U-1480, 1:500; RGS9: U-1784, 1:1000; RGS11: U-1717, 1:1000; Gβ5: SGS-1, 1:2000)]. A commercially available mouse monoclonal poly-histidine antibody (Qiagen #34660, 1:2000) is also used to track tagged Gβ5. Immunodetection is particularly useful in monitoring protein expression, solubility, and binding to Ni-NTA. After Ni-NTA binding and elution, the dimer is usually sufficiently enriched to be tracked by Coomassie staining of proteins resolved by SDS–PAGE (Fig. 2B). RGS (47–55 kDa) and Gβ5 (44 kDa) proteins migrate as predicted by their molecular masses in 12.5% SDS–polyacrylamide gels.

Cell Culture

Sf9 cells (Invitrogen) are maintained in Grace's medium supplemented with 10% fetal bovine serum (FBS) in spinner flasks (Bellco Biotechnology) at 27°. Cell densities are maintained between 0.5 and 3 × 10^6 cells/ml. Cell density doubles in approximately 24 h, and this growth rate should be monitored regularly. On the day before an infection, cells at a density of approximately 3 × 10^6 cells/ml are diluted to 0.8 × 10^6 cells/ml in Grace's medium without FBS, resulting in a final concentration of FBS of approximately 3%. The surfactant Pluronic F-68 (Gibco BRL) is added to the culture medium to a final concentration of 0.1% to protect cells from lysis. Cells are grown overnight in a shaking incubator at 130 rpm and 27°. The infection is begun the following day at a cell density of approximately 1.5 × 10^6 cells/ml.

Cell Infection and Cytosol Harvest

Comments. Lipid modification of R7 family RGS proteins has been reported (Rose *et al.*, 2000), but these signaling proteins are generally considered to exist as soluble proteins. We purified Gβ5/R7 dimers from the cytosolic fraction; however, the majority of R7 immunoreactivity occurred in sedimentable fractions. The nature of this R7 immunoreactive protein is not known, although its resistance to solubilization by detergent treatment of membranes suggests that it reflects unfolded protein.

Procedure. Optimal infection conditions for the generation of soluble Gβ5/RGS7 heterodimers are somewhat variable and should be determined empirically. In our experience, soluble heterodimers with RGS6, 7, and 9 were expressed to high levels at a multiplicity of infection (MOI, number of virus particles per cell) of 1 for the RGS virus and a MOI of 0.5 for the Gβ5 virus and when harvested 48 h postinfection. The expression of soluble Gβ5/RGS11 heterodimers was improved at lower MOI values (0.5 and 0.05 for RGS11 and Gβ5 viruses, respectively) and longer infection times (60 h).

Coinfect 4 liters of Sf9 cells (1.5 × 10^6 cells/ml) with working stock viruses encoding the RGS protein and hexa-histidine-tagged Gβ5. At 48–60 h postinfection, collect cells by centrifugation at 1000g for 15 min in a JA10 rotor at 4° in a J2 Beckman centrifuge. All subsequent steps are carried out at 4°. Resuspend cells in 600 ml of ice-cold buffer A [20 mM KPO$_4$, pH 8, 150 mM NaCl, 2 mM MgCl$_2$, 5 mM β-mercaptoethanol, protease inhibitors (PIs: 500 nM aprotinin, 10 μM leupeptin, 200 μM phenylmethylsulfonyl fluoride, 1 nM pepstatin, 10 μM N-Tosyl-1-phenyl-alanyl Chloromethyl ketone TPCK)] and lyse with 25 strokes of a 40-ml tight dounce homogenizer on ice. (Alternatively, cells are disrupted by nitrogen cavitation with similar results.) Clear the lysate by low-speed centrifugation (500g, 15 min, JA10 rotor in a J2 Beckman centrifuge) to remove intact cells and nuclei. Combine the supernatants and centrifuge at 150,000g for 35 min in an ultracentrifuge (Beckman). Gβ5/R7 heterodimers are purified from the supernatant fraction.

Ni-NTA/Hexa-Histidine Affinity Chromatography

Comments. The first step of purification takes advantage of the hexa-histidine tag of the expressed Gβ5 protein, which binds tightly to nickel ions at slightly basic pH via the imidazole ring of histidine. The soluble fraction of the cell lysate is passed over a column packed with Ni-conjugated beads, and Gβ5, along with its associated R7 protein, is retained on the column. The dimer is then specifically eluted with an excess of free imidazole. This step results in significant purification and also concentrates the sample 20- to 50-fold, which simplifies further purification via fast performance liquid chromatography (FPLC).

Procedure. Perform the following steps in a 4° cold room and chill all buffers on ice prior to use. To prepare the nickel agarose column, pour 10 ml of a 50% slurry of nickel-nitrilotriacetic acid (Ni-NTA) agarose resin (Qiagen, Germany) in a disposable 0.75 × 8-cm column (final bed volume of 5 ml). Wash the column with 20 ml buffer A. Load the soluble protein fraction (600 ml) onto the column over 3 h by gravity flow and collect the flow through. Reapply the flow through to the column by gravity flow and retain the final flow through. Wash the column with three column volumes (15 ml) high salt wash (20 mM KPO$_4$, pH 8, 400 mM NaCl, 2 mM MgCl$_2$, 5 mM β-mercaptoethanol, and protease inhibitors), followed by 5 ml of low salt wash (20 mM KPO$_4$, pH 8, 25 mM NaCl, 2 mM MgCl$_2$, 5 mM β-mercaptoethanol, and protease inhibitors). Wash with 5 ml of low salt wash plus 5 mM imidazole to further remove nonspecifically bound contaminant proteins. Elute the Gβ5/R7 dimer with four washes (5 ml each) of low salt wash plus 20–50 mM imidazole; the dimer elutes in the second and third column volume fractions. A final wash with 10 ml low salt wash plus 300 mM imidazole removes all proteins bound to the Ni-NTA resin (several Sf9 proteins interact with Ni-NTA and elute at imidazole concentrations greater than or equal to that required to elute Gβ5/R7 dimers). Retain all wash and elution fractions for western blot analysis (Fig. 2A).

Ion-Exchange Chromatography

Comments. Gβ5/RGS7 dimer-containing fractions from the Ni-NTA purification are purified further by binding to ion-exchange columns using an automated Amersham Pharmacia Biotech FPLC (Fig. 2B). Gβ5/RGS6 and Gβ5/RGS7 are best retained and separated from contaminants by Q-Sepharose anion-exchange resin, whereas better purification of Gβ5/RGS9 and Gβ5/RGS11 is achieved using a S-Sepharose cation-exchange resin. This is consistent with differences in their pI values, approximately pH 7.5 for RGS6 and −7 and approximately pH 8.5 for RGS9 and −11.

Several protein peaks eluted from the anion-exchange column during purification of Gβ5/RGS6 and Gβ5/RGS7 (Fig. 3A, B), whereas Gβ5/RGS9 was the only major peak to elute from the cation-exchange column during its purification (Fig. 3C). Although a cation-exchange step was also used for the purification of Gβ5/RGS11, the low levels of expression resulted in an increase in the relative level of contaminants following this step (Fig. 3D). Therefore, a third purification step, described in the next section, may be used to increase the purity of Gβ5/RGS11.

Procedure. Wash and equilibrate a 1-ml HiTrap FPLC column (Amersham Pharmacia, Sweden) of either Q-Sepharose (Gβ5/RGS6 and Gβ5/RGS7) or S-Sepharose (Gβ5/RGS9 and Gβ5/RGS11) according to the manufacturer's recommendations. Briefly, connect the column to the FPLC and wash the column with 5–10 column volumes of sterile water to remove the ethanol in which the column is stored. Equilibrate the column with the following series of washes: 10 ml starting buffer, 10 ml elution buffer, 10 ml starting buffer [starting buffer for Q-Sepharose: 50 mM Tris, pH 8, 2 mM dithiothreitol (DTT), PIs; starting buffer for S-Sepharose: 50 mM HEPES, pH 8, 1 mM EDTA, 2 mM DTT, 10% glycerol, PIs; elution buffer for Q or S: starting buffer containing 1 M NaCl]. Dilute the dimer-containing Ni-NTA fractions (typically the second and third 5-ml fractions of imidazole elution) 5:1 in starting buffer to a final volume of 50 ml and a final salt concentration of 5 mM NaCl. Load the diluted sample onto the equilibrated column at a flow rate of 0.5 ml/min, maintaining a column pressure of <0.5 mPa, and retain the flow through. Wash the column with 2 ml starting buffer prior to initiation of the elution gradient. Ultraviolet absorbance levels (A_{280}) should return to baseline

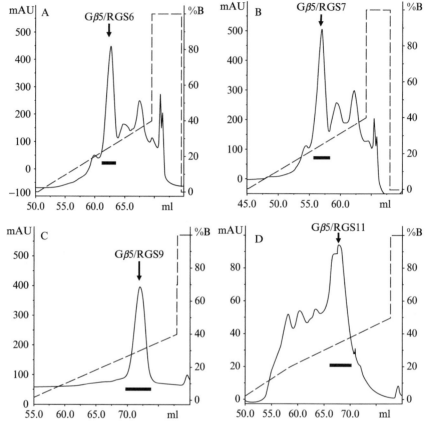

FIG. 3. The elution of Gβ5/RGS6, Gβ5/RGS7, Gβ5/RGS9, and Gβ5/RGS11 from ion-exchange columns. (A) Elution of Gβ5/RGS6 from Q-Sepharose. (B) Elution of Gβ5/RGS7 from Q-Sepharose. (C) Elution of Gβ5/RGS9 from S-Sepharose. (D) Elution of Gβ5/RGS11 from S-Sepharose. The left-hand y axis and solid line indicate milliabsorbance units (A_{280}), and the right-hand y axis and hatched line indicate percentage of elution buffer (1 M NaCl). The bold line indicates which fractions were pooled.

during this wash. Elute bound proteins using a linear gradient from 100% starting buffer, 0% elution buffer to 60% starting buffer, 40% elution buffer (0 to 400 mM NaCl) over 20–30 column volumes. Collect the elution in 0.5-ml volumes. Wash the column with 5 ml of 100% elution buffer.

Gβ5/RGS6 and Gβ5/RGS7 dimers elute at approximately 200 mM NaCl, and Gβ5/RGS9 and Gβ5/RGS11 elute at approximately 300 mM NaCl. Confirm the location of the dimer with Coomassie-stained gels (Fig. 2B) and determine the concentration of purified Gβ5/R7 dimers in the pooled peak fractions by Coomassie staining of an SDS–PAGE gel of the purified product and a standard curve of protein standard (e.g., bovine serum albumin). Concentrate the purified Gβ5/R7 dimer to a final concentration of at least 10 μM using a Centricon centrifugal filter device (Millipore, Bedford, CT) and store in 25- to 50-μl aliquots at $-80°$. Repeated freeze–thaw of the purified sample should be avoided. For optimal activity, the purification process should be completed with less than 48 h between cell lysis and freezing of purified protein, and the protein should be kept on ice or in the cold room at all times.

Gβ5/RGS11 Purification via Chelating FPLC Column

Comments. As described earlier, Gβ5/RGS11 dimers are insufficiently pure for certain applications following Ni-NTA affinity chromatography and cation-exchange chromatography. Therefore, the dimer-containing fractions eluted from the S-Sepharose column are pooled and reloaded onto a chelating Sepharose column charged with nickel ions and eluted with an imidazole gradient using an FPLC.

Procedure. Manually wash and charge a 1-ml HiTrap chelating Sepharose column (Amersham Pharmacia, Sweden) with Ni ions according to the manufacturer's instructions. Briefly, using a disposable syringe with luer fittings, apply 5 ml sterile distilled water to the column followed by 0.5 ml of 0.1 M NiSO$_4$ at a flow rate of approximately 1 ml/min. Connect the column to an automated FPLC and wash with 5 ml water. Equilibrate the column with 10 ml starting buffer (20 mM Tris, pH 8, 300 mM NaCl, PIs), followed by 10 ml elution buffer (20 mM Tris, pH 8, 50 mM NaCl, 500 mM imidazole, PIs), and again 10 ml starting buffer. Load the pooled S-Sepharose fractions (Fig. 3D, bar) at a flow rate of 1 ml/min and elute with a gradient of 0–500 mM imidazole (0% elution buffer to 100% elution buffer) over 20 column volumes (Fig. 4A). Collect the eluate in 0.5-ml volumes and analyze by Coomassie-stained SDS–PAGE gels (Fig. 4B). The Gβ5/RGS11 dimer elutes at approximately 250 mM imidazole. Concentrate and store the final purified product as described earlier.

Vesicle Reconstitution

Comments. The following protocol for vesicle reconstitution uses the M2 muscarinic cholinergic receptor reconstituted with Gα_o and G$\beta_1\gamma_2$ as an example; other purified receptors and G-protein subunits may be reconstituted using the same protocol.

The reconstitution of receptors and G proteins in lipid vesicles was originally developed by Elliot Ross and colleagues and has been discussed extensively (e.g., Asano *et al.*, 1984). Gα and G$\beta\gamma$ subunits and

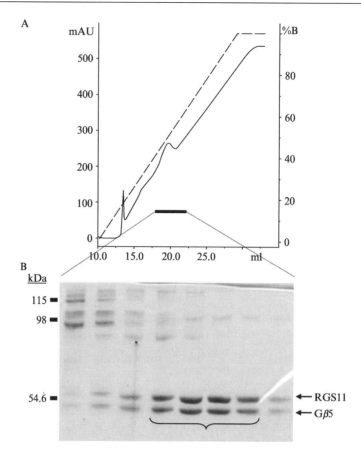

Fig. 4. Purification of Gβ5/RGS11 with a chelating FPLC column. (A) Elution of a protein peak (A_{280}) from a Ni-charged chelating column is shown. Imidazole, used in the elution gradient, also absorbs light at this wavelength and accounts for the linear increase in absorbance that is superimposed on the elution of protein. The left-hand y axis and solid line indicate mAU (A_{280}), and the right-hand y axis and hatched line indicate percentage of elution buffer (1 M NaCl). The bold line indicates fractions analyzed by SDS–PAGE. (B) The purity of the heterodimer eluting from the chelating column is shown on a Coomassie-stained SDS–PAGE gel. The bracket indicates which fractions were pooled prior to concentration. (See color insert.)

G-protein-coupled receptors are purified from Sf9 insect cells using the baculovirus expression system and previously described methodology (Bodor *et al.*, 2003; Kozasa and Gilman, 1995; Parker *et al.*, 1991). Briefly, the membrane fraction of Sf9 cells expressing hexa-histidine-tagged M2 muscarinic cholinergic receptors is isolated, and the receptors are extracted from the membrane fraction using 1% digitonin. The tagged receptor is purified by a two-step nickel-binding protocol. First, the extract is applied to Ni-NTA resin and bound proteins are eluted with 150 mM imidazole. This fraction is then diluted and passed over a FPLC chelating column charged with Ni ions and eluted with an imidazole gradient (Bodor *et al.*, 2003). G-protein subunits are purified after 1% cholate extraction of membranes prepared from Sf9 cells heterologously expressing tagged heterotrimeric G proteins. Gα_o subunits are purified after coexpression with Gβ and hexa-histidine-tagged Gγ subunit. The extract is passed over a Ni-NTA column, and the Gα_o subunit is eluted with aluminum tetrafluoride. Conversely, G$\beta_1\gamma_2$ dimers are purified after coexpression with a histidine-tagged Gα subunit. The extract is passed over a Ni-NTA column, and the G$\beta_1\gamma_2$ subunits are eluted with aluminum tetrafluoride (Kozasa and Gilman, 1995). The G proteins may be purified further using a Q-Sepharose column.

Procedure. Prepare detergent/phospholipid mixed micelles by drying L-α-phosphatidylethanolamine (PE), L-α-phosphatidylserine (PS), and cholesteryl hemisuccinate (CH) and resuspending in detergent buffer (0.4% deoxycholate, 20 mM HEPES, pH 8, 1 mM EDTA, 100 mM NaCl). Brain PS (Avanti Polar Lipids) is stored as a 10-mg/ml solution in chloroform, brain PE (Avanti Sigma,) as a 10-mg/ml chloroform solution, and CH (Sigma) as a 2 mM stock in chloroform. For a 200-μl batch of lipid resuspension, combine 11 μl (110 μg) PE, 7 μl (70 μg) PS, and 4 μl (3.9 μg) CH stocks in a 12 × 75-mm glass test tube, and dry under a gentle stream of argon. To the dried lipids add 200 μl of detergent buffer, pipetting up and down several times. Cover the tube with parafilm and place in a room temperature bath sonicator for 15 min. Vortex briefly and place on ice.

Prepare a G-50 size-exclusion column by packing a 0.75 × 8-cm column with a 3.5-ml bed volume of Sephadex G-50 resin and equilibrate with 30 ml of buffer B (20 mM HEPES, pH 8, 100 mM NaCl, 1 mM EDTA, 2 mM MgCl$_2$). Allow the buffer to completely enter the gel, and seal the column to stop the flow. Combine 50 μl of the detergent/phospholipid-mixed micelles with 150 pmol G$\beta_1\gamma_2$, 50 pmol Gα_o, and 15 pmol M2 muscarinic cholinergic receptor (generally less than 10 μl each, depending on the concentration of purified protein preparation). Add the purified proteins in the order listed, vortex briefly between each addition, and then return to ice. Bring the final volume to 100 μl with buffer B, vortex, and immediately load the complete mixture onto the equilibrated G-50 column, making sure to pipette directly into the center of the surface of the gel. Carefully pipette 200 μl washes of buffer B onto the top of the column without disrupting the surface of the gel bed. Collect the eluate in 200-μl fractions by hand in 0.6-ml Eppendorf tubes. The column preparation, loading, and elution should be performed in a 4° cold room.

The location of the column void volume, which contains the reconstituted vesicles, should be determined empirically by binding of the radiolabeled antagonist to the reconstituted receptor. Assay fractions for the presence of the M2 muscarinic cholinergic receptor by incubating 5 μl of each 200-μl fraction with 20 nM ($\sim$200,000 cpm) of [^{3}H]quinuclidinyl benzilate (NEN, Boston, MA) in a volume of 100 μl (buffer: 20 mM HEPES, pH 8, 3 mM MgCl$_2$, 1 mM EDTA) for 90 min at 30°. Dilute the reaction with 4 ml of cold stop buffer (20 mM Tris, pH 7.5, 150 mM NaCl, 2 mM MgCl$_2$) and filter over GF/F filters (Whatman, England) using a vacuum filtration manifold. Wash the filters twice with 4 ml of cold stop buffer and add filters to scintillant for scintillation counting. For a 3.5-ml G-50 column, the void volume containing the vesicles is typically in the sixth and seventh 200-μl fractions. Pool the peak fractions and use immediately or snap freeze in aliquots for storage at $-80°$.

Steady-State GTPase Assays

Comments. Gα subunits of heterotrimeric G proteins exhibit an intrinsic GTPase activity that may be measured by the liberation of [^{32}P]Pi from [γ-^{32}P]GTP. The rate of this enzymatic activity is determined by measuring GTP hydrolysis as a function of time. Once a linear relationship of product generation versus time has been established, a single time point may be used to determine GTPase rates (15 pmol GTP hydrolyzed in a 15-min assay under linear kinetics = 1 pmol/min). Steady-state GTPase activity of a receptor-regulated Gα subunit reflects multiple cycles of nucleotide exchange and hydrolysis and is influenced by both agonist-promoted activation of receptor-mediated nucleotide exchange activity and RGS protein-promoted stimulation of catalysis by the intrinsic GTPase activity of the Gα subunit (Fig. 5A). Concentration–effect relationships for the activation of receptors by agonists may be determined in the presence of a maximal concentration of RGS protein. Likewise, a concentration–effect relationship for RGS protein may be determined in the presence of a maximum concentration of agonist (Fig. 5B).

Procedure. Steady-state GTPase assays are performed in duplicate in conical 12 × 75-mm polypropylene tubes (Sarstedt) in a 25-μl reaction volume. Thaw aliquots of purified Gβ5/RGS dimer on ice. Radiolabeled [γ-^{32}P]GTP (10 mCi/mL, NEN Life Science Products, BLU-004Z) may be thawed at room temperature. For 50 reactions, prepare 1 ml buffer B and 1 mL 2X GTP buffer [buffer B plus GTP (4 μM GTP + $\sim$25 μCi [γ-^{32}P]GTP)]. If assaying a frozen vesicle preparation, thaw on ice. When assaying G$\alpha_{i/o}$-containing vesicles, 1 μl of 400 μl pooled vesicles (as described earlier) is sufficient for robust GTPase activity. Because of the level

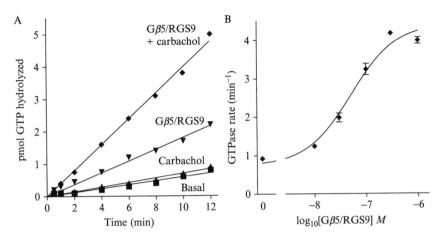

FIG. 5. Steady-state GTPase activity of $G\alpha_o\beta_1\gamma_2$ heterotrimers reconstituted in proteoliposomes with M2 muscarinic receptors. (A) The GTPase activity of vesicles alone or in the presence of 100 μM carbachol, 100 nM Gβ5/RGS9, or both was measured over 12 min. (B) The GTPase rate of vesicles in the presence of 100 μM carbachol and various concentrations of the Gβ5/RGS9 dimer was determined in 15-min assays.

of error associated with pipetting small volumes, dilute the vesicles 1:7.5 in buffer B. Prepare 10X purified RGS protein and agonist stocks; 10 μM stocks should be prepared for assays of maximal R7 family RGS activity, diluting in buffer B if necessary. To assay lower concentrations, dilute the RGS protein in buffer B. In the case of muscarinic receptors, prepare 1 mM carbachol stocks (in water) for a final assay concentration of 100 μM. A stock of 5% activated charcoal (Sigma) in 20 mM NaH_2PO_4 should be prepared and kept on ice. Pre-cool reaction tubes in an ice bath.

Add 7.5 μl of the diluted vesicles to each reaction tube and add 2.5 μl 10X RGS protein and/or 2.5 μl 10X carbachol according to the experimental design. Bring the volume in each tube to 12.5 μl with buffer B. Pipette up and down to mix between each addition and return to the ice bath. Next add 12.5 μl of cold 2X GTP reaction mix (~400,000 cpm/reaction) to each tube and return to ice. Vortex each tube briefly and return to the ice bath. Move tubes simultaneously to a 30° water bath and incubate for the desired length of time, typically 15 min for $G\alpha_i$ family G proteins and 30 min for $G\alpha_q$ family G proteins. The reaction is linear for at least 30 min under these conditions (Fig. 5A). The charcoal solution should be placed on a stir plate to suspend the charcoal evenly in advance of the reaction termination. To terminate the reaction, move the reaction tubes back to the ice bath and add 975 μl of the suspended activated charcoal (final volume

1 ml). Cut off the end of a 1-ml pipette tip to allow easier pipetting of the charcoal suspension. Centrifuge the reaction tubes in a Beckman J6 swinging bucket centrifuge for 10 min at 3000g. Transfer 600 μl of the cleared supernatant to scintillation vials and add scintillation fluid for counting.

Data Preparation and Calculations. A buffer blank, with only 12.5 μl buffer B and 12.5 μl 2X GTP mix, is used to determine the background level of free [^{32}P]Pi in the GTP mix; this value is subtracted from all data points. The GTPase activity of the purified Gβ5/R7 preparation in the absence of vesicles should be measured to determine if background GTP-hydrolyzing activity is present. This activity should be subtracted from RGS protein-promoted activities measured in the presence of G-protein-containing vesicles.

Determine the total radioactivity added to each reaction by directly pipetting 12.5 μl of the 2X GTP reaction mix into scintillation fluid (Scintisafe gel, Fisher Scientific) and counting. Less than 10% of the total GTP should be hydrolyzed during the reaction, and the amount of vesicles per reaction and/or the reaction time should be altered if necessary to maintain initial rate assay conditions. GTPase rates are typically reported in units of picomoles GTP hydrolyzed per picomoles Gα per minute, or min^{-1}. The concentration of Gα protein in the vesicle preparation is measured by [^{35}S]GTPγS binding (1 μM GTPγS, 90 min at 30$°$). Determine total radioactivity per reaction and use this value to calculate the conversion factor to convert cpm values to pmol GTP hydrolyzed (cpm/RXN $\div$ 50 pmol GTP/RXN = cpm/pmol).

Additional Comments. The methods described in this article have been used to purify Gβ5/R7 dimers with both long and short forms of Gβ5. The Gβ5/R7 *C. elegans* homologs GPB-2/EAT-16 and GPB-2/EGL-10 have also been purified by our laboratory using these methods.

Acknowledgments

This work is supported by NIH Grants F32 GM66561 and P01GM065533. The authors thank Gary Waldo for technical advice and expertise.

References

Asano, T., Pedersen, S. E., Scott, C. W., and Ross, E. M. (1984). Reconstitution of catecholamine-stimulated binding of guanosine 5′-O-(3-thiotriphosphate) to the stimulatory GTP-binding protein of adenylate cyclase. *Biochemistry* **23**, 5460–5467.

Bodor, E. T., Waldo, G. L., Hooks, S. B., Corbitt, J., Boyer, J. L., and Harden, T. K. (2003). Purification and functional reconstitution of the human P2Y12 receptor. *Mol. Pharmacol.* **64**, 1210–1216.

Chen, C. K., Eversole-Cire, P., Zhang, H., Mancino, V., Chen, Y. J., He, W., Wensel, T. G., and Simon, M. I. (2003). Instability of GGL domain-containing RGS proteins in mice lacking the G protein β-subunit Gβ5. *Proc. Natl. Acad. Sci. USA* **100,** 6604–6609.

Hooks, S. B., Waldo, G. L., Corbitt, J., Bodor, E. T., Krumins, A. M., and Harden, T. K. (2003). RGS6, RGS7, RGS9, and RGS11 stimulate GTPase activity of Gi family G-proteins with differential selectivity and maximal activity. *J. Biol. Chem.* **278,** 10087–10093.

Kozasa, T., and Gilman, A. G. (1995). Purification of recombinant G proteins from Sf9 cells by hexahistidine tagging of associated subunits: Characterization of alpha 12 and inhibition of adenylyl cyclase by alpha z. *J. Biol. Chem.* **270,** 1734–1741.

Makino, E. R., Handy, J. W., Li, T., and Arshavsky, V. Y. (1999). The GTPase activating factor for transducin in rod photoreceptors is the complex between RGS9 and type 5 G protein β subunit. *Proc. Natl. Acad. Sci. USA* **96,** 1947–1952.

Martemyanov, K. A., Lishko, P. V., Calero, N., Keresztes, G., Sokolov, M., Strissel, K. J., Leskov, I. B., Hopp, J. A., Kolesnikov, A. V., Chen, C. K., Lem, J., Heller, S., Burns, M. E., and Arshavsky, V. Y. (2003). The DEP domain determines subcellular targeting of the GTPase activating protein RGS9 *in vivo*. *J. Neurosci.* **23**(32), 10175–10181.

Parker, E. M., Kameyama, K., Higashijima, T., and Ross, E. M. (1991). Reconstitutively active G protein-coupled receptors purified from baculovirus-infected insect cells. *J. Biol. Chem.* **266,** 519–527.

Patikoglou, G. A., and Koelle, M. R. (2002). An N-terminal region of *Caenorhabditis elegans* RGS proteins EGL-10 and EAT-16 directs inhibition of Gαo versus Gαq signaling. *J. Biol. Chem.* **277,** 47004–47013.

Posner, B. A., Gilman, A. G., and Harris, B. A. (1999). Regulators of G protein signaling 6 and 7: Purification of complexes with Gβ5 and assessment of their effects on G protein-mediated signaling pathways. *J. Biol. Chem.* **274,** 31087–31093.

Rose, J. J., Taylor, J. B., Shi, J., Cockett, M. I., Jones, P. G., and Hepler, J. R. (2000). RGS7 is palmitoylated and exists as biochemically distinct forms. *J. Neurochem.* **75,** 2103–2112.

Siderovski, D. P., Snow, B. E., Chung, S., Brothers, G. M., Sondek, J., and Betts, L. (2002). Assays of complex formation between RGS protein G gamma subunit-like domains and G beta subunits. *Methods Enzymol.* **344,** 702–723.

Snow, B. E., Betts, L., Mangion, J., Sondek, J., and Siderovski, D. P. (1999). Fidelity of G protein β-subunit association by the G protein γ-subunit-like domains of RGS6, RGS7, and RGS11. *Proc. Natl. Acad. Sci. USA* **96,** 6489–6494.

Snow, B. E., Krumins, A. M., Brothers, G. M., Lee, S. F., Wall, M. A., Chung, S., Mangion, J., Arya, S., Gilman, A. G., and Siderovski, D. P. (1998). A G protein γ subunit-like domain shared between RGS11 and other RGS proteins specifies binding to Gβ5 subunits. *Proc. Natl. Acad. Sci. USA* **95,** 13307–13312.

Zhang, J. H., and Simonds, W. F. (2000). Copurification of brain G-protein β5 with RGS6 and RGS7. *J. Neurosci.* **20,** RC59 1–5.

[12] Characterization of R9AP, a Membrane Anchor for the Photoreceptor GTPase-Accelerating Protein, RGS9-1

By Guang Hu and Theodore G. Wensel

Abstract

The proper recovery of photoreceptor light responses requires timely inactivation of the G-protein transducin (G_t) by GTP hydrolysis. It is now well established that the GTPase-accelerating protein (GAP) RGS9-1 plays an important role in determining the recovery kinetics of photoresponses. RGS9-1 has been found to be anchored to photoreceptor disk membranes by a novel photoreceptor protein, R9AP. R9AP has a single transmembrane domain at its C-terminal region. Membrane tethering by R9AP enhances RGS9-1 GAP activity *in vitro* and has been hypothesized to be important for the regulation of RGS9-1 function *in vivo*. In addition, R9AP shows structural similarity to the SNARE complex protein syntaxin and has been shown to be required for the correct targeting and localization of the RGS9-1 protein in photoreceptors. Therefore, R9AP may have additional functions other than that in the phototransduction pathway.

This article presents methods and protocols developed for the functional characterization of R9AP in phototransduction, including the immunoprecipitation of the endogenous protein, the expression and purification of recombinant proteins, the reconstitution of proteoliposomes, and assays for its interaction with RGS9-1 and its effects on RGS9-1 GAP activity. These methods may also be applied to the study of R9AP function in other pathways or other cell types or to the studies of other membrane proteins that are structurally similar to R9AP.

Introduction

In phototransduction, deactivation in a timely fashion of the G-protein α subunit, transducin-α ($G_{t\alpha}$), by GTP hydrolysis is important in the regulation of light responses in vertebrate photoreceptors (Lyubarsky and Pugh, 1996; Pugh, 2000; Sagoo and Lagnado, 1997). A member of the RGS protein family, RGS9-1 accelerates GTP hydrolysis on $G_{t\alpha}$ and plays a central role in determining the inactivation kinetics of phototransduction (Chen *et al.*, 2000; He *et al.*, 1998; Lyubarsky *et al.*, 2001; Makino *et al.*, 1999; Zhang *et al.*, 2003). RGS9-1 is the endogenous GTPase-accelerating protein (GAP) for $G_{t\alpha}$ and is the mammalian RGS protein

METHODS IN ENZYMOLOGY, VOL. 390

whose physiological function has been characterized most thoroughly (reviewed in Cowan *et al.*, 2001). It is a peripheral membrane protein on the photoreceptor outer segment membranes (Cowan *et al.*, 1998) and is tethered to the membranes by a newly identified retinal protein, R9AP (Hu and Wensel, 2002). R9AP was discovered by its interaction with RGS9-1 in detergent extracts from rod outer segment membranes (ROS) and was found to bind directly to the N-terminal region of RGS9-1. It is expressed predominantly in the retina and localizes largely in the photoreceptor outer segment layer. It has a single transmembrane helix in its C-terminal region and serves as a membrane anchor for RGS9-1 (Hu and Wensel, 2002).

Studies have shown that membrane anchoring of RGS9-1 by R9AP enhances its GAP activity toward $G_{t\alpha}$ *in vitro* (Hu *et al.*, 2003), and this membrane anchoring can be regulated by the phosphorylation state of RGS9-1 (Sokal *et al.*, 2002). Therefore, R9AP may play a role in the modulation of phototransduction recovery kinetics. In addition, R9AP also seems to play a role in the sorting of RGS9-1 protein in photoreceptors. Expression of a mutant RGS9-1 that has impaired interaction with R9AP but intact GAP activity resulted in mislocalization of the mutant protein and complete loss of GTPase acceleration in phototransduction in mice.

R9AP orthologs have been identified in human, cattle, mouse, rat, *Xenopus*, zebrafish, and puffer fish (Hu and Wensel, 2002; Hu *et al.*, 2003). Expressed sequence tags of R9AP have been found in cDNA from human fetal eye (BQ187216), human kidney (AW302149), adult mouse retina (BB591662, BI730314), pig (BX665878, BX667653), *Xenopus laevis* brain (CD328477, CD328533, CD303540, CD256685), whole 31–32 stage embryo (CF288302) and lung (BG515592), *Xenopus* (*Silurana*) *tropicalis* (BX667653), zebrafish adult retina (BE015922), and three-spined stickleback fish (CD507448).

Although there are two distinct R9AP-like sequences among *Xenopus* expressed sequence tags (ESTs), there is only one intronless R9AP gene and no close homolog in available human and mouse genome sequences. There are many distant homologs, as the structure of R9AP predicted by computational methods is similar to the syntaxin family of SNARE proteins involved in vesicle trafficking.

This article describes the experimental procedures developed to characterize the function of R9AP in phototransduction, including protocols for isolation of the endogenous protein from retinal extract, expression and purification of recombinant proteins, reconstitution of proteoliposomes, and assays for its interaction with RGS9-1 and its effect on RGS9-1 GAP activity. An interesting feature of R9AP is that it is one of a very few mammalian transmembrane proteins that can be expressed in *Escherichia coli*, purified in detergent, and reconstituted in active form in lipid vesicles. It is hoped that these methods will not only be helpful in further investigations of the function of R9AP in photoreceptors, but in future studies on its potential functions in other organs or cell types as well.

Immunoprecipitation of RGS9-1 and R9AP from ROS Extract

To understand how the function of RGS9-1 is regulated, it is useful to know the proteins with which it interacts. We took a proteomics approach and developed an affinity purification protocol to isolate the RGS9-1 complex efficiently from rod outer segment membranes. The protocol allowed us to extract the majority of the ROS membrane proteins into solution, including the major signaling components in phototransduction, and to obtain a sufficiently large quantity of pure RGS9-1 or R9AP protein complex for sequence analysis and further study.

Buffers and Reagents

GAPN buffer: 100 mM NaCl, 10 mM HEPES, pH 7.4, 2 mM MgCl$_2$, 1 mM dithiothreitol (DTT), solid phenylmethylsulfonyl fluoride (PMSF), ~20 mg/liter). In this and other buffers, PMSF crystals are added at an amount above its solubility in water and maintained in suspension as a continuous source of PMSF, which is poorly soluble and fairly rapidly hydrolyzed once in solution. The buffers are filtered routinely through nitrocellulose filters with 0.2- or 0.45-μm pores before use, and the PMSF is added after filtration

Solubilization buffer: 1% NP40 (Sigma) in GAPN, without DTT n-Octyl-β-D-glucopyranoside (Anatrace)

Antibodies: Rabbit anti-RGS9-1c polyclonal antiserum, anti-RGS9-N polyclonal antiserum, monoclonal anti-RGS9-1 antibody D7, and polyclonal goat anti-R9AP antiserum (Cowan et al., 1998; He et al., 1998, 2000; Hu and Wensel, 2002)

CNBr-activated Sepharose 4B (Amersham Biosciences)

Protein A-Sepharose 4 fast flow (Amersham Biosciences)

Procedures. ROS are isolated from frozen bovine retina by a standard procedure (Papermaster and Dreyer, 1974; Papermaster, 1982) and stored at −80°. Purified ROS are pelleted by centrifugation at 84,000g at 4° and washed with GAPN buffer using the same centrifugation conditions and a rhodopsin concentration of 10–15 μM. The membranes are then dissolved in detergent at a rhodopsin concentration of 60 μM by homogenization in syringes with 18- or 23-gauge needles and incubation at 0–4° for 30 min with gentle agitation to solubilize ROS membrane proteins. Both 40 mM

octyl-glucoside and 1% NP40 in GAPN are found to solubilize RGS9-1 efficiently, and NP40 is chosen in routine preparations due to its lower cost. Insoluble proteins (the major one is tubulin) are removed by centrifugation at 84,000g, and the supernatant (containing the majority of ROS proteins) is collected for immunoprecipitation.

To prepare antibody-coupled Sepharose beads for immunoprecipitation, IgG is first purified from rabbit anti-RGS9-1c antiserum using protein A—Sepharose 4 fast flow affinity chromatography (Harlow and Lane, 1988), or from goat anti-R9AP antiserum following a protocol of Gray *et al.* (1969) and Ibrahimi *et al.* (1980). Purified IgG is then coupled to CNBr-activated Sepharose 4B at a ratio of 10 mg IgG to 1 ml beads according to the manufacturer's instructions (Amersham Biosciences). IgG that is not coupled covalently to the beads is removed by washing the beads alternately using 0.1 M Tris–HCl, pH 8.0, 0.5 M NaCl, and 0.1 M CH$_3$COONa, pH 4.0, 0.5 M NaCl for four to five times. Beads are stored in 0.1 M Tris–HCl, pH 8.0, 0.5 M NaCl at 4° as ~50% slurry and can be stored for up to 1 year (sodium azide can be added to prevent contamination by microorganisms, but is not necessary) without a significant decrease in its efficiency in immunoprecipitation. However, some detachment of IgG from the beads is observed after long–term storage. Before every immunoprecipitation experiment, antibody-coupled beads need to be washed again more stringently with 0.1 M glycine, pH 2.5, two to three times and then equilibrated in ROS solubilization buffer. This step removes loosely attached IgG light chains or any residual detached IgGs from the beads and therefore is critical in reducing IgG contamination in the final immunoprecipitation product. IgG light chains and heavy chains appear on SDS–PAGE as diffuse bands at apparent sizes of ~25 and ~55 kDa, respectively, and interfere with R9AP and RGS9-1 detection on Coomassie-stained gels.

Small-scale RGS9-1 immunoprecipitation from detergent-solubilized ROS membranes is performed by incubating 300 μl solubilized ROS with 40 μl IgG-coupled beads for 2.5 h at 4° with mixing on a shaker. Beads are then separated from the supernatant by a brief centrifugation and are washed three times with the solubilization buffer. Bound proteins are eluted from the beads by 0.1 M glycine at pH 3.0, concentrated by trichloroacetic acid precipitation, and redissolved in the SDS–PAGE sample buffer. Under these conditions, more than 90% of the RGS9-1 (~0.5–1 μg) in the ROS can be immunoprecipitated and recovered in the eluted proteins. To measure the efficiency of immunoprecipitation, eluted proteins are resolved by SDS–PAGE, and RGS9-1 is detected by immunoblotting using the monoclonal antibody D7. Under these conditions, ~90% or more of the RGS9-1 protein can be depleted from the retinal extract and eluted from the beads (see Hu *et al.*, 2001). For large-scale immunoprecipitation of RGS9-1, 2 ml of anti-RGS9-1c IgG-coupled Sepharose beads is packed into a column, and detergent extracts of ROS membranes (equivalent to 50–100 mg rhodopsin) are incubated with the beads in the column with shaking at 4° for 3–4 h. The column is washed with 40–60 ml solubilization buffer, and proteins are eluted stepwise with 0.1 M glycine, pH 3.0, at 1 ml per sample for a total of 20 ml. Eluted proteins are precipitated by 10% trichloroacetic acid (TCA), washed with acetone, resuspended in SDS–PAGE sample buffer, resolved by SDS–PAGE, and visualized by Coomassie staining. Milligrams of the RGS9-1 complex can be isolated using this method in a heterotetrameric complex with $G_{\beta5L}$, $G_{t\alpha}$, and R9AP with minimal residual contamination from other retinal proteins. The stoichiometry of the proteins in the complex is in an approximate molar ratio of 1:1:1:1 (Hu and Wensel, 2002) (Fig. 1), although the relative amount of $G_{t\alpha}$ is somewhat variable.

Similar procedures are used for the immunoprecipitation of R9AP. However, because R9AP is a membrane protein and precipitates in buffers without detergent, the SDS–PAGE sample buffer is used to elute it from the antibody-coupled beads. The concentration of the reducing reagents in the sample buffer has to be minimized and elution has to be done at room temperature in 2–5 min in order to reduce the elution of IgG light chains from the beads.

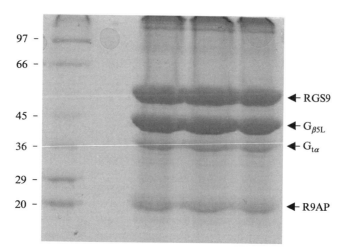

FIG. 1. Preparative immunoprecipitation for R9AP sequence analysis. Detergent-solubilized ROS proteins were immunoprecipitated as described in the text, separated on a preparative SDS–PAGE, and stained with Coomassie blue dye.

Localization of R9AP in the Retina

Immunofluorescence Microscopy

Buffers and Reagents

PBST: 155 mM NaCl, 1.5 mM KH$_2$PO$_4$, 2.7 mM Na$_2$HPO$_4$, pH 7.2, 0.1% Triton X-100

Antibodies: polyclonal goat anti-R9AP antiserum, Texas Red-labeled rabbit antigoat antibody (Vector Laboratories)

Sheep serum (Sigma)

Vectashield mounting solution (Vector Laboratories)

C57BL-6 mice

Procedures. Mouse eyes are dissected, fixed, cryoprotected with sucrose buffer, frozen, and then cryosectioned (Lyubarsky *et al.*, 2001). Sections are blocked with 10% sheep serum (Sigma) in PBST for 1 h and are then incubated with goat anti-R9AP antiserum at a 1:500 dilution in PBST for 2 h. For control experiments, the 1:500 diluted goat anti-R9AP antiserum is neutralized with 0.1 mg/ml recombinant His-R9AP-ΔC (see later for preparation) in PBST before being applied to the sections. After a 2-h incubation, sections are washed three times for 5 min in PBST to remove excess antibodies, and the Texas Red-labeled rabbit antigoat antibody is added to the sections at a dilution ratio of 1:100 and incubated for 1 h. The sections are washed again in PBST three times for 5 min, mounted using Vectashield mounting solution, and imaged using an LSM-510 confocal microscope (Zeiss). R9AP staining is largely detected in the photoreceptor outer segments, with stronger signals in the cones than rods. Staining in the inner segments is also detectable, but with less intensity. Note that the strong labeling of the outer plexiform layer is very likely caused by the cross-reactivity of the antibody toward syntaxin-like proteins, which share structural similarity to R9AP and are highly concentrated in the synaptic junctions between photoreceptors and bipolar cells (see later for more supporting evidence) (Hu and Wensel, 2002).

Localization by Fractionation of Retinal Membranes

The R9AP signal is largely detected in the photoreceptor layer of the retina and is enriched in the photoreceptor outer segments by immunofluorescence, suggesting its function in phototransduction. However, strong labeling is also observed at the outer plexiform layer where RGS9-1 labeling is not obvious. Therefore, to find out if there is a distinct population of R9AP that is not associated with RGS9-1, we also use biochemical methods to investigate its localization and interaction with RGS9-1 in the retina.

Reagents

Antibodies: polyclonal goat anti-R9AP antiserum, polyclonal anti-RGS9-1c antiserum

Procedures. Because RGS9-1 is only observed in the photoreceptor outer segment in immunofluorescence studies, we separate retinal membranes into ROS membranes and ROS-depleted membranes and test whether there is R9AP present in ROS-depleted membranes and whether it exists in the RGS9-1-free form. Forty bovine retinas are extracted twice with 20 ml 34% sucrose buffer by gentle vortexing. The resulting crude ROS and ROS-depleted retinal membranes are separated by a 20-min low-speed centrifugation at 6000 rpm in a Beckman JA 25.50 rotor. The supernatants are pooled, diluted four to five times in GAPN, and centrifuged at 15,000 rpm in the same rotor to collect ROS membranes. The ROS or ROS-depleted membranes are solubilized in 1% NP40 in GAPN, and RGS9-1 is immunodepleted from the detergent-solubilized proteins as described in the previous section. R9AP in the samples before and after RGS9-1 depletion is determined by immunoblotting. In both ROS or ROS-depleted fractions, complete RGS9-1 depletion results in 100% R9AP depletion, indicating that all the R9AP protein must be in complex with RGS9-1 in the retina (Hu and Wensel, 2002).

To test whether R9AP localizes predominantly to fractionated ROS membranes, ROS membranes from frozen bovine retinas are prepared in dim red light and separated by sucrose gradient centrifugation using a standard technique (Papermaster and Dreyer, 1974). Samples in the sucrose gradient are fractionated into 1-ml aliquots using an automatic gradient puller (Auto-Densi-Flow from Labconco) and stored at $-80°$. Proteins in these fractions are resolved by SDS–PAGE, and R9AP is detected by Western blotting. The concentration of the major rod outer segment marker protein rhodopsin in these fractions is determined by measuring the difference of absorbance at 500 nm before and after light bleaching in 1.5% *N,N*-dimethyldodecylamine *N*-oxide. We observed faithful copurification of R9AP and rhodopsin, strongly supporting the conclusion that R9AP is localized within ROS membranes (Hu and Wensel, 2002).

Expression and Purification of Recombinant R9AP

To characterize the biochemical properties of R9AP, it is necessary to generate recombinant proteins for *in vitro* functional assays and antibody production. Although *E. coli* has been the most commonly used heterologous expression system, there are few examples of purification and reconstitution of functional mammalian transmembrane proteins from bacteria (Quick and Wright, 2002). We have developed a successful strategy for preparation of R9AP in *E. coli*, and this strategy may also be applicable to other membrane proteins of similar type. We also developed a protocol to express R9AP in insect cells. These highly efficient expression and purification methods may facilitate future structural studies of the GAP complex.

Buffers and Reagents

Lysis buffer: 300 mM NaCl, 25 mM Tris, pH 8.0, 20 mM imidazole, 2 mM DTT and solid PMSF (~20 mg/liter)

Dialysis buffer: 10 mM Tris–HCl, 1 mM DTT, pH 8.0

Sodium cholate (3α, 7α, 12α-trihydroxy-5β-cholanic acid sodium salt), Sigma

Isopropyl-β-D-thiogalactopyranoside (IPTG)

Plasmids: pET14b (Novagen) and pVL1392 (BD Biosciences).

E. coli strain BL21 (DE3) LysS (Novagen)

BD BaculoGold baculovirus expression vector system (BD Biosciences)

Sf9 cells (serum-free medium adapted) (Invitrogen)

SF900 II serum-free insect cell medium (Invitrogen)

Ni-NTA superflow beads (Qiagen)

POROS HQ strong anion-exchange column (Applied Biosystems)

Procedures. Bovine R9AP fragment (His-R9AP-ΔC, amino acid 1–212), or His-R9AP-ΔC2, amino acid 1–191) and mouse R9AP full-length cDNA (His-mR9AP) are amplified by polymerase chain reaction from bovine retina cDNA library and mouse genomic DNA, respectively, and cloned into the pET14b vector using *Bam*HI and *Nde*I sites (Hu and Wensel, 2002). N-terminal His$_6$-tagged bovine R9AP domains or mouse full-length R9AP protein is expressed in *E. coli* [BL21(DE3) LysS] by growing freshly transformed cells to OD$_{600}$ ~0.6–0.8 and inducing with 0.3 mM IPTG at 30° for 4–5 h. Cells are harvested by centrifugation at 3000g and can be stored at $-80°$ until use. Under these conditions, ~10 mg protein can be produced in every liter of culture. Bovine R9AP fragments are largely soluble, as they do not contain the C-terminal transmembrane helix, and the mouse full-length protein is insoluble.

For insect cell expression, the N-terminal His$_6$-tagged mouse full-length R9AP was cut out of pET14b using *Xba*I and *Nde*I sites and cloned into pVL1392. The resulting plasmid is used to generate baculoviruses that can infect and express R9AP in insect cells using the BaculoGold system (BD biosciences). $Sf9$ cells are seeded at 1.0×10^6 cells/ml in serum-free SF900 II medium and are infected with the R9AP viruses at MOI (multiplicity of infection) of 3 after 24 h of growth at 28°. Cells are grown for another 50–54 h postinfection and are then harvested by centrifugation. About 0.5 mg protein can be obtained from every liter of culture.

Because $E.\ coli$ expression is much more economical and time-saving, it is used routinely to obtain large quantities of recombinant R9AP. To purify His-R9AP-ΔC or His-R9AP-ΔC2, $E.\ coli$ cells are resuspended in lysis buffer and lysed by sonication. Cell debris and insoluble materials are removed by centrifugation at 24,000g in a Beckman JA 25.50 rotor for 30 min. R9AP-soluble fragments in the supernatant are purified by affinity chromatography using nickel nitrilotriacetate (Ni-NTA superflow beads, Qiagen) according to the manufacturer's protocol. Proteins are eluted with 200 mM imidazole in the lysis buffer, and the purest fractions are combined and dialyzed against dialysis buffer. After dialysis, proteins are loaded onto a POROS HQ anion-exchange HPLC column (Applied Biosystems), washed with 20 mM NaCl in dialysis buffer, and eluted with a linear gradient from 20 to 200 mM NaCl in the same buffer. The flow rate is kept at 2 or 3 ml/min because a faster flow rate decreases resolution. The purest fractions are combined, concentrated to $\sim$2–4 mg/ml, and stored at 4° for short-term usage or at -20° after adding glycerol to 40% for long-term usage (Hu and Wensel, 2002).

To purify His-mR9AP, $E.\ coli$ cells are lysed, and insoluble proteins including R9AP are separated from soluble proteins in a similar way. The insoluble proteins, including His-mR9AP, are washed once in lysis buffer and are then extracted with 4% sodium cholate in lysis buffer for $\sim$30–60 min at 4° with gentle agitation, and the detergent-solubilized proteins in the supernatant are separated from the pellet by centrifugation. The extraction is repeated a total of three to four times, yielding >70% of total His-mR9AP extracted in soluble form (Hu et al., 2003). Being in the insoluble fraction actually facilitates purification, as most contaminating $E.\ coli$ proteins are soluble and are removed in the beginning. The detergent supernatants are pooled together and applied to Ni-NTA beads, and His-mR9AP is purified in 4% sodium cholate in lysis buffer according to the manufacturer's protocol. From 2 liters of $E.\ coli$ culture, at least 5 mg R9AP with 90–95% purity can be obtained routinely. It should be noted that discarding the supernatant from the first detergent extraction can improve the purity of the final product, although at the expense of lowering the yield, as the first extract contains some $E.\ coli$ proteins that copurify with R9AP on Ni-NTA beads.

Vesicle Reconstitution

Because R9AP is a transmembrane protein, it is necessary to reconstitute it into lipid vesicles to study its function in its natural conformation. The protocol we developed to reconstitute R9AP and rhodopsin into lipid vesicles allowed us to reconstitute much of the vertebrate visual cascade *in vitro*. It will facilitate further studies on the effect of specific lipids and other molecules of ROS membranes in regulating RGS9-1 activity. It may also be useful in the characterization of other phototransduction components or be applied to the studies of membrane proteins of similar types in other systems.

Reagents

L-α-Phosphatidylserine (PS: brain, porcine-sodium salt) (Avanti Polar Lipids)
L-α-Phosphatidylcholine (PC: egg, chicken) (Avanti Polar Lipids)
L-α-Phosphatidylethanolamine (PE: egg, chicken) (Avanti Polar Lipids)
Rhodamine-labeled PE [*N*-(6-tetramethylrhodaminethiocarbamoyl)-1,2-dihexadecanoyl-*sn*-glycero-3-phosphoethanolamine, triethylammonium salt] (Molecular Probes Inc.)
Sodium cholate (Sigma)

Procedures. Lipids are mixed at a ratio (w/w) of PC:PE:PS = 50:35:15 to mimic the ROS phospholipid composition (Daemen, 1973). They are mixed in chloroform, dried under argon flow, and redissolved in 4% sodium cholate in lysis buffer to a final concentration of 20 mg/ml. To facilitate the use of vesicles in binding assays, those used for this purpose contain rhodamine-labeled PE at a final concentration of 0.5% (w/w) to help in visualizing the lipids. Lipids in detergent solution are mixed with purified rhodopsin, His-mR9AP, or rhodopsin plus His-mR9AP, each in the same detergent concentration, using a lipid to protein (w/w) ratio of 20–40:1. The mixture is dialyzed against GAPN buffer (>100 × volume) with buffer changes every 20–24 h for a total of approximately 120 h (Parmar *et al.*, 1999). Dialyzed samples are centrifuged (15 min, 18,000*g*, Beckman TL100 rotor) to remove any aggregates. Reconstituted proteoliposomes are in the supernatant at this point in the procedure and contain the majority of the protein added initially to the reaction mixture in a roughly random orientation with respect to the inner and outer membrane leaflets. The resulting proteoliposomes are unilamellar and have a fairly narrow size distribution: radius = 240 (± 50) Å (Fig. 2).

R9AP   Rhodopsin + R9AP

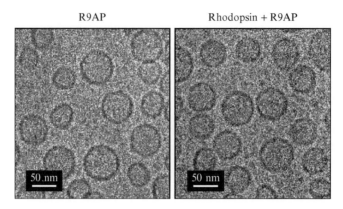

FIG. 2. Electron micrographs of reconstituted vesicles, imaged in vitreous ice. The reconstituted vesicles (in solution) were frozen on carbon-coated holey grids by rapid plunging into liquid ethane. Images were recorded at a nominal magnification of 40,000 × with a dose of ~20 electrons per Å^2 on a JEOL1200EX electron microscope. R9AP, R9AP-only vesicles; rhodopsin + R9AP, rhodopsin and R9AP vesicles. Reprinted, with permission, from Hu *et al.* (2003).

Two different methods are used to concentrate the vesicles. R9AP-only vesicles are concentrated by ultracentrifugation at 88,000g. The easily pelleted vesicles are collected to be used in binding assays, and vesicles that stay in the supernatant are discarded. Pelleted vesicles are resuspended in GAPN buffer to the desired concentration by repeated extrusion through a 26-gauge needle. Lipid-only vesicles used as controls in the binding assays are made similarly in parallel, except that the proteins are left out. However, lipid-only vesicles are less dense than R9AP vesicles and do not pellet as well. Therefore, disturbance of the pellet should be minimized to avoid mixing the pellet with the supernatant. Rhodopsin or rhodopsin and R9AP vesicles (Fig. 2), which do not pellet as efficiently as the R9AP-only vesicles, are concentrated in a protein concentrator (50 kDa cutoff, Millipore) by centrifugation to collect all the vesicles and obtain the maximum yield. Concentrated samples are then diluted to the desired concentrations in GAPN buffer.

Lipid concentrations of all the samples are determined by measuring the total phosphate concentrations colorimetrically (Chen *et al.*, 1956). The concentrations of vesicles expressed in terms of "moles" are calculated as follows: molar concentration of vesicles = (formal concentration of phospholipids × s)/$4\pi r^2$, where s is the average surface area of a phospholipid head group in Å^2 (70 Å^2) and r is the average vesicle radius in Å. Rhodopsin concentrations are determined from bleachable absorbance at 500 nm

($\varepsilon = 40{,}600 \ M^{-1}cm^{-1}$) in lauryldimethylamine oxide (1.5%, v/v). R9AP concentrations are determined by the densitometry of Coomassie-stained SDS–PAGE gels using bovine serum albumin as the standard. The molar ratios of R9AP, rhodopsin, and vesicles are fairly constant across different batches of the vesicles, as they are largely determined by the amount of R9AP, rhodopsin, and lipids used in the reconstitution.

Binding of R9AP and RGS9-1

Because RGS9-1 forms a heterotetramer with $G_{\beta 5L}$, $G_{t\alpha}$, and R9AP, binding assays between RGS9-1 and R9AP are carried out to test the direct interaction between RGS9-1 and R9AP and to identify the domains on RGS9-1 responsible for this interaction. Binding assays between RGS9-1 and the R9AP-soluble domain are carried out using the His$_6$-tagged R9AP soluble domain immobilized on Ni^{2+}-NTA beads. Binding to full-length R9AP is measured by reconstituting detergent-solubilized recombinant R9AP into vesicles and separating vesicle-bound proteins by centrifugation.

Buffers and Reagents

Binding buffer: 100 mM NaCl, 10 mM HEPES, pH 7.4, 2 mM MgCl$_2$, 0.5% NP40, 10 mM imidazole, and 0.5 mg/ml chicken egg ovalbumin
Ni-NTA superflow beads (Qiagen)

Procedures. GST-RGS9-1 (amino acid 1–484)·Gβ_5, GST-RGS9-1-NG (amino acid 1–280), GST-RGS9-1-NGD (amino acid 1–431)·Gβ_5, and His-RGS9-1-IGRC (amino acid 112–484)·Gβ_5 complexes are expressed and purified from Sf9 cells as described previously (He *et al.*, 2000; Martemyanov and Arshavsky, 2002). GST-tagged RGS9-1-D (amino acid 276–431) is expressed and purified from *E. coli* (Slep *et al.*, 2001). The His-RGS9-1-IGRC·Gβ_5 complex cannot be used for assays of binding to the Ni^{2+}-NTA-immobilized R9AP soluble domain, but can be used in vesicle-binding assays.

For a typical binding reaction between the RGS9-1 and R9AP-soluble domain, 0.5 μM of GST-RGS9-1·Gβ_5 or another GST-tagged RGS9-1 fragment is mixed with 2.5 μM His-R9AP-ΔC2 in 200 μl binding buffer and incubated at 4° for 1 h. The mixtures are then incubated with 20 μl Ni-NTA superflow beads on a shaker for another 1.5 h. The presence of detergent minimizes the nonspecific binding of RGS9-1 to the beads and the microtubes. The beads are separated from the supernatant by brief centrifugation and are washed three times with the binding buffer. Bound proteins are eluted from the beads with SDS–PAGE sample buffer, and RGS9-1 proteins or Gβ_5 in the elution is detected by immunoblotting using anti-RGS9-N antiserum or anti-Gβ_5 antiserum. Control experiments are performed similarly, except that His-R9AP-ΔC2 is not added. Both full-length RGS9-1 · Gβ_5 and RGS9-1NG · Gβ_5 are found to bind to R9AP, whereas RGS9-1G · Gβ_5 does not, suggesting that the NG domain (which includes the DEP domain) of RGS9-1 is responsible for the interaction (Hu and Wensel, 2002).

For binding of RGS9-1 proteins to R9AP vesicles, purified GST-RGS9-1 proteins (in GAPN buffer) are first centrifuged at 88,000g for 40 min (TLA 100.3 rotor) to remove possible aggregates. Soluble proteins in the supernatant are diluted and mixed with R9AP vesicles or lipid-only vesicles in a volume of 300 μl and are incubated at 4° with very gentle vortexing for ~3 h in GAPN buffer containing 0.2 mg/ml ovalbumin. One hundred microliters of the reaction mixture is then transferred to a new polypropylene tube (Beckman), and unbound proteins are separated from bound proteins by ultracentrifugation at 88,000g for 40 min. The pink (due to the rhodamine tag) vesicle pellet is redissolved in SDS–PAGE sample buffer and is subjected to electrophoresis. Again, the pellet of lipid-only vesicles is not as tight as that of the R9AP vesicles and has to be handled very gently to avoid disturbance. Typical final concentrations in binding reactions are RGS9-1 proteins, 0.1–0.5 μM; R9AP (total), 3.0 μM; and lipids (as phosphate), 0.8 mM. For SDS–PAGE, equal proportions of each fraction, including the starting reaction mixture, should be loaded on SDS–PAGE for easy comparison. RGS9 proteins are detected by immunoblotting with anti-RGS9-1c antiserum (see Hu et al., 2003). RGS9-1 · Gβ_5, RGS9-1NGD · Gβ_5, and RGS9-1GDC · Gβ_5 all bound to R9AP vesicles, with significantly less binding detected for RGS9-1GDC · Gβ_5. RGS9-1D did not bind at all. Again, these results indicate that the NG domain is important for the interaction, in agreement with the findings from the aforementioned binding assays using the R9AP-soluble domain (Hu et al., 2003) and with work by others (Lishko et al., 2002).

Effect of R9AP on RGS9-1 GAP Activity

Two kinetic approaches for measuring GTPase activity have been developed previously: the multiple turnover and the single turnover method (reviewed in Cowan et al., 2000). The multiple turnover method monitors the steady-state GTP hydrolysis rate during cycles of GTPase activation by the receptor and inactivation by GTP hydrolysis. The use of this assay requires that GTP hydrolysis must be the rate-limiting step. However, in experiments containing the other proteins that may interfere with GTPase recycling, such as PDEγ, this assay could yield ambiguous results. Therefore, to avoid any possible interference caused by G-protein recycling, we used single turnover G$_{\alpha t}$ GTPase assays (Cowan *et al.*, 2000; He and Wensel, 2001) to determine the effect of R9AP on the GAP activity of RGS9-1. We first tested the effect on the reconstituted vesicles and then tested it on urea-treated ROS membranes in which the endogenous R9AP was made accessible to recombinant RGS9-1. Note that the procedures used here employ GST-tagged RGS9-1. Lower GAP activity in the absence of R9AP and higher GAP activity in its presence were measured for a His_6-tagged form of the RGS9-1/$G\beta5$ complex (Martemyanov and Arshavsky, 2002).

Enhancement of RGS9-1 GAP Activity by R9AP Vesicles

Procedures. Rhodopsin vesicles or rhodopsin and R9AP vesicles are mixed with purified G_t [purified from bovine retina (Struthers *et al.*, 2000)], His_6-tagged PDEγ [expressed and purified from *E. coli* (He *et al.*, 2000)], and individual GST-tagged RGS9-1 proteins (GST-RGS9-1-D or GST-RGS9-1·$G\beta5$) in GAPN buffer and incubated on ice for ~30 min to allow for the binding between RGS9-1 and R9AP. The reaction is started by the $[\gamma^{-32}P]GTP$ addition and is stopped by the addition of trichloroacetic acid (5% w/v) at different times. Released phosphate is isolated by binding to activated charcoal, and radioactivity is measured using scintillation counting (for detailed procedures, see Cowan *et al.*, 2000; He *et al.*, 2000). Basal GTP hydrolysis rates by $G_{t\alpha}$ are measured similarly by leaving out the RGS9-1 proteins in the aforementioned reactions. Typical final protein concentrations in the reactions are 4.0 μM total rhodopsin, 0.5 μM $G_{t\alpha}$, 0.1 μM RGS9-1, 0.2 μM His_6-PDEγ, and 1.2 μM R9AP; the final GTP concentration is 25 nM. Data are fitted to a single exponential curve to derive the GTP hydrolysis rate constant k_{inact}. RGS9-1 GAP activity (Δk_{inact}) is calculated as $\Delta k_{inact} = k_{inact}$ (of $G_{t\alpha}$ in the presence of RGS9-1) – k_{inact} (of $G_{t\alpha}$).

At a given RGS9-1 concentration, the apparent RGS9-1 GAP activity is expected to be a function of the vesicle concentration. On the one hand, an increasing vesicle concentration increases the total R9AP concentration, and therefore increases the fraction of RGS9-1 that can be anchored to the vesicles, which will eventually be manifested as increased RGS9-1 activity. However, such an increase of activity will saturate when the vesicle concentration exceeds that of RGS9-1. On the other hand, as the vesicle concentration approaches and exceeds that of RGS9-1, the probability of $G_{\alpha t}$-GTP being formed on a vesicle lacking RGS9-1 also increases. If $G_{\alpha t}$-GTP exchange between vesicles is very slow on the timescale of GTP hydrolysis, then at higher vesicle concentrations the apparent GAP activity must decline. Therefore, in order to determine the maximal enhancement of GAP activity by R9AP-containing vesicles, we performed a vesicle titration, holding RGS9-1 concentration constant. RGS9-1 GAP activity (Δk_{inact}) on rhodopsin-only or rhodopsin and R9AP vesicles were measured similarly as described earlier over a series of vesicle concentrations. Final concentrations in the reactions were vesicles: 9.0, 18, 36, 72, 144, 288, and 576 nM; rhodopsin: 0.38, 0.75, 1.5, 3.0, 6.0, 12.0, and 24.0 μM; R9AP: 0.12, 0.24, 0.48, 0.96, 1.9, 3.8, and 7.7 μM; GST-RGS9-1·Gβ_5, 0.1 μM, G$_t$: 0.2 μM (for vesicle concentrations from 9.0 to 36.0 nM) or 0.5 μM (for vesicle concentrations from 72.0 to 576.0 nM). Enhancement on RGS9-1 GAP activity by R9AP membrane anchoring was defined as fold enhancement = Δk_{inact} (on rhodopsin and R9AP vesicles)/Δk_{inact} (on rhodopsin only vesicles) and was plotted as mean ± standard deviation against vesicle concentration and R9AP concentration (on a log scale) (Fig. 3A).

   Mathematical Model for Data Analysis. Because G$_t$ and RGS9-1·Gβ_5 should bind to rhodopsin- and R9AP-containing vesicles independently of one another, the apparent GAP activity should decline in a way predicted

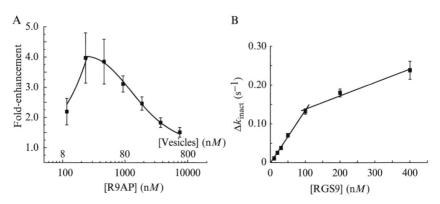

Fig. 3. Activation of GAP activity of RGS9-1 by reconstituted R9AP. (A) RGS9-1 GAP activity (Δk_{inact}) toward transducin was measured on rhodopsin-only or rhodopsin and R9AP vesicles at different vesicle concentrations and compared to yield fold enhancement = Δk_{inact} (on rhodopsin and R9AP vesicles)/Δk_{inact} (on rhodopsin only vesicles), plotted as mean ± standard deviation. The final GST-RGS9-1·Gβ_5 concentration was held constant at 0.1 μM, and the G$_t$ concentration was 0.2 μM for vesicle concentrations from 9.0 to 36.0 nM and 0.5 μM for vesicle concentrations from 72.0 to 576.0 nM. The *continuous curve* was calculated using Eq (1), assuming a maximal enhancement of fourfold. (B) RGS9-1 GAP activity (Δk_{inact}) on urea-washed ROS membranes was measured by single turnover assays as described in the text. Data were plotted as mean ± standard deviation and fit to two linear functions (*straight lines*) with slopes of 1.36 and 0.34 $\mu M^{-1} \cdot s^{-1}$. Reprinted, with permission, from Hu *et al.* (2003).

by Poisson statistics. Therefore, the probability of a vesicle having exactly x molecules of RGS9-1 when the average number of RGS9-1 molecules per vesicle is μ can be calculated by the Poisson distribution [Eq. (1)]. Based on this assumption, together with the assumption that the dissociation constant K_d between RGS9-1 and R9AP on the vesicles was much lower than their concentrations used in the experiments and that each $G\alpha_t \cdot$ GTP stays on the vesicle in which it was originally activated until its bound GTP is hydrolyzed, we were able to build the following mathematical model for the effect of R9AP on RGS9-1 GAP activity: Eq. (2) (Fig. 3A, smooth line).

$$P(x, \mu) = \mu^x e^{-\mu}/x! \tag{1}$$

$$
\begin{aligned}
E - 1 = (E_{max} - 1) &\times [(\text{total concentration of accessible R9AP})/ \\
&(\text{total concentration of RGS9})] \times [1 - \exp(-\mu)], \\
&\text{when accessible R9AP} \leq \text{RGS9, and by } E - 1 \\
= (E_{max} - 1) \times [1 &- \exp(-\mu)], \text{when accessible R9AP} \geq \text{RGS9.} \\
&E, \text{the expected fold enhancement } (\textit{smooth line}), \\
&E_{max}, \text{ maximal fold enhancement of RGS9-1 activity}
\end{aligned}
$$

$$\tag{2}$$

Enhancement of RGS9-1 GAP Activity by Urea-Washed ROS Membranes

It was reported previously that urea washing can expose endogenous binding sites for RGS9-1 on purified ROS, and binding of RGS9-1 to these sites enhances its activity. To compare our results on reconstituted vesicles to that on urea-washed ROS, we titrated different concentrations of recombinant RGS9-1 protein into urea-washed membranes held at a constant concentration and measured the GAP activity by single turnover assays.

Buffers and Reagents

Hypotonic buffer: 5 mM Tris–HCl, pH 7.4, 3 mM EDTA
4 M urea buffer: 4 M urea, 5 mM Tris–HCl, pH 7.4
6 M urea buffer: 6 M urea, 5 mM Tris–HCl, pH 7.4
GAPN buffer: 100 mM NaCl, 10 mM HEPES, pH 7.4, 2 mM MgCl$_2$
    1 mM DTT and solid PMSF are added to all these buffers before
    use
Proteins: GST-tagged RGS9-1·Gβ_5 and His$_6$-PDEγ
Purified bovine ROS membranes (see earlier discussion)

Procedures. ROS membranes are washed sequentially with hypertonic buffer (twice), hypotonic buffer (twice), 4 M urea buffer (once), and GAPN buffer (twice). For each sequential washing step, ROS membranes are diluted, typically to $\sim$10 μM rhodopsin, in the indicated buffer, homogenized extensively on ice using a glass/Teflon homogenizer, and centrifuged in a Ti-45 rotor (Beckman) at 44,000 rpm for 45 min. Washed membranes are resuspended in GAPN buffer and stored at $-80°$.

For GAP assays, urea-washed ROS are incubated with purified G_t, His6-PDEγ, and GST-RGS9-1 $\cdot$ Gβ_5 in GAPN buffer on ice for $\sim$30 min. GTP hydrolysis is initiated by adding [γ-^{32}P]GTP, and the GTP hydrolysis rate constant is determined using the single turnover method described earlier. Typical final concentrations of proteins are rhodopsin, 72 μM; G_t, 0.5 μM; His$_6$-PDEγ, 0.2 μM, and GST-RGS9-1 $\cdot$ Gβ_5, 10, 20, 30, 50, 100, 200, and 400 nM. RGS9-1 GAP activity (Δk_{inact}) is plotted against its concentration as mean $\pm$ standard deviation and is fitted with straight lines (Fig. 3B). GAP assays using 6 M urea buffer-washed ROS yields similar results to those using 4 M urea buffer-washed membranes, except that there is a slightly decreased k_{cat}/K_m value of RGS9-1 $\cdot$ Gβ_5 on 6 M urea-treated membranes.

Acknowledgments

This work was supported by National Eye Institute Grant R01-EY11900, the Robert A. Welch Foundation, and the National Center for Macromolecular Imaging, P41-RR02250.

References

Chen, C. K., Burns, M. E., He, W., Wensel, T. G., Baylor, D. A., and Simon, M. I. (2000). Slowed recovery of rod photoresponse in mice lacking the GTPase accelerating protein RGS9-1. *Nature* **403,** 557–560.

Chen, P. S., Toribara, T. Y., and Warner, H. (1956). Microdetermination of phosphorous. *Anal. Chem.* **28,** 1756–1758.

Cowan, C. W., Wensel, T. G., and Arshavsky, V. Y. (2000). Enzymology of GTPase acceleration in phototransduction. *Methods Enzymol.* **315,** 524–538.

Cowan, C. W., He, W., and Wensel, T. G. (2001). RGS proteins: Lessons from the RGS9 subfamily. *Prog. Nucleic Acid Res. Mol. Biol.* **65,** 341–359.

Cowan, C. W., Fariss, R. N., Sokal, I., Palczewski, K., and Wensel, T. G. (1998). High expression levels in cones of RGS9, the predominant GTPase accelerating protein of rods. *Proc. Natl. Acad. Sci. USA* **95,** 5351–5356.

Daemen, F. J. M. (1973). Vertebrate rod outer segment membranes. *Biochim. Biophys. Acta* **300,** 255–288.

Gray, G. D., Mickelson, M. M., and Crim, J. A. (1969). The demonstration of two gamma-globulin subclasses in the goat. *Immunochemistry* **6,** 641–644.

Harlow, E., and Lane, D. (1988). "Antibodies: A Laboratory Manual." Cold Spring Harbor Laboratories, Cold Spring Harbor, NY.

He, W., Cowan, C. W., and Wensel, T. G. (1998). RGS9, a GTPase accelerator for phototransduction. *Neuron* **20,** 95–102.

He, W., Lu, L., Zhang, X., El-Hodiri, H. M., Chen, C. K., Slep, K. C., Simon, M. I., Jamrich, M., and Wensel, T. G. (2000). Modules in the photoreceptor RGS9-1.Gbeta 5L GTPase-accelerating protein complex control effector coupling, GTPase acceleration, protein folding, and stability. *J. Biol. Chem.* **275,** 37093–37100.

He, W., and Wensel, T. G. (2002). RGS function in visual signal transduction. *Methods Enzymol.* **344,** 724–740.

Hu, G., Jang, G. F., Cowan, C. W., Wensel, T. G., and Palczewski, K. (2001). Phosphorylation of RGS9-1 by an endogenous protein kinase in rod outer segments. *J. Biol. Chem.* **276,** 22287–22295.

Hu, G., and Wensel, T.G. (2002). R9AP, a membrane anchor for the photoreceptor GTPase accelerating protein, RGS9-1. *Proc. Natl. Acad. Sci. USA* **99,** 9755–9760.

Hu, G., Zhang, Z., and Wensel, T. G. (2003). Activation of RGS9-1 GTPase acceleration by its membrane anchor, R9AP. *J. Biol. Chem.* **278,** 14550–14554.

Ibrahimi, I. M., Eder, J., Prager, E. M., Wilson, A. C., and Arnon, R. (1980). The effect of a single amino acid substitution on the antigenic specificity of the loop region of lysozyme. *Mol. Immunol.* **17,** 37–46.

Lishko, P. V., Martemyanov, K. V., Hopp, J. A., and Arshavsky, V. Y. (2002). Specific binding of RGS9-Gbeta 5L to protein anchor in photoreceptor membranes greatly enhances its catalytic activity. *J. Biol. Chem.* **276,** 24376–24381.

Lyubarsky, A. L., Naarendorp, F., Zhang, X., Wensel, T., Simon, M. I., and Pugh, E. N., Jr. (2001). RGS9-1 is required for normal inactivation of mouse cone phototransduction. *Mol. Vis.* **7,** 71–78.

Lyubarsky, A. L., and Pugh, E. N., Jr. (1996). Recovery phase of the murine rod photoresponse reconstructed from electroretinographic recordings. *J. Neurosci.* **16,** 563–571.

Makino, E. R., Handy, J. W., Li, T., and Arshavsky, V. Y. (1999). The GTPase activating factor for transducin in rod photoreceptors is the complex between RGS9 and type 5 G protein beta subunit. *Proc. Natl. Acad. Sci. USA* **96,** 1947–1952.

Martemyanov, K. A., and Arshavsky, V. Y. (2002). Noncatalytic domains of RGS9-1. Gbeta 5L play a decisive role in establishing its substrate specificity. *J. Biol. Chem.* **277,** 32843–32848.

Papermaster, D. S. (1982). Preparation of retinal rod outer segments. *Methods Enzymol.* **81,** 48–52.

Papermaster, D. S., and Dreyer, W. J. (1974). Rhodopsin content in the outer segment membranes of bovine and frog retinal rods. *Biochemistry* **13,** 2438–2444.

Parmar, M. M., Edwards, K., and Madden, T. D. (1999). Incorporation of bacterial membrane proteins into liposomes: Factors influencing protein reconstitution. *Biochim. Biophys. Acta* **1421,** 77–90.

Pugh, E. N., Jr., and Lamb, T. D. (2000). "Phototransduction in Vertebrate Rods and Cones: Molecular Mechanisms of Amplification, Recovery and Light Adaptation." Elsevier, New York.

Quick, M., and Wright, E. M. (2002). Employing *Escherichia coli* to functionally express, purify, and characterize a human transporter. *Proc. Natl. Acad. Sci. USA* **99,** 8597–8601.

Sagoo, M. S., and Lagnado, L. (1997). G-protein deactivation is rate-limiting for shut-off of the phototransduction cascade. *Nature* **389,** 392–395.

Slep, K. C., Kercher, M. A., He, W., Cowan, C. W., Wensel, T. G., and Sigler, P. B. (2001). Structural determinants for regulation of phosphodiesterase by a G protein at 2.0 Å. *Nature* **409,** 1071–1077.

Sokal, I., Hu, G., Liang, Y., Mao, M., Wensel, T. G., and Palczewski, K. (2002). Identification of protein kinase C isozymes responsible for the phosphorylation of photoreceptor-specific RGS9-1 at Ser475. *J. Biol. Chem.* **278,** 8316–8325.

Struthers, M., Yu, H., and Oprian, D. D. (2000). G protein-coupled receptor activation: Analysis of a highly constrained, "straitjacketed" rhodopsin. *Biochemistry* **39,** 7938–7942.

Zhang, X., Wensel, T. G., and Kraft, T. W. (2003). GTPase regulators and photoresponses in cones of the eastern chipmunk. *J. Neurosci.* **23,** 1287–1297.

[13] Kinetic Approaches to Study the Function of RGS9 Isoforms

By Kirill A. Martemyanov *and* Vadim Y. Arshavsky

Abstract

The experimental strategies developed in kinetic studies of interactions between RGS9 isoforms with G proteins of the G_i subfamily provide a useful framework for conducting similar studies with essentially any regulator of G-protein signaling (RGS) protein – G-protein pair. This article describes two major kinetic approaches used in the studies of RGS9 isoforms: single turnover and multiple turnover GTPase assays. We also describe pull-down assays as a method complementary to the kinetic assays. The discussion of the strengths and limitations of each individual assay emphasizes the importance of combining multiple experimental approaches in order to obtain comprehensive and internally consistent information regarding the mechanisms of RGS protein action.

Introduction

Regulator of G-protein signaling (RGS) proteins regulate the duration of G-protein signaling by stimulating GTP hydrolysis on the G-protein α subunits ($G\alpha$), thus determining the lifetime of G proteins in their active states. For this reason, RGS proteins can be considered enzymes that catalyze the conversion of the active state of $G\alpha$, $G\alpha \cdot GTP$, to its inactive state, $G\alpha \cdot GDP$. In all RGS proteins, this catalysis is performed by the $\sim$120 amino acid-long RGS homology domain. RGS homology domains of different RGS proteins have different catalytic efficiencies and different spectra of G-protein specificity. Furthermore, most RGS proteins are equipped with additional noncatalytic domains that can drastically modify the catalytic properties of RGS homology domains or couple RGS proteins to their regulators. This versatility in modulating the catalytic activity of

RGS proteins allows for the precise regulation of signal duration in a broad range of G-protein pathways.

RGS9 provides an instructive example of how specific needs of the timing of individual signaling pathways are met through a complex organization of an RGS protein. There are two alternative splice variants of RGS9, both expressed in the nervous system (reviewed in Burns and Wensel, 2003) and both existing as constitutive complexes with the type 5 G-protein β subunit (Gβ5). The short splice isoform RGS9-1, complexed with the long splice isoform of Gβ5 (Gβ5L), regulates signal duration in the visual pathway of vertebrate photoreceptors (reviewed in Arshavsky et al., 2002; Cowan et al., 2000a). In this cascade, a G-protein-coupled receptor (photoexcited rhodopsin) activates the G-protein transducin (Gα_t). Activated transducin then stimulates the activity of the effector enzyme, cGMP phosphodiesterase (PDE), by binding to the PDEγ-subunits (PDEγ) and releasing the inhibitory constraint that PDEγ imposes on the PDE catalytic subunits (for reviews, see Burns and Baylor, 2001; Arshavsky et al., 2002). GTP hydrolysis on Gα_t results in PDE inactivation and termination of the light-evoked signal. The physiologically rapid rate of signal inactivation is achieved via the stimulation of Gα_t GTPase activity by RGS9-1 · Gβ5L (Chen et al., 2000; He et al., 1998). In photoreceptors, the activity of RGS9-1 · Gβ5L is regulated by two proteins, PDEγ and RGS9 anchor protein (R9AP). PDEγ increases the affinity of RGS9-1 · Gβ5L for Gα_t by over 20-fold (Skiba et al., 2000), which allows RGS9-1 · Gβ5L to target the Gα_t · PDE complex specifically and prevents RGS9-1 · Gβ5L from inactivating Gα_t prior to PDE activation. R9AP holds RGS9-1 · Gβ5L on the surface of photoreceptor membranes and potentiates its catalytic activity dramatically (Hu and Wensel, 2002; Hu et al., 2003; Lishko et al., 2002; Martemyanov et al., 2003b).

The long splice isoform RGS9-2 exists as a complex with the short splice isoform of Gβ5 (Gβ5S) and regulates signaling via dopamine and/or μ-opioid receptors in the central nervous system (Chen et al., 2003; Granneman et al., 1998; Rahman et al., 1999, 2003; Zachariou et al., 2003). RGS9-2 interacts with all members of the G$_i$ subfamily, with Gα_o being its most likely physiological target (Martemyanov et al., 2003a). The affinity of RGS9-2 · Gβ5S for Gα subunits from this subfamily is enhanced by the C-terminal domain of RGS9-2, which is absent in RGS9-1 (Martemyanov et al., 2003a). This C-terminal domain shares several structural and functional similarities with PDEγ (Martemyanov et al., 2003a). The detailed mechanisms of RGS9-2 activity regulation in the brain are presently unknown.

This article first discusses general kinetic approaches used to study G-protein GTPase activity regulation by RGS proteins. It then describes the applications of two major GTPase assays, single turnover and multiple turnover, to the studies of RGS9 isoforms. Finally, it demonstrates the utility of pull-down assays as a method complementary to the kinetic assays in the studies of RGS9 function.

Kinetic Approaches to the Studies of RGS Protein Catalytic Activity

Since RGS proteins act as enzymes, the efficiency of a given RGS protein in stimulating the GTPase activity of any of its $G\alpha$ targets can be described in terms of the Michaelis kinetics. In general, the action of RGS proteins can be represented by the following reaction scheme, modified from Ross (2002):

$$RGS + G\alpha \cdot GTP \underset{k_{-1}}{\overset{k_{+1}}{\rightleftharpoons}} RGS \cdot G\alpha \cdot GTP \overset{k_{cat}}{\longrightarrow} RGS + G\alpha \cdot GDP + P_i$$

$$\Big\downarrow k_h$$

$$G\alpha \cdot GDP + P_i$$

$$(1)$$

In this scheme, the RGS protein is considered to be the enzyme and $G\alpha \cdot GTP$ the substrate. In this approach, the catalytic efficiency of an RGS protein is expressed by two kinetic parameters: k_{cat}, reflecting the maximal number of G protein molecules "served" by a single molecule of an RGS protein per second; and K_m (equal to $(k_{-1} + k_{cat})/k_{+1}$), reflecting the affinity of RGS for $G\alpha \cdot GTP$. The identification of these parameters for any given RGS–$G\alpha$ pair provides insights to understanding how signal duration is regulated in corresponding signaling pathways. The hallmark of this reaction scheme is that the rate at which the substrate, $G\alpha \cdot GTP$, turns into the products $G\alpha \cdot GDP$ and P_i "nonenzymatically" (determined by the rate constant k_h) usually makes a significant contribution to the overall rate of product formation. As illustrated later, this should be taken into account while performing the Michaelis analysis of the RGS catalytic activity.

Single Turnover GTPase Assays

Single turnover GTPase assays monitor a single synchronized round of GTP hydrolysis on a G-protein α subunit. This is achieved by forming the $G\alpha \cdot GTP$ complex before initiating GTP hydrolysis and by preventing any additional GTP binding to $G\alpha$ during the course of the reaction. The most significant advantage of this approach is that the rate of GTP hydrolysis measured in single turnover assays is not influenced by other reactions of the G-protein activation/inactivation cycle, the reactions which can be rate limiting for the overall rate of GTP hydrolysis under other experimental conditions.

There are two experimental modifications of single turnover GTPase assays. The first involves a two-step procedure that takes advantage of the fact that Mg^{2+} is a crucial cofactor in the GTP hydrolysis on G-protein α subunits (Krumins and Gilman, 2002). During the first step of the reaction, the GDP/GTP exchange on $G\alpha$ is conducted in the absence of Mg^{2+}. The $G\alpha \cdot GTP$ complex is then separated from excess GTP and the GTPase is initiated by the addition of Mg^{2+} on the second step. A detailed discussion of how this modification of single turnover assays can be used for many RGS–G protein pairs is provided by Ross (2002).

The second modification of single turnover assays involves a one-step procedure where the GTPase reaction is initiated by the addition of a substoichiometric amount of GTP to the preformed complex between the G-protein heterotrimer and its activated receptor (Cowan et al., 2000b). Because GTP binding to $G\alpha$ in this case is initiated simultaneously with GTP hydrolysis, this approach is applicable only under experimental conditions where the rate of GTP binding is much faster than the rate of GTP hydrolysis. In practice, this condition could be met only in studies of G proteins from the G_i subfamily, which can be activated by rhodopsin, the GPCR receptor available in essentially unlimited amounts (Arshavsky et al., 1987; Martemyanov et al., 2003a; Skiba et al., 1999; Yamanaka et al., 1985).

Analysis of RGS9 Activity in Single Turnover GTPase Assays

Because both splice isoforms of RGS9 regulate the activity of the G_i subfamily members, all reported single turnover assays using RGS9 were based on the relatively simple one-step approach. A detailed description of the single turnover assay for the measurements of transducin GTPase activity can be found in Cowan et al. (2000b). This section discusses the application of this assay for studying the catalytic activity of RGS9 and its regulation by other proteins.

In a typical experiment, the G protein of interest in the form of heterotrimer is combined with bleached urea-treated photoreceptor membranes serving as the source of photoexcited rhodopsin. The mixture is also supplied by RGS9 and any of its regulators. The GTPase reaction is initiated by the addition of $[\gamma\text{-}^{32}P]GTP$ and is quenched by the addition of 6% perchloric acid. The amount of $[^{32}P]P_i$ formed in the reaction is then determined using activated charcoal (Cowan et al., 2000b). The time course of the reaction is determined by collecting data from at least five individual assays of various durations; data are fitted by a single exponent. In studies from this laboratory, the most commonly used concentrations of the reacting components are 200 nM GTP, 20 μM rhodopsin, and 2 μM G-protein heterotrimer; the buffer contains 10 mM Tris–HCl (pH 7.8), 100 mM NaCl, 8 mM MgCl$_2$, and 1 mM dithiothreitol. In most published studies, these assays are performed with transducin purified from rod outer segment membranes. It is also possible to conduct these assays with native or recombinant Gα_t, Gα_i, and Gα_o reconstituted with G$\beta_1\gamma_1$ (Martemyanov et al., 2003a). This is possible because, in the presence of G$\beta_1\gamma_1$, rhodopsin effectively catalyzes the nucleotide exchange not only on Gα_t, but also on Gα_i (Skiba et al., 1996, 1999) and Gα_o (Martemyanov et al., 2003a).

The experiment shown in Fig. 1 illustrates the utility of this approach in comparing the abilities of different RGS9 splice isoforms and their mod-ulators to stimulate the GTPase activity of the α subunit of the same G protein, transducin. The rate of transducin GTPase was increased by both RGS9-1 and RGS9-2, with the latter being more efficient. However, the addition of PDEγ had opposite effects on the activities of RGS9-1 and RGS9-2. The activity of RGS9-1 was potentiated by PDEγ, whereas the

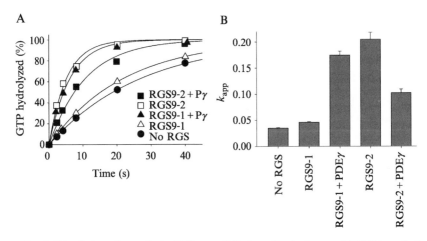

FIG. 1. Stimulation of transducin GTPase activity by splice isoforms of RGS9 in a single turnover GTPase assay. (A) GTPase activities of transducin α subunits in the absence or presence of 1 μM of RGS9 splice isoforms and 2 μM PDEγ. The reactions were conducted for various times under conditions described in the text. The amounts of P$_i$ formed at specified time points were measured and normalized to total amount of P$_i$ formed after the completion of GTP hydrolysis. The resulting values were fitted with single exponents to derive the apparent rate constants k_{app} plotted in the B.

activity of RGS9-2 was inhibited. As discussed in Martemyanov *et al.* (2003a), the former effect is explained by PDEγ increasing the affinity between RGS9-1 and Gα_t, whereas the latter effect is explained by the competition between PDEγ and the RGS9-2 C terminus.

An example of somewhat more advanced analysis is shown in Fig. 2, where the potentiation of RGS9-1 activity was analyzed by its membrane anchor R9AP. In a set of single turnover assays, we monitored transducin GTPase activity at a fixed concentration of R9AP and various concentrations of RGS9-1 (Fig. 2A). R9AP in this experiment was an intrinsic component of the urea-treated photoreceptor membranes used in the assays. The rate constant of nonactivated transducin GTPase (k_h) determined from the single exponential analysis of data obtained without RGS9 (lower curve in Fig. 2A) was then subtracted from the rate constants (k_{app}) determined by the exponential fits of data obtained in the presence of RGS9-1. The resulting values were plotted as the function of RGS9-1 concentration (Fig. 2B) and showed a strongly biphasic dependency. The initial phase, reflecting a steep rise in transducin GTPase activity with a slope of 10.5 s^{-1} μM^{-1}, was followed by one with a slope of only 0.15 s^{-1} μM^{-1}. Based on independent measurements indicating that the amount of R9AP available for RGS9-1 binding in these membranes corresponds to the braking point between the two slopes (Lishko *et al.*, 2002), we concluded that the two phases represent the activities of R9AP-bound RGS9-1 and free RGS9-1, respectively. The $\sim$70-fold ratio between the two slopes indicates that R9AP causes an $\sim$70-fold increase in the overall ability of

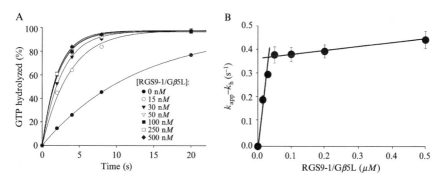

FIG. 2. The effect of R9AP on the ability of RGS9-1 to stimulate transducin GTPase. (A) Single turnover GTPase assays conducted in the presence of increasing concentrations of RGS9-1. Data points were fitted with single exponents to derive the k_{app} values. (B) The rate constant of Gα GTPase measured in the absence of RGS9 (k_h) was subtracted from the k_{app} values obtained in A, and the resulting value was plotted as a function of RGS9 concentration. Data were fitted with two straight lines with the slopes of 10.5 and 0.15 s^{-1} μM^{-1}.

RGS9-1 to stimulate transducin GTPase. Importantly, the same degree of RGS9-1 regulation by R9AP present in urea-treated membranes was obtained in published experiments using multiple turnover assays described in the next section (Lishko *et al.*, 2002).

These examples illustrate that single turnover assays provide a convenient means of directly comparing overall catalytic activities of various splice variants of RGS9 (Martemyanov *et al.*, 2003a) or RGS9 mutants (He *et al.*, 2000; Martemyanov and Arshavsky, 2002). They are also very useful in studying RGS9 modulators (Arshavsky and Bownds, 1992; Hu *et al.*, 2003). Although we are not aware of any published work where single turnover assays were used for the determination of K_m and k_{cat} of RGS9, it is plausible that this task could be accomplished by the methodology described for other RGS proteins (Ross, 2002). Such experiments would involve conducting a series of single turnover GTPase assays at a fixed concentration of RGS9 in the presence of increasing concentrations of $G\alpha \cdot GTP$ followed by the Michaelis analysis of the initial reaction rates. It should be noted, however, that such an analysis may be complicated by the strong nonlinearity of GTP hydrolysis in single turnover assays. In addition, it might be difficult to achieve a reliable difference between GTP hydrolysis rates measured with and without RGS9 while still satisfying the Michaelis requirement of $[RGS] \ll [G\alpha \cdot GTP]$.

Multiple Turnover GTPase Assays

Multiple turnover GTPase assays are conducted under steady-state conditions, where the concentration of $G\alpha \cdot GTP$ remains constant during the assay because the $G\alpha \cdot GDP$ formed in the reaction is recycled rapidly into $G\alpha \cdot GTP$ by the activated GPCR. The major advantage of these assays is that the rate of GTP hydrolysis remains linear as long as the GTP concentration remains saturating for the reaction. This makes multiple turnover assays optimal for conducting Michaelis analysis of the RGS activity. A potential complication of these assays, however, is that under many experimental conditions, various reactions of the G-protein activation/inactivation cycle other than the GTP hydrolysis itself may limit the rate of GTP hydrolysis. Examples of such reactions are the GDP release from $G\alpha$ or the reassociation of $G\alpha$ and $G\beta\gamma$ subunits following GTP hydrolysis (for a more detailed discussion, see Cowan *et al.*, 2000b; Ross, 2002). Furthermore, some of the activators of the GTP hydrolysis, such as PDEγ, may also inhibit G-protein activation by the receptor. These issues should be specifically addressed in appropriate control experiments and, whenever possible, multiple turnover assays should be complemented by single turnover assays.

Analysis of RGS9 Activity in Multiple Turnover GTPase Assays

Multiple turnover GTPase assays have been used successfully in determining the catalytic parameters of RGS9-1 and its mutants (Skiba *et al.*, 2000, 2001; Martemyanov and Arshavsky, 2002). The experimental protocol for a typical experiment is similar to that of a single turnover experiment, except that $[\gamma\text{-}^{32}\text{P}]\text{GTP}$ is taken in large excess over the G-protein α subunit. The concentrations of the reacting components should be optimized as follows.

1. The levels of GTP should be saturating, i.e., shown experimentally to not affect the rate of GTP hydrolysis at the maximal concentrations of rhodopsin and $G\alpha$ used in the assay.

2. The levels of photoexcited rhodopsin should also be saturating, i.e., also shown experimentally to not affect the rate of GTP hydrolysis at saturating GTP.

3. The minimal concentration of the G-protein heterotrimer used in the assays should be sufficiently high to ensure that the rate of the α and $\beta\gamma$ subunit reassociation after the GTP hydrolysis does not limit the overall GTPase rate. This condition is met when the rate of GTP hydrolysis is dependent linearly on the G-protein concentration.

In studies of transducin GTPase regulation by RGS9, these conditions are usually met at $[G_t] = 0.5\ \mu M$, [rhodopsin] $= 20\ \mu M$, and [GTP] $= 200\ \mu M$.

In the experiment shown in Fig. 3, we studied the activation of transducin GTPase by the catalytic domain of RGS9 (RGS9d) and by its mutant bearing two amino acid substitutions (corresponding to L353E and R360P of the full-length RGS9). The rates of GTP hydrolysis were measured at a range of transducin concentrations with and without RGS9d variants. Data obtained without RGS9d (the lower data set in Fig. 3A) represented the basal GTPase activity of transducin. Note that the linearity requirement was met throughout the entire range. The slope of this line was subtracted from data obtained with RGS9d and its mutant (Fig. 3B), and the corresponding catalytic parameters were determined from the hyperbolic fits of data. This analysis revealed that the RGS9d mutant has an ~3-fold higher catalytic activity than the wild-type RGS9d. This was reflected by a 3-fold difference in the V_{max} value (the k_{cat} values could be calculated by dividing the V_{max} value by the concentration of RGS9d in the assays). The effect of the mutations on the K_m value was smaller, with the RGS9d mutant having ~1.6-fold lower K_m than the wild-type RGS9d.

In another application of multiple turnover assays, illustrated in Fig. 4, the kinetic mechanism of RGS9-1·Gβ5L regulation by PDEγ was analyzed. Note that PDEγ in this assay was substituted by its PDEγ63–87

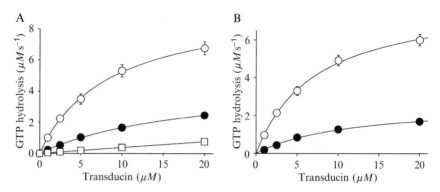

FIG. 3. Comparison of catalytic properties of RGS9 homology domains in multiple turnover GTPase assays. (A) Multiple turnover assays were performed as described in the text. GTP hydrolysis rates at increasing transducin heterotrimer concentrations were measured without RGS proteins ($\square$), with 1 μM RGS9d ($\bullet$), or with 1 μM RGS9d mutant ($\bigcirc$). Data obtained in the absence of RGS were fitted with a straight line with a slope of 0.038 s^{-1}. Data obtained in the presence of RGS were fitted with the sum of the same straight line and a hyperbola (from B). (B) Data from A replotted after the subtraction of the basal activity of transducin and fitted with hyperbolas. The values of catalytic parameters were V_{max} = 2.7 ± 0.2 $\mu M \cdot s^{-1}$ and K_m = 11.8 ± 1.6 μM for RGS9d and V_{max} = 8.2 ± 0.5 $\mu M \cdot s^{-1}$ and K_m = 7.2 ± 1.0 μM for the RGS9d mutant. *Error bars* represent the scatter of the points. Reproduced from Martemyanov and Arshavsky (2002), with permission.

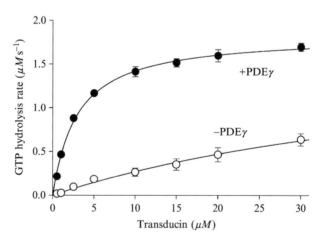

FIG. 4. The effect of PDEγ on the catalytic properties of RGS9-1 · Gβ5L in multiple turnover GTPase assays. The catalytic activity of 1 μM RGS9-1 · Gβ5L was determined in the absence ($\bigcirc$) or presence ($\bullet$) of 50 μM of PDEγ peptide 63–87 as described in the text. The values of kinetic constants were V_{max} = 1.84 ± 0.03 $\mu M \cdot s^{-1}$ and K_m = 2.9 ± 0.2 μM with PDEγ and V_{max} = 1.79 ± 0.39 $\mu M \cdot s^{-1}$ and K_m = 56 ± 17 μM without PDEγ. *Error bars* represent the scatter of the points. Reprinted from Skiba *et al.* (2001), with permission.

peptide, which is sufficient for regulation of RGS9 activity, but does not interfere with the cycle of transducin activation and inactivation, as does the full-length PDEγ (Skiba et al., 2000). Three data sets were obtained: (1) basal GTPase activity of transducin measured without RGS9; (2) transducin GTPase activity measured in the presence of RGS9; and (3) transducin GTPase activity measured in the presence of both RGS9 and PDEγ. Values obtained from the linear fit of basal GTPase activity were subtracted from data obtained in the two latter sets, and the resulting values were fitted by hyperbolas. The catalytic parameters obtained in this experiment indicate that PDEγ acts by reducing the RGS9-1 K_m for transducin by $\sim$20-fold, with the V_{max} value being practically unaffected.

Analysis of RGS9 Interactions with G Proteins in Pull-Down Assays

Pull-down assays are used for qualitative assessments of binding affinities between RGS proteins and their G-protein α subunit partners. They are also used in studies of how regulatory proteins affect RGS–G protein interactions. In these assays, the binding equilibrium between an RGS protein and a Gα subunit is achieved in a system where one of these proteins is immobilized on beads and the other one is soluble. The extent of complex formation is determined after washing the beads and measuring the amount of the retained nonimmobilized protein. The degree of retention of the nonimmobilized protein is considered a measure of RGS–Gα mutual affinity. Because the affinities of Gα subunits in their GDP- or GTP-bound states for RGS proteins are usually too low to retain significant amounts of protein after washing the beads, Gα subunits in pull-down assays are usually activated by AlF$_4$$^-$. The binding of AlF$_4$$^-$ to Gα subunits mimics their transition state during GTP hydrolysis in which they interact with RGS proteins with highest affinity (Berman et al., 1996; Tesmer et al., 1997).

Pull-down assays are relatively simple, straightforward, and useful in providing independent qualitative confirmations of results obtained in kinetic assays. However, they have at least two limitations. First, the actual measurements of protein association are conducted after extensive washes of the beads so that kinetics of protein complex dissociation influence results obtained in pull-down assays. Second, it is possible that the relative affinities of various RGS proteins for AlF$_4$$^-$-bound species of G$\alpha$ do not necessarily reflect their affinities for GTP-bound Gα species.

In an illustrative experiment shown in Fig. 5, we used the pull-down assay to compare affinities of Gα_o and Gα_t for both splice isoforms of RGS9, with and without PDEγ. Recombinant RGS9-1 · Gβ5L or RGS9-2 · Gβ5S (250 pmol each) are immobilized on Ni-NTA agarose

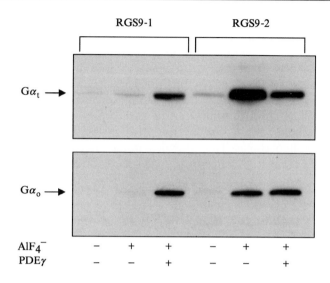

FIG. 5. The binding of RGS9 splice isoforms to $G\alpha_o$ and $G\alpha_t$ in pull-down assays. Pull-down assays were conducted as described in the text. The amounts of $G\alpha_o$ or $G\alpha_t$ retained on beads containing immobilized RGS9 isoforms were determined by western blotting using antibodies specific for each subunit. Reprinted from Martemyanov *et al.* (2003a), with permission.

beads (10 μl) via hexahistidine tags located on the N termini of RGS9 molecules. Immobilization is performed on ice for 20 min in a buffer containing 20 mM Tris–HCl (pH 8.0), 300 mM NaCl, 2 mM MgCl$_2$, 10 μM GDP, 0.25% lauryl sucrose, and 50 μg/ml bovine serum albumin. The beads are then washed once with 500 μl buffer to remove excess unbound RGS9 and are resuspended in 50 μl of the same buffer containing 250 nM of the G-protein α subunit. Ten millimolar NaF and 30 μM AlCl$_3$ (yielding AlF$_4^-$) and/or PDEγ at the final concentration of 1 μM are added when necessary. The samples are incubated on ice for 20 min with occasional shaking. The beads are sedimented and washed twice with 1 ml of the binding buffer supplemented with 20 mM imidazole to minimize non-specific protein binding to the beads. Ten millimolar NaF and 30 μM AlCl$_3$ are included in the washing buffer when required. Gα subunits retained on the immobilized RGS9 are eluted by 40 μl of SDS–PAGE sample buffer. Ten-microliter aliquots of the eluates are then subjected to SDS–PAGE followed by western blot detection using specific anti-Gα antibodies.

Results obtained with $G\alpha_t$ (Fig. 5, top) indicate that appreciable $G\alpha_t$ binding to RGS9-1 occurred only in the presence of PDEγ. The binding of

$G\alpha_t$ to RGS9-2, however, occurred in both cases, with $PDE\gamma$ reducing the amount of bound $G\alpha_t$. These data are entirely consistent with the results of GTPase assays from Figs. 1 and 4, and further confirm our conclusion that $PDE\gamma$ increases the affinity between RGS9-1 and $G\alpha_t$, but competes with the RGS9-2 C terminus for $G\alpha_t$ binding. Results obtained with $G\alpha_o$ were very similar except that in this case $PDE\gamma$ did not inhibit $G\alpha_o$ binding to RGS9-2.

Concluding Remarks

The experimental strategies developed in kinetic studies of interactions between RGS9 isoforms with transducin and other G proteins of the G_i subfamily provide a useful framework for conducting similar studies with essentially any RGS protein – G protein pair. In addition, these studies allow one not only to appreciate the strengths of any individual experimental approach, but also to realize that every approach has its specific limitations. These studies thus emphasize the importance of combining multiple experimental approaches in order to obtain comprehensive and internally consistent information regarding kinetic mechanisms of RGS protein action.

Materials

Detailed descriptions of obtaining protein materials described in this chapter can be found in the following publications. The preparation of urea-treated photoreceptor membranes serving as the source of rhodopsin and/or R9AP is described by Nekrasova *et al.* (1997). Full-length recombinant RGS9 proteins were obtained as constitutive complexes with $G\beta5$ splice isoforms ($G\beta5L$ for RGS9-1 and $G\beta5S$ for RGS9-2) by expression in Sf9 cells and isolation using the introduced affinity tag (He *et al.*, 2000; Skiba *et al.*, 2001). The RGS9d homology domain of RGS9 and its mutants was obtained by expressing them in *Escherichia coli* as fusion proteins with glutathione transferase (He *et al.*, 2000). The GST tag was then removed proteolytically. The transducin heterotrimer was isolated from bovine rod outer segments as described by Ting *et al.* (1993). The $\beta_1\gamma_1$ subunits of transducin were isolated from the transducin heterotrimer by Blue Sepharose chromatography (Ting *et al.*, 1993). The α subunits of G_o and G_i were expressed in *E. coli* as N-terminal fusions with hexahistidine tags and were subsequently purified by metal-chelating chromatography (Lee *et al.*, 1994). $PDE\gamma$ was obtained by expression and purification in *E. coli* (Slepak *et al.*, 1995). $PDE\gamma$ peptide 63–87 used in the multiple turnover GTPase assays was obtained by chemical peptide synthesis.

Acknowledgments

We thank Dr. V. I. Govardovski for many helpful discussions and I.B. Leskov for critical reading of the manuscript. This work was supported by NIH Grant EY 12859 to V.Y.A. and the American Heart Association Award to K.A.M. V.Y.A. is a recipient of the Jules and Doris Stein Professorship from Research to Prevent Blindness.

References

Arshavsky, V. Y., Antoch, M. P., and Philippov, P. P. (1987). On the role of transducin GTPase in the quenching of a phosphodiesterase cascade of vision. *FEBS Lett.* **224**, 19–22.

Arshavsky, V. Y., and Bownds, M. D. (1992). Regulation of deactivation of photoreceptor G protein by its target enzyme and cGMP. *Nature* **357**, 416–417.

Arshavsky, V. Y., Lamb, T. D., and Pugh, E. N., Jr. (2002). G proteins and phototransduction. *Annu. Rev. Physiol.* **64**, 153–187.

Berman, D. M., Kozasa, T., and Gilman, A. G. (1996). The GTPase-activating protein RGS4 stabilizes the transition state for nucleotide hydrolysis. *J. Biol. Chem.* **271**, 27209–27212.

Burns, M. E., and Baylor, D. A. (2001). Activation, deactivation, and adaptation in vertebrate, photoreceptor cells. *Annu. Rev. Neurosci.* **24**, 779–805.

Burns, M. E., and Wensel, T. G. (2003). From molecules to behavior: New clues for RGS function in the striatum. *Neuron* **38**, 853–856.

Chen, C. K., Burns, M. E., He, W., Wensel, T. G., Baylor, D. A., and Simon, M. I. (2000). Slowed recovery of rod photoresponse in mice lacking the GTPase accelerating protein RGS9-1. *Nature* **403**, 557–560.

Chen, C. K., Eversole-Cire, P., Zhang, H. K., Mancino, V., Chen, Y. J., He, W., Wensel, T. G., and Simon, M. I. (2003). Instability of GGL domain-containing RGS proteins in mice lacking the G protein β-subunit Gβ5. *Proc. Natl. Acad. Sci. USA* **100**, 6604–6609.

Cowan, C. W., He, W., and Wensel, T. G. (2000a). RGS proteins: Lessons from the RGS9 subfamily. *Prog. Nucleic Acid Res. Mol. Biol.* **65**, 341–359.

Cowan, C. W., Wensel, T. G., and Arshavsky, V. Y. (2000b). Enzymology of GTPase acceleration in phototransduction. *Methods Enzymol.* **315**, 524–538.

Granneman, J. G., Zhai, Y., Zhu, Z., Bannon, M. J., Burchett, S. A., Schmidt, C. J., Andrade, R., and Cooper, J. (1998). Molecular characterization of human and rat RGS9L, a novel splice variant enriched in dopamine target regions, and chromosomal localization of the RGS9 Gene. *Mol. Pharmacol.* **54**, 687–694.

He, W., Cowan, C. W., and Wensel, T. G. (1998). RGS9, a GTPase accelerator for phototransduction. *Neuron* **20**, 95–102.

He, W., Lu, L. S., Zhang, X., El Hodiri, H. M., Chen, C. K., Slep, K. C., Simon, M. I., Jamrich, M., and Wensel, T. G. (2000). Modules in the photoreceptor RGS9-1·Gβ5L GTPase-accelerating protein complex control effector coupling, GTPase acceleration, protein folding, and stability. *J. Biol. Chem.* **275**, 37093–37100.

Hu, G., and Wensel, T. G. (2002). R9AP, a membrane anchor for the photoreceptor GTPase accelerating protein, RGS9-1. *Proc. Natl. Acad. Sci. USA* **99**, 9755–9760.

Hu, G., Zhang, Z., and Wensel, T. G. (2003). Activation of RGS9-1 GTPase acceleration by Its membrane anchor, R9AP. *J. Biol. Chem.* **278**, 14550–14554.

Krumins, A. M., and Gilman, A. G. (2002). Assay of RGS protein activity *in vitro* using purified components. *Methods Enzymol.* **344**, 673–685.

Lee, E., Linder, M. E., and Gilman, A. G. (1994). Expression of G-protein α subunits in *Escherichia coli*. *Methods Enzymol.* **237**, 146–164.

Lishko, P. V., Martemyanov, K. A., Hopp, J. A., and Arshavsky, V. Y. (2002). Specific binding of RGS9-Gβ5L to protein anchor in photoreceptor membranes greatly enhances its catalytic activity. *J. Biol. Chem.* **277**, 24376–24381.

Martemyanov, K. A., and Arshavsky, V. Y. (2002). Noncatalytic domains of RGS9-1·Gβ5L play a decisive role in establishing its substrate specificity. *J. Biol. Chem.* **277**, 32843–32848.

Martemyanov, K. A., Hopp, J. A., and Arshavsky, V. Y. (2003a). Specificity of G protein-RGS protein recognition is regulated by affinity adapters. *Neuron* **38**, 857–862.

Martemyanov, K. A., Lishko, P. V., Calero, N., Keresztes, G. T., Sokolov, M., Strissel, K. J., Leskov, I. B., Hopp, J. A., Kolesnikov, A. V., Chen, C.-K., Lem, J., Heller, S., Burns, M. E., and Arshavsky, V. Y. (2003b). The DEP domain determines subcellular targeting of the GTPase activating protein RGS9 in vivo. *J. Neurosci.* **23**, 10175–10181.

Nekrasova, E. R., Berman, D. M., Rustandi, R. R., Hamm, H. E., Gilman, A. G., and Arshavsky, V. Y. (1997). Activation of transducin guanosine triphosphatase by two proteins of the RGS family. *Biochemistry* **36**, 7638–7643.

Rahman, Z., Gold, S. J., Potenza, M. N., Cowan, C. W., Ni, Y. G., He, W., Wensel, T. G., and Nestler, E. J. (1999). Cloning and characterization of RGS9-2: A striatal-enriched alternatively spliced product of the RGS9 gene. *J. Neurosci.* **19**, 2016–2026.

Rahman, Z., Schwarz, J., Gold, S. J., Zachariou, V., Wein, M. N., Choi, K. H., Kovoor, A., Chen, C. K., DiLeone, R. J., Schwarz, S. C., Selley, D. E., Sim-Selley, L. J., Barrot, M., Luedtke, R. R., Self, D., Neve, R. L., Lester, H. A., Simon, M. I., and Nestler, E. J. (2003). RGS9 modulates dopamine signaling in the basal ganglia. *Neuron* **38**, 941–952.

Ross, E. M. (2002). Quantitative assays for GTPase-activating proteins. *Methods Enzymol.* **344**, 601–617.

Skiba, N. P., Bae, H., and Hamm, H. E. (1996). Mapping of effector binding sites of transducin α-subunit using Gαt/Gαi1 chimeras. *J. Biol. Chem.* **271**, 413–424.

Skiba, N. P., Hopp, J. A., and Arshavsky, V. Y. (2000). The effector enzyme regulates the duration of G protein signaling in vertebrate photoreceptors by increasing the affinity between transducin and RGS protein. *J. Biol. Chem.* **275**, 32716–32720.

Skiba, N. P., Martemyanov, K. A., Elfenbein, A., Hopp, J. A., Bohm, A., Simonds, W. F., and Arshavsky, V. Y. (2001). RGS9-Gβ5 substrate selectivity in photoreceptors: Opposing effects of constituent domains yield high affinity of RGS interaction with the G protein-effector complex. *J. Biol. Chem.* **276**, 37365–37372.

Skiba, N. P., Yang, C. S., Huang, T., Bae, H., and Hamm, H. E. (1999). The α-helical domain of Gαt determines specific interaction with regulator of G protein signaling 9. *J. Biol. Chem.* **274**, 8770–8778.

Slepak, V. Z., Artemyev, N. O., Zhu, Y., Dumke, C. L., Sabacan, L., Sondek, J., Hamm, H. E., Bownds, M. D., and Arshavsky, V. Y. (1995). An effector site that stimulates G-protein GTPase in photoreceptors. *J. Biol. Chem.* **270**, 14319–14324.

Tesmer, J. J. G., Berman, D. M., Gilman, A. G., and Sprang, S. R. (1997). Structure of RGS4 bound to AlF4−-activated Giα1: Stabilization of the transition state for GTP hydrolysis. *Cell* **89**, 251–261.

Ting, T. T., Goldin, S. B., and Ho, Y.-K. (1993). Purification and characterization of bovine transducin and its subunits. *Methods Neurosci.* **15**, 180–195.

Yamanaka, G., Eckstein, F., and Stryer, L. (1985). Stereochemistry of the guanyl nucleotide binding site of transducin probed by phosphorothioate analogues of GTP and GDP. *Biochemistry* **24**, 8094–8101.

Zachariou, V., Georgescu, D., Sanchez, N., Rahman, Z., DiLeone, R., Berton, O., Neve, R. L., Sim-Selley, L. J., Selley, D. E., Gold, S. J., and Nestler, E. J. (2003). Essential role for RGS9 in opiate action. *Proc. Natl. Acad. Sci. USA* **100**, 13656–13661.

[14] Assays of Nuclear Localization of R7/Gβ_5 Complexes

By WILLIAM F. SIMONDS, GEOFFREY E. WOODARD, and JIAN-HUA ZHANG

Abstract

Heterodimeric complexes between individual members of the R7 subfamily of regulators of G-protein signaling proteins and the Gβ_5 isoform of the heterotrimeric G-protein β subunit family are strongly expressed in the cell nucleus in neurons and brain, as well as in the cytoplasm and plasma membrane. Native and recombinant Gβ_5 and/or R7 expression have been studied in model systems like rat pheochromocytoma PC12 cells where their nuclear localization can be studied by fluorescence microscopy and/or subcellular fractionation. Nucleic acid counterstains chosen for compatibility with the fluorescent tags on secondary antibodies can facilitate the assay of R7/Gβ_5 nuclear localization by epifluorescence or confocal laser microscopy. Subcellular fractionation allows isolation of a purified nuclear fraction that can be probed for the presence of Gβ_5 and/or R7 subunits by immunoblots or immunoprecipitation and compared to other subcellular fractions. While the function of nuclear R7/Gβ_5 complexes is unknown, comparison with the properties of other RGS proteins that localize to the cell nucleus may suggest modes of action. Models are offered in which the reversible post-translational modification of R7/Gβ_5 complexes regulates their nuclear localization and signaling activity, whether the target of such signaling activity is in the nucleus, at the plasma membrane, or both.

Introduction

Complexes between members of the R7 subfamily of regulators of G-protein signaling (RGS) protein and the Gβ_5 isoform of the heterotrimeric G-protein β subunit family, R7/Gβ_5 complexes, are strongly expressed in the cell nucleus in neurons and brain, as well as in the cytoplasm and plasma membrane (Chatterjee *et al.*, 2003; Zhang *et al.*, 2001). Understanding the significance of R7/Gβ_5 complex nuclear localization and assaying such localization require consideration of their known biochemical, signaling and other functional properties. In mammals, R7/Gβ_5 complexes are expressed primarily in the brain and nervous system, including the photosensitive cells of the retina (Cabrera *et al.*, 1998; Witherow and Slepak,

2003). The function of the R7/Gβ_5 complex in the retina, which consists of the RGS9-1 splice variant bound as a heterodimer with the long splice variant of Gβ_5, Gβ_5L, is perhaps the best understood of all R7/Gβ_5 complexes. In photoreceptor cells, RGS9-1/Gβ_5L complexes accelerate the deactivation of transducin in an effector-regulated fashion (Arshavsky *et al.*, 2002). Because RGS9-1/Gβ_5L complexes are anchored to the photoreceptor disk membrane by a special anchoring protein, they are likely not free to traffic to the nucleus (Hu and Wensel, 2002). For this reason the remainder of this article focuses on the understanding and assaying of nuclear localization of R7/Gβ_5 complexes in the brain and non-photosensitive neurons.

Members of the R7 subfamily of RGS proteins possess a similar domain structure, which consists of an N-terminal DEP domain (Ponting and Bork, 1996), a Gγ-like (GGL) domain (Snow *et al.*, 1998; Sondek and Siderovski, 2001), an RGS core domain (Ross and Wilkie, 2000), and a C-terminal region of variable length. Their ability to complex with Gβ_5 subunits requires the GGL domain on the R7 subunit and a hydrophobic pocket on the Gβ_5 subunit that can accommodate a conserved tryptophan residue near the C terminus of the GGL domain (Snow *et al.*, 1999). The R7 domain structure is conserved even in the phylogenetically remote members EGL-10 (Koelle and Horvitz, 1996) and EAT-16 (Hajdu-Cronin *et al.*, 1999) in *Caenorhabditis elegans* and drosRGS7 (Elmore *et al.*, 1998) and CG7095 (Sondek and Siderovski, 2001) in *Drosophila*.

Alternative R7 gene splicing can affect the nuclear expression and subcellular localization of the resulting R7 proteins and R7/Gβ_5 complexes, as shown for the human RGS6 gene by Chatterjee *et al.* (2003). Multiple transcripts arising from alternative splicing of the RGS6 gene encode proteins with or without the N-terminal DEP domain, with a complete or partial GGL domain, and with one of several C-terminal sequences. The variable presence of these sequences, as well as the absence or presence of coexpressed Gβ_5, can alter the nuclear localization of RGS6 and Gβ_5 dramatically. For example, coexpression of Gβ_5 can alter the expression of a 472 residue RGS6 isoform containing the DEP domain from exclusively cytoplasmic (without Gβ_5) to predominantly nuclear (with Gβ_5) (Chatterjee *et al.*, 2003).

Post-translational modification has been demonstrated for R7 proteins that could, in theory, affect the nuclear localization of R7/Gβ_5 complexes, although the regulation of nuclear localization has not yet been shown for any such modification. RGS7 and perhaps other R7 subfamily proteins can be phosphorylated (Benzing *et al.*, 1999) and/or palmitoylated (Rose *et al.*, 2000); the former modification may govern binding of 14-3-3 proteins (Benzing *et al.*, 2000, 2002).

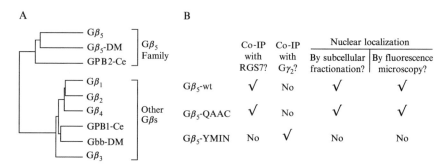

Fig. 1. (A) The Gβ_5 subunit, unique among heterotrimeric Gβ subunits for its ability to bind R7 subfamily RGS proteins, is the most highly conserved Gβ subunit in metazoans. Included in this phylogenetic tree are mammalian Gβ subunits Gβ_1 thru Gβ_5, *Drosophila* Gβ_5 and Gbb, and *C. elegans* GPB-1 and GPB-2. Not shown is the *Drosophila* Gbe subunit, which forms a distinct branch by itself. Figure generated from v. 8 of the PileUp algorithm of the Genetics Computer Group (Madison, WI). (B) Nuclear localization of wild-type and mutant Gβ_5 subunits correlates with its ability to bind to R7 subfamily RGS proteins. The wild-type Gβ_5 is compared to the Gβ_5-QAAC mutant (Gβ_5-Lys25Gln/Glu29Ala/Lys32Ala/Leu33Cys) and the Gβ_5-YMIN mutant (Gβ_5-Phe93Tyr/Thr338Met/Val351Ile/Ala353Asn) for its ability to coimmunoprecipitate (Co-IP) with RGS7 or Gγ_2 and for its ability to localize to the cell nucleus, as assayed by subcellular fractionation followed by immunoblotting or laser confocal fluorescence microscopy. Mutations in Gβ_5-QAAC map to the Gβ coiled-coil region, and mutations in the Gβ_5-YMIN mutant map to a putative hydrophobic pocket on the side of the β-propeller region of Gβ (Snow *et al.*, 1999). Data taken from Rojkova *et al.* (2003).

Gβ_5 is the most highly conserved metazoan Gβ subunit isoform (Watson *et al.*, 1994), with homologs present in the nervous systems of worms, flies, and humans (Fig. 1A). From native mammalian tissues, Gβ_5 has only been purified in complex with an R7 subfamily RGS protein (Cabrera *et al.*, 1998; Witherow *et al.*, 2000; Zhang and Simonds, 2000), never as part of a G$\beta\gamma$ complex, even though a functional association of Gβ_5 with Gγ subunits can be demonstrated *in vitro* (Lindorfer *et al.*, 1998; Watson *et al.*, 1994; Zhang *et al.*, 1996). Mutational studies suggest that the ability of Gβ_5 to undergo nuclear localization depends on its ability to heterodimerize with R7 proteins (Fig. 1B) (Rojkova *et al.*, 2003).

Assays of R7/Gβ_5 Nuclear Localization

A variety of biochemical and microscopic methods can be used to assess and monitor the nuclear localization of native and recombinant R7/Gβ_5 complexes. The three principal assays used in our laboratory to assay the nuclear localization of R7/Gβ_5 complexes are presented here. Two methods employ fluorescence microscopy. These are indirect immunofluorescence microscopy employing anti-R7 or anti-Gβ_5 primary antibodies and secondary antibodies tagged with fluorescent chemical moieties, and direct fluorescence microscopy of recombinant protein representing a fusion between *Aequorea victoria* green fluorescent protein (GFP) or a related fluorescent protein with either an R7 subfamily RGS protein or Gβ_5. The final method used to assay R7/Gβ_5 nuclear localization is subcellular fractionation and nuclear isolation in combination with immunoblotting with an anti-R7 or anti-Gβ_5 primary antibody. The method of subcellular fractionation is modification of a previously described technique (Schilling *et al.*, 1999). Each method is discussed in more detail, drawing on model systems studied successfully in our laboratory.

Fluorescence Microscopy of Nuclear R7/Gβ_5 Complexes

Materials

Cell Culture Media and Buffers: Dulbecco's modified Eagle's medium (DMEM) supplemented with 10% horse serum, 5% fetal bovine serum (FBS), 4 mM L-glutamine, 100 U/ml penicillin, and 100 μg/ml streptomycin ("supplemented DMEM"), nerve growth factor-β (NGF) [Sigma N-2513; reconstitute in 0.2-μm-filtered phosphate-buffered saline (PBS) containing 0.1% bovine serum albumin (BSA) at >0.1 μg/ml concentration; store in aliquots at $-20°$], PBS, and FBS

Slides: Covered chamber slides (Lab-Tek II Chamber Slide System, Nunc; for epifluorescence) or chambered coverglass (Lab-Tek II Chambered Coverglass, Nunc; for laser confocal microscopy). Coat the chamber slides or coverglass with poly-D-lysine using the following method: dilute stock poly-D-lysine (1 mg/ml) to a concentration of 50 μg/ml in unsupplemented DMEM. Add 1 ml of the dilution into the chamber and incubate at room temperature for 1 h. Wash twice with sterile dH$_2$O. Discard the wash and let the slides air dry for 1 h in a laminar flow (cell culture) hood.

Paraformaldehyde, 16% solution (Electron Microscopy Sciences)

Saponin (Calbiochem)

Nuclear Counterstaining Dyes: YOYO-1 (Molecular Probes; a dimer of oxazole yellow; stock solution is 1 mM in dimethyl sulfoxide; for use dilute 1:25,000 into PBS); DAPI (diamidino-2-phenylindole; stock solution of 25 mg/ml in dH$_2$O; for use dilute ~1:300 into PBS); propidium iodide (stock solution of 0.5 mg/ml in PBS; for use dilute 1:10 into PBS); Hoechst 33342 (stock solution 10 μg/ml in dH$_2$O; for use dilute 1:100 into dH$_2$O)

Transfection Reagents: high-quality plasmid DNA encoding desired construct(s), prepared by Qiagen maxi-prep or endotoxin-free Qiagen maxi-prep; LipofectAmine 2000 transfection reagent (Invitrogen); OptiMEM I reduced serum medium (Invitrogen)

Primary Antibodies: goat anti-RGS7 C-19 IgG (Santa Cruz sc-8139), rabbit anti-Gβ_5 ATDG polyclonal (Zhang and Simonds, 2000) (although not tested, Santa Cruz goat anti-Gβ_5 N-14 IgG [sc-14942] applications and immunogen appear similar to ATDG), rabbit anti-TFIID (TBP) N-12 IgG (Santa Cruz sc-204)

Indirect Immunofluorescence Microscopy of R7 Subfamily RGS Protein and/or Gβ_5 in Rat Pheochromocytoma PC12 Cells

1. Maintain PC12 cells in culture flasks at 37° in supplemented DMEM.

2. When cells are 80–90% confluent, dislodge from the feeder flasks by tapping, and plate PC12 cells onto poly-D-lysine precoated covered chamber slides or chambered coverslips at ~60–70% of initial density.

3. Grow cells at 37° for 16 h in supplemented DMEM containing 50 ng/ml NGF.

4. Discard the spent DMEM. Wash cells twice with PBS at room temperature and then fix in 4% (w/v) paraformaldehyde in PBS (dilute 1:4 from stock 16% paraformaldehyde solution) at room temperature for 15 min.

5. Incubate slides with one or more primary antibodies in PBS/10% FBS/0.1% (w/v) saponin for 1.5 h at room temperature.

6. Wash slides twice with PBS and then incubate with appropriate secondary antibody labeled covalently with fluorescein isothiocyanate or rhodamine red in PBS/10% FBS/0.1% (w/v) saponin for 45 min at room temperature.

7. For imaging of cell nuclei, counterstain cells with the desired DNA-binding fluorescent dye for 5 to 30 min at room temperature to achieve desired intensity (Table I).

8. Rinse slides three times with PBS.

9. Mount slides with 1 or 2 drops of Permount (Vector Labs) applied to the slide surface.

10. For epifluorescent imaging, view cells in Zeiss Axiophot or comparable inverted microscope with appropriate excitation and barrier filters (Fig. 2A).

11. For confocal imaging, view cells with a Zeiss LSM 510 laser-scanning microscope with an appropriate wavelength laser source (Fig. 2B).

TABLE I
PROPERTIES OF DYES USEFUL AS NUCLEAR COUNTERSTAINS IN THE ASSAY OF Gβ_5/R7 COMPLEX
NUCLEAR LOCALIZATION BY FLUORESCENCE MICROSCOPY

Nucleic acid stain	Excitation λ maximum (nm)	Emission λ maximum (nm)	Visible fluorescence	General comments
Hoechst 33342	343 (UV)	483	Blue	Membrane permeant, stains nuclei of living cells (fixation/ permeabilization not required)
DAPI	358 (UV)	461	Blue	Semipermeant of membranes in living cells
YOYO-1	491	509	Green	Bright nuclear staining, but weakly stains RNA so may need to treat cells/tissues with RNase to reduce cytoplasmic staining
Propidium iodide	535	617	Red	May need to treat cells/tissues with RNase to reduce cytoplasmic staining; membrane impermeant— without fixation/permeabilization indicator of cell death

Direct Fluorescence Microscopy of a Recombinant Fusion Protein Combining GFP with either an R7 Subfamily RGS Protein or Gβ_5 Expressed in PC12 Cells

1. Clone the cDNA encoding a Gβ_5 or the R7 subfamily RGS protein into a mammalian expression vector containing GFP (or a related fluorescent derivative) followed by a multiple cloning site. Ensure that the insert is in frame with the GFP and, if possible, omit or substitute the starting methionine of Gβ_5 or the R7 subfamily RGS protein with alanine to minimize the possibility of expressing translational products lacking the GFP moiety. To make a GFP–Gβ_5 construct, we employed the pEGFP-C2 vector (Clontech) encoding a red-shifted variant of GFP (Zhang *et al.*, 2001). Note that other laboratories have successfully employed expression constructs in which the R7 protein entity comprises the N-terminal part of the fusion followed by the GFP moiety (Chatterjee *et al.*, 2003). Confirm the DNA sequence of the GFP-insert junction and the remainder of the insert, and prepare a sufficient quantity of high-quality purified plasmid to allow subsequent transfections.

2. One day prior to transfection, harvest PC12 cells at 80–90% confluence by tapping gently to dislodge from the feeder flasks, and plate PC12 cells in 2 ml of supplemented DMEM containing 50 ng/ml NGF into

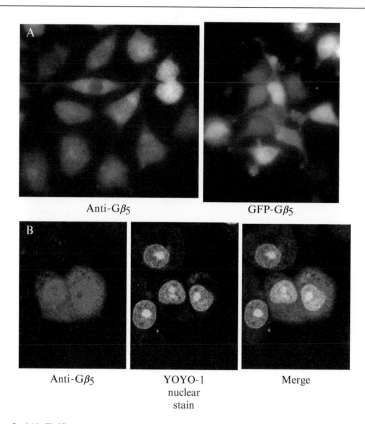

Anti-Gβ_5 GFP-Gβ_5

Anti-Gβ_5 YOYO-1 nuclear stain   Merge

FIG. 2. (A) Epifluorescence microscopic images (black and white) of rat pheochromo-cytoma PC12 cells. Untransfected PC12 cells processed for immunohistochemistry with affinity-purified ATDG anti-Gβ_5 primary antibody and fluorescein isothiocyanate (FITC)-labeled antirabbit secondary antibody (left). PC12 cells transfected with the pEGFP-C2 construct encoding a GFP–Gβ_5 fusion protein (right). Modified from Zhang *et al.* (2001), with permission. (B) Laser confocal immunofluorescence analysis of stably transfected PC12 cells. Cells transfected stably with HA epitope-tagged Gβ_5 were analyzed by laser confocal microscopy after dual staining with affinity-purified ATDG anti-Gβ_5 antibody (red signal) and the nuclear counterstain YOYO-1 (green signal). The yellow–orange signal in the merge image indicates colocalization of probes. Modified from Zhang *et al.* (2001), with permission. (See color insert.)

poly-D-lysine precoated covered chamber slides or chambered coverslips. The number of cells is adjusted to 60–70% of their density at harvest.

3. Let the cells incubate overnight at 37°. Combine 2 μg of plasmid DNA and 8 μl of LipofectAmine 2000 transfection reagent and dilute into 200 μl of Opti-MEM I medium, mix gently, and incubate at room

temperature for 20 min. The transfection mixture should then be added directly to each chamber.

4. After mixing gently, incubate the cultures for an additional 24 to 48 h. Although we usually do not use this step, if desired, the medium containing the DNA and transfection reagent can be discarded after 5 h and replaced with fresh supplemented DMEM for the remainder of the 24 to 48 h.

5. Discard the spent DMEM. Wash cells twice with PBS at room temperature and then fix in 4% (w/v) paraformaldehyde in PBS (dilute 1:4 from stock 16% paraformaldehyde solution) at room temperature for 15 min.

6. Rinse the slides in PBS. For imaging of cell nuclei, counterstain cells with the desired DNA-binding fluorescent dye for 5 to 30 min at room temperature to achieve desired intensity (Table I).

7. Rinse slides three times with PBS.

8. Mount slides with 1 or 2 drops of Permount (Vector Labs) applied to the slide surface.

9. For epifluorescent imaging, view cells in Zeiss Axiophot or comparable inverted microscope with appropriate excitation and barrier filters (Fig. 2A).

10. For confocal imaging, view cells with a Zeiss LSM 510 laser-scanning microscope with an appropriate wavelength laser source (Fig. 2B).

Subcellular Fractionation of Mouse Brain or PC12 Cells Followed by Immunoblotting or Immunoprecipitation to Assay for Nuclear Localization of R7/Gβ₅ Complexes

The methods used here for subcellular fractionation and nuclear isolation, modified from a report by Schilling *et al.* (1999), have been employed by us with good results (Zhang *et al.*, 2001). The reader should be aware that several commercially available kits are available to aid in nuclear isolation and subcellular fractionations, which, although not rigorously compared by us to the method presented here, appear to achieve qualitatively similar results (FractionPREP kit, BioVision, Inc.; NE-PER nuclear and cytoplasmic extraction kit, Pierce Chemical Co.).

Materials and Buffers

Buffer A: 50 mM triethanolamine HCl, pH 7.5, 25 mM KCl, 5 mM MgCl$_2$, 0.5 mM dithiothreitol, 17 μg/ml 4-(2-aminoethyl)-benzenesulfonylfluoride HCl (AEBSF), 2 μg/ml each aprotinin, leupeptin, and pepstatin, and 1 μg/ml soybean trypsin inhibitor Homogenization buffer: 0.25 M sucrose in buffer A, kept on ice prior to use.

Membrane extraction buffer: 0.5% (w/v) Genapol C-100 (or C$_{12}$E$_9$ Lubrol) in buffer A.

Buffer B: 2.3 M sucrose in buffer A

Trypan blue solution 0.4% (w/v) in sterile dH_2O

Dounce homogenizer with type B (loose-type) pestle

Ultracentrifuge with fixed-angle and swinging bucket rotors (e.g., Beckman L8-70M ultracentrifuge, with SW50.1 swinging bucket rotor utilizing 13 × 51-mm tubes)

Polypropylene disposable pellet pestle (Kimble/Kontes; Fisher)

Cordless motor for pellet pestle (Kimble/Kontes; Fisher)

1. All steps should be performed at 4° or on ice.

2. Homogenize a mouse brain (1 g wet weight) or a PC12 cell pellet ($\sim$1 × 10^7 cells, freshly harvested by tapping to dislodge cells after a gentle PBS rinse to remove traces of spent medium) in 5 ml homogenization buffer with 30 strokes in a Dounce homogenizer on ice.

3. For PC12 cells, pass the homogenate through a blunt 25-gauge needle five times on ice to ensure lysis of the majority of cells.

4. Remove 50 μl of homogenate and mix with 50 μl of trypan blue solution and confirm thoroughness of cell lysis and viability of nuclei under a light microscope.

5. Centrifuge the homogenate at 500g for 5 min to remove unbroken cells and tissue debris (first pellet). Save an aliquot of the supernatant as the "total" fraction to compare with subsequent subcellular fractions.

6. Recentrifuge the first supernatant at 2000g for 15 min. The pellet is the second pellet (P2) and should be put aside on ice.

7. Recentrifuge the second supernatant at 100,000g for 1 h in a fixed-angle rotor in an ultracentrifuge. The pellet is the third pellet (P3) and should be put aside on ice.

8. The third supernatant is taken as the cytosolic fraction.

9. Fraction P3 is taken as the membrane fraction and can be taken up in 4 ml membrane extraction buffer by rocking at 4° for 2 h.

10. Resuspend P2 in 1 ml of buffer A and mix with 2 ml of buffer B.

11. Layer the 3-ml mixture carefully on top of 1 ml of buffer B and then centrifuge at 12,400g for 1 h in a swinging bucket rotor.

12. Discard the supernatant carefully and wash the pellet by resuspension in 4 ml homogenization buffer and recentrifugation at 2000g for 15 min.

13. Discard the wash. The washed pellet is taken as the nuclear fraction. The pellet can be resuspended in 2 ml buffer A and a 0.5-ml portion homogenized in a 1.5-ml polypropylene microtube with a motor-driven pellet pestle to reduce viscosity after lysis, in preparation for SDS–PAGE.

14. Analyze expression of Gβ_5 and/or the R7 subfamily RGS protein by SDS–PAGE followed by immunoblotting with specific antibodies (to

native protein or to engineered epitopes if PC12 cells transfected with recombinant cDNAs). The nuclear fraction can be analyzed for Gβ_5 and/or R7 expression alone or in comparison to the "total" starting fraction and/ or to the other subcellular fractions generated. Immunoblotting for expression of the TATA-binding protein can validate the identity of the nuclear fraction and detect possible cross-contamination of other fractions. Immunoprecipitation with specific antibodies can also verify the identity of Gβ_5 and/or R7 expressed in different subcellular fractions.

Nuclear Localization of Other RGS Proteins

In addition to members of the R7 subfamily, other RGS proteins have been demonstrated in the cell nucleus under various conditions using assays of nuclear localization similar to those presented here for R7/ Gβ_5 complexes. General aspects of the nuclear localization of RGS proteins have been reviewed by Burchett (2003). An N-terminally truncated RGS3 variant is expressed strongly in the cell nucleus and induces the apoptosis of transfected CHO cells (Dulin *et al.*, 2000). Fluorescence microsopy of transfected RGS-GFP fusion proteins demonstrated strong nuclear localization of full-length RGS2 and RGS10 (Chatterjee and Fisher, 2000b). The nuclear localization of RGS2 requires a sequence in its N-terminal region that causes nuclear retention after passive inward diffusion and not active nuclear import (Heximer *et al.*, 2001). The nuclear localization of RGS8 also requires an N-terminal sequence for nuclear retention (Saitoh *et al.*, 2001). The nuclear localization of RGS10 and diminution of its negative effect on G-protein-regulated inwardly rectifying potassium channel function both correlated with its protein kinase A (PKA)-mediated phosphorylation (Burgon *et al.*, 2001). This suggests the PKA-regulated nuclear localization of RGS10 represents, at least in part, a functional sequestration resulting in the disinhibition of G-protein pathways at the cell membrane. Three splice variants of RGS12 with alternative N-terminal regions were all strongly localized to the cell nucleus (Chatterjee and Fisher, 2000a). One of the RGS12 splice variants associates with the nuclear matrix, represses DNA transcription, and inhibits cell cycle progression, properties all associated with an N-terminal domain independent of the RGS core homology region (Chatterjee and Fisher, 2002).

Models of R7/Gβ_5 Complex Function in the Nucleus

Canonical models of heterotrimeric G-protein signal transduction do not explain the presence of R7/Gβ_5 complexes in the cell nucleus. The function of R7/Gβ_5 complexes best defined from *in vitro* studies (Posner *et al.*, 1999), accelerating the GTP hydrolysis of $G\alpha \cdot GTP$ complexes [GTPase-activating protein (GAP) activity], would seem best accomplished by localization at the plasma membrane, not in the nucleus. Nor has it been shown, for example, that changes in the activity of G-protein-coupled receptors or heterotrimeric G proteins can cause the relative accumulation or depletion of nuclear $R7/G\beta_5$ complexes or changes in nucleocytoplasmic shuttling. Mild heat stress and treatment with proteasome inhibitors induce the nucleolar accumulation of RGS6, but such redistribution does not seem to involve G-protein-regulated stress-activated protein kinases (Chatterjee and Fisher, 2003).

Despite the current gaps in our understanding, it is worthwhile to consider models that assume the intracellular distribution and nuclear localization of $R7/G\beta_5$ complexes are not constitutive, but are instead dynamic and regulated by signaling events (at the plasma membrane and/or nucleus). In all of these models, reversible post-translational modification of $R7/G\beta_5$ complexes could directly or indirectly cause retention in, or export from, one or another intracellular compartment. Phosphorylation of RGS7 that enables binding of 14-3-3-proteins, for example, might provide such a mechanism (Benzing *et al.*, 2000), as might palmitoylation or depalmitoylation of an R7 protein (Rose *et al.*, 2000). A model that nuclear localization or $R7/G\beta_5$ complexes simply sequesters the heterodimer and reduces GAP activity at the plasma membrane, enhancing signaling through heterotrimeric G-protein pathways normally deactivated by R7/$G\beta_5$ complexes, is an extrapolation of the observations for RGS10 cited earlier (Burgon *et al.*, 2001). More complex models could allow for information transfer between the intracellular compartments mediated by the translocation of $R7/G\beta_5$ complexes, such as between the plasma membrane and nucleus or vice versa. One or more signals incoming to the nucleus could mark nuclear $R7/G\beta_5$ complexes for export, resulting in enhanced GAP activity at the plasma membrane and more rapid deactivation of targeted heterotrimeric G-protein pathways. Alternatively, the incoming signals could mark nuclear $R7/G\beta_5$ complexes for nuclear retention with opposite effects at the plasma membrane. Changes in the activity of G-protein-mediated, or other signaling pathways at the plasma membrane, could mark $R7/G\beta_5$ complexes and trigger the centripetal flow of information with translocation of the heterodimers to the nucleus. In the nucleus it is conceivable that $R7/G\beta_5$ complexes could regulate gene transcription and/or alter progression through the cell cycle, as already speculated for an RGS12 isoform (Chatterjee and Fisher, 2002), or induce apoptosis, as shown for the N-terminally truncated RGS3 isoform (Dulin *et al.*, 2000). Assays for the nuclear localization of $R7/G\beta_5$ complexes such as those presented here should help develop and distinguish these models as functional correlates of the intracellular distribution of R7/Gβ_5 in different signaling environments are established.

Acknowledgments

The authors thank Drs. Valarie A. Barr, Alexandra M. Rojkova, and Christian A. Combs for their collaboration, instruction, and guidance during the development of the assays described here.

References

Arshavsky, V. Y., Lamb, T. D., and Pugh, E. N., Jr. (2002). G proteins and phototransduction. *Annu. Rev. Physiol.* **64,** 153–187.

Benzing, T., Brandes, R., Sellin, L., Schermer, B., Lecker, S., Walz, G., and Kim, E. (1999). Upregulation of RGS7 may contribute to tumor necrosis factor-induced changes in central nervous function. *Nature Med.* **5,** 913–918.

Benzing, T., Yaffe, M. B., Arnould, T., Sellin, L., Schermer, B., Schilling, B., Schreiber, R., Kunzelmann, K., Leparc, G. G., Kim, E., and Walz, G. (2000). 14–3–3 interacts with regulator of G protein signaling proteins and modulates their activity. *J. Biol. Chem.* **275,** 28167–28172.

Benzing, T., Kottgen, M., Johnson, M., Schermer, B., Zentgraf, H., Walz, G., and Kim, E. (2002). Interaction of 14-3-3 protein with regulator of G protein signaling 7 is dynamically regulated by tumor necrosis factor-alpha. *J. Biol. Chem.* **277,** 32954–32962.

Burchett, S. A. (2003). In through the out door: Nuclear localization of the regulators of G protein signaling. *J. Neurochem.* **87,** 551–559.

Burgon, P. G., Lee, W. L., Nixon, A. B., Peralta, E. G., and Casey, P. J. (2001). Phosphorylation and nuclear translocation of a regulator of G protein signaling (RGS10). *J. Biol. Chem.* **276,** 32828–32834.

Cabrera, J. L., De Freitas, F., Satpaev, D. K., and Slepak, V. Z. (1998). Identification of the Gβ_5-RGS7 protein complex in the retina. *Biochem. Biophys. Res. Commun.* **249,** 898–902.

Chatterjee, T. K., and Fisher, R. A. (2000a). Novel alternative splicing and nuclear localization of human RGS12 gene products. *J. Biol. Chem.* **275,** 29660–29671.

Chatterjee, T. K., and Fisher, R. A. (2000b). Cytoplasmic, nuclear, and golgi localization of RGS proteins: Evidence for N-terminal and RGS domain sequences as intracellular targeting motifs. *J. Biol. Chem.* **275,** 24013–24021.

Chatterjee, T. K., and Fisher, R. A. (2002). RGS12TS-S localizes at nuclear matrix-associated subnuclear structures and represses transcription: Structural requirements for subnuclear targeting and transcriptional repression. *Mol. Cell. Biol.* **22,** 4334–4345.

Chatterjee, T. K., and Fisher, R. A. (2003). Mild heat and proteotoxic stress promote unique subcellular trafficking and nucleolar accumulation of RGS6 and other RGS proteins: Role of the RGS domain in stress-induced trafficking of RGS proteins. *J. Biol. Chem.* **278,** 30272–30282.

Chatterjee, T. K., Liu, Z., and Fisher, R. A. (2003). Human RGS6 gene structure, complex alternative splicing, and role of N terminus and G protein gamma-subunit-like (GGL) domain in subcellular localization of RGS6 splice variants. *J. Biol. Chem.* **278,** 30261–30271.

Dulin, N. O., Pratt, P., Tiruppathi, C., Niu, J. X., Voyno-Yasenetskaya, T., and Dunn, M. J. (2000). Regulator of G protein signaling RGS3T is localized to the nucleus and induces apoptosis. *J. Biol. Chem.* **275,** 21317–21323.

Elmore, T., Rodriguez, A., and Smith, D. P. (1998). dRGS7 encodes a Drosophila homolog of EGL-10 and vertebrate RGS7. *DNA Cell Biol.* **17,** 983–989.

Hajdu-Cronin, Y. M., Chen, W. J., Patikoglou, G., Koelle, M. R., and Sternberg, P. W. (1999). Antagonism between G(o)alpha and G(q)alpha in *Caenorhabditis elegans*: The RGS protein EAT-16 is necessary for G(o)alpha signaling and regulates G(q)alpha activity. *Genes Dev.* **13,** 1780–1793.

Heximer, S. P., Lim, H., Bernard, J. L., and Blumer, K. J. (2001). Mechanisms governing subcellular localization and function of human RGS2. *J. Biol. Chem.* **276,** 14195–14203.

Hu, G., and Wensel, T. G. (2002). R9AP, a membrane anchor for the photoreceptor GTPase accelerating protein, RGS9-1. *Proc. Natl. Acad. Sci. USA* **99,** 9755–9760.

Koelle, M. R., and Horvitz, H. R. (1996). EGL-10 regulates G protein signaling in the *C. elegans* nervous system and shares a conserved domain with many mammalian proteins. *Cell* **84,** 115–125.

Lindorfer, M. A., Myung, C. S., Savino, Y., Yasuda, H., Khazan, R., and Garrison, J. C. (1998). Differential activity of the G protein $\beta 5\gamma 2$ subunit at receptors and effectors. *J. Biol. Chem.* **273,** 34429–34436.

Ponting, C. P., and Bork, P. (1996). Pleckstrin's repeat performance: A novel domain in G-protein signaling? *Trends Biochem. Sci.* **21,** 245–246.

Posner, B. A., Gilman, A. G., and Harris, B. A. (1999). Regulators of G protein signaling 6 and 7. Purification of complexes with G$\beta 5$ and assessment of their effects on G protein-mediated signaling pathways. *J. Biol. Chem.* **274,** 31087–31093.

Rojkova, A. M., Woodard, G. E., Huang, T. C., Combs, C. A., Zhang, J. H., and Simonds, W. F. (2003). Ggamma subunit-selective G protein beta 5 mutant defines regulators of G protein signaling protein binding requirement for nuclear localization. *J. Biol. Chem.* **278,** 12507–12512.

Rose, J. J., Taylor, J. B., Shi, J., Cockett, M. I., Jones, P. G., and Hepler, J. R. (2000). RGS7 is palmitoylated and exists as biochemically distinct forms. *J. Neurochem.* **75,** 2103–2112.

Ross, E. M., and Wilkie, T. M. (2000). GTPase-activating proteins for heterotrimeric G proteins: Regulators of G protein signaling (RGS) and RGS-like proteins. *Annu. Rev. Biochem.* **69,** 795–827.

Saitoh, O., Masuho, I., Terakawa, I., Nomoto, S., Asano, T., and Kubo, Y. (2001). Regulator of G protein signaling 8 (RGS8) requires its NH2 terminus for subcellular localization and acute desensitization of G protein-gated K+ channels. *J. Biol. Chem.* **276,** 5052–5058.

Schilling, G., Wood, J. D., Duan, K., Slunt, H. H., Gonzales, V., Yamada, M., Cooper, J. K., Margolis, R. L., Jenkins, N. A., Copeland, N. G., Takahashi, H., Tsuji, S., Price, D. L., Borchelt, D. R., and Ross, C. A. (1999). Nuclear accumulation of truncated atrophin-1 fragments in a transgenic mouse model of DRPLA. *Neuron* **24,** 275–286.

Snow, B. E., Betts, L., Mangion, J., Sondek, J., and Siderovski, D. P. (1999). Fidelity of G protein β-subunit association by the G protein gamma-subunit-like domains of RGS6, RGS7, and RGS11. *Proc. Natl. Acad. Sci. USA* **96,** 6489–6494.

Snow, B. E., Krumins, A. M., Brothers, G. M., Lee, S. F., Wall, M. A., Chung, S., Mangion, J., Arya, S., Gilman, A. G., and Siderovski, D. P. (1998). A G protein gamma subunit-like domain shared between RGS11 and other RGS proteins specifies binding to G$\beta 5$ subunits. *Proc. Natl. Acad. Sci. USA* **95,** 13307–13312.

Sondek, J., and Siderovski, D. P. (2001). Ggamma-like (GGL) domains: New frontiers in G-protein signaling and β-propeller scaffolding. *Biochem. Pharmacol.* **61,** 1329–1337.

Watson, A. J., Katz, A., and Simon, M. I. (1994). A fifth member of the mammalian G-protein β-subunit family: Expression in brain and activation of the $\gamma2$ isotype of phospholipase C. *J. Biol. Chem.* **269,** 22150–22156.

Witherow, D. S., and Slepak, V. Z. (2003). A novel kind of G protein heterodimer: The Gβ5-RGS complex. *Receptors Channels* **9,** 205–212.

Witherow, D. S., Wang, Q., Levay, K., Cabrera, J. L., Chen, J., Willars, G. B., and Slepak, V. Z. (2000). Complexes of the G protein subunit Gβ5 with the regulators of G protein signaling RGS7 and RGS9: Characterization in native tissues and in transfected cells. *J. Biol. Chem.* **275,** 24872–24880.

Zhang, J. H., Barr, V. A., Mo, Y., Rojkova, A. M., Liu, S., and Simonds, W. F. (2001). Nuclear localization of Gβ5 and regulator of G protein signalling 7 in neurons and brain. *J. Biol. Chem.* **276,** 10284–10289.

Zhang, J. H., and Simonds, W. F. (2000). Copurification of brain G-protein β5 with RGS6 and RGS7. *J. Neurosci.* **20**(RC59), 1–5.

Zhang, S. Y., Coso, O. A., Lee, C. H., Gutkind, J. S., and Simonds, W. F. (1996). Selective activation of effector pathways by brain-specific G protein β5. *J. Biol. Chem.* **271,** 33575–33579.

Subsection D

R12 Subfamily

[15] Mapping of RGS12-Ca$_v$2.2 Channel Interaction

By RYAN W. RICHMAN and MARÍA A. DIVERSÉ-PIERLUISSI

Abstract

The α1 (pore-forming) subunit of the Ca$_v$2.2 (N-type) channel is tyrosine phosphorylated by Src kinase upon activation of GABA$_B$ receptors. The tyrosine-phosphorylated form of the α1 subunit of the Ca$_v$2.2 channel becomes a target for the binding of RGS12, a GTPase-accelerating protein. Binding of the phosphotyrosine-binding domain of RGS12 to the tyrosine-phosphorylated channel alters the kinetics of the termination of GABA-mediated inhibition of the calcium current. Using a combination of biochemical and electrophysiological approaches, we have determined that the SNARE binding or "synprint" region of the Ca$_v$2.2 binds to RGS12. This article describes the protocols used to map the interaction using primary neuronal cultures.

Introduction

Regulation of the timing of voltage-dependent calcium channel activity is traditionally thought to involve heterotrimeric G-protein signaling pathways (Hille, 1994; Ikeda and Dunlap, 1999; Wickman and Clapham, 1995). Data from our laboratory suggest that the timing of GABA-induced inhibition is a complex process that involves several signals. The α (pore-forming) subunit of the Ca$_v$2.2 (N-type) channel is tyrosine phosphorylated by Src kinase in a GABA-dependent manner (Schiff et al., 2000). This phosphorylation makes the channel a target for the binding of a GTPase-accelerating protein, RGS12. The phosphotyrosine-binding (PTB) domain of RGS12 binds to the tyrosine-phosphorylated channel, altering the kinetics of the termination of GABA-mediated inhibition of the calcium current (Schiff et al., 2000).

Methods presented in this article have been restricted to those using the endogenous cellular machinery of primary neurons. These protocols measure acute changes in signaling without exogenous gene expression. The reader is referred to reconstitution experiments of calcium channel modulation for methods for heterologous expression of calcium channels (Ewald *et al.*, 1989; Jeong and Ikeda, 2000; Jones *et al.*, 1997).

Neuronal Cultures

Solutions and Materials

Dorsal Root Ganglia (DRG) Dissociation. DRG are dissected from 10- to 12-day-old chick embryos incubated at 99.5 °F (Schiff *et al.*, 2000). Fertilized eggs are obtained from Spafas Corp. (Charles River, MA). Puck's saline solution contains 150 mM NaCl, 5 mM KCl, 1.2 mM Na$_2$HO$_4$, 4 mM KH$_2$PO$_4$, and phenol red (approximately 0.001%, w/v) in deionized H$_2$O; adjust pH to 7.4 with 1 M NaOH. The 1% (w/v) glucose solution is prepared in H$_2$O. Add the glucose solution to Puck's saline to a final concentration of 0.1% just prior to use. All reagents are obtained from Sigma (St. Louis, MO). Forceps used for dissection are Dumont #5 forceps obtained from Fine Science Tools (Foster City, CA).

Trituration. Collagenase A (Boehringer Mannheim, Indianapolis, IN) is dissolved in H$_2$O to make a stock of 1% (w/v), filtered using a 0.22-μm syringe filter, and set aside in 250-μl aliquots at $-20°$. The DRG culture medium is prepared from the following reagents: 83 ml of Dulbecco's modified Eagle's medium (low glucose, with L-glutamine, sodium pyruvate, and pyridoxine hydrochloride), 10 ml of heat-inactivated horse serum, 5 ml of chick embryo extract, 1 ml of nerve growth factor, and 1 ml of penicillin–streptomycin/glutamine. Nerve growth factor (mouse, 7S) is prepared by dissolving 10 μg in DMEM to a final concentration of 1 μg/ml. Warm the medium to 37° before using. All the reagents are obtained from GIBCO (Rockville, MD).

Plating DRG Neurons. Collagen, type VII from rat tail (Sigma), is dissolved in 1 mM acetic acid to a final concentration of 3 mg/ml. Falcon culture dishes (35 mm) are obtained from Fisher (Pittsburgh, PA). To prepare the dishes for plating cells, spread one drop of collagen on each dish that will be used. Then allow the plates to dry (at least 2 h) before using.

Protocols

DRG Neuron Isolation. It is recommended to consult Bellairs (1997) before performing this isolation procedure. Open the egg at the broad end with spatula forceps and discard shell lid. Remove the embryo via the limbs and place in a large dissection dish. Decapitate immediately, grasping the chick with spatula forceps. Switching to fine forceps, lay the embryo on its back and spread its limbs. With a clipping motion, cut through the sternum from the rostral end. Open the first layers of skin and spread the skin and ribs laterally. Wash the cavity with Puck's/glucose medium. Next, remove intestines and wash cavity.

Move the embryo to the dissection microscope. Remove lungs after detaching them from the abdominal base and wash the cavity. Remove kidneys and wash the cavity. Remove blood vessels and wash the cavity. Remove superficial nerves and wash the cavity. Remove the connective tissue on either side of the spinal cord and wash the cavity. Using the finest forceps, detach the DRG from peripheral nerves and remove them by cutting their attachments to the spinal cord. Place the DRG in Puck's/ glucose medium in an untreated 35-mm culture dish.

Triturating DRG Neurons. Perform the rest of the procedure in a sterile laminar flow hood. Add 250 μl of collagenase A (1% solution) to 9 ml of Puck's/glucose solution in a sterile 15-ml Corning tube (Acton, MA). Transfer dissociated DRG from the 35-mm dish to the Puck's/glucose solution + collagenase A medium using a glass Pasteur pipette (in a minimal volume). Incubate at 37° for 15 min. This should be an adequate time for the connective tissue holding the bundled DRG together to begin to deteriorate. Meanwhile, set aside two more 15-ml tubes, each with 5 ml of the Puck's/glucose medium. Preheat these to 37° in the incubator. They will be used for washing away the collagenase from the DRG, since the dishes are treated with collagen and any collagenase carried over would break down the collagen coating of the dishes.

Washes. Run the end of a sterile, autoclaved glass Pasteur pipette over a gas flame. Allow it to cool and then transfer the DRG from the first 15-ml Corning tube (containing the Puck's/glucose + collagenase A medium) to the next (containing 5 ml of the Puck's/glucose medium only) using a Pasteur pipette. Be careful when transferring DRG not to draw cells up past the narrow portion of the Pasteur pipette, as they have a tendency to stick to the sides of the glass. After the DRG have settled to the bottom of the tube, repeat the procedure, transferring them to the third 15-ml tube (containing 5 ml of the Puck's/glucose medium only). Finally, transfer the DRG from the third 15-ml tube to a fourth containing 2-ml of DRG medium, preheated to 37°. Again, allow the DRG to settle to the bottom of the tube. Next, fire polish the tip of a glass Pasteur pipette until its diameter is approximately one-third of the original tip diameter. Allow it to cool. Then triturate the DRG by pipetting up and down approximately 15 times. Continue to triturate until the solution is cloudy and large pieces of ganglia are not seen.

Plating DRG Neurons. After triturating, bring up the volume of tritu- rated DRG neurons in complete medium so that you will be able to plate

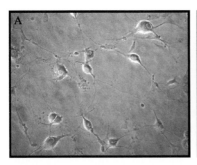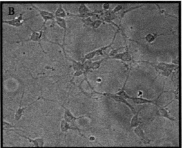

Fig. 1. Neuronal primary cultures. (A) Embryonic chick DRG neurons 24 h after plating. (B) E18 rat hippocampal neurons 14 days after plating. Images were taken with a Hamamatsu Orca CCD camera mounted on a Zeiss Axiophot 2 microscope. Cells were visualized with a 40× objective lens using DIC II filters.

approximately 2 ml per collagen-treated 35-mm dish (the desired dissection-to-plating ratio is approximately two embryos per dish, depending on DRG viability and dissection efficiency). After pipetting the appropriate volume of medium into each culture dish, gently move it up and down, and left to right, to disperse the DRG neurons. Return the culture dishes to a 37° incubator.

Feeding DRG Neurons. The day after plating the DRG neurons, and every 2 days subsequent to that, feed the cells (Fig. 1A). Gently remove the lid of the 35-mm dish and place it face down on the surface of the sterile hood. Prop the dish up against the lid so it is at approximately a 20–30° angle to horizontal. Being careful not to touch the bottom of the dish, remove 1 ml of the DRG medium. Then place the dish flat with the surface of the hood. Add 1 ml of new DRG medium to the culture by pipetting gently against the side of the dish. Return the 35-mm dishes to the incubator.

Embryonic Rat Hippocampal Neuron Cultures

This procedure is adapted from Goslin *et al.* (1998).

Solutions and Materials

Balanced saline solution (BSS)
Stock solution 1: 3.0 g dextrose,
0.6 g $KH_2PO_2 \cdot (H_2O)_7$, 1.07 g Na_2HPO_4 $(H_2O)_7$, and
6 ml 0.5% phenol red solution. Dissolve and bring to 300 ml with deionized water.

Stock solution 2: 558 mg $CaCl_2 \cdot (H_2O)_2$,
  1.2 g KCl,
  24 g NaCl,
  0.6 g $MgCl_2 \cdot (H_2O)_6$, and
  0.6 g $MgSO_4 \cdot (H_2O)_7$.

Dissolve and bring to 300 ml with double distilled water. Let both stock solutions stand overnight at $4°$. They may be stored at $4°$ or at room temperature. The pH of stock solutions should be around 7.2.

To prepare 3 liters of BSS, add 300 ml of stock solution 1 to 2.4 liters of water and mix. Add 300 ml of stock solution 2 to the dilute stock solution 1 and mix. Filter under sterile conditions. Incubate for 1 week to check for contamination. Store at room temperature.

Coverslip Preparation. Round-glass coverslips are obtained from Fisher and arranged in ceramic racks. Groups of 48 coverslips (12 per rack) are treated overnight in 70% nitric acid. The next day these racks are transferred to distilled water and washed for 1 h, after which they are transferred to fresh distilled water and washed again. Coverslips are dry heat sterilized for at least 6 h at 250 °C and are placed into 12-well plates obtained from Fisher.

Poly-L-lysine is obtained from Sigma and is dissolved in borate buffer (1 mg/ml borate buffer, pH 8.5; filter sterilized). This solution is filtered using a 0.22-μm syringe filter, and 1 ml is placed over each glass coverslip. The coverslips are incubated overnight at room temperature. Before plating cells, the poly-L-lysine coverslips are washed twice in autoclaved water.

Hippocampal Dissection. Pregnant rats at embryonic day 17 are obtained from Charles River. Instruments used in the dissection are obtained from Fine Science Tools and autoclaved prior to use.

Trituration, Plating, and Maintenance. The hippocampal growth medium is prepared in stages, starting with 485 ml of Dulbecco's modified Eagle's medium (low glucose, with L-glutamine, sodium pyruvate, and pyridoxine hydrochloride). NMEM solution is prepared by adding 15 ml of 20% glucose solution (w/v in H_2O) followed by 0.22 μm Nalgene filter sterilization. HS-NMEM growth medium is prepared by adding 88 ml of NMEM to 10 ml of heat-inactivated horse serum, 1 ml of penicillin–streptomycin, and 1 ml of 100 mM sodium pyruvate. B27 growth medium is prepared by adding 48.5 ml of NMEM medium to 1 ml of B27 supplement and 0.5 ml of L-glutamate. The 25% trypsin solution used is divided into 0.5-ml aliquots and stored at $-20°$. All cell culture reagents, supplements, and solutions are obtained from GIBCO/Invitrogen.

Protocols

Coverslip Preparation. In a laminar flow hood, fill ceramic racks with 12 coverslips each with clean tweezers. Place the racks 4 at a time (48 coverslips) inside a 2000-ml beaker. In the fume hood, pour just enough 70% nitric acid into the beaker to cover the entire racks. Cover the beaker with a large glass plate and incubate in the hood overnight.

The next day, carefully transfer racks to another 2000-ml beaker filled with enough distilled water to cover the racks. Incubate coverslips in distilled water for 2 h, changing water four times in that time. Place racks into a dry beaker, cover with aluminum foil, and sterilize in dry heat (250 °C for 6 h). Cool to room temperature, and transfer the coverslips to sterile 12-well plates in a laminar flow hood.

Dissolve poly-L-lysine hydrobromide in 0.1 M borate buffer for a final concentration of 1 mg/ml. Using a 0.22-μm syringe filter, deliver approximately 1.5 ml of poly-L-lysine solution to each coverslip. Incubate the coverslips overnight at room temperature in a laminar flow hood.

After overnight incubation, wash coverslips with autoclaved deionized water for 1 h. Repeat wash a second time for another hour. Apply 1.5 ml of HS-NMEM medium to each coverslip and incubate at 37°, just prior to plating.

Hippocampal Dissection. For optimal cultures, dissect embryos at embryonic day 17. Sacrifice a pregnant rat with CO_2 for 5 min. Check for a negative tail pinch response, eye response, and lack of chest movement before proceeding with dissection.

Wipe down the dissecting work area with a chlorinated antiseptic solution followed by 70% ethanol. Dissection should be performed in full surgical attire. Place an absorbent bench pad on a dissecting bench, and lay the rat out in an anatomical position. Sterilize the absorbent bench pad, gloves, and rat with 70% ethanol, making sure to drench the fur around the belly thoroughly.

Using sterilized instruments, make a small lateral incision approximately 1 in. from the base of the tail through the skin only, 2 in. across. After opening the skin covering the belly, make a small incision through the muscle layer. Widen this incision 2 in. across. Locate the fetuses above the ovaries, in two uterine horns. Remove fetuses and uterus together by cutting connective fat as you pull. Place embryos in tube with a small amount of ice-cold BSS. Seal this tube and place on ice.

In a laminar flow hood, remove embryos from the tube into a large plate filled with ice-cold BSS. Cut each fetus out of the uterus and out of their amniotic sacks individually. Place them into a second dish filled with ice-cold BSS.

To remove brains, begin by placing the point of a small, sharp pair of scissors at the occipital base of the skull. Pierce the skull and angle the scissor blade up. Keeping the scissors angled up so that they do not cut the brain, extend the incision all the way to the nose. Using #7 curved forceps, sever the brain stem and angle the curve of the forceps up and under the brain. Lift the brain out in one motion and place in a dish filled with ice-cold BSS.

Place the plate under the microscope and position the brain such that the top is oriented toward the ceiling. Using a scalpel blade to cut and #5 forceps to hold, cut gently half way in between the lobes and then at a 45° angle to sever off each of the two hemispheres. Using sharp #5 forceps, carefully remove the pink-colored meninges by holding onto a flap and tearing the covering apart. Being careful not to pull the brain apart, remove all the meninges, leaving behind just the purely white-colored brain. The hippocampus is a densely white crescent of tissue. With a fiber-optic light source placed on the side of the dish to create a better distinction, cut sideways into the brain with the scalpel. Locate the area on the edge of the flap that is separated from the rest of the lobe by a dark line and carefully cut on that dark line to separate the hippocampus.

After separating the entire hippocampus, use a Pasteur pipette to transfer it into a 15-ml conical tube on ice. Place all other isolated hippocampus in this tube as well.

Trituration, Plating, and Maintenance. Fill the 15-ml tube with the hippocampal tissue to the 4.5-ml mark with fresh BSS. Add 0.5 ml of 25% trypsin, invert, and mix. Incubate for 15–20 min at 37°. After incubation, remove most of the trypsin/BSS medium with a Pasteur pipette, making sure to avoid the tissue at the bottom of the tube. Refill the tube to the top with fresh BSS, invert, mix, and incubate for 5 min at 37°. Repeat removal and refilling with BSS two more times to remove as much trypsin as possible. On the last refill, use HS-NMEM medium instead of BSS and fill up to the 3.5-ml mark. Avoiding bubbles, pipette up and down to separate and mix cells with a regular 10-ml plastic pipette 15 times. Wait until the large pieces of tissue settle to the bottom and remove most of the cloudy supernatant to a separate tube. Add another 2–3 ml fresh HS-NMEM to the large pieces of tissue in the first tube and repeat pipetting up and down 15 times. Combine the old supernatant from the first tube with the contents of the second tube.

Flame a Pasteur pipette such that its opening is narrowed to about half of its original diameter. Using this, pipette the cell suspension up and down another 15 times, avoiding bubbles. Drop about 0.6 ml of cell suspension to each well of the 12-well plates. For proper cell density, use approximately 0.5–1 embryonic brain equivalents per coverslip.

Incubate cultures overnight and then change the medium to B27 medium. Feed cells once every week with B27 medium (Fig. 1B).

Biochemical Approaches

Materials and Solutions

Lysis Buffers. Lysis buffer A contains 1% NP-40 (v/v), 1 mM sodium orthovanadate, 1 mM EDTA, 17 μg/ml calpain inhibitor I, 7 μg/ml calpain inhibitor II, 10 μg/ml pepstatin, 10 μg/ml soybean trypsin inhibitor, 2 μg/ml aprotinin, 1 mg/ml Pefabloc SC, and 10 μg/ml leupeptin in phosphate-buffered saline (PBS). Lysis buffer B contains 50 mM HEPES, pH 7.4, 5 mM EDTA, 1 mM dithiothreitol (DTT), 17 μg/ml calpain inhibitor I, 7 μg/ml calpain inhibitor II, 10 μg/ml pepstatin, 10 μg/ml soybean trypsin inhibitor, 2 μg/ml aprotinin, 1 mg/ml Pefabloc SC, and 10 μg/ml leupeptin in PBS. For lysis buffer A, the detergent NP-40 is obtained from Pierce (Rockford, IL); all other reagents are obtained from Roche/Boehringer Mannheim (Indianapolis, IN).

Electrophoresis. Separating and stacking gels are prepared following Laemmli (1970) using Bio-Rad reagents (Hercules, CA). For spacer thickness 1.00 mm, the total monomer solution volume is 10 ml for the stacking gel and 25 ml for the separating gel portions. The 7.5% separating gel is prepared using 6.25 ml of acrylamide/bisacrylamide (30% T, 2.67 C stock), 12.125 ml of deionized water, 6.25 ml of 1.5 M Tris–HCl, pH 8.8, 250 μl of 10% (w/v) SDS, 125 μl of 10% ammonium persulfate, and 12.5 μl of TEMED. The 4% stacking gel is prepared using 1.3 ml of acrylamide/bisacrylamide (30% T, 2.67 C stock), 6.1 ml of deionized water, 2.5 ml of 0.5 M Tris–HCl, pH 6.8, 100 μl of 10% (w/v) SDS, 50 μl of 10% ammonium persulfate, and 10 μl of TEMED. Running (25 mM Tris, pH 8.3, 192 mM glycine, 0.1% SDS) and transfer (Tris/glycine buffer containing methanol) buffers are obtained as 10× stocks from Bio-Rad. The gel is run on a Protean II xi cell system using the Power Pac 300 power supply, also obtained from Bio-Rad. The Rainbow protein marker is obtained from Amersham Pharmacia Biotech (Piscataway, NJ).

Antibodies. Antisera directed against the N-terminal PDZ/PTB tandem (amino acids 1–440) and RGS box (amino acids 664–885) of human and rat RGS12 (kindly provided by Dr. David Siderovski, University of North Carolina, Chapel Hill, NC) are generated and affinity purified using previously described methods (Snow *et al.*, 1998). The monoclonal antibody against phosphotyrosine, 4G10, is from Upstate Biotechnology (Lake Placid, NY). Horseradish peroxidase-conjugated protein A/G is from Pierce (Rockford, IL).

Peptides. Sequences of the peptides used in this study are based on the $Ca_v2.2$ $\alpha1$ sequence from chick DRG neuron (CDB1, GenBank AAD51815). Peptides are synthesized by FastMoc chemistry at the Tufts University Core Facility and are purified by HPLC with >97% purity as determined by mass spectrometry. An amino-terminal biotin is included in every peptide. Peptides are dissolved in 5 mM acetic acid at 1 mg/ml and diluted into the internal solution for electrophysiological experiments or HEPES-buffered saline solution HBSS for biochemical experiments.

Protocols

Preparation of Membrane-Associated Fractions. Cells are homogenized in lysis buffer B using a mechanized, pellet pestle homogenizer (Kontes, Vineland, NJ) using about 20 strokes. The homogenates are then centrifuged at 1000g for 10 minutes. Transfer the supernatants from this spin to fresh prechilled Eppendorf tubes and spin again at 100,000g for 1 h. Drain off the supernatant and resuspend the pellet in ice-cold lysis buffer A [1% NP-40 (v/v)].

In Vitro *Phosphorylation Reaction. In vitro* phosphorylation of recombinant channel loop proteins is performed using a recombinant active form of Src kinase (Upstate Biotechnologies, New York). Ten to 20 units of recombinant Src per assay are used. The kinase reaction buffer contains 100 mM Tris–HCl, pH 7.2, 125 mM $MgCl_2$, 25 mM $MnCl_2$, 2 mM EGTA, 0.25 mM sodium orthovanadate, 2 mM DTT, and 100 μM ATP. Reactions are run for 40 min at 30° and terminated by the addition of Laemmli sample buffer. Proteins are resolved in a 12% SDS–polyacrylamide gel, and tyrosine phosphorylation is detected using immunoblotting with the antiphosphotyrosine antibody 4G10 (1:1000, Upstate Biotechnologies).

Immunoblot Quantitation. For quantitation of bands in immunoblots visualized using chemiluminescence, images are obtained using a Molecular Dynamics personal densitometer SI. ImageQuant 1.11 (for Macintosh) is used to analyze data. Data are reported as a Microsoft Excel table. Values are pooled for all the experiments, and mean values ± standard error of the mean (SEM) are plotted.

All experiments were repeated two to five times with similar results. Representative examples are shown for the biochemical experiments. For electrophysiological experiments, data derived using cells from three independent platings are shown.

Overlay Assay (Far Western). After transferring proteins from the SDS–PAGE gel to the nitrocellulose membrane, block for 1 h in blocking solution containing 5% nonfat dry milk and PBS–0.1% Tween 20 (PBS-T). Incubate the membrane in this blocking buffer containing the test inter-acting protein to a final concentration of 500 ng/ml at 25° or room temper-ature for 4 h. Ideally, this will encourage the direct association of proteins, which can subsequently be assayed using an antibody to the protein that was overlain. Wash the membrane three times in 5% milk/PBS-T before incubating it overnight at 4° with an antibody (directed against the overlain protein) diluted 1:2000 in blocking buffer. Wash the membrane three more times and incubate it for 1 h at 25° in the appropriate horseradish per-oxidase-conjugated secondary antibody (Pierce) diluted 1:5000 in blocking buffer. Wash the membrane three times in PBS-T and once in PBS. Bound protein is detected using enhanced chemiluminescence (Amersham Pharmacia Biotech, Piscataway, NJ).

Binding of Endogenous RGS12 to Channel Peptides. Affinity columns are prepared by incubating 100 μg of N-terminally biotinylated, synthetic peptides with streptavidin-Sepharose beads as per the manufacturer's in-structions (Pierce). Two milligrams of DRG lysate is loaded onto the biotinylated peptide-avidin column and incubated for 4 h at 4°. Negative controls using (a) beads without peptide and (b) lysate from HEK293T cells are run in parallel. Elution is performed as per the manufacturer's instructions (Pierce). The eluate is mixed with the Laemmli sample buffer and resolved by 7.5% SDS–polyacrylamide gel electrophoresis. Immuno-detection of RGS12 is carried out as descubed previously using an antibody raised against the RGS box (1:1000) (Schiff *et al.*, 2000).

Experimental Data

Studies of RGS12 Binding to the Calcium Channel as Assessed by Overlay Assays

To determine which cytoplasmic regions of the Ca$_v$2.2 channel α1 subunit could be potential sites for phosphorylation by Src kinase, we perform *in vitro* phosphorylation assays using recombinant proteins containing the N termi-nus, cytoplasmic loops, and C terminus of the Ca$_v$2.2 channel α1 subunit. These recombinant proteins were expressed in bacteria as His$_6$- and S-tagged proteins. The pattern of the bands detected is shown in Fig. 2. Both the synprint region of loop II–III and the C terminus are tyrosine phosphorylated.

As reported previously that RGS12 and the tyrosine-phosphorylated form of the N-type calcium channel interact *in vivo* (Schiff *et al.*, 2000), we

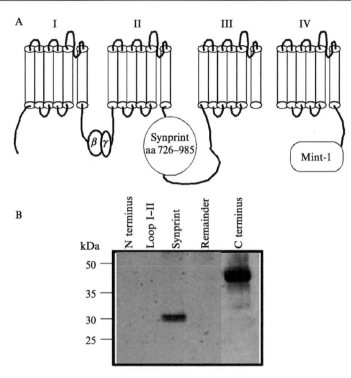

FIG. 2. (A) Topology of the α1 subunit of the Ca$_v$2.2 (N-type) calcium channel. Loop I–II binds to G protein βγ subunits ("βγ"). The "synprint" region (amino acids 726–984) is the SNARE-binding domain found in loop II–III. Mint-1 is a munc-18-interacting protein that binds to the C terminus of the calcium channel though a PDZ domain. (B) *In vitro* phosphorylation of channel loop peptides using active Src kinase was performed as described in the text. Loop II–III was divided into a "synprint" region (amino acids 726–984) and "remainder" (amino acids 985–1170). Samples were resolved by 12% SDS–polyacrylamide gel electrophoresis, and tyrosine phosphorylation was detected by immunoblotting with the antiphosphotyrosine antibody 4G10. Experiments were performed five independent times with similar results.

were interested to see whether this interaction was direct. Using tyrosine-phosphorylated forms of the recombinant proteins used in the *in vitro* phosphorylation reaction, we determined which region of the N-type calcium channel could bind to the N-terminal PDZ/PTB domains of RGS12. While both the C terminus and the loop II–III synprint region of the Ca$_v$2.2 channel α1 subunit are tyrosine phosphorylated, the RGS12 N terminus binds only to the synprint (amino acids 726–984) region of loop II–III. Because this region of loop II–III is the site where SNARE proteins bind to the channel, this interaction could potentially have functional consequences for the regulation of exocytosis (Sheng *et al.*, 1994) (Fig. 3).

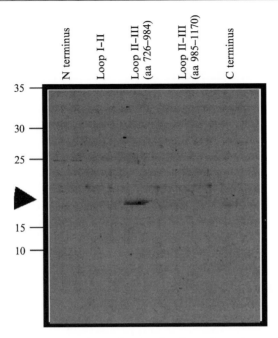

Fig. 3. RGS12 N terminus binds the phosphorylated synprint region peptide (loop II–III; amino acids 726–984) in a protein overlay assay. Channel loop peptides were resolved by SDS–PAGE and transferred to nitrocellulose membrane. Membranes were incubated with 500 ng/ml of recombinant GST-fusion protein comprising the N-terminal PDZ and PTB domains of RGS12. Data are representative of four experiments. The anti-GST antibody (1:2000) was used to detect binding.

The region of loop II–III to which RGS12 binds to contains two tyrosine residues: Y804 and Y815. We designed biotinylated peptides that would contain one of these two tyrosine residues (aa 793–808 and 808–819) as phosphotyrosine. As control peptides, we designed peptides in which the tyrosine residues were left unphosphorylated or mutated to phenylalanine. These peptides were used to test the ability of this region of the calcium channel to interact with the full-length endogenous RGS12 derived from lysates of chick dorsal root ganglion neurons and rat hippocampal neurons. To avoid proteolysis, all the steps (once the peptide is bound to the beads) are performed at 4°. RGS12 from both types of neurons binds to the peptide containing phosphotyrosine at position 804 (e.g., Fig. 4). While some binding is observed to the nonphosphorylated form of the peptide (Y804; Fig. 4), the binding to the phosphorylated form of the peptide is significantly stronger.

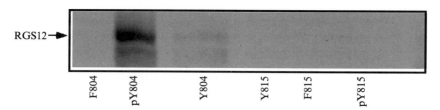

RGS12 →

| F804 | pY804 | Y804 | Y815 | F815 | pY815 |

FIG. 4. Binding of endogenous RGS12 from rat hippocampal lysates to channel peptides. Biotinylated peptides from the chick synprint site were bound to strepavidin–Sepharose beads. Immunodetection was performed using the anti-RGS12 antibody (1:1000).

Electrophysiological Approaches

Solutions for Electrophysiology

The external solution contains (in mM) NaCl 133, CaCl$_2$ 1, MgCl$_2$ 0.8, HEPES 25 (pH 7.4), NaOH 12.5, glucose 5, tetraethylammonium chloride 10 (to block K$^+$ channels), and tetrodotoxin 0.3 μM (to block voltage-dependent Na$^+$ channels). The internal recording solution contains (in mM) CsCl 150, BAPTA 5, MgATP 5, and HEPES 10 (pH 7.2).

Protocols

Electrophysiological Recordings. Whole cell recordings are performed as described previously (Schiff *et al.*, 2000). For extracellular application, agents are diluted into standard extracellular saline and applied via a wide-bore (140 μm i.d.) "sewer pipe," which exchanges solutions with subsecond kinetics. For the experiments presented in this report, the calcium current has been corrected for rundown by measuring calcium current as a function of time in control cells without a neurotransmitter. Cells used for experiments exhibited a rundown of the current of less than 1%/min.

If the reader is not familiarized with the technique of whole cell electrophysiology, details of the technique and equipment have been described by Hescheler (1994).

Data Analysis

Data are filtered at 3 kHz, acquired at 10–20 kHz, and analyzed using PulseFit (HEKA) and Igor (WaveMetrics) on a Macintosh computer. Strong depolarizing conditioning pulses (to +80 mV) that precede test pulses (to 0 mV) reverse neurotransmitter-induced voltage-dependent inhibition of the Ca$_v$2.2 channel. Such conditioning pulses have no effect on control currents recorded in the absence of the neurotransmitter. During application of the neurotransmitter, test pulse currents measured before and after the conditioning pulse are subtracted to yield the voltage-dependent inhibitory component. Test pulses measured following the conditioning pulse are subtracted from control currents (measured in the absence of neurotransmitter) to yield the voltage-independent component. Integration of these two currents gives measurements of total charge entry carried by the two modulatory components.

Introduction of Peptides into the Cytosolic Environment through the Recording Pipette

One of the advantages of whole cell electrophysiology is that reagents can be introduced into the cellular environment through the recording pipette. Peptides, small molecules (<60 kDa), and pharmacological reagents can be included in the internal recording solution. Experimental and theoretical observations have identified factors that determine the rate of exchange between the recording pipette solution and the cytosol such as the diffusion coefficient of the reagent, pipette access resistance, and cell volume.

In our experiments using embryonic chick DRG neurons, the variability in cell volume is minimized using cells with a diameter of 30–40 μm. The pipette resistance is usually 1–2.5 megaohms. Slow seal formation is observed when the internal solution contains a high protein concentration (micromolar range). Once the cell membrane is ruptured, gaining whole cell access, the capacitive transients should be faster than 1 ms. Sometimes, during the equilibration time, the transients become slower, indicating that one is losing access; suction can be applied to improve cell access. If this problem persists, the exchange will be poor and the time to obtain optimal equilibration with the cytosolic environment will vary widely from cell to cells. Under these conditions, the investigator cannot obtain reliable data on the time course of the effects of the reagent and negative data become questionable.

Cells will become granular in appearance under phase-contrast optics as the internal solution containing the reagent equilibrates with the cytosolic environment. We use an internal solution that contains 0.1% fluoresceinated-dextran (70,000 size, Molecular Probes, Oregon), which allows visualization under epifluorescence optics. In DRG neurons, it takes 3–5 min for the soma to become completely fluorescent.

The investigator has to pay particular attention to the choice of vehicle or buffer used in the experiments. Some organic solvents, such

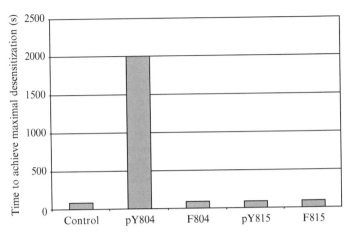

FIG. 5. Effect of channel loop peptides on the rate of desensitization of GABA-mediated voltage-independent inhibition of the $Ca_v2.2$ channel. Inhibitory components were determined as described in the text.

as chloroform and DMSO, have adverse effects on the membrane in concentrations greater than 0.1%.

Experimental Data

Loop II–III synprint region peptides (in concentrations from 1 nM to 1 mM) were introduced into the cytosolic environment through the recording pipette. Calcium current as a function of time was monitored to determine any agonist-independent effect that the peptides may have on the basal calcium current. DRG neurons were exposed to 100 μM GABA (in the presence of 100 μM bicuculline to block $GABA_A$ receptors), and both the magnitude and the time course of the GABA-mediated voltage-independent inhibition was determined. Under these experimental conditions, the peptides did not cause a significant change in the magnitude of voltage-independent inhibition.

At 1 μM, the pY804 peptide slowed the rate of desensitization by a factor of 48 (Fig. 5). Mutation of Y804 to phenylalanine (F804) abolished the effect. Desensitization rates in cells equilibrated with peptides containing the second tyrosine Y815 were not significantly different from control cells.

Acknowledgments

This work was supported by a grant from the National Institutes of Health (NINDS 37443) and a Hirschl Trust Fund Career Award.

References

Bellairs, R. (1997). "The Atlas of Chick Development." Academic Press, San Diego.

Ewald, D. A., Pang, I. H., Sternweis, P. C., and Miller, R. J. (1989). Differential G protein-mediated coupling of neurotransmitter receptors to Ca^{2+} channels in rat dorsal root ganglion neurons *in vitro*. *Neuron* **2**, 1185–1193.

Goslin, K., Asmusson, H., and Banker, G. (1998). "Culturing Nerve Cells," 2nd Ed., Chapter 13. MIT Press, Boston, MA.

Hescheler, J. (1994). Whole-cell clamp analysis for G-protein regulation of channels. *Methods Enzymol.* **238**, 365–375.

Hille, B. (1994). Modulation of ion-channel function by G-protein-coupled receptors. *Trends Neurosci.* **17**, 531–536.

Ikeda, S. R., and Dunlap, K. (1999). Voltage-dependent modulation of N-type calcium channels: Role of G protein subunits. *Adv Second Messenger Phosphoprotein Res.* **33**, 131–134.

Jeong, S. W., and Ikeda, S. R. (2000). Effect of G protein heterotrimer composition on coupling of neurotransmitter receptors to N-type Ca(2+) channel modulation in sympathetic neurons. *Proc. Natl. Acad. Sci. USA* **97**, 907–912.

Jones, L. P., Patil, P. G., Snutch, T. P., and Yue, D. T. (1997). G-protein modulation of N-type calcium channel gating current in human embryonic kidney cells (HEK 293). *J. Physiol.* **498**, 601–610.

Laemmli, U. K. (1970). Cleavage of structural proteins during the assembly of the head of bacteriophage T4. *Nature* **227**, 680–685.

Schiff, M. L., Siderovski, D. P., Jordan, D. J., Brothers, G., Snow, B., DeVries, L., Ortiz, D. F., and Diversé-Pierluissi, M. A. (2000). Tyrosine kinase-dependent recruitment of RGS12 to the N-type calcium channels. *Nature* **408**, 723–726.

Sheng, Z. H., Rettig, J., Takahashi, M., and Catterall, W. A. (1994). Identification of a syntaxin-binding site on N-type calcium channels. *Neuron* **13**, 1303–1313.

Snow, B. E., Hall, R. A., Krumins, A. M., Brothers, G. M., Bouchard, D., Brothers, C. A., Chung, S., Mangion, J., Gilman, A. G., Lefkowitz, R. J., and Siderovski, D. P. (1998). GTPase activating specificity of RGS12 and binding specificity of an alternatively spliced PDZ (PSD-95/Dlg/ZO-1) domain. *J. Biol. Chem.* **273**, 17749–17755.

Wickman, K., and Clapham, D. E. (1995). Ion channel regulation by G proteins. *Physiol. Rev.* **75**, 865–885.

[16] Analysis of Interactions between Regulator of G-Protein Signaling-14 and Microtubules

By Luke Martin-McCaffrey, Francis S. Willard, Agnieszka Pajak, Lina Dagnino, David P. Siderovski, and Sudhir J. A. D'Souza

Abstract

Microtubules are dynamic polymers essential for mitosis and cell division, intracellular transport, and maintaining cell organization and structure. Microtubule dynamics are tightly controlled in a context-specific manner by a myriad of microtubule-associated proteins. We have identified regulator of G-protein signaling-14 (RGS14) as a microtubule-associated protein. RGS14 is a component of the mitotic apparatus that binds directly to and stabilizes microtubules *in vitro* and is essential for the first cell division in the mouse embryo. This article describes methods used for examining the impact of the RGS14/microtubule interaction *in vivo* and *in vitro*.

Introduction

Mitosis is the phase of the cell cycle when newly replicated chromosomes assemble on the mitotic spindle for distribution to the daughter cells. The mitotic spindle is principally composed of microtubules, which assemble from α/β tubulin heterodimers. α/β tubulin dimers are guanine nucleotide-binding proteins that associate in a head-to-tail orientation to produce a polarized tubule in a GTP-dependent fashion. The dynamic polymerization/depolymerization of microtubules is essential for chromosome segregation during mitosis and is tightly regulated by a variety of proteins, including heterotrimeric G proteins (Roychowdhury and Rasenick, 1997; Roychowdhury *et al.*, 1999).

Heterotrimeric guanine nucleotide-binding proteins (G proteins) are composed of a GTP-binding α subunit and a membrane-associated $\beta\gamma$ dimer and have well-described functions as plasma membrane-delimited signal transducers for seven transmembrane receptors. It has become apparent that heterotrimeric G proteins also play a key regulatory role in mitotic cell division, as shown by studies of *Drosophila* neuroblast and *C. elegans* embryo asymmetric cell division (Gotta and Ahringer, 2001; Knoblich, 2001; Willard *et al.*, 2004). Microtubule assembly and disassembly are modulated by $G\alpha_{i1}$, $G\alpha_o$, $G\alpha_s$, and $G\beta_1\gamma_2$ *in vitro* (Roychowdhury and Rasenick, 1997; Roychowdhury *et al.*, 1999). Furthermore, regulator of G-protein signaling (RGS) proteins RGS6 and RGSZ1, which act as GTPase-accelerating proteins (GAPs) for $G\alpha$ proteins, have been implicated in the regulation of microtubule assembly/disassembly through their interaction with the neuronal microtubule-destabilizing protein SCG10 (Liu *et al.*, 2002; Nixon *et al.*, 2002) (see also Nixon and Casey, 2004).

We have demonstrated that RGS14 is an essential regulator of mitotic spindle dynamics in mammalian cells (Martin McCaffrey *et al.*, 2004). RGS14 colocalizes with mitotic spindles in cultured cells, as well as in the preimplantation mouse embryo, and associates directly with microtubules to promote assembly. Furthermore, RGS14 is required for formation of the mitotic apparatus and is essential for the first cell division of mouse embryogenesis. The methods described here are used to examine the interaction of RGS14 with microtubules and its impact on the dynamics of the mitotic spindle both *in vitro* and *in vivo*.

Materials

Unless otherwise noted, all chemicals were purchased from Sigma (St. Louis, MO).

Preimplantation Embryo Collection and Manipulations

To make 100 ml of flushing medium I (Spindle, 1980), dissolve 1.407 g NaCl, 0.08 g KCl, 0.04 g $MgSO_4 \cdot 7H_2O$, 0.04 g $MgCl_2 \cdot 6H_2O$, 0.038 g $Na_2H\text{-}PO_4$, 0.012 g KH_2PO_4, 0.2 g D-glucose, 0.05 g calcium lactate, 0.01 g sodium pyruvate, 0.3 g bovine serum albumin, and 0.001 g phenol red in sterile water and adjust the pH to 7.4 with 1 M NaOH. Filter sterilize through a 0.2-μm filter (Pall Gelman, Ann Arbor, MI).

Make a flushing needle by using a fine metal file (a nail file works well) to shave down the end of a 25-gauge syringe needle (BD Biosciences, Palo Alto, CA), tapering the end to round off the sharp tip. If the sharp tip is not removed, the flushing needle will puncture the oviduct, making it very difficult to flush embryos.

Microtubule stabilization/fixation buffer (MTF) (Albertini and Eppig, 1995; Hunt *et al.*, 1995) is prepared freshly before use by adding 5 μl of 10% (v/v) Triton X-100 and 27 μl of 37% (w/v) formaldehyde (Bioshop, Burlington, Ontario, Canada) to 500 μl of microtubule stabilization buffer (MTSB) [20 mM PIPES, pH 6.9, 1 mM $MgCl_2$, 0.5 mM EGTA, 1 μg/ml aprotinin, 1 mM dithiothreitol (DTT)] with 10 μg/ml taxol. Note that PIPES will not go into solution until the pH is between 6.1 and 7.4.

To make a micropipette, flame the tip of a glass Pasteur pipette and pull the ends apart to a length of 5–7 cm and the approximate thickness of a capillary tube, before breaking the tip off at the thinnest part. Connect a piece of rubber tubing to the wide end of the pipette and fit a small rubber bulb on the other end of the tubing for suction. Using a glass micropipette enables one to handle embryos in a small volume, which is important during wash steps and embryo mounting. Embryos are manipulated under a dissecting microscope, with one hand guiding the micropipette and the other regulating suction.

Fixing Cultured Cells for Microtubule Visualization

Coverslips are washed in acid (1 M HCl, 55°, 16 h) to increase cell adherence. Let the coverslips cool and then wash five times with sterile phosphate-buffered saline (PBS). Immerse and store coverslips in 70% ethanol.

Microtubules are stabilized during fixation using MTSB with 4 M glycerol (Gaglio *et al.*, 1995). Store the solution at 4°, but warm it to room temperature before use. The high concentration of glycerol is used to stabilize microtubules (Keates, 1981). Taxol (10 μM) is a good substitute for the glycerol, but it must be added to the buffer immediately before use from a 2 mM stock dissolved in dimethyl sulfoxide (DMSO). Enzyme grade glycerol and DMSO are from Fisher Scientific, (Nepean, Ontario, Canada).

GST-RGS14-His$_6$ Fusion Protein

Isolation and purification of full-length RGS14 as a soluble GST fusion protein from *Escherichia coli* can be problematic. Here, we describe an alternative approach involving refolding of insoluble RGS14. To produce GST-RGS14 fusion protein, transform the chemically competent *E. coli* strain BL21-DE3 (Stratagene, La Jolla, CA) with a pGEX4T1-based vector (Amersham Biosciences, Uppsala, Sweden) containing the full-length rat RGS14 open reading frame fused to a C-terminal hexahistidine tag (Kimple *et al.*, 2001). We routinely use the BL21-DE3 strain to produce GST fusion proteins because it is deficient in proteases and was established solely for the purpose of generating recombinant protein. After an overnight incubation at 37° on an LB agar plate containing 100 μg/ml ampicillin, pick a single colony, inoculate 5 ml of NYZ medium (5 g yeast extract, 5 g NaCl, 2 g MgSO$_4$, 10 g NZ-amine or tryptone dissolved in 1 liter), and incubate with vigorous shaking at 37° overnight. Dilute cultures 1:100 (500 ml) in a 2-liter flask and grow at 30° until the OD$_{600}$ reaches 1.0–1.2. Add 100 mM isopropylthio-β-D-galactoside (IPTG, Invitrogen, Carlsbad, CA) to a final concentration of 0.5 mM and continue incubating at 30° for another 1 h. Collect the bacteria by centrifugation (6000g, 10 min, 4°).

Under these fermentation conditions, the bulk of the full-length recombinant GST-RGS14 segregates into highly dense aggregates of insoluble protein known as inclusion bodies (Fig. 1A). Standard lysis via a French press or a nondenaturing buffer and sonication fails to solubilize protein within inclusion bodies. GST-RGS14-His$_6$ protein can be purified from inclusion bodies by solubilization under denaturing conditions and dialysis into nondenaturing buffer to induce refolding. Specifically, we use reagents from Novagen (Madison, WI) that are designed for isolating and refolding proteins in inclusion bodies. First, resuspend the bacterial pellet in 4 ml of ice-cold buffer A [140 mM NaCl, 10 mM Na$_2$HPO$_4$, 1.8 mM KH$_2$PO$_4$, pH 7.4, 1 mM phenylmethylsulfonyl fluoride (PMSF), 4 μg/ml leupeptin, 2 μg/ml aprotinin, 2 μg/ml pepstatin], supplemented with 1 μg/ml DNase (Invitrogen) and 1% (v/v) Triton X-100 per gram of bacterial pellet. To isolate inclusion bodies, load 4-ml samples of suspended bacteria in a 50-ml capacity French pressure cell and break the cell walls (10,000 psi), chilling the pressure cell on ice between runs. A single pass through the pressure cell is usually sufficient to ensure adequate lysis of the bacterial cell wall. The ensuing extract is often viscous; consequently, sonication or an additional pass through the French press is necessary to disrupt the DNA. Next, add lysozyme from a freshly prepared stock to a final concentration of 100 mg/ml and let stand at 30° for 15 min before swirling to mix and sonicating on ice until the solution is no longer viscous. Keep the sample cold during sonication to avoid heat-induced denaturation of proteins. Because conditions must be optimized to account for the amount of heat generated by the specific sonicator used, consult the manufacturer for recommended settings for your equipment. To collect the inclusion bodies, centrifuge the crude extract (10,000g for 10 min, 4°). Remove the supernatant and resuspend the pellet thoroughly in 5 ml of wash buffer [200 mM Tris–HCl, pH 7.5, 100 mM EDTA, 10% (v/v) Triton X-100]. Transfer the suspension to a clean centrifuge tube with a known tare weight. Collect the inclusion bodies by centrifugation (10,000g for 10 min). Decant the supernatant and remove the last traces of liquid by tapping the inverted tube on a paper towel. Weigh the tube and subtract the tare weight to obtain the wet weight of the inclusion bodies.

To solubilize the inclusion bodies, use a 3-(cyclohexylamino)-1-propane sulfonic acid (CAPS) buffer (50 mM CAPS, pH 11) with freshly added 0.3% (w/v) N-lauroylsarcosine (solubilization buffer). A combination of CAPS buffer with 4–4.5 M urea can also be used to solubilize the protein. This is of use if proteins are sensitive to detergents. Based on the wet weight of the inclusion bodies, resuspend the inclusion bodies at a concentration of 10–20 mg/ml in 1X solubilization buffer. Disrupt large debris by repeated trituration, incubate at room temperature for 15 min, and clarify

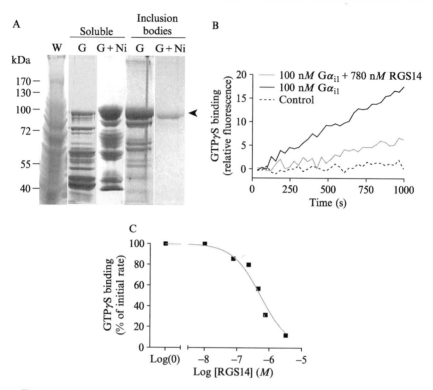

FIG. 1. Purification and functional assessment of refolded GST-RGS14-His$_6$ protein. (A) Refolding and purification of full-length GST-RGS14-His$_6$ fusion protein. Following induced expression of GST-RGS14-His$_6$ protein in the BL21 DE3 strain of *E. coli*, the majority of full-length GST-RGS14-His$_6$ fusion protein is found in the insoluble fraction, representing inclusion bodies. After induction of protein production at 30°, protein was extracted from inclusion bodies using 4 M urea, dialyzed, refolded (protein refolding kit, Novagen, WI), and purified sequentially with glutathione Sepharose (G) and Ni^{2+} (Ni) columns (Novagen, WI), which bind to the GST and His tags on the N and C-termini of the fusion protein, respectively. This procedure extracts approximately 95% of the full-length GST-RGS14-His$_6$ fusion protein (denoted by arrowhead) produced by the bacteria. W, whole cell extract. (B) Time course of 1 μM BODIPY FL GTPγS binding to 100 nM Gα_{i1}·GDP preincubated with 780 nM purified GST-RGS14-His$_6$ fusion protein. Fluorescence measurements (λ_{ex} = 485 nm, λ_{em} = 530 nm) in the presence of only BODIPY FL GTPγS ("control") and Gα_{i1}·GDP are also shown. (C) Dose–response analysis of GST-RGS14-His$_6$ guanine nucleotide dissociation inhibitor activity toward Gα_{i1}. Experiments were performed as in B but with GST-RGS14-His$_6$ protein concentrations of between 0 and 3.26 μM. See Kimple *et al.* (2004b) for further details on the BODIPY FL GTPγS assay used.

by centrifugation (10,000g, 10 min, room temperature). Transfer the supernatant containing the solubilized protein into a clean tube. Prepare the required volume of buffer for dialysis (20 mM Tris–HCl, pH 8.5) of solubilized protein. Dialysis should be performed with at least two buffer changes of >50 times the volume of the sample. Dialyze at 4° for at least 3 h using Spectra/Por dialysis tubing with a molecular weight cutoff of 6000–8000 Da (Spectrum Laboratories, Inc., Gardena, CA). Change the buffer and dialyze for at least 3 h. Continue the dialysis through two additional changes (3 h each) with the dialysis buffer lacking DTT. Centrifuge the dialyzed protein solution (10,000g, 10 min 4°) to remove visible insoluble material evident following dialysis.

To purify the refolded GST-RGS14, add 250 μl of glutathione Sepharose 4B (Amersham Pharmacia Biosciences) (as a 50% slurry in buffer A) to the supernatant. Incubate with gentle rocking for 30 min at room temperature. Wash the gel with 2.5 ml of buffer A three times for 5 min. Elute bound GST fusion proteins from the glutathione Sepharose by incubating three times with 250 μl of elution buffer (50 mM Tris–HCl, pH 8.0, 10 mM reduced glutathione) for 30 min at 25°.

Following elution from the glutathione Sepharose beads, if needed the fusion protein can be purified further by the carboxyl-terminal His$_6$-tag using a nickel-charged 5 ml HiTrap HP column (Amersham Pharmacia Biosciences) attached to a 20-ml syringe (BD Biosciences). The column is washed with 20 ml deionized water and is then charged with 2.5 ml 0.1 M NiSO$_4$. The column is washed sequentially with 20 ml each of deionized water, His-binding buffer (20 mM sodium phosphate, 0.5 M NaCl, pH 7.4), imidazole buffer (20 mM sodium phosphate, 0.5 M NaCl, 0.6 M imidazole, pH 7.4), and then His-binding buffer again. For binding to the Ni-charged column, dilute the GST-RGS14-His$_6$ protein 25-fold in His-binding buffer and apply it to the column. Wash the column with 20 ml of His wash buffer (20 mM sodium phosphate, 0.5 M NaCl, 30 mM imidazole, pH 7.4) to remove unbound protein. The bound protein is then eluted in 20 ml of imidazole buffer.

For the microtubule-binding assay and the tubulin polymerization assay, we prefer to solubilize the test proteins in PEM buffer (100 mM PIPES, pH 6.9, 1 mM EGTA, 2 mM MgSO$_4$). To exchange the buffer and concentrate the protein, centrifuge the Ni column-eluted GST-RGS14-His$_6$ protein at 10,000g in a Nanosep 30 K Omega concentrating column (Pall Gelman) down to 10% of the original volume. Add 1 volume of PEM and spin the column again. Repeat for a total of five times to exchange the buffer approximately 99.9%. Finally, measure the protein concentration using the Bradford method. For use in tubulin polymerization assays, we concentrate the GST-RGS14 protein to at least 7 mg/ml.

Aliquots of the lysed bacterial pellet (before and after addition of IPTG), the lysate (before and after binding to gluthathione Sepharose), and the eluted protein should be examined by SDS–PAGE and Coomassie blue staining to assess the efficiency of purification (e.g., Fig. 1).

Sources of Purified Tubulin

Purified, polymerization-competent tubulin is available from Cytoskeleton, Inc. (Denver, CO). Tubulin greater than 99% pure is suitable for analysis of direct protein–microtubule interactions. Alternatively, tubulin can be purified easily from pig brain, using a method described by E.D. Salmon (http://www.bio.unc.edu/faculty/salmon/lab/salmonprotocols.html). The GTP stock is prepared as a 100 mM Mg salt, (100 mM GTP with 100 mM $MgSO_4$) and can be stored in aliquots at −20°.

Antibodies

We have generated two rabbit polyclonal antibodies against rat RGS14 fragments (amino acids 60–544 and 263–544). The anti-β-tubulin monoclonal antibody (E7) developed by Dr. Michael Klymkowsky was obtained from the Developmental Studies Hybridoma Bank (The University of Iowa, Iowa City, IA).

Methods

Colocalization of RGS14 with Microtubules in Embryos and Cultured Cells

Isolating Preimplantation Embryos. Superovulated mice release more eggs, and therefore produce more embryos than naturally ovulating mice; in addition, when mice are induced to superovulate, the experimentalist has control on the day and time of ovulation and, to some degree, of conception. To superovulate mice, inject intraperitoneally 10 international units (IU) of pregnant mare serum gonadotropin (PMSG; Sigma) in a 0.1-ml volume into 3- to 5-week-old female CD-1 mice (Hogan *et al.*, 1994). Forty-six hours later, inject the mice with 5 IU of human chorionic gonadotropin (hCG; Sigma) in a 0.1-ml volume and mate singly with a male. The next morning, check for a vaginal plug, indicative of copulation.

Isolate embryos at specific times after injection of hCG for the desired stage of zygotic development [18–28 h post-hCG for one-cell embryos and 36–52 h post-hCG for two-cell embryos (Hogan *et al.*, 1994)]. Euthanize the dam by carbon dioxide inhalation and place on its back with limbs spread. Cut through the abdomen from the rostral end up the midline to just below the sternum, taking care to only cut skin and muscle layers. Make a second cut across the bottom of the abdomen just above the hind legs. Open the layers of skin and muscle and follow the uterine tubes up to the ovaries. With fine sharp scissors and fine forceps, cut the oviducts from the ovaries and uterine tubes and place in a drop of flushing medium I in a 35-mm tissue culture dish. The rest of the procedure is performed with a dissecting microscope (4× objective, 10× oculars). With fine forceps, remove the oviduct from flushing medium I and place it in another 35-mm culture dish without flushing medium I. While holding the oviduct with fine forceps, insert a rounded 25-gauge needle with a 1-ml syringe containing flushing medium I into one end, clamp the oviduct around the needle with forceps, and inject approximately 100 μl of flushing medium I. The oviduct will balloon out and embryos will flow out the opposite end. Typically, one can expect at least 20–30 embryos from a single female. Using a glass micropipette, transfer the embryos to a watch glass well (we use a nine-well watch glass) and wash them three times in PBS.

Immunostaining of Microtubules. For fixation of the microtubule cytoskeleton or mitotic spindle apparatus in preimplantation embryos, incubate embryos in MTF at 37° for 30 min and wash three times with PBS containing 5% (v/v) normal goat serum (Albertini and Eppig, 1995; Hunt *et al.*, 1995). We use goat serum because all our secondary antibodies are raised in goat. Embryos can be stored in this solution for up to 2 weeks at 4° before staining. For cultured cells, seed cells on acid-washed coverslips placed inside the wells of a 24-well plate on the day before staining cells. To preserve the microtubule cytoskeleton, wash the cells with MTSB plus 4 M glycerol three times for 1 min each at room temperature. Permeabilize cells in MTSB plus 4 M glycerol containing 0.5% (v/v) Triton X-100 for 2 min at room temperature. Wash cells with MTSB plus 4 M glycerol three times for 1 min each at room temperature. Fix cells for 10 min at $-20°$ with cold methanol. Aspirate off the methanol and wash three times with PBST [PBS, 0.05% (v/v) Triton X-100] at room temperature.

Block the cells or embryos in 10% normal goat serum (diluted in PBS) for 2 h at room temperature or overnight at 4°. Wash the samples over the period of 1 h with at least four changes of PBST. Incubate with the anti-β-tubulin primary antibody [1:10 dilution in PBS with 5% (v/v) normal goat serum] overnight at 4°. Wash the cells with PBST at 37° for 1 h. From this point onward, cover the cells or embryos with aluminum foil to shield them from the light to minimize photobleaching. Incubate cells for 1 h with a fluorescent-labeled (FITC, or Cy3) goat antimouse antibody (Jackson ImmunoResearch Laboratories, Inc., West Grove, PA) diluted 1:1000 in PBS with 5% (v/v) normal goat serum. Nuclei are labeled with Hoescht 33258 (10 μg/ml) (Sigma) for 15 min and are then washed for 45 min with PBST.

Colocalization of Proteins with Microtubules Using Fluorescence Microscopy and Deconvolution. Embryos are mounted in a fibrin clot to immobilize them prior to microscopy (Hunt *et al.*, 1995). Several embryos can be mounted on a single microscope slide. Spot a 1-μl drop of fibrinogen solution (12.5 mg/ml fibrinogen, Sigma; 154 mM NaCl, 5.6 mM KCl, 1.7 mM CaCl$_2$) on a poly-L-lysine-coated slide and overlay the fibrinogen with 5 μl of mineral oil (Sigma). The mineral oil prevents evaporation of the fibrinogen drop. Using a glass micropipette, pick up a single embryo in a minimal amount of liquid and transfer it to the fibrinogen drop. Special care needs to be taken so that embryos are placed in the fibrinogen and not the mineral oil. Add 1 μl of thrombin solution (100 U/ml thrombin, Sigma; dissolved in PBS) to the fibrinogen drop. Allow 3 min for the clot to form. Dip the slide in 2% (v/v) Triton X-100 in PBS for 1 min to remove the mineral oil. Wash the slide three times in PBS to remove the detergent. Spread a 40-μl bead of DAKO antifade reagent (DAKO Diagnostics, Mississauga, Ontario, Canada) along the edge of the slide and lay a 24 $\times$ 50-mm glass coverslip over the slide such that the DAKO antifade reagent is spread evenly. Allow the DAKO reagent to set at room temperature for at least 2 h.

For cultured cells, place a 5-μl drop of DAKO antifade reagent on a microscope slide. Remove the coverslip from the 24-well plate using a pair of forceps and place cell side down on the drop of DAKO antifade reagent.

Before viewing under the microscope, gently clean off the slide and coverslip with distilled water to remove dried salts and use 95% (v/v) ethanol to dry. We use either a 40$\times$ objective or a 63$\times$ oil immersion objective. At least six serial images are captured at 5-μm intervals in the Z plane using an automated inverted epifluorescent microscope (DMIRBE, Leica Microsystems, Richmond Hill, Ontario, Canada) equipped with a digital camera (Hammatsu Orca II). Our images are processed on a G4 Macintosh computer running OpenLab 3.0 (Improvision, Coventry, UK). Identical sections are captured individually for Cy3, FITC, and DAPI signals. Z sections are deconvolved, pseudo-colored, and overlayed using OpenLab software. The image appears yellow where microtubules (green) and RGS14 (orange) colocalize (Fig. 2). Colocalization of two proteins does not imply interaction, but solely proximity. To determine whether RGS14 and microtubules interact directly, an *in vitro* cosedimentation assay is used.

Determining Direct Binding between Microtubules and RGS14

To form microtubules, mix pure tubulin (20 μl of 5 mg/ml stock; this is enough tubulin for five samples or controls) with 2.5 μl of cushion buffer [50% (v/v) 100 mM PIPES, pH 6.9, 1 mM EGTA, 2 mM MgSO$_4$, 50%

β-tubulin RGS14 Overlay

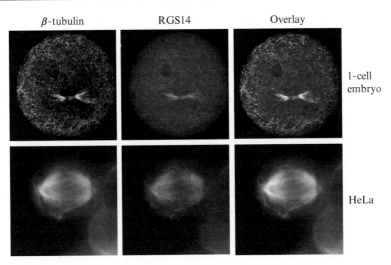

1-cell embryo

HeLa

FIG. 2. Colocalization of RGS14 and β-tubulin in preimplantation mouse embryos (one-cell stage; top) and mitotic HeLa cells (bottom). Microtubules (green) were detected with a monoclonal antibody raised to β-tubulin and an FITC-conjugated goat anti-mouse secondary antibody. RGS14 (orange) was detected using an affinity-purified rabbit anti-RGS14-60 antiserum (antigen: rat RGS14 amino acids 60–544) and a Cy3-conjugated goat anti-rabbit secondary antibody. Colocalization between RGS14 and microtubules appears yellow in the overlay images because of the combination of the orange and green pseudocolors. (See color insert.)

glycerol, 1 mM GTP, 10 μM taxol] in an appropriately labeled microfuge tube and incubate at 35° for 20 min. Add 180 μl of prewarmed (35°) G-PEM (1 mM GTP, 100 mM PIPES, 1 mM EGTA, 2 mM MgSO$_4$) buffer containing 10 μM taxol to the microtubules to stabilize them. For example, dispense five 20-μl aliquots of the tubulin into five tubes containing 1 μg of the GST-RGS14 (tube 1); 2–30 μg of GST-RGS14 protein (tube 2); 5 μg microtubule-associated protein 2 (MAP2), positive control (tube 3) (Manso-Martinez *et al.*, 1980); 5 μg GST, negative control (tube 4); and empty (tubulin alone), negative control (tube 5). Five additional tubes are similarly labeled and filled with 20 μl of G-PEM buffer. These tubes do not contain microtubules, but are treated identically to act as controls to determine whether the sample protein will sediment in the absence of microtubules due to the formation of protein aggregates. Dilute 1–2 μg of the sample (e.g., GST-RGS14) or control (e.g., GST alone) protein in G-PEM buffer to a final volume of 30 μl and mix this with a tube containing microtubules, taking care not to introduce bubbles into the mixture. Incubate for 30 min at room temperature and overlay on 100 μl of cushion buffer in a 1.5-ml conical ultracentrifuge tube (Beckman). Centrifuge for 40 min (100,000g, 25°). Put the hinge of the centrifuge tube toward the outside of the rotor to identify the side of the tube on which the pellet will form. The high molecular weight microtubules will sediment through the buffer, along with any bound proteins. Unbound proteins will remain in the supernatant.

Carefully remove the supernatant to a labeled microfuge tube without disturbing the pellet. Resuspend the pellet in Laemmli sample buffer (Laemmli, 1970), boil for 5 min, and resolve by denaturing sodium dodecyl sulfate–polyacrylamide gel electrophoresis (SDS-PAGE) on a 10% gel. Once the dye front reaches the end of the gel, stop the electrophoresis, remove the gel, and place it in a dish. Fix the gel in Coomassie fix solution [25% (v/v) propan-2-ol, 10% (v/v) acetic acid, diluted in deionized water] for 60 min with gentle mixing on a rotary mixer. Wash the gel three times with water and then stain with Coomassie stain solution [10% (w/v) acetic acid, 0.006% (w/v) Coomassie brilliant blue] for 2–5 h, until the protein is clearly visible (e.g., Fig. 3A). Wash the gel with Coomassie wash solution [10% (v/v) acetic acid] for several hours through several changes of wash solution until the bands are clear. It is useful to put a KimWipe (Kimberly-Clark Limited, West Malling, UK) in with the gel during the wash steps to soak up residual Coomassie stain.

If desired, the direct association between microtubules and GST fusion protein may be further characterized by repeating the assay using a range of protein concentrations to generate a saturation-binding curve to determine the dissociation constant (K_D) of the interaction (e.g., Fig. 3B). Because the K_D will be different for each microtubule-associating protein, the concentration range will have to be determined empirically. A good starting point would be 0.1 to 30 μM, as many microtubule-associated proteins fall within this range (Goode and Feinstein, 1994). After Coomassie staining of the gel, measure the band intensity of the bound GST fusion protein by densitometry (Alphaimager 1200, Alphaease Image Analysis, AlphaInnotech Corporation, San Leandro, CA). The binding of GST is also examined to estimate the nonspecific component of binding. Specific binding at each concentration can be determined by subtracting nonspecific binding of GST from binding of the GST fusion protein. Plot the curve as specific binding on the ordinate and GST fusion protein concentration on the abscissa (e.g., Fig. 3C).

Determining the Effect of RGS14 on Microtubule Dynamics

The effect of tubulin ligands on polymerization can be quantified by measuring the turbidity of the tubulin solution by optical density (Gaskin *et al.*, 1974). Kits for this turbidometric assay are also available from

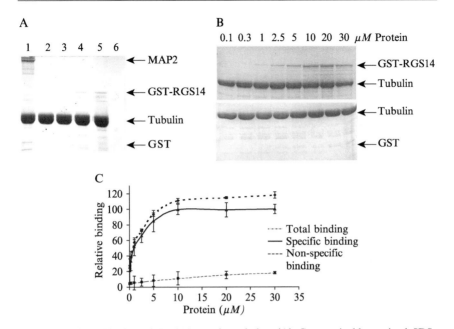

FIG. 3. Direct binding of RGS14 to microtubules. (A) Coomassie blue-stained SDS–polyacrylamide gel showing that the GST-RGS14-His$_6$ fusion protein binds to microtubules *in vitro*. Sample protein was incubated with 100 μM tubulin that was polymerized, and the reaction was then centrifuged through a glycerol buffer to sediment nascent microtubules and associated proteins. The pellet was then separated by SDS–PAGE and stained with Coomassie brilliant blue. Lane 1, positive control (1 μg MAP2); lane 2, tubulin alone; lane 3, negative control (2 μg GST); lane 4, 1 μg GST-RGS14-His$_6$; lane 5, 2 μg GST-RGS14-His$_6$; and lane 6, 2 μg GST-RGS14-His$_6$ without microtubules. (B) Various amounts of GST-RGS14-His$_6$ (0.2–30 μM; top) or GST alone (bottom) were incubated with microtubules as in A. (C) Graph of GST-RGS14-His$_6$ protein binding to microtubules demonstrating that RGS14 binds in a saturable fashion to microtubules.

Cytoskeleton Inc. Set up a 96-well microplate reader (*e.g.*, Safire Microplate Reader, Xfluor software, Tecan Systems Inc.) to read absorbance (340 nm) every minute for 60 min. Warm the plate reader to 37° for at least 30 min. For this assay, we use a Corning Costar 96-well plate (Corning Inc. Life Sciences). Add 10 μl of the GST fusion protein or GST (at a concentration 10-fold higher than the final concentration in the assay) to the warm plate and allow to stand for 1 min to come to the desired temperature. All proteins used must be in PEM buffer for accurate results. As a positive control for tubulin polymerization, set up wells containing 10 μl of a 20 μM taxol solution for a final concentration of 2 μM during the assay. For the negative control, add 10 μl of a 20 μM nocodazole solution to prevent

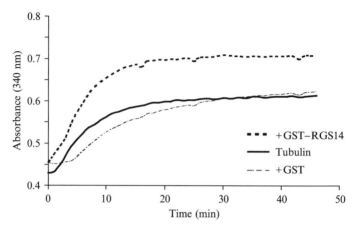

Fig. 4. Effect of RGS14 on microtubule dynamics. RGS14 (10 μM) was incubated with 100 μM of MAP-rich (>10%) tubulin, which was subsequently induced to polymerize. The turbidity of the solution (absorbance 340 nm), which correlates to the formation of nascent microtubules, was measured every minute over 45 min. RGS14 was found to promote microtubule stability with MAP-rich tubulin, but not with MAP-depleted tubulin (data not shown).

microtubule polymerization. Add 90 μl of microtubule-associated protein-rich tubulin (1.1 mg/ml stock in G-PEM buffer) to each well and, with a pipette tip, mix thoroughly. This amount of tubulin is a good starting point, but the concentration may need to be either increased for proteins that slow microtubule polymerization or decreased for proteins that enhance microtubule polymerization. It is of note that, to initiate *in vitro* tubulin polymerization, a so-called "critical concentration" of dimeric tubulin (approximately 1 mg/ml) must be reached. This is not an absolute requirement, as the apparent critical concentration can be lowered by the addition of microtubule-stabilizing compounds such as glycerol and taxol. It is critical that no bubbles are introduced into the sample because they will affect the readings. The onset of polymerization is very rapid (∼ 1–2 min). Therefore, it is best to use a multichannel pipetter to introduce the tubulin synchronously to each well. Start reading the absorbance at 340 nm immediately with one reading every minute for 60 min. Plot data with absorbance on the abscissa and time on the ordinate (e.g., Fig. 4).

Determining the Effect of RGS14 on Mitotic Aster Formation

To determine whether proteins that localize to the mitotic apparatus also affect the formation of the spindle, an *in vitro* mitotic aster assay may be used (see flowchart of procedure in Fig. 5A). Mitotic extracts may be

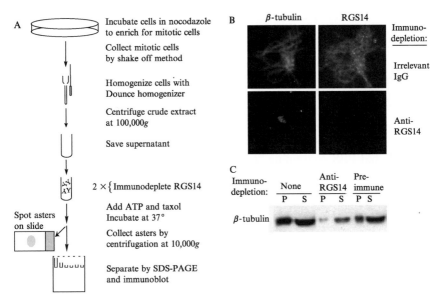

FIG. 5. Immunodepletion of RGS14 from HeLa cell mitotic extracts disrupts aster formation. (A) Flow diagram of method for preparing and immunodepleting mitotic cell extracts. (B) RGS14 was immunodepleted from 50 μl of extract (3×10^7 cells/ml) twice with 100 μg affinity-purified anti-RGS14-263 antiserum (antigen: rat RGS14 amino acids 263–544) conjugated to a Seize-IP matrix (Pierce). Induction of mitotic aster formation on RGS14-depleted and control extracts (depleted with irrelevant IgG) was accomplished by adding 3 μM ATP and 10 μM taxol to the extracts and incubating at 30° for 1 h. RGS14-depleted asters are very small compared to control asters. (C) Asters were collected by centrifugation and analyzed by SDS–PAGE. The relative amounts of asters formed were compared by immunoblotting for β-tubulin. There was less tubulin present in the pellet ("P") formed by RGS14-immunodepleted mitotic extracts induced to form asters than those formed by mitotic extracts depleted with preimmune serum or nondepleted, untreated mitotic extracts, indicating that RGS14 is required for normal mitotic aster formation. The content of β-tubulin in the supernatent ("S") postimmunodepletion of RGS14 is not altered grossly. (See color insert.)

isolated from HeLa cells (Gaglio *et al.*, 1995). Culture HeLa cells in Dulbecco's modified Eagle's medium (DMEM) with 8% (v/v) fetal bovine serum (FBS). When the cells are approximately 40% confluent, incubate them in DMEM/8% FBS containing 2 mM thymidine for 15–18 h, remove the medium, and wash five times with PBS to release the thymidine block. Incubate the cells in DMEM/10% (v/v) FBS for 9 h before subjecting them to a second 16-h thymidine block. A double thymidine block will place >95% of the cells into the G_1/S transition point of the cell cycle. Release the cells from the block by rinsing five times with PBS and culture with

DMEM/10% (v/v) FBS supplemented with 40 ng/ml of nocodazole. Noco-dazole inhibits microtubule polymerization and blocks cells during mitosis at metaphase (Zieve and Solomon, 1982; Zieve *et al.*, 1980). After 9–12 h, more than 80% of the cells will be in mitosis and appear rounded. Tap the tissue culture dish on the bench five times to release poorly adherent mitotic cells. Transfer the suspended cells to a conical tube, add cytochala-sin B (20 μg/ml final), and incubate at 37° for 30 min. Cytochalasin B is added to exclude actin microfilament contamination. It also improves cell lysis. Collect the cells by centrifugation (600g, 5 min, 4°), wash three times with ice-cold PBS containing 20 μg/ml cytochalasin B by centrifuging the suspended cells between each wash and resuspend in 1 ml of KHM buffer (78 mM KCl, 50 mM HEPES, pH 7.0, 4 mM MgCl$_2$, 2 mM EGTA, add DTT 1 mM final immediately prior to use). Take a 20-μl sample of the resuspended cells, add 20 μl of trypan blue [0.4% (v/v) in PBS] and count the number of intact cells (exclusion of trypan blue) using a hemocytome-ter. Centrifuge the remaining cell suspension (600g, 5 min, 4°), aspirate the supernatant, and resuspend the cell pellet in KHM buffer to a final con-centration of approximately 3 × 10^7 cells/ml. Homogenize the cells with a Dounce homogenizer (tight pestle) on ice with five sets of 10 strokes. Please note that the number of strokes and sets is important. Overdoing the isolation results in poor-quality extracts that will perform suboptimally. Transfer the crude mitotic extract to a 1.5-ml conical ultracentrifuge tube and centrifuge (100,000g, 15 min, 4°). The supernatant can be divided into 50-μl aliquots for immunodepletion experiments.

To immunodeplete the extracts, incubate the mitotic extract with either 100 μg of affinity-purified anti-RGS14 antiserum or 100 μg of preimmune serum immobilized on a Seize-IP matrix column (Pierce Biotechnology, Rockford, IL). Prior to adding the mitotic extract, the column is washed three times with KHM buffer. The advantage of using these columns is that the antibody is covalently coupled to the matrix; consequently, each col-umn may be used for several immunodepletion experiments with reduced nonspecific binding. The mitotic extract is rocked continuously at 4° for 2 h to ensure proper mixing. Then, the extract is eluted by centrifugation (3000g, 1 min, 4°). Release the immunoprecipitated protein from the col-umn using three 50-μl washes with IgG elution buffer (Pierce) and save the fractions. Wash the column sequentially three times with PBS and three times with KHM buffer. Add the immunodepleted mitotic extract back onto the column and incubate it for an additional 2 h at 4°. The second incubation depletes the extract of nearly all the endogenous RGS14 (as determined by gel electrophoresis analysis).

Retrieve the depleted mitotic extract by centrifugation (3000g, 1 min, 4°) and then add MgATP and taxol to the extract to a final concentration of

3 and 10 μM, respectively. Incubate the mitotic extracts, MgATP, and taxol at 30° for 1 h, spot 5 μl of the solution on a poly-L-lysine-coated microscope slide, and let stand on the slide for 5 min, prior to adding 1 μl of 20% paraformaldehyde to the mitotic extract drop on the slide and fixing for 15 min. Once fixed, wash the slides in PBS and stain for tubulin and RGS14 as described previously.

Centrifuge the remainder of the mitotic extract reaction (14,000g, 15 min, room temperature) to sediment the nascent asters and associated proteins. Resuspend the pellet in Laemmli sample buffer and resolve the proteins by SDS–PAGE (10% gel). Transfer the separated proteins to a PVDF membrane. Rinse the membrane with TBST [50 mM Tris, pH 7.5, 150 mM NaCl, 0.1% (v/v) Tween 20] before incubating in 5% (w/v) skim milk powder in TBST for 1 h at room temperature to block non-specific binding sites. Wash the membrane for 3 h with TBST, ensuring frequent solution changes. Then incubate the membrane for 2 h with the appropriate primary antibody. Before incubating the membrane with a secondary antibody coupled to horseradish peroxidase [1:5000 in 5% (w/v) dried milk powder in TBST], wash with TBST with frequent solution exchanges for at least 1 h. Incubate the membrane with the secondary antibody for 1 h and rewash with TBST for 1 h. Proteins labeled with the secondary antibody are detected by enhanced chemiluminescence (Amersham Pharmacia Biotech) and BioMax film (Kodak, Rochester, NY). To reprobe, strip the membranes by incubating them for 30 min at 50° in stripping buffer [0.5 M Tris–HCl, 0.4% (w/v) SDS, 100 mM 2-mercaptoethanol; 2-mercaptoethanol should be handled in a fume hood with gloves] and then washing in TBST for 1 h with frequent changes. Membranes may be safely stripped and reprobed three times.

Applications

Fluorescence-based measurement of refolded GST-RGS14-His$_6$ guanine nucleotide dissociation inhibitor (GDI) activity is performed at 30° using a LS55 luminescence spectrofluorimeter (Perkin Elmer, Boston, MA), with excitation and emission wavelengths of 485 and 530 nm, respectively. The assay is conducted as described previously (Kimple *et al.*, 2003) using buffer containing 10 mM Tris–HCl, pH 8.0, 1 mM EDTA, 10 mM MgCl$_2$, and 1 μM BODIPY FL-GTPγS (Molecular Probes, Eugene, OR). Recombinant purified human G$\alpha_{i1}\Delta$N30 (100 nM) (Kimple *et al.*, 2004a) is preincubated for 5 min with various concentrations of GST-RGS14-His$_6$ protein (between 0 and 3.26 μM) and then added to the cuvette. Figure 1B illustrates that a 7.8-fold molar excess of refolded GST-RGS14-His$_6$ to Gα_{i1} decreases the initial rate of BODIPY FL-GTPγS binding to Gα_{i1} by

70%, consistent with the reported effects of RGS14 as a GDI for $G\alpha_{i1}$ (Kimple *et al.*, 2001). Initial rate data from a range of concentrations were converted into percentage form and were then fit to a dose–response curve using Graph Pad Prism version 4.0 (GraphPad Software, San Diego, CA). Figure 1C illustrates the concentration–response relationship of refolded GST-RGS14-His$_6$ GDI activity for $G\alpha_{i1}$. The calculated IC_{50} for inhibition of GTPγS binding by GST-RGS14-His$_6$ protein was 540 nM. An IC_{50} of 144 nM has been reported previously, in an identical assay, for a synthetic peptide derived from the GoLoco motif region of RGS14 (Kimple *et al.*, 2001). The observed ~4-fold difference in potency could be due to inhibitory constraints only present within the full-length RGS14 protein or, more likely, due to a lower specific activity of the 88-kDa refolded recombinant protein versus a synthetic GoLoco motif peptide. Nonetheless, these data illustrate the biochemical functionality of refolded GST-RGS14-His$_6$ protein.

We have found that RGS14 colocalizes with the mitotic spindle in preimplantation mouse embryos, as well as in cell lines and primary cell cultures (Fig. 2). To determine whether RGS14 binds directly to microtubules, GST-RGS14-fusion protein or GST controls were incubated with tubulin, which was induced to polymerize, and the nascent microtubules were isolated from free tubulin by centrifugation. Whereas binding of GST was minimal, GST-RGS14 bound robustly to the nascent microtubules, demonstrating that RGS14 is a microtubule-associated protein (Fig. 3A). Defining the interaction further, we found that RGS14 exhibits high affinity for microtubules, with a $K_D = 1.3 \pm 0.3$ μM (Fig. 3B and C), which is in the range of other known microtubule-associated proteins (Goode and Feinstein, 1994).

We examined whether RGS14 altered microtubule dynamics. We chose to look at its effect using MAP-rich (>10%) tubulin. We reasoned that this preparation would permit us to identify direct or indirect actions. We incubated 10 μM GST-RGS14 with 100 μM tubulin and found that RGS14 was capable of stabilizing microtubules (Fig. 4). When we repeated this assay using tubulin with a low percentage (<3%) of microtubule-associated proteins, we found that RGS14 did not affect the polymerization of tubulin. Therefore, we concluded that RGS14 acts indirectly to promote microtubule polymerization.

Because RGS14 colocalizes with the mitotic apparatus and binds and regulates microtubule assembly *in vitro*, we asked whether RGS14 could also regulate the formation of mitotic asters. We immunodepleted RGS14 from HeLa cell mitotic extracts and found that normal mitotic asters were not detected by immunofluorescence microscopy (Fig. 5B). When we examined the asters by immunoblot and probed for β-tubulin, we found that there was much less tubulin in the insoluble fraction (Fig. 5C). Therefore, RGS14 appears to have a role in regulating the formation of microtubules during formation of the mitotic apparatus.

In conclusion, multiple methodologies and techniques need to be used to understand the interaction between RGS14 and microtubules and the impact of this relationship on the formation and dynamics of the mitotic spindle.

Acknowledgments

Mr. Martin McCaffrey is an Ontario Graduate Studies Scholar. Dr. Willard is a postdoctoral fellow of the American Heart Association. Drs. Dagnino and D'Souza are New Investigators of the Canadian Institutes of Health Research. The Heart and Stroke Foundation of Ontario (Grant 5051 to SJAD) and NIH Grant GM062338 (to DPS) supported the work described. The authors acknowledge Dr. Patricia Hunt for sharing her protocol for the mounting and immunostaining of the cytoskeleton and chromatin of preimplantation embryo and Dr. Duane Compton for clarifying the nuances of HeLa cell mitotic extract isolation.

References

Albertini, D. F., and Eppig, J. J. (1995). Unusual cytoskeletal and chromatin configurations in mouse oocytes that are atypical in meiotic progression. *Dev. Genet.* **16**, 13–19.

Gaglio, T., Saredi, A., and Compton, D. A. (1995). NuMA is required for the organization of microtubules into aster-like mitotic arrays. *J. Cell Biol.* **131**, 693–708.

Gaskin, F., Cantor, C. R., and Shelanski, M. L. (1974). Turbidimetric studies of the *in vitro* assembly and disassembly of porcine neurotubules. *J. Mol. Biol.* **89**, 737–755.

Goode, B. L., and Feinstein, S. C. (1994). Identification of a novel microtubule binding and assembly domain in the developmentally regulated inter-repeat region of tau. *J. Cell Biol.* **124**, 769–782.

Gotta, M., and Ahringer, J. (2001). Axis determination in *C. elegans*: Initiating and transducing polarity. *Curr. Opin. Genet. Dev.* **11**, 367–373.

Hogan, B., Costantini, F., Beddington, R., and Lacy, E. (1994). "Manipulating the Mouse Embryo: A Laboratory Manual," 2 Ed., Cold Spring Harbor Laboratory Press, Plainview, NY.

Hunt, P., LeMaire, R., Embury, P., Sheean, L., and Mroz, K. (1995). Analysis of chromosome behavior in intact mammalian oocytes: Monitoring the segregation of a univalent chromosome during female meiosis. *Hum. Mol. Genet.* **4**, 2007–2012.

Keates, R. A. (1981). Stabilization of microtubule protein in glycerol solutions. *Can. J. Biochem.* **59**, 353–360.

Kimple, R. J., De Vries, L., Tronchere, H., Behe, C. I., Morris, R. A., Gist Farquhar, M., and Siderovski, D. P. (2001). RGS12 and RGS14 GoLoco motifs are G alpha(i) interaction sites with guanine nucleotide dissociation inhibitor activity. *J. Biol. Chem.* **276**, 29275–29281.

Kimple, R. J., Jones, M. B., Shutes, A., Yerxa, B. R., Siderovski, D. P., and Willard, F. S. (2003). Established and emerging fluorescence-based assays for G-protein function: Heterotrimeric G-protein alpha subunits and regulator of G-protein signaling (RGS) proteins. *Comb. Chem. High Throughput Screen.* **6**, 399–407.

Kimple, R. J., Willard, F. S., Hains, M. D., Jones, M. B., Nweke, G. K., and Siderovski, D. P. (2004a). Guanine nucleotide dissociation inhibitor activity of the triple GoLoco motif protein G18: Alanine-to-aspartate mutation restores function to an inactive second GoLoco motif. *Biochem. J.* **378,** 801–808.

Kimple, R. J., Willard, F. S., and Siderovski, D. P. (2004b). Purification and *in vitro* functional analyses of RGS12 and RGS14 GoLoco motif peptides. *Methods Enzymol.* **390,** 419–436.

Knoblich, J. A. (2001). Asymmetric cell division during animal development. *Nature Rev. Mol. Cell Biol.* **2,** 11–20.

Laemmli, U. K. (1970). Cleavage of structural proteins during the assembly of the head of bacteriophage T4. *Nature* **227,** 680–685.

Liu, Z., Chatterjee, T. K., and Fisher, R. A. (2002). RGS6 interacts with SCG10 and promotes neuronal differentiation: Role of the G-gamma subunit-like (GGL) domain of RGS6. *J. Biol. Chem.* **277,** 37832–37839.

Manso-Martinez, R., Villasante, A., and Avila, J. (1980). Incorporation of the high-molecular-weight microtubule-associated protein 2 (MAP2) into microtubules at steady state *in vitro. Eur. J. Biochem.* **105,** 307–313.

Martin McCaffrey, L., Willard, F. S., Oliveira-dos-Santos, A. J., Natale, D. R., Snow, B., Kimple, R. J., Watson, A. J., Dagnino, L., Penninger, J. M., Siderovski, D. P., and D'Souza, S. J. (2004). RGS14 is a mitotic spindle protein essential from the first division of the mammalian zygote. Submitted for publication.

Nixon, A. B., and Casey, P. J. (2004). Analysis of the regulation of microtubule dynamics by interaction of RGSZ1 (RGS20) with the neuronal stathmin, SCG10. *Methods Enzymol.* **390,** 53–64.

Nixon, A. B., Grenningloh, G., and Casey, P. J. (2002). The interaction of RGSZ1 with SCG10 attenuates the ability of SCG10 to promote microtubule disassembly. *J. Biol. Chem.* **277,** 18127–18133.

Roychowdhury, S., Panda, D., Wilson, L., and Rasenick, M. M. (1999). G protein alpha subunits activate tubulin GTPase and modulate microtubule polymerization dynamics. *J. Biol. Chem.* **274,** 13485–13490.

Roychowdhury, S., and Rasenick, M. M. (1997). G protein beta1gamma2 subunits promote microtubule assembly. *J. Biol. Chem.* **272,** 31576–31581.

Spindle, A. (1980). An improved culture medium for mouse blastocysts. *In Vitro Cell Dev. Biol. Anim.* **16,** 669–674.

Willard, F. S., Kimple, R. J., and Siderovski, D. P. (2004). Return of the GDI: The GoLoco motif in cell division. *Annu. Rev. Biochem.* **73,** 925–951.

Zieve, G., and Solomon, F. (1982). Proteins specifically associated with the microtubules of the mammalian mitotic spindle. *Cell* **28,** 233–242.

Zieve, G. W., Turnbull, D., Mullins, J. M., and McIntosh, J. R. (1980). Production of large numbers of mitotic mammalian cells by use of the reversible microtubule inhibitor nocodazole: Nocodazole accumulated mitotic cells. *Exp. Cell Res.* **126,** 397–405.

Subsection E
Rhogef Subfamily

[17] Modular Architecture and Novel Protein–Protein
Interactions Regulating the RGS-Containing Rho
Guanine Nucleotide Exchange Factors

By José Vázquez-Prado, John Basile, and J. Silvio Gutkind

Abstract

The regulator of G-protein signaling (RGS)-containing RhoGEFs, including p115RhoGEF, PDZ-RhoGEF, and LARG, represent a novel family of guanine nucleotide exchange factors for RhoA that are regulated by the $G\alpha_{12/13}$ family of heterotrimeric G proteins. Experimental evidence indicates that the complex architecture of these RhoGEFs provides the structural basis for novel regulatory mechanisms mediated by protein–protein interactions. These include the direct association of their RGS domain with GTP-bound forms of $G\alpha_{12/13}$ and the binding of the PDZ domain present in PDZ-RhoGEF and LARG to plexins, which are receptors for semaphorins. The carboxyl-terminal region of these GEFs also exerts regulatory properties, including the ability to form dimers, which is inhibitory to their *in vivo* GEF activity and, in the case of PDZ-RhoGEF, to associate with PAK4, a downstream target of Cdc42. This carboxyl-terminal region also acts as the target for tyrosine kinases, which have a positive effect on the long-term activity of these GEFs. This article describes the experimental strategies that have been utilized to begin unraveling the molecular mechanisms regulating the functional activity of RGS-containing RhoGEFs.

Introduction

Small GTP-binding proteins of the Rho family play a central role in transducing signals from cell surface receptors, including those that are coupled to heterotrimeric G proteins, into rapid changes in the actin-based cytoskeleton (Bar-Sagi and Hall, 2000; Narumiya, 1996; Takai *et al.*, 2001). By doing so, Rho GTPases such as RhoA, Rac1, and Cdc42, orchestrate the remodeling of actin-containing cytoskeletal structures and regulate the cell contractile machinery, which are required for many cellular processes, including polarized cell motility that occurs during embryogenesis, wound healing, immune response, vascular development, and axonal growth (Etienne-Manneville and Hall, 2002; Fukata *et al.*, 2003). In addition, RhoA can control the expression of growth-promoting genes, such as those regulated by the serum response factor, thereby promoting cell cycle progression and normal or aberrant cell growth (Marshall, 1999; Sahai and Marshall, 2002).

The activity of RhoA is regulated by guanine nucleotide exchange factors (GEFs) that promote the exchange of GDP for GTP, thereby activating this small GTPase, and by GTPase-activating proteins (GAPs) and guanine nucleotide dissociation inhibitors (GDIs) that limit or inhibit its function (Burridge and Wennerberg, 2004; Schmidt and Hall, 2002). Recent efforts have focused on the identification of the nature of the GEFs by which each class of cell surface receptors stimulate RhoA. In particular, for G-protein-coupled receptors (GPCRs), a family of closely related GEFs has been identified that mediate the activation of RhoA in response to GPCRs linked to the $G\alpha_{12}/G\alpha_{13}$ family of heterotrimeric G proteins, such as the receptors for lysophosphatidic acid (LPA) and thrombin (Fukuhara *et al.*, 2001; Kozasa, 2001). Of interest, this GEF family shares a structural domain similar to regulators of G-protein signaling (RGS). This RGS-like domain establishes direct interactions with active $G\alpha_{12/13}$ and thus can accelerate the hydrolysis of GTP bound to $G\alpha_{12}/G\alpha_{13}$ (Kozasa *et al.*, 1998), as well as provide a structural feature by which these active heterotrimeric α subunits can bind to RGS-containing GEFs, thereby causing their activation (Fukuhara *et al.*, 1999, 2001; Hart *et al.*, 1998).

The family of RGS-RhoGEFs is composed of three members: PDZ-RhoGEF, LARG, and p115RhoGEF (Fukuhara *et al.*, 2001). Interestingly, these RGS-RhoGEFs also contain a series of structural elements, thus broadening the array of extracellular signals that they can transduce and conferring specific regulatory properties. We and others have demonstrated that the multidomain nature of PDZ-RhoGEF and LARG provides elements for their regulation by $G\alpha_{12/13}$ proteins, plexin receptors, and a variety of tyrosine and serine/theonine kinases (Aurandt *et al.*, 2002; Barac *et al.*, 2004; Chikumi *et al.*, 2002, 2004; Fukuhara *et al.*, 1999, 2000; Perrot *et al.*, 2002; Swiercz *et al.*, 2002). Each activity is attributable to a distinct structural element. Specifically, the amino-terminal PDZ module is key for the interaction with plexins (Aurandt *et al.*, 2002; Perrot *et al.*, 2002; Swiercz *et al.*, 2002), which are receptors for a large family of soluble and membrane-bound semaphorins (Tamagnone and Comoglio, 2000). The RGS domain connects to $G\alpha_{12/13}$ signaling (Fukuhara *et al.*, 1999, 2000; Kozasa *et al.*, 1998). The DH catalytic domain, followed by a PH domain, is necessary to promote the exchange of GDP for GTP, leading to formation of the active RhoA-GTP form, whereas the carboxyl-terminal region determines the ability of the GEF to dimerize and establishes direct interactions with PAK4, suggesting a regulatory loop connecting RhoA with other members of the Rho family of small GTPases (Fig. 1) (Barac *et al.*, 2004; Chikumi *et al.*, 2004).

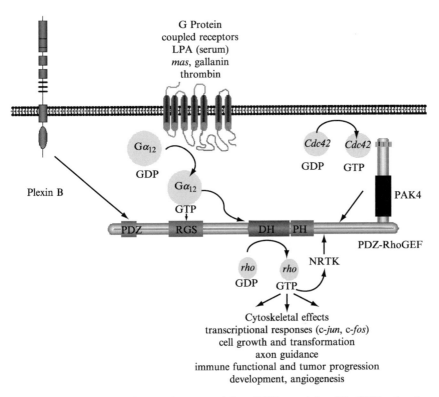

FIG. 1. Diverse signaling pathways modulate RGS-containing RhoGEFs, thereby stimulating or preventing the activation of RhoA. The RGS-like domain, which defines this family of GEFs, establishes direct interactions with $G\alpha_{12/13}$ family members; this interaction has a reciprocal effect, activating the GEF and modulating the activity of the $G\alpha$ subunit. In the case of PDZ-RhoGEF (shown) and LARG, the amino-terminal PDZ domain serves to link plexins to RhoA activation. The DH and PH domains are responsible for the catalytic activity and membrane interactions, respectively, whereas the carboxyl-terminal region provides regulatory properties, including the ability to form dimers and to act as the target of non-receptor tyrosine kinases (NRTKs), which have a positive effect on its long-term activity. PAK4, an effector of Cdc42 and very likely other Rho family GTPases, interacts with this region, phosphorylating it, and preventing the activation of RhoA. Further details and the experimental strategies that have served to reveal the diverse interactions and effects of RGS-containing RhoGEFs are provided in the text.

This article describes the experimental strategies used to identify novel protein–protein interactions by which diverse signaling pathways regulate the activity of RGS-RhoGEF and the methods used to characterize their functional consequences. We will not assume previous experience from the reader and emphasize practical simplifications that we have developed in some of the protocols, which we expect can be useful for the characterization of other multidomain-containing proteins, in addition to RGS-RhoGEFs.

Methods

$G\alpha_{12/13}$ Interact with the RGS Domain of RGS-RhoGEFs and Promote Their Activity

The activation of RhoA by $G_{12/13}$ protein-coupled receptors can occur though direct interactions of the $G\alpha_{12/13}$ subunit with the RGS domain characteristic of the family of RhoGEFs that includes PDZ-RhoGEF (Fukuhara et al., 2001). In fact, a catalytically inactive mutant of PDZ-RhoGEFs acts as a dominant-negative interfering molecule decreasing the activation of RhoA by $G\alpha_{12/13}$ active mutants (Fukuhara et al., 1999). Interestingly, the interaction between $G\alpha_{12/13}$ and PDZ-RhoGEF not only determines the activation of the GEF, but limits the extent of the activity achieved; accordingly, an RGS-deletion mutant shows an increased, obviously nonregulated activity (Fukuhara et al., 1999). All the members of this family of GEFs establish direct interactions with $G\alpha_{12/13}$, which promote the activation of RhoA downstream pathways. This section describes the technical details used to reveal those interactions and some of the RhoA-dependent pathways that are stimulated by the $G_{12/13}$-dependent activation of RhoA.

Coimmunoprecipitation of G Protein and RGS-RhoGEF

Overview

The interaction between activated $G\alpha_{13}$ and PDZ-RhoGEF is determined in HEK293T cells transfected with the corresponding plasmids. In this case, an HA-tagged, constitutively active (GTPase-deficient) point mutant of $G\alpha_{13}$ (glutamine 226 to leucine or "$G\alpha_{13}QL$") is expressed using the pCEFL-HA-$G\alpha_{13}QL$ plasmid upon transfection into HEK293T cells in the presence or absence of an AU1-tagged PDZ-RhoGEF expressed from pCEFL-AU1-PDZ-RhoGEF. Two days after transfection the cells are lysed and immunoprecipitations are carried out using an anti-AU1 tag monoclonal antibody. Coimmunoprecipitated $G\alpha_{13}QL$ is detected by Western blotting using anti-HA monoclonal antibodies. As a reference, the expression of $G\alpha_{13}QL$ in total lysates is detected.

Procedure

Transfection. The day before transfection, divide one 10-cm confluent plate of HEK293T cells into five poly-D-lysine-coated 10-cm dishes [poly-D-lysine: 5 μg/ml in phosphate-buffered saline (PBS), incubated in the plate for 10 min at room temperature and washed out with PBS]. Be sure to resuspend the cells well and distribute them evenly into the plates. Use a 10-cm plate for each condition to be tested. Transfect the cells with either Polyfect (Qiagen) or Lipofectamine Plus (Invitrogen) or a similar DNA transfection reagent. Follow the instructions provided by the manufacturer but reduce the amount of reagent indicated: for Polyfect, use 25 μl of reagent, and for Lipofectamine Plus, use 10 μl of Plus reagent and 12 μl of Lipofectamine. Transfect 4 μg of DNA (prepared with Qiagen QIAprep spin miniprep kit, Qiagen plasmid midikit, or equivalent). A typical experiment will include 1 μg of AU1-tagged PDZ-RhoGEF (pCEFL-AU1-PDZ-RhoGEF) or a control expression vector (pCEFL), 1 μg of HA-tagged $G\alpha_{13}QL$ (pCEFL-HA-$G\alpha_{13}QL$), and 2 μg of empty vector (pCEFL). All the indicated DNAs are available from our laboratory.

Coimmunoprecipitation. Two days after transfection, wash the cells with PBS and lyse them on ice for 20–30 min with 1 ml/plate of ice-cold lysis buffer (50 mM Tris, pH 7.5, 0.15 M NaCl, 5 mM EDTA, 1% Triton X-100 and protease inhibitors added just before use as described later). Prepare enough lysis buffer with protease inhibitors considering at least 4 ml per sample (1 ml for lysis and 3 ml for washes). To prepare 50 ml of lysis buffer, add 50 μl aprotinin, 50 μl leupeptin (both from a 10-mg/ml stock prepared in water and kept at $-20°$ in 50-μl aliquots) and 500 μl phenylmethylsulfonyl fluoride (PMSF; from a 0.1 M stock prepared in ethanol and kept at room temperature). An alternative is to use a commercially available protease inhibitor cocktail (Roche or equivalent). Scrape the cells and transfer the lysates to microcentrifuge tubes on ice. Centrifuge at 14,000 rpm for 10 min at 4°. Transfer supernatants to new tubes (save 75 μl for total cell lysate analysis). Add 1 μl of AU1 monoclonal antibody (Covance) and incubate for 1 h to overnight at 4°. Add 25 μl of Gamma-bind Sepharose (Amersham Pharmacia Biotech; to use, dilute 1:2 with PBS; any equivalent antibody-binding resin can be used). Incubate for 30 min at 4° with constant shaking. Spin down briefly and wash three times (1 ml/each using lysis buffer containing protease inhibitors). Vortex and spin down briefly. For resolving coimmunoprecipitated proteins by sodium dodecyl sulfate–polyacrylamide gel electrophoresis (SDS–PAGE), boil the beads in 50 μl of 1× sample buffer containing 5% 2-mercaptoethanol. Boil for 5 min and centrifuge at 14,000 rpm for 5 min at room temperature. Run the immunoprecipitates and the respective total cell lysates in a 10% SDS–PAGE gel and transfer to Immobilon-P (Millipore or equivalent). Block the membrane with 5% nonfat dry milk in T-TBS (Tris-buffered saline containing 0.05% Tween 20) for 1 h at room temperature, wash with T-TBS, and determine the interaction between $G\alpha_{13}$ and PDZ-RhoGEF by western blotting. The HA-tagged $G\alpha_{13}$ is detected using an anti-HA monoclonal antibody (Covance) diluted 1:1000 in 0.5% bovine serum albumin (BSA) in T-TBS, incubated for 1 h at room temperature or overnight at 4° with constant shaking. Wash the blot four to five times, 5 min each, with T-TBS. Incubate with a secondary horseradish peroxidase (HP)-conjugated goat antimouse antibody diluted 1:10,000 in 0.5% BSA in T-TBS. Incubate for 1 h at room temperature, wash, and develop using the ECL Plus system (Amersham Biosciences), SuperSignal WestPico (Pierce), or similar enhanced chemiluminescence system following the instructions provided by the manufacturer. For an example of a coimmunoprecipitation experiment, see Fukuhara *et al.* (1999).

RhoA-Dependent Transcription Factor Assays

Overview

$G\alpha_{12/13}$ protein-coupled receptors or other stimuli acting on RhoA promote the activation of transcription factors stimulating the serum response element (SRE) (Fromm *et al.*, 1997). The assay used to monitor this activity measures the expression and function of reporter genes such as luciferase or chloramphenicol O-acetyltransferase (CAT), which are cloned in a plasmid downstream of the SRE that is mutated to respond just to serum response factor and not to tertiary complex factors ("SRE.L"; Hill *et al.*, 1995). The synthesis and activity of these reporter enzymes reflect the activity of the serum response factor. This section describes the SRE-CAT assay as it is performed in HEK293T cells.

Procedure

The day before transfection, split the cells into poly-D-lysine-coated 24-well plates. Divide one 10-cm confluent plate into three 24-well plates and scale up or down as required. The next day, wash the cells with PBS and incubate with 200 μl of serum-free medium. Transfect the cells with a total of 0.5 μg DNA per well, including 0.15 μg of reporter SRE-CAT vector, 0.1 μg of the DNA to be tested, and adjust with an empty vector. Dilute the DNA in a final volume of 25 μl in serum-free medium, add 4 μl Plus reagent mix, and incubate for 15 min. In parallel, prepare serum-free medium with Lipofectamine; consider 25 μl of serum-free medium and 1 μl of Lipofectamine per well. Mix with the tubes containing DNA-medium-Plus reagent and incubate for 15 min. Add the mixture to the cells that have been washed and incubated with 200 μl of serum-free medium. Leave for 2 h, aspirate the medium, and replace with 300 μl of serum-free Dulbecco's modified Eagle's medium (DMEM) containing 10 mM HEPES, pH 7.4. The next day, process the experiment. If the effect of some transfected DNA is being tested, aspirate the medium and lyse with 150 μl/well of reporter lysis buffer (Promega, prepared fresh in water from a 5$\times$ stock). If the effect of some stimuli is being tested (e.g., the application of thrombin, lysophosphatidic acid, and carbachol), incubate the cells with the stimulant for 4–6 h and proceed to the lysis as indicated earlier. Leave the plate at room temperature for 20 min with constant shaking. Transfer the supernatant to microcentrifuge tubes. Vortex briefly and centrifuge for 10 min at maximum speed. Transfer 5 μl of the supernatant to new tubes for the CAT assay.

Prepare the following substrate mixture for the reaction (per sample): 112.5 μl 0.25 M Tris–HCl (pH 7.5), 2.5 μl [^{14}C]chloramphenicol (ICN, 100 mCi/mmol), and 5 μl n-butyryl CoA (C4:0, Sigma, 10 mg/ml water).

Add 120 μl of the substrate mixture to each tube containing 5 μl of lysate, mix, and incubate at 37° for 30 min. Add 300 μl of extraction buffer (one part of xylene and two parts of 2,6,10,14-tetramethylpentadecane kept at room temperature in a dark glass bottle). Vortex for 30 s and centrifuge. Transfer 150 μl of the upper phase to scintillation vials; add 5 ml of liquid scintillation cocktail and count for ^{14}C (see Fig. 2).

G$_{13}$-Coupled Receptor-Stimulated Activation of RhoA

Overview

The principle of the assay depends on the ability of the active, GTP-bound form of RhoA to interact with the Rho-binding domain (RBD) of rhotekin expressed as a glutathione S-transferase (GST) fusion protein (Reid et al., 1996). The GST-RBD protein is first expressed in bacteria and the recombinant protein is isolated with glutathione-Sepharose beads. These beads are then used to capture the active form of RhoA from lysates of stimulated HEK293T cells, and the fraction of active RhoA is detected by western blotting using the total amount of RhoA detected in parallel in total cell lysates as a reference.

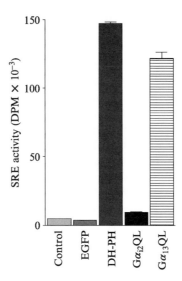

Fig. 2. SRE assay. Typical example of an SRE-CAT assay in HEK293T cells transfected with pSRE-CAT, together with EGFP (negative control), the DH-PH domain of PDZ-RhoGEF (positive control), and activated mutants of the $G\alpha_{i2}$ and $G\alpha_{13}$ G-protein α subunits.

Procedure

Preparation of Rhotekin Beads. The expression vector pGEX-rhotekin-RBD, which encodes the GST fusion protein with the isolated GTP-dependent binding domain of the RhoA effector rhotekin (rhotekin-RBD), was provided by Dr. S. Narumiya (Reid *et al.*, 1996). Transform the protease-deficient *Escherichia coli* BL21 strain with pGEX-rhotekin-RBD and plate the bacteria on LB agar containing 50 μg/ml ampicillin. Pick a colony of pGEX-transformed BL21 cells, and start a liquid culture in 5 ml of LB ampicillin and leave overnight at 37° with constant shaking. The next day, transfer the culture into a 1-liter flask containing 250 ml of LB ampicillin and leave 2–3 h at 37° with constant shaking. Add 0.2 mM isopropyl β-D-thiogalactoside (IPTG; Sigma; prepare a 200 mM stock in water and keep it frozen in small aliquots), transfer the culture to room temperature, and leave it overnight with constant shaking. Harvest bacteria by centrifugation and resuspend in 10 ml of ice-cold PBS-Triton-EDTA including protease inhibitors (1% Triton X-100, 1 mM EDTA, 1 mM PMSF, 10 μg/ml each aprotinin and leupeptin). Transfer the lysates to a 50-ml polypropylene centrifuge tube and freeze–thaw three times by immersing in a bath with ethanol-dry ice followed by cold water. Keep the lysates on ice and sonicate three times for 10 s each. Centrifuge at 14,000 rpm (Sorvall, SS34 rotor or equivalent). In the meantime, prepare 250 μl of glutathione-Sepharose beads (Amersham Pharmacia or similar) by washing them with PBS-Triton-EDTA. Transfer the supernatant of the lysates to a 15-ml tube and incubate with the beads for 30 min at 4° with constant shaking. Centrifuge at 4° for 1 min at 3000 rpm. Resuspend the beads in PBS-Triton-EDTA-containing protease inhibitors, transfer to a microcentrifuge tube and wash three times, resuspend the beads each time by vortexing, and spin down briefly. Wash three times with PBS-containing protease inhibitors and resuspend in 500 μl of this buffer. For the experiment, use 50 μl of resuspended beads per sample. We prefer to keep the beads at 4° and use them within a week.

Experiment. Transfect HEK293T cells with wild-type or GTPase-deficient mutant Gα_{13} or any cDNA to be tested. The day after transfection, leave the cells in serum-free medium and process the experiment the following day, stimulating with agonists when required. Wash once with ice-cold PBS and keep the plates on ice from that point on, using ice-cold buffers. Lyse the cells at 4° with 1 ml of a buffer containing 50 mM Tris, pH 7.5, 0.15 M NaCl, 1% Triton X-100, 5 mM EDTA, 10 mM MgCl$_2$, 10 μg/ml aprotinin, 10 μg/ml leupeptin, and 1 mM PMSF. Transfer the lysates to microcentrifuge tubes, centrifuge at 14,000 rpm for 5 min at 4°, and transfer the supernatants to new tubes for the isolation of GTP-RhoA and 75 μl to a second set of tubes for total cell lysates. Incubate the cell lysates with 50 μl of GST-RBD beads (vortex briefly before taking the indicated volume), leave at 4° for 30–45 min with constant shaking, wash three times with lysis buffer, vortex, and spin down briefly for each wash. Detect the active GTP-bound form of RhoA associated with GST-rhotekin-RBD and the total RhoA in cell lysates by western blot analysis using a monoclonal antibody against RhoA (Santa Cruz Biotechnology). For examples of RhoA activation assays, see Barac *et al.* (2004), Chikumi *et al.* (2002), and Fukuhara *et al.* (1999).

Focus Formation Assay

Overview

One of the most remarkable cellular effects of the persistent activation of RhoA is acquisition of a transformed phenotype, which can be induced experimentally by transfecting mouse fibroblasts with a GTPase-deficient RhoA, which acts as a constitutively active mutant or with an upstream activator of this small GTPase. A very convenient way to detect and characterize such activity, and therefore to characterize molecular mediators of RhoA activity, is by transfecting cells with a DNA vector expressing a putative RhoA activator and then monitor for the formation of foci. When the transforming activity is dependent on RhoA, it is sensitive to treatment with *Clostridium botulinum* C3 exotoxin, which is a very specific bacterial toxin inhibiting RhoA. If required, RhoA mutants insensitive to the effect of the C3 toxin (RhoI41) can be used as an additional specificity control (Fromm *et al.*, 1997).

Procedure

Transfect NIH3T3 cells by the calcium phosphate precipitation technique with different expression plasmids together with 1 μg of pcDNAIII-β-gal, a plasmid expressing the enzyme β-galactosidase, adjusting the total amount of plasmid DNA with an empty vector, and maintaining the cells in DMEM supplemented with 10% calf bovine serum. The day after transfection, wash the cells in medium supplemented with 5% calf serum and then maintain them in the same medium until foci are scored, 2–3 weeks later. Change the medium twice weekly. Fix the plates with PBS containing 2% (v/v) formaldehyde and 0.2% (v/v) glutaraldehyde and stain at 37° for β-galactosidase activity with a PBS solution containing 2 mM MgCl$_2$, 5 mM K$_3$Fe(CN)$_6$, 5 mM K$_4$Fe(CN)$_6$, and 0.1% 5-bromo-4-chloro-3-indolyl-β-D-galactopyranoside (X-gal) to evaluate the transfection efficiency. For examples of focus formation assays, see Chikumi *et al.* (2004).

Tyrosine Phosphorylation

Overview

Thrombin and activated focal adhesion kinase (FAK) promote the tyrosine phosphorylation of PDZ-RhoGEF, which enhances RhoA activation *in vivo* during the more sustained and delayed phase of the response (Chikumi *et al.*, 2002), representing a novel biochemical route regulating the signaling pathway from GPCRs to RhoA. Similarly, tyrosine phosphorylation can regulate the ability of the RGS-RhoGEF LARG to be stimulated by Gα_{12} (Suzuki *et al.*, 2003). Based on these observations, it can be hypothesized that tyrosine kinases may play an important regulatory role in the biochemical route linking GPCRs to RhoA.

Procedure

Transfect HEK293T cells with the expression vector for AU1-PDZ-RhoGEF, together with expression plasmids carrying cDNAs for enhanced green fluorescent protein (EGFP) or myristoylated FAK (myr-FAK). Two days after transfection, prepare lysates at 4° in a buffer containing 25 mM

HEPES, pH 7.5, 0.3 M NaCl, 1.5 mM MgCl$_2$, 0.2 mM EDTA, 0.5 mM dithiothreitol (DTT), 1% Triton X-100, 0.5% sodium deoxycholate, 0.1% SDS, 20 mM β-glycerophosphate, 1 mM sodium vanadate, 1 mM PMSF, 20 μg/ml aprotinin, and 20 μg/ml leupeptin. Isolate PDZ-RhoGEF from lysates by immunoprecipitating with the anti-AU1 antibody and determine its tyrosine phosphorylation status by western blotting with the antiphosphotyrosine antibody. For example, use a mixture of monoclonal antiphosphotyrosine (4G10) and monoclonal antiphosphotyrosine (PY99) from Upstate Biotechnology and Santa Cruz Biotechnology, respectively. Determine the expression level of PDZ-RhoGEF in total cell lysates by western blotting using an anti-AU1 antibody. To determine the agonist-induced tyrosine phosphorylation of PDZ-RhoGEF, transfect HEK293T cells with the plasmid expressing AU1-PDZ-RhoGEF. The next day, change the medium to serum-free medium and, 24 h later, stimulate the cells with 5 units/ml thrombin for different times. Immunoprecipitate the lysates with the anti-AU1 antibody and subject them to western blot analysis with the antiphosphotyrosine antibodies (see Chikumi et al., 2002).

Yeast Two-Hybrid Interaction Assay

Overview

The yeast two-hybrid system is a powerful method used to identify novel protein–protein interactions (Fields and Song, 1989). The principle is based on the reconstitution of a transcription factor that promotes the proliferation of yeast under restrictive conditions and the expression of reporter genes. This system selects for those cells carrying the cDNAs for two interacting proteins, contained in two independent plasmids, fused to either the DNA-binding or the transactivation domain of the transcription factor, permitting the isolation of the cDNA for the novel interacting protein. One of the fusion proteins is used as the bait with which novel interacting proteins will be "fished out" and the other includes the novel interacting protein contributed by the cDNA library, which will be identified and characterized further. We have used the yeast two-hybrid system successfully to clone cDNAs coding for proteins able to interact with specific domains of PDZ-RhoGEF (Barac et al., 2004; Perrot et al., 2002). There are different commercially available yeast two-hybrid systems; we have had very good experience with a system based on the reconstitution of the Gal4 transcription factor (MatchMaker III system, Clontech). We introduced some modifications and simplifications to the instructions provided by the manufacturer that, in our opinion, make the system easier to use. The whole procedure is outlined here.

a. Define the limits, at the protein and cDNA levels, of the structural domain that will be used as the bait to identify and clone cDNAs for novel interacting proteins.

b. Design the strategy to clone the cDNA for that domain into the plasmid pGBKT7, specific for the bait. As an example, cloning of the PDZ domain from PDZ-RhoGEF is described here.

c. Determine that the bait is not toxic for the yeast and that it does not promote proliferation under restrictive conditions.

d. Choose a library and perform the screening.

e. Isolate cDNAs for the prey and determine the specificity of the interaction.

f. Sequence the positive clones and determine their identity by comparing their DNA (and predicted protein) sequence against a public sequence database available at the National Center for Biotechnology Information (NCBI; http://www.ncbi.nlm.nih.gov); based on sequence analyses, decide which clones are relevant for further studies.

Define Structural Domains Present in RGS-RhoGEFs

1. Obtain the cDNA and protein sequences of RGS-RhoGEFs and identify the limits of their structural domains. The complete cDNA and protein sequences of PDZ-RhoGEF and LARG can be obtained from the NCBI's PubMed web page: *http://www.ncbi.nlm.nih.gov/entrez/query.fcgi*; once in this page, search for nucleotide using the accession numbers *NM_014784* and *NM_198236* (accession numbers for the two splice variants of human PDZ-RhoGEF, also known as ARHGEF11). Obtain the sequence corresponding to the open reading frame by clicking on the hyperlink for CDS ("coding sequence"). The page with the CDS displays the nucleotide sequence starting from the initial ATG codon and finishing with the stop codon. For LARG, use the accession number *NM_015313*; LARG is also known as ARHGEF12.

It is useful to keep all the information contained in these NCBI web pages organized in a single file. Different commercial or freely available programs are useful for this purpose. As an example, we use Bioedit, which is a software for sequence analysis for PCs developed by Tom Hall from the Department of Microbiology, North Carolina State University that can be downloaded freely from http://www.mbio.ncsu.edu/BioEdit/bioedit.html. To organize the sequences in a Bioedit file, simply open a new Bioedit file and paste all the information from the NCBI web page into it for the sequence to be analyzed. To start working with the sequence, obtain a printed sheet with the cDNA and the corresponding protein translation; to prepare constructs for each different domain, it will be very convenient to have this sheet available to highlight the sequence of the domains at both protein and cDNA levels. To analyze the activities and putative protein–protein interactions attributable to the specific domains of PDZ-RhoGEF, LARG, or any other multidomain protein of interest, it is very likely that each domain will need to be cloned in different constructs to be studied independently, e.g., for yeast two-hybrid screens, proteomic approaches, protein–protein interaction assays, and functional assays. Start by defining the limits of each domain. The prediction is done by analyzing the protein sequence with SMART (Simple Modular Architecture Research Tool) at the following web page: *http://smart.embl-heidelberg.de/*. The protein sequence can be obtained directly from the NCBI web page by searching for protein instead of nucleotide. Obtain a graphical representation of the protein, which in the SMART website will be presented with a table indicating the polypeptide limits of the different domains. Highlight the sequence corresponding to the different domains in the sheet containing the cDNA and protein sequences. To determine the limits for the constructs, we usually extend a few amino acids toward both extremes, up to a proline when available, which is not included in the construct to avoid undesired changes in the secondary structure.

2. Prepare the constructs for the yeast two-hybrid system. Using the polymerase chain reaction (PCR), amplify the cDNA corresponding to the specific domain that will be used to search for novel interacting proteins. Design primers, including restriction sites absent in the cDNA of the bait and present in the multicloning site of pGBKT7; be extremely careful to avoid frameshifts.

a. SMART site analysis of the protein sequence of PDZ-RhoGEF indicated that the PDZ domain spans histidine-55 to serine-123. Clone the cDNA coding for the first 123 amino acids (from methionine-1 to serine-123) corresponding to the first 369 bp at the cDNA level. Amplify by PCR using the primers

PDZ5'*Nde*I 5'-ataCATATGatgagtgtaaggttaccccag-3' and
PDZ3'*Not*I 5'-ataGCGGCCGCtcatgaagagcccaggagggtg-3'

and clone the PCR product into pGBKT7.

b. Transform AH109 yeast cells with the bait pGBKT7-PDZ from PDZ-RhoGEF. From the frozen stock of AH109 yeast cells, provided with the yeast two-hybrid system (MatchMaker III Clontech), start a culture by streaking a few microliters with a sterile bacteriological loop into YPDA plates (YPD Agar, Q-Biogene or equivalent, supplemented with 0.003% adenine hemisulfate, Sigma). Incubate at 30° until colonies are 3–5 mm in diameter; they must be white and healthy. Transfer a colony to a 50-ml tube containing 10 ml of liquid medium (YPD broth, Q-Biogene or equivalent supplemented with adenine hemisulfate as indicated for the plates) and grow overnight (12–16 h) at 30° with constant shaking. The next day, recover the yeasts by centrifugation (2000 rpm, 5 min), wash once with water, and proceed to transform with the bait vector using the lithium acetate/single-stranded carrier DNA/polyethylene glycol method (Gietz and Woods, 2002).

Materials for yeast transformation:

AH109 yeast (overnight culture in adenine-supplemented YPD medium)

pGBKT7-PDZ from PDZ-RhoGEF (200–1000 ng/μl)

pGADT7 AD vector (from Clontech, provided with the Matchmaker III two-hybrid system)

100 mM lithium acetate (prepared fresh from a 1 M stock)

50% polyethylene glycol 3350 (Sigma; dissolved with sterile water; it can be prepared in a microwave oven by giving several pulses of heating at medium power)

Single-stranded DNA from salmon testes (Sigma; to dissolve, leave overnight in sterile water at room temperature and, the next day, incubate for 5 min in a boiling water bath, cool down on ice, and keep frozen in aliquots)

Sterile MilliQ water

Agar plates with minimal medium and supplements lacking tryptophan; leucine and tryptophan; histidine, leucine, and tryptophan; or adenine, histidine, leucine, and tryptophan [minimal medium agar, DOBA, Qbiogene; "dropout" (DO) supplements, all from Clontech: -Trp, -Leu/-Trp, -Leu/-Trp/-His, and -Ade/-His/-Leu/-Trp].

Note. The DO supplements are necessary to select for transformants and to determine that yeast expressing the bait are unable to activate the system in the absence of a specific interacting protein; otherwise, the bait cannot be used for screening. The medium lacking tryptophan (-T) will select for yeast containing the bait plasmid; the medium lacking tryptophan and leucine (-LT) will select for yeast containing two plasmids, pGBKT7-PDZ and pGADT7 AD; and media lacking tryptophan, leucine, and histidine (-HLT) and tryptophan, leucine, histidine, and adenine (-AHLT) are intended to select for protein–protein interactions and are used in this step to demonstrate that the bait does not autoactivate the system in the absence of a relevant binding partner.

Procedure

1. Centrifuge the yeast cells for 5 min at 2000 rpm. Discard the supernatant and resuspend the cells in 20 ml of sterile water. Centrifuge

again and resuspend in 3 ml of 100 mM lithium acetate. Transfer to 1.5-ml microcentrifuge tubes.

2. Spin down for 3–5 s, just enough to pellet the yeast. Discard the supernatant with a pipette tip connected to a vacuum source. Add 4 volumes of 100 mM lithium acetate (with respect to the volume of pelleted cells). For example, add 400 μl of lithium acetate and transfer 500 μl of the cell suspension to a new tube.

3. Distribute 50 μl of the cell suspension to each tube to be used in the transformation protocol.

4. Prepare the transformation mixture as follows: for (n + 1) number of tubes, mix (n + 1) times (240 μl PEG 3350, 36 μl 1 M lithium acetate, 50 μl of salmon testes single-stranded DNA). Add 326 μl to each tube containing the yeasts. Add the plasmids to different tubes as follows:
   i. "bait": 2 μl of pGBKT7-PDZ.
   ii. "bait" + "control plasmid": 2 μl of pGBKT7-PDZ plus 2 μl of pGADT7 AD.

5. Mix gently with the pipette tip (use a 1-ml pipette tip that can be cut with scissors to facilitate the job).

6. Leave at 30° for 30 min. Mix every 10 min by inverting the tubes. Transfer to 42° for 30 min.

7. Centrifuge for 5 s at 6000 rpm and discard the supernatant.

8. Resuspend the yeast in 1 ml of sterile water. Plate 20 or 200 μl of the cell suspension as indicated for the different tubes:
   i. Tube 1, "bait" into medium -T. This plate is intended to select yeast with the bait for the subsequent screen.
   ii. Tube 2, "bait + control" into medium −LT.

9. Incubate at 30° until the colonies reach at least 3 mm in diameter.

10. Scrape a colony from plate 2 and resuspend it with 500 μl of sterile water. Drop 10–20 μl into -LT, -HLT, and -AHLT plates. Incubate at 30° for several days. Yeast should grow only on -LT plates. In parallel, transform the positive control bait and prey plasmids provided with the system, which should grow in the three different auxotrophic conditions.

c. Use yeast two-hybrid screening for PDZ-interacting proteins. Once it is demonstrated that the "bait," in this case the PDZ domain from PDZ-RhoGEF, is not toxic to yeast and does not activate the system in the absence of specific protein–protein interactions, it can be used to screen a library to identify and clone the cDNA for novel interacting proteins. For that purpose, it is important to select a cDNA library prepared from a tissue where the protein under study is known to be expressed. We have successfully screened cDNA libraries from human brain (Clontech) and bone marrow (Clontech) (Barac et al., 2004; Perrot et al., 2002). The use of pretransformed libraries facilitates the work. The procedure consists of mating AH109 yeast previously transformed with the "bait" (the PDZ domain of PDZ-RhoGEF) with yeast pretransformed with the selected cDNA library. After overnight mating, diploid yeast are plated onto -HLT or -AHLT plates, which represent medium and high stringency conditions for the growth of yeast containing the cDNA of the bait and the cDNA of a putative interacting protein.

Materials for screening:

PDZ-bait plasmid-pretransformed AH109 yeast (pretransformed with pGBKT7-PDZ from PDZ-RhoGEF)

cDNA library transformed into Y187 yeast (available from Clontech)

2X liquid YPDA (prepare 100 ml)

Agar plates (15 cm diameter) with restrictive medium lacking histidine, leucine, and tryptophan (10 plates) and lacking adenine, histidine, leucine, and tryptophan (30 plates)

Agar plates (10 cm diameter) with restrictive medium lacking tryptophan, leucine, or tryptophan and leucine (prepare 10 of each to determine the titer of the library and the efficiency of the mating)

X-α-Gal (Clontech): prepare a 20-mg/ml stock in dimethyl formamide, keep at $-20°$

Liquid minimal medium, both -HLT and -AHLT (prepare 1 liter of each)

Procedure

1. Obtain 1.0 to 1.5 ml of PDZ-bait-transformed yeast by growing them in liquid restrictive medium lacking tryptophan (-T medium). Start a culture with a colony from the -T plate, prepared as described in the previous section, into 10 ml of -T liquid medium in a 50-ml tube with constant shaking at 30°. The next day, transfer to a 250-ml Erlenmeyer plastic flask and grow in 50 ml of -T medium with constant shaking. If required, divide into two flasks and continue the culture one more day.

2. Prepare the agar plates with minimal medium. The plates can be prepared in advance; leave them overnight at room temperature and keep at 4° until required. Alternatively, they can be prepared just after starting the mating. Use 100 ml of medium per 15-cm plate.

3. Start the mating by mixing the PDZ-bait-pretransformed AH109 yeast with the cDNA library-pretransformed Y187 yeast resupended in 40 ml of 2X YPDA. Transfer to a 50-ml tube, centrifuge at 2000 rpm for 5 min, and leave the pellet of yeast for 1–2 h at room temperature. Add 2X YPDA up to 50 ml. Transfer to a 15-cm plate and leave for 24 h at 30° with gentle shaking (40 rpm).

4. The next day, recover yeasts by centrifugation and resuspend in sterile water up to 10 ml. Distribute 250 μl of the yeast suspension on each plate with either -HLT or -AHLT restrictive medium. Distribute gently and evenly with the help of sterile glass beads and incubate upside down at 30° for 1 to 3 weeks until the colonies are about 5 mm in diameter (Fig. 3, left).

5. Determine the titer of the bait and the cDNA library and the efficiency of mating as indicated in the procedures provided with the Matchmaker III system.

After the colonies have reached a minimum size of 5 mm, start liquid cultures by inoculating each of them into 1 ml of the respective restrictive medium (-HLT or -AHLT). Using sterile loops to pick the colonies, cut the agar at the same time in order to indicate that a colony has been picked. Leave the liquid cultures growing at 30° with constant shaking. One or 2 days later, when growth is clearly visible, transfer 1 μl into new tubes containing restrictive medium, leave for 1 or 2 days until the growth is visible, and repeat once more. The growth must be clearly visible; discard tubes where no clear growth is detected. After the third passage, drop 5–10 μl on agar plates with restrictive medium supplemented with X-α-Gal (prepare plates with X-α-Gal just a few minutes before they are needed; use 50–100 μl of X-α-Gal per 15-cm plate, distributed evenly with the help of folded plastic loop and left drying out at room temperature).

FIG. 3. The yeast two-hybrid system is a powerful tool used to reveal novel protein–protein interactions regulating RGS-containing RhoGEFs. This figure illustrates different steps involved in the cloning of cDNAs of PDZ-RhoGEF-interacting proteins. As explained in the text, a structural domain of PDZ-RhoGEF is used as bait to fish out interacting proteins expressed from a cDNA library. The interaction reconstitutes a transcription factor, which allows the yeast to grow under restrictive conditions (left) and turn blue in the presence of X-αGal (middle). To confirm the interactions, cDNA of the putative-interacting protein is isolated and transformed in yeast containing the bait or an unrelated control (right). Those yeast strains carrying the bait and a positive-interacting protein grow and turn blue under restrictive conditions (-AHLT + XαGal). The efficiency of the transformation is verified by the growth in medium only restricting for the presence of the plasmids (-LT). (See color insert.)

Leave at 30° for 1–2 days or until yeast start to grow. With the tip of a loop, transfer those that are dark blue (Fig. 3, middle) into new tubes containing restrictive medium and leave the cultures at 30° with constant shaking. Isolate the plasmids from the yeast 1 or 2 days later.

To isolate plasmids from yeast cells, use a yeast plasmid miniprepara-tion kit (Zymoprep, Zymo Research or equivalent). Follow the standard protocol provided with the kit and transform 4 μl from the miniprep into 10 μl of GC5 competent cells (high efficiency, 10^9 transformants/μg, Gene-Choice or equivalent). Plate on ampicillin-containing LB plates and pre-pare minipreps as indicated previously. In some cases, it is impossible to obtain plasmids from some clones; discard those clones from which no positive transformation is obtained from two different yeast minipreps. Plasmids prepared from bacteria are used for further studies.

d. Determine the specificity of the interactions and sequence the posi-tive clones in yeast. To determine the specificity of the interactions be-tween the PDZ domain from PDZ-RhoGEF and any of the proteins encoded by the plasmids that were isolated from the screening, it is neces-sary to transform these plasmids (known as "prey") into PDZ-pretrans-formed yeast and pGBKT7 control pretransformed yeast. After transformation, all yeast must be able to grow in -LT plates, but only those in which the bait and prey interact should grow under restrictive conditions and turn blue in the presence of X-α-Gal (Fig. 3, right).

Procedure

1. Leave an overnight culture of PDZ- or pGBKT7-pretransformed AH109 yeast in -T liquid medium.

2. The next day, recover yeast by centrifugation and transform each of them with each cDNA isolated from the screening; use the plasmids recovered from bacteria, not those that were originally isolated from yeast. Follow the lithium acetate transformation protocol indicated earlier. Plate on -LT plates.

3. Scrape colonies from each transformation and resuspend in 1 ml of sterile water. Drop 5–10 μl of each on X-α-Gal-containing restrictive plates (-HLT for those clones recovered in the screening with that level of selective medium and -AHLT for all of them). Clones that grow and turn blue in X-α-Gal-containing restrictive medium only when cDNA from the library is transformed in yeast containing the PDZ domain from PDZ-RhoGEF, and not in yeast containing the control plasmid, are considered positive.

Obtain the sequence from those clones that show a positive phenotype. Usually, 5 μl from the bacteria minipreps obtained for each prey is enough for sequencing. Use the primer 5' AD SeqPri for pACT2 (5'-TACCAC-TACAATGGATG-3') for automated dye termination sequencing. The sequence obtained will include part of the vector and the 5' end of the cDNA of the prey. Determine the identity of the prey by sequence homology searches using the BLAST algorithm at *http://www.ncbi.nlm.nih.gov/BLAST/*, selecting for the nucleotide–nucleotide search program "blastn." Prepare a database with all the cDNAs obtained from the screening and decide which of them are going to be studied further based on BLAST-generated sequence identities/similarities.

Novel Protein–Protein Interactions Regulating RGS-RhoGEFs in Mammalian Cells

Overview

The significance of the protein–protein interactions detected in the yeast two-hybrid system must be determined by confirmatory studies in mammalian cells. To confirm that the PDZ domain from PDZ-RhoGEF used as the bait interacts with the prey and that this interaction correlates with a potential physiological activity, it is important to assess this interaction and its effects in mammalian cells.

Procedure

Subclone the cDNA for the putative PDZ-interacting protein into a mammalian expression vector. Available plasmids include pCMV-HA or pCMV-Myc (Clontech); both plasmids allow expression in mammalian cells and the restriction sites in the multicloning site are compatible with those present in pACT2, the plasmid from which the cDNA of the putative PDZ-interacting protein is recovered. As described for cloning of the PDZ domain from PDZ-RhoGEF, use restriction sites that are not internal in the cDNA to be cloned. *Sfi*I and *Xho*I are frequently good candidates for these vectors. If a different vector is selected, the relevant restriction sites can be introduced in the cDNA to be cloned by PCR amplification.

One of the PDZ-interacting proteins that we identified by this method was plexin B (Perrot *et al.*, 2002). We cloned the cDNA corresponding to the carboxyl-terminal domain of plexin B into pCEFL-EGFP (available from the Gutkind laboratory). In this construct, the fusion protein contains EGFP at the amino-terminal end. It is found frequently that cDNAs obtained from yeast two-hybrid screens correspond only to small stretches of the full cDNA. In that case, it is convenient to obtain the full cDNA for functional studies. The sources from which a full cDNA can be obtained vary and include laboratories working with the cDNA identified in the screening or organizations such as the National Institutes of Health through the Mammalian Gene Collection at *http://mgc.nci.nih.gov/Info/Buy* and the Kazusa DNA Research Institute at *http://www.kazusa.or.jp/en/*. This section details the methods used for characterization of the interaction between plexin and other proteins with PDZ-RhoGEF.

Plexin–PDZ-RhoGEF Interactions in HEK293T Cells. The whole procedure is similar to that described earlier to assess the interaction between $G\alpha_{12/13}$ and PDZ-RhoGEF. A typical experiment will include 1 μg of AU1-tagged PDZ-RhoGEF (pCEFL-AU1-PDZ-RhoGEF) or a control expression vector (pCEFL), 1 μg of the EGFP-tagged carboxyl-terminal domain of plexin B (pCEFL-EGFP-cter-Plexin B), and 2 μg of empty vector (pCEFL). All the indicated DNAs are available from our laboratory. Transfect the cells and immunoprecipitate PDZ-RhoGEF as indicated previously. Detect the interactions by western blotting using the anti-GFP monoclonal antibody (Covance) diluted 1:5000 before immunoblotting (see Perrot *et al.*, 2002).

Plexin Stimulation of RhoA Activation in HEK293T Cells. RhoA-dependent activities can be determined as indicated earlier for G-protein-coupled receptor-mediated activation of RhoA. To determine the functional significance of the interactions detected between the carboxyl-terminal domain of plexin and the PDZ domain of PDZ-RhoGEF, we prepared a construct for a fusion protein in which the extracellular domain corresponds to TrkA, the receptor for nerve growth factor (NGF), and the intracellular domain corresponds to the carboxyl-terminal domain of plexin B2, which is the fragment isolated as a PDZ-interacting protein in the yeast two-hybrid system. This chimeric receptor can be activated by soluble NGF and is able to transduce plexin-dependent activation of RhoA (see Perrot *et al.*, 2002).

Cytoskeletal Changes

Overview

A reliable indicator for the activation of RhoA-mediated pathways in a cell is the formation of actin stress fibers (Hall, 1998). Actin polymerization in response to agonist activation of a cell surface receptor implies the recruitment of proteins capable of modulating RhoA activation, such as GEFs, to the cytoplasmic portion of the receptor at the cell membrane. Such is the case for G-protein-coupled receptors—specific domains of the GEFs are able to interact with $G\alpha_{12/13}$ to generate an appropriate response from the Rho family of GTPases (Schmidt and Hall, 2002). GEF activity is similarly regulated by the B family of plexins though protein–protein interactions initiated by a short stretch of amino acids called the PDZ-binding motif.

PDZ/PDZ-binding motif interactions are involved in the sorting, targeting, and assembly of supramolecular complexes at the cytoplasmic, C-terminal segment of transmemembrane proteins (Sheng and Sala, 2001). It has been suggested that the treatment of endothelial cells with semaphorin 4D activates RhoA signaling by recruiting PDZ-RhoGEF and LARG *via* their PDZ domains to the PDZ-binding motif at the C-terminal cytoplasmic portion of the plexin-B1 receptor, which results in actin polymerization (Perrot *et al.*, 2002; Swiercz *et al.*, 2002). Detection of these PDZ-binding domain interactions and the cytoskeletal changes that follow can be achieved using fluorescent phalloidin to stain polymerized actin (An *et al.*, 1998). Phalloidin is a convenient probe for identifying stress fiber formation in cells because it has a particular affinity for F-actin, especially at lower pHs, and works well at very low concentrations (Adams and Pringle, 1991). In addition, phalloidin is also water soluble and binds to polymerized actins from many different species (Adams and Pringle, 1991).

Procedure

Detection of Stress Fibers in PAECs Using Rhodamine-Labeled Phalloidin

1. Grow porcine aortic endothelial cells (PAECs) in HAM/F-12 medium in 35-mm, six-well plates on sterile glass coverslips in a 37° incubator. Serum starve the cells for 36 h before treatment.

2. For a positive control, stimulate cells with 0.5 μM LPA (Sigma-Aldrich, Inc) or serum (10%) for 20 min. Fibroblast growth factor (FGF) or hepatocyte growth factor (HGF) (R&D Systems, Minneapolis, MN) can also be used as a positive control, titrated from 100 to 400 ng/ml for maximum effect. Treat experimental populations with semaphorin 4D over a time course using different concentrations or, if using cells expressing Trk-A/plexin B1 fusions, treat with 50–100 ng/ml of NGF (R&D Systems) to activate plexin-B1 signaling.

3. After growth factor treatment, wash with PBS and fix cells with 4% paraformaldehyde in PBS for 10 min at room temperature.

4. Leaving the glass coverslips in the six-well plates, wash three times in PBS and treat with 0.1% Triton-X in PBS for 3–5 min.

5. Wash twice with PBS and incubate coverslips in 1% BSA in PBS for 20–30 min at room temperature.

6. Wash in PBS and stain with Texas Red-X phalloidin (Molecular Probes, Inc.) for 20 min at room temperature following the manufacturer's

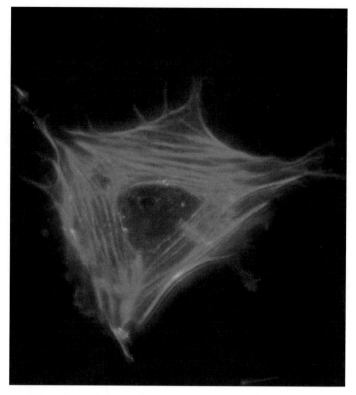

FIG. 4. Texas Red-conjugated phalloidin was used to label polymerized actin stress fibers in this LPA-treated cell. (See color insert.)

instructions. A dilution of labeled phalloidin of 1:200 to 1:500 in PBS is generally sufficient to label polymerized actin.

7. Remove coverslips from the six-well plates and mount on glass slides with Vectashield mounting medium for fluorescence containing DAPI to counterstain cell nuclei (Vector Laboratories, Inc., CA). Stress fibers can be viewed with fluorescence microscopy with a TRITC filter (see Fig. 4).

Migration Assays

Overview

Chemotaxis is defined as cell movement toward a gradient of increasing chemical concentration (Lauffenburger and Zigmond, 1981). This is different from chemokinesis, which is movement in response to a compound that lacks a specific directional component (Lauffenburger and Zigmond, 1981). Many eukaryotic cells have the capacity to migrate toward a chemoattractant and it is this ability that underlies cell migration observed in such diverse processes as growth and development, inflammation and wound healing, and angiogenesis. The standard assay used to test if a compound acts as a chemoattractant for a specific cell type is the Boyden chamber assay (Boyden, 1962).

In this assay, a solution of cells suspended in buffer or medium is placed in a small chamber or well separated from a lower reservoir of chemoattractant by a porous filter that is often coated with extracellular matrix proteins. A gradient forms in the lower chamber by diffusion and it is this difference in concentration, combined with the intrinsic qualities of the compound and cell type used, that determines the direction and strength of the chemotactic response.

The readout for this assay is redistribution of the cell population over time as they migrate though the pores in the filter and adhere onto the opposite side. Results are quantified by physically removing the cells on the outer part of the membrane that failed to migrate and measuring the number of cells attached to the bottom of the filter; however, it is possible that some cells may migrate completely though the filter and fall into the lower chamber, depending on the cell type used (Baggiolini *et al.*, 1997). Cells that migrate to the lower surface of the filter can be counted using light microscopy or, more objectively, though the use of an image analyzer. Increased migration toward a test material in these assays suggests that the material might be a chemoattractant, although these results can be masked by using too high a concentration of a compound, which either desensitizes the receptors to further responses from the ligand or becomes so high as to prevent the establishment of a proper gradient (Tranquillo *et al.*, 1988).

The Boyden chamber assay can be employed to determine if semaphorin 4D acts as a chemoattractant for cells expressing the plexin-B1 receptor. The actin polymerization observed in PAECs treated with semaphorin 4D suggested that this plexin-B1/PDZ domain interaction may be important for chemotaxis, as RhoA-mediated remodeling of the actin cytoskeleton is essential for cell polarization and directional cell movement (Hall, 1998). Therefore, we generated soluble semaphorin 4D and used it as the chemoattractant in the lower portion of the chamber to determine its role in PAEC migration and angiogenic signaling. Semaphorin 4D did, in fact, induce chemotaxis in PAECs equal to that seen for FGF and HGF, both strong inducers of angiogenesis (Cross and Claesson-Welsh, 2001; Rosen and Goldberg, 1997).

Procedure

Migration Assay

1. Place 28 μl of serum-free HAM/F-12 medium containing 0.1% BSA and the appropriate chemoattractant in the bottom well of a Boyden chamber. One hundred to 400 ng/ml FGF or HGF (R&D Systems) can be used as a positive control and serum-free medium containing 0.1% BSA alone as the negative control.

2. Place a PVDF membrane (Osmonics, GE Water Technologies, Trevose, PA, 8 μm pore size) coated with 10 μg/ml fibronectin (GIBCO, Invitrogen, Carlsbad, CA) on top of the chemoattractant wells, avoiding the creation of bubbles that might block diffusion of the chemoattractant. Place the rubber gasket over the membrane and seal the chamber with the lid.

3. Trypsinize cells and neutralize trypsin with HAM/F-12 medium containing 10% serum. Centrifuge at 1000 rpm for 4–5 min and remove the medium, adding back serum-free HAM/F-12 medium to wash the cells. Recentrifuge at 1000 rpm for 4–5 min, remove the medium, and resuspend the cell pellet in serum-free HAM/F-12 medium with 0.1% BSA using a Coulter counter or hemocytometer to create a solution of 1×10^6 cells per milliliter.

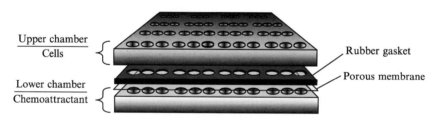

FIG. 5. The Boyden chamber illustrated here consists of two plexiglass members separated by an extracellular matrix protein-coated membrane. The upper portion contains 48 holes, arranged in four rows of threes that pass completely through the lid. The base contains the same pattern of holes closed at the bottom end to form wells capable of holding dilutions of the chemoattractant. The two members are secured together, usually by tightening small screws on the lid, after insertion of the rubber gasket between them to prevent leaks. A solution of cells is then added to the upper wells, directly on top of the exposed membrane. At the appropriate time, the chamber is disassembled and the membrane is processed as indicated in the text. The final stained membrane is digitized, and the intensity of staining at each spot is quantified with the NIH Image or other appropriate software. The intensity of spots is determined in triplicate and is illustrated numerically and as a bar graph. (See color insert.)

4. Add 50 μl of this solution, which is approximately 50,000 cells, to the top chamber. Incubate at 37° in a container kept humid with a damp paper towel for 6 to 8 h. Longer incubation times may require that the assay be set up under sterile conditions.

5. Disassemble the chamber and remove the membrane, being careful not to disrupt the cells on the portion of the membrane that had been facing down toward the chemoattractant wells. Immerse in 100% methanol for 2 min.

6. Stain the membrane with the Diff-Quick stain kit (Diff-Quick, Dade Behing, Deerfield, IL) according to the manufacturer's instruction.

7. Place the membrane on a glass slide with the surface that had been facing the chemoattractant face down on the slide. Wipe the top surface with a wet Kim-Wipe or rubber policeman. Cover with another glass slide or coverslip secured with Permount (Fisher Scientific).

8. Individual cells remaining on the membrane can be counted under the microscope or the slide can be scanned and densitometric quantitation performed with NIH Image freeware or other appropriate quantitation software. See Fig. 5 for a scheme of the Boyden chamber.

Acknowledgments

J.V.-P. acknowledges support by the Fogarty International Center, NIH Research Grant D43 TW06664, and Fundación Miguel Alemán.

References

Adams, A. E., and Pringle, J. R. (1991). Staining of actin with fluorochrome-conjugated phalloidin. *Methods Enzymol.* **194**, 729–731.

An, S., Goetzl, E. J., and Lee, H. (1998). Signaling mechanisms and molecular characteristics of G protein-coupled receptors for lysophosphatidic acid and sphingosine 1-phosphate. *J. Cell Biochem. Suppl.* **30–31**, 147–157.

Aurandt, J., Vikis, H. G., Gutkind, J. S., Ahn, N., and Guan, K. L. (2002). The semaphorin receptor plexin-B1 signals through a direct interaction with the Rho-specific nucleotide exchange factor, LARG. *Proc. Natl. Acad. Sci. USA* **99**, 12085–12090.

Baggiolini, M., Dewald, B., and Moser, B. (1997). Human chemokines: An update. *Annu. Rev. Immunol.* **15**, 675–705.

Barac, A., Basile, J., Vazquez-Prado, J., Gao, Y., Zheng, Y., and Gutkind, J. S. (2004). Direct interaction of p21-activated Kinase 4 with PDZ-RhoGEF, a G protein-linked Rho guanine exchange factor. *J. Biol. Chem.* **279**, 6182–6189.

Bar-Sagi, D., and Hall, A. (2000). Ras and Rho GTPases: A family reunion. *Cell* **103**, 227–238.

Boyden, S. (1962). The chemotactic effect of mixtures of antibody and antigen on polymorphonuclear leucocytes. *J. Exp. Med.* **115**, 453–466.

Burridge, K., and Wennerberg, K. (2004). Rho and Rac take center stage. *Cell* **116**, 167–179.

Chikumi, H., Barac, A., Behbahani, B., Gao, Y., Teramoto, H., Zheng, Y., and Gutkind, J. S. (2004). Homo- and hetero-oligomerization of PDZ-RhoGEF, LARG and p115RhoGEF by their C-terminal region regulates their in vivo Rho GEF activity and transforming potential. *Oncogene* **23**, 233–240.

Chikumi, H., Fukuhara, S., and Gutkind, J. S. (2002). Regulation of G protein-linked guanine nucleotide exchange factors for Rho, PDZ-RhoGEF, and LARG by tyrosine phosphorylation: Evidence of a role for focal adhesion kinase. *J. Biol. Chem.* **277**, 12463–12473.

Cross, M. J., and Claesson-Welsh, L. (2001). FGF and VEGF function in angiogenesis: Signalling pathways, biological responses and therapeutic inhibition. *Trends Pharmacol. Sci.* **22**, 201–207.

Etienne-Manneville, S., and Hall, A. (2002). Rho GTPases in cell biology. *Nature* **420**, 629–635.

Fields, S., and Song, O. (1989). A novel genetic system to detect protein-protein interactions. *Nature* **340**, 245–246.

Fromm, C., Coso, O. A., Montaner, S., Xu, N., and Gutkind, J. S. (1997). The small GTP-binding protein Rho links G protein-coupled receptors and Galpha12 to the serum response element and to cellular transformation. *Proc. Natl. Acad. Sci. USA* **94**, 10098–10103.

Fukata, M., Nakagawa, M., and Kaibuchi, K. (2003). Roles of Rho-family GTPases in cell polarisation and directional migration. *Curr. Opin. Cell Biol.* **15**, 590–597.

Fukuhara, S., Chikumi, H., and Gutkind, J. S. (2000). Leukemia-associated Rho guanine nucleotide exchange factor (LARG) links heterotrimeric G proteins of the G(12) family to Rho. *FEBS Lett.* **485**, 183–188.

Fukuhara, S., Chikumi, H., and Gutkind, J. S. (2001). RGS-containing RhoGEFs: The missing link between transforming G proteins and Rho? *Oncogene* **20**, 1661–1668.

Fukuhara, S., Murga, C., Zohar, M., Igishi, T., and Gutkind, J. S. (1999). A novel PDZ domain containing guanine nucleotide exchange factor links heterotrimeric G proteins to Rho. *J. Biol. Chem.* **274**, 5868–5879.

Gietz, R. D., and Woods, R. A. (2002). Transformation of yeast by lithium acetate/single-stranded carrier DNA/polyethylene glycol method. *Methods Enzymol.* **350**, 87–96.

Hall, A. (1998). Rho GTPases and the actin cytoskeleton. *Science* **279**, 509–514.

Hart, M. J., Jiang, X., Kozasa, T., Roscoe, W., Singer, W. D., Gilman, A. G., Sternweis, P. C., and Bollag, G. (1998). Direct stimulation of the guanine nucleotide exchange activity of p115 RhoGEF by Galpha13. *Science* **280**, 2112–2114.

Hill, C. S., Wynne, J., and Treisman, R. (1995). The Rho family GTPases RhoA, Rac1, and CDC42Hs regulate transcriptional activation by SRF. *Cell* **81**, 1159–1170.

Kozasa, T. (2001). Regulation of G protein-mediated signal transduction by RGS proteins. *Life Sci.* **68**, 2309–2317.

Kozasa, T., Jiang, X., Hart, M. J., Sternweis, P. M., Singer, W. D., Gilman, A. G., Bollag, G., and Sternweis, P. C. (1998). p115 RhoGEF, a GTPase activating protein for Galpha12 and Galpha13. *Science* **280**, 2109–2111.

Lauffenburger, D. A., and Zigmond, S. H. (1981). Chemotactic factor concentration gradients in chemotaxis assay systems. *J. Immunol. Methods* **40**, 45–60.

Marshall, C. (1999). How do small GTPase signal transduction pathways regulate cell cycle entry? *Curr. Opin. Cell Biol.* **11**, 732–736.

Narumiya, S. (1996). The small GTPase Rho: Cellular functions and signal transduction. *J. Biochem. (Tokyo)* **120**, 215–228.

Perrot, V., Vazquez-Prado, J., and Gutkind, J. S. (2002). Plexin B regulates Rho through the guanine nucleotide exchange factors leukemia-associated Rho GEF (LARG) and PDZ-RhoGEF. *J. Biol. Chem.* **277**, 43115–43120.

Reid, T., Furuyashiki, T., Ishizaki, T., Watanabe, G., Watanabe, N., Fujisawa, K., Morii, N., Madaule, P., and Narumiya, S. (1996). Rhotekin, a new putative target for Rho bearing homology to a serine/threonine kinase, PKN, and rhophilin in the rho-binding domain. *J. Biol. Chem.* **271**, 13556–13560.

Rosen, E. M., and Goldberg, I. D. (1997). Regulation of angiogenesis by scatter factor. *Exs* **79**, 193–208.

Sahai, E., and Marshall, C. J. (2002). RHO-GTPases and cancer. *Natl. Rev. Cancer* **2**, 133–142.

Schmidt, A., and Hall, A. (2002). Guanine nucleotide exchange factors for Rho GTPases: Turning on the switch. *Genes Dev.* **16**, 1587–1609.

Sheng, M., and Sala, C. (2001). PDZ domains and the organization of supramolecular complexes. *Annu. Rev. Neurosci.* **24**, 1–29.

Suzuki, N., Nakamura, S., Mano, H., and Kozasa, T. (2003). Galpha 12 activates Rho GTPase through tyrosine-phosphorylated leukemia-associated RhoGEF. *Proc. Natl. Acad. Sci. USA* **100**, 733–738.

Swiercz, J. M., Kuner, R., Behrens, J., and Offermanns, S. (2002). Plexin-B1 directly interacts with PDZ-RhoGEF/LARG to regulate RhoA and growth cone morphology. *Neuron* **35**, 51–63.

Takai, Y., Sasaki, T., and Matozaki, T. (2001). Small GTP-binding proteins. *Physiol. Rev.* **81**, 153–208.

Tamagnone, L., and Comoglio, P. M. (2000). Signalling by semaphorin receptors: Cell guidance and beyond. *Trends Cell Biol.* **10**, 377–383.

Tranquillo, R. T., Zigmond, S. H., and Lauffenburger, D. A. (1988). Measurement of the chemotaxis coefficient for human neutrophils in the under-agarose migration assay. *Cell Motil. Cytoskel.* **11**, 1–15.

[18]   Regulation of RGS-RhoGEFs by Gα12 and Gα13 Proteins

By SHIHORI TANABE, BARRY KREUTZ, NOBUCHIKA SUZUKI, and TOHRU KOZASA

Abstract

Three mammalian Rho guanine nucleotide exchange factors (Rho-GEFs), leukemia-associated RhoGEF (LARG), p115RhoGEF, and PDZ-RhoGEF, contain regulator of G-protein signaling (RGS) domains within their amino-terminal regions. These RhoGEFs link signals from heterotrimeric G12/13 protein-coupled receptors to Rho GTPase activation, leading to various cellular responses, such as actin reorganization and gene expression. The activity of these RhoGEFs is regulated by Gα12/13 through their RGS domains. Because RhoGEFs stimulate guanine nucleotide exchange by Rho GTPases, RhoGEF activation can be measured by monitoring GTP binding to or GDP dissociation from Rho GTPases. This article describes methods used to perform reconstitution assays to measure the activity of RhoGEFs regulated by Gα12/13.

Introduction

Rho GTPases regulate various cellular responses such as cell cycle progression, chemotaxis, and axonal guidance by controlling actin reorganization and gene expression (Hall, 1998). Rho GTPases are activated by Rho guanine nucleotide exchange factors (RhoGEFs) through their Dbl homology (DH)/Pleckstrin homology (PH) domain, which is responsible for catalyzing GDP-GTP exchange on Rho GTPase.

In mammalian cells, three RhoGEFs, which contain regulator of G-protein signaling (RGS) domains (RGS-RhoGEF), have been identified as p115RhoGEF, leukemia-associated RhoGEF (LARG), and PDZ-RhoGEF [Fukuhara et al., 1999; Hart et al., 1998; Kozasa et al., 1998; Suzuki et al., 2003 (Fig. 1)]. These RGS-RhoGEFs are direct links between Gα12/13 and Rho GTPase proteins. Heterotrimeric Gα12 and Gα13 specifically interact with RGS-RhoGEFs through their RGS domains and regulate their GEF activity. The RGS domains of p115RhoGEF and LARG function as GTPase-activating proteins (GAPs) for Gα12 and Gα13 to facilitate inactivation of these Gα subunits. Thus, RGS-RhoGEFs relay signals from heterotrimeric G12/13 protein-coupled receptors to Rho GTPase activation. This article describes methods used to assay the RhoGEF activity of p115RhoGEF and LARG proteins regulated by Gα12/13.

Proteins Used for Reconstitution Assays

Sf9 Culture

Sf9 cells are cultured in IPL-41 medium (JRH)/10% heat-inactivated fetal bovine serum (treated at 57° for 30 min)/0.1% (v/v) Pluronic F-68 (Invitrogen)/25 μg/ml gentamicin at 27° with constant shaking (140 rpm). Glass culture flasks with steel closures of various sizes (10 ml to 2 liters) (Bellco) are used for suspension cultures. Stock cultures (50 ml) of cells in

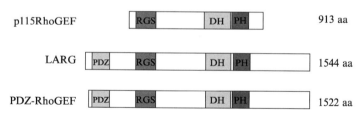

Fig. 1. Domain structure of p115RhoGEF, LARG, and PDZ-RhoGEF. LARG and PDZ-RhoGEF contain PDZ, RGS, DH, and PH domains, whereas p115RhoGEF lacks the PDZ domain.

a 100-ml flask are subcultured every 3 days with a density between 0.5 and 4×10^6 cells/ml. Sf9 cells are grown up from stock culture to 250 ml $(0.5 \times 10^6/\text{ml})$ in IPL-41 medium containing 10% fetal bovine serum in two 250-ml flasks or a 500-ml flask. After 2 or 3 days, they are further expanded to 1 liter (about $0.5–1 \times 10^6/\text{ml}$). After 2 or 3 days, cells are transferred to four 2-liter flasks containing 750 ml of scale-up medium [IPL-41/1% fetal bovine serum/1% chemically defined lipid concentrate (Invitrogen)/0.1% (v/v) Pluronic F-68/25 μg/ml gentamicin]. The cells (usually $1.5–2 \times 10^6/\text{ml}$) are infected with amplified, recombinant baculo-viruses the next day. Cells are harvested 48 h after infection by centrifuga-tion at 2000 rpm at 4° for 15 min in a JLA 10,500 rotor (Beckman). Cell pellets are frozen in liquid nitrogen and stored at $-80°$ until purification.

Reagents

Cholate is purified using a DEAE-Sephacel column and acid precipita-tion. Five hundred grams of cholic acid (Sigma) is dissolved in 3 liters of dH_2O and adjusted to pH 7.5–8.0 with 10 N NaOH. Cholate is diluted to 20 liters with dH_2O and purified over a 200- to 300-ml DEAE-Sephacel column. Cholate is precipitated from a flow-through fraction by decreasing the pH to 2–3 with 10 N HCl. The stock cholate solution (20%) is filtered through a Whatman #4 paper filter and stored at 4°. Stock solutions (1000×) of proteinase inhibitors (Sigma) are prepared as follows. Phenyl-methylsulfonyl fluoride (PMSF), L-1-p-tosylamino-2-phenylethyl chloro-methyl ketone (TPCK), and 1-chloro-3-tosylamido-7-amino-2-heptanone hydrochloride (TLCK) (800 mg of each) are dissolved in 50 ml of 50% dimethyl sulfoxide (DMSO)/50% isopropanol. Leupeptin and lima bean trypsin inhibitor (160 mg each) are dissolved in 50 ml of H_2O. They are stored at $-20°$ and added to purification buffers before use.

Purification of LARG Protein

LARG was originally isolated as KIAA0382, which encodes a partial cDNA lacking N-terminal PDZ and RGS domains. Full-length cDNA was obtained by 5′-RACE using the cDNA as a template and the human brain cDNA library (Clontech). The constructs of LARG are subcloned into the pFastBacHT transfer vector encoding a six-histidine tag at the amino terminus (Invitrogen), and their recombinant baculoviruses are generated. Sf9 cells (200–500 ml) are infected with the full-length LARG or ΔPDZ-LARG (LARG lacking the amino-terminal PDZ domain) virus and harvested after 48 h. Cells are resuspended in lysis buffer (20 mM HEPES, pH 8.0/50 mM NaCl/1 mM MgCl$_2$/0.1 mM EDTA/10 mM 2-mercaptoethanol, and proteinase inhibitors) and lysed by nitrogen cavitation (Parr bomb) or homogenized with a homogenizer (Kontes). The lysates are centrifuged at 35,000 rpm in a Ti45 rotor (Beckman) at 4° for 30 min. The supernatants are diluted fivefold with buffer A (20 mM HEPES, pH 8.0/100 mM NaCl/1 mM MgCl$_2$/10 mM 2-mercaptoethanol, and proteinase inhibitors) and are loaded onto a Ni-NTA column equilibrated with buffer A. The column is washed with 10 column volumes of buffer B (buffer A containing 400 mM NaCl and 10 mM imidazole). Recombinant LARG is eluted with 10 column volumes of buffer C (buffer A containing 150 mM imidazole). The elution fractions are concentrated, and the buffer is exchanged to buffer D [20 mM HEPES, pH 8.0/50 mM NaCl/ 1 mM EDTA/1 mM dithiothreitol (DTT)/10% glycerol, and proteinase inhibitors] using a Centricon YM-50. The full-length LARG and ΔPDZ-LARG have apparent molecular masses of about 200 and 180 kDa on SDS–PAGE, respectively. The yield of ΔPDZ-LARG from 200 ml of Sf9 culture is about 500 μg.

Purification of p115RhoGEF

p115RhoGEF is purified using an anti-glu-glu immunoaffinity column as described, with some modifications (Hart *et al.*, 1996). Pellets from 500 ml of the Sf9 cell culture infected with tagged p115RhoGEF are lysed by homogenization in 2.5 volumes of lysis buffer (50 mM Tris, pH 8.0/1 mM EDTA/1 mM DTT/1% NP-40, and proteinase inhibitors). The cell lysates are centrifuged at 500g for 15 min followed by centrifugation at 35,000 rpm in a Ti45 rotor for 30 min at 4°. The supernatant is loaded onto a 0.5-ml anti-glu-glu antibody-conjugated column (Covance), equilibrated with 5 ml of lysis buffer. The column is washed with 10 ml of wash buffer I (50 mM Tris, pH 8.0/1 mM EDTA/1 mM DTT/0.5% NP-40, and proteinase inhibitors) and then with 10 ml of wash buffer II (50 mM Tris, pH 8.0/1 mM EDTA/1 mM DTT/400 mM NaCl, and proteinase inhibitors). The p115RhoGEF is eluted with 5 ml of elution buffer (0.5 ml per fraction) (50 mM Tris, pH 8.0/1 mM EDTA/1 mM DTT/400 mM NaCl/100 μg/ml EYMPME peptide, and proteinase inhibitors). The peak fractions are concentrated with a Centricon YM-50. Glycerol is added to a final concentration of 10%.

Purification of RhoA

cDNA for RhoA was ligated into the pFastBacHTb vector, and the baculovirus encoding RhoA with an amino-terminal hexahistidine tag was generated following the Bac-to-Bac baculovirus expression system protocol (Invitrogen). Recombinant His$_6$-RhoA used in the GTPγS binding assay is purified from Sf9 cells infected with this virus. Cell pellets from a

1-liter culture are resuspended in 200 ml of lysis buffer (20 mM HEPES, pH 7.4/50 mM NaCl/1 mM MgCl$_2$/10 mM 2-mercaptoethanol/10 μM GDP, and proteinase inhibitors) and are lysed by nitrogen cavitation at 4° for 30 min. The lysates are centrifuged at 2500 rpm in a JLA 10,500 rotor at 4° for 15 min, and the supernatant is extracted with 1% cholate on ice for 1 h, followed by centrifugation at 35,000 rpm at 4° for 30 min in a Ti45 rotor. The supernatant is loaded onto a 1-ml Ni-NTA column equili- brated with buffer A (20 mM HEPES, pH 7.4/50 mM NaCl/1 mM MgCl$_2$/ 10 mM 2-mercaptoethanol/10 μM GDP/1% cholate and proteinase inhibi- tors). The column is washed with 10 ml of buffer B (20 mM HEPES, pH 7.4/400 mM NaCl/1 mM MgCl$_2$/10 mM 2-mercaptoethanol/10 μM GDP/ 10 mM imidazole, pH 8.0/0.5% cholate). His$_6$-RhoA is eluted with 5 ml of buffer C (20 mM HEPES, pH 7.4/100 mM NaCl/1 mM MgCl$_2$/1% cholate/10 mM 2-mercaptoethanol/100 mM imidazole/10 μM GDP). The eluted fractions are analyzed by SDS–PAGE and Coomassie brilliant blue staining. The peak fractions are concentrated, and the buffer is exchanged to buffer D (20 mM HEPES, pH 7.5/100 mM NaCl/1 mM MgCl$_2$/1% cholate/1 mM DTT/1 μM GDP) using a Centricon YM-10. Octyl β-D- glucopyranoside (OG) is added to a final concentration of 1%. Purified protein is divided into small aliquots (5–10 μl), frozen in liquid nitrogen, and stored at −80°. The typical yield of the protein from 1 liter of Sf9 cells is about 5 mg. Nontagged RhoA used in the GDP dissociation assay is purified from Sf9 cells by coexpressing GST-RhoGDI as an affinity tag as described (Wells *et al.*, 2002).

Purification of Gα12 and Gα13

Gα12 and Gα13 are purified using the Sf9 baculovirus system as de- scribed, with some modifications (Kozasa, 1999, 2003). This section describes procedures used for the purification of Gα13. Sf9 cell cultures (4 liters) are infected with 15 ml of Gα13, 10 ml of Gβ1, and 7.5 ml of His$_6$- Gγ2 viruses per liter. Cell pellets are resuspended in 600 ml of ice-cold lysis buffer (20 mM HEPES, pH 8.0/100 mM NaCl/0.1 mM EDTA/2 mM MgCl$_2$/10 mM 2-mercaptoethanol/10 μM GDP, and proteinase inhibitors). The remaining procedures are carried out at 4° unless otherwise specified. Cells are lysed by nitrogen cavitation (Parr bomb) at 500 psi for 30 min. Cell lysates are centrifuged at 2000 rpm for 10 min in a JLA 10,500 rotor. The supernatants are centrifuged further at 35,000 rpm for 30 min in a Ti45 rotor. The pellets are resuspended in 300 ml of wash buffer (20 mM HEPES, pH 8.0/100 mM NaCl/1 mM MgCl$_2$/10 mM 2-mercaptoethanol/ 10 μM GDP, and proteinase inhibitors) and centrifuged again as described earlier. The resulting pellets (cell membranes) are resuspended in 200 ml of wash buffer. The yield of membrane protein from 4 liters of Sf9 cells is about 1 g. Cell membranes are thawed and stirred at 4° with fresh proteinase inhibitors. Sodium cholate is added to a final concentration of 1% (w/v), and the mixture is stirred on ice for 1 h prior to centrifugation at 35,000 rpm in a Ti45 rotor for 30 min. The supernatants (membrane extract) are collected, diluted fourfold with buffer A, and centrifuged at 4000 rpm for 10 min. The supernatant is loaded onto a 2-ml Ni-NTA column equilibrated with buffer A (20 mM HEPES, pH 8.0/100 mM NaCl/1 mM MgCl$_2$/ 10 mM 2-mercaptoethanol/10 μM GDP/0.5% C$_{12}$E$_{10}$). The Ni-NTA column is washed with 50 ml of buffer B (20 mM HEPES, pH 8.0/300 mM NaCl/3 mM MgCl$_2$/10 mM 2-mercaptoethanol/50 μM GDP/10 mM imidazole/0.5% C$_{12}$E$_{10}$). The column is warmed to room temperature for 15 min and washed with 6 ml of buffer C [20 mM HEPES, pH 8.0/ 100 mM NaCl/1 mM MgCl$_2$/10 mM 2-mercaptoethanol/50 μM GDP/ 10 mM imidazole/0.06% n-dodecyl-β-D-maltoside (DDM)], 6 ml of buffer D (buffer C containing 0.2% DDM), 16 ml of buffer E (20 mM HEPES, pH 8.0/50 mM NaCl/50 mM MgCl$_2$/10 mM 2-mercaptoethanol/50 μM GDP/10 mM imidazole/0.2% DDM/10 mM NaF/30 μM AlCl$_3$/10% glycerol), and 12 ml of buffer F (20 mM HEPES, pH 8.0/50 mM NaCl/1 mM MgCl$_2$/10 mM 2-mercaptoethanol/50 μM GDP/150 mM imidazole/0.2% DDM). Peak fractions of Gα13 eluted with buffer E are combined and loaded onto a 1-ml ceramic hydroxyapatite column equilibrated with buffer G (20 mM HEPES, pH 8.0/50 mM NaCl/1 mM MgCl$_2$/1 mM DTT/50 μM GDP/10% glycerol/0.2% DDM). The flow through is collected and reapplied to the column. The column is then washed with 4 ml of buffer G, 4 ml of buffer G150 (buffer G containing 150 mM potassium phosphate, pH 8.0), and 4 ml of buffer G300 (buffer G containing 300 mM potassium phosphate, pH 8.0). Gα13 is eluted during washing with buffer G150. The peak fractions are combined, and the buffer is exchanged to buffer H (20 mM HEPES, pH 8.0/100 mM NaCl/0.5 mM EDTA/2 mM MgCl$_2$/1 mM DTT/10 μM GDP/10% glycerol/0.2% DDM) using a Centricon YM-30. The yield of Gα13 from 4 liters of Sf9 culture is about 200–400 μg.

Methods

GEF Assay (GDP Dissociation from RhoA)

The guanine nucleotide exchange activity of a RhoGEF protein can be assayed by monitoring the dissociation of GDP from RhoA. RhoA (2 μM) is mixed with loading buffer (50 mM Tris, pH 7.5/1.5 mM MgCl$_2$/ 4.5 mM EDTA/1 mM DTT/0.1% C$_{12}$E$_{10}$/10 μM GDP/~2000 cpm/pmol

[^{3}H]GDP), and the mixture is incubated at 30° for 1 min. The reaction is stopped by adding MgCl$_2$ to a final concentration of 10 mM, and the mixture is incubated at 30° for another 15 min. RhoA (100 nM) loaded with [^{3}H]GDP is mixed with the indicated proteins in GEF assay buffer (50 mM Tris, pH 7.5/50 mM NaCl/1 mM EDTA/1 mM DTT/10 mM MgCl$_2$/5 μM GTPγS/0.1% C$_{12}$E$_{10}$) in a final volume of 20 μl. G-protein α subunits are preincubated with AMF (30 μM AlCl$_3$/10 mM NaF) in GEF assay buffer and added to the GEF reaction mixture. The mixture is incubated at 20°. Reactions are terminated by the addition of 2 ml of washing buffer (20 mM Tris–HCl, pH 7.5/40 mM MgSO$_4$/100 mM NaCl), followed by filtration through nitrocellulose filters [BA-85; pore size 0.45 μm (Schleicher & Schuell)]. The amount of [^{3}H]GDP that remains on the filters is determined by liquid scintillation counting.

GEF Assay (GTPγS Binding to RhoA)

The GDP-GTP exchange activity of a RhoGEF can also be assayed by monitoring the production of GTPγS-bound Rho GTPase. G-protein α subunits are activated by incubating with 2×AMF in buffer II for 10 min on ice. G-protein α subunits are added to RhoGEF protein diluted with buffer I (total volume 10 μl) and are incubated for 10 min on ice. Finally, 25 μl of 2×binding buffer and 5 μl of His$_6$-RhoA (diluted to 5 μM in buffer I) are added to the mixture. The final concentration of RhoA is 500 nM. The reaction mixture is incubated at 30°. The reaction is terminated by adding 3 ml of washing buffer, and the solution is filtered through nitrocellulose. The filters are washed three times with 3 ml of washing buffer and are dried under an infrared heat lamp. The amount of [^{35}S]GTPγS on the filters is determined in 5 ml of scintillation fluid (Bio-Safe II; RPI). For calculation of specific activity, 2–5 μl of 2×binding buffer is spotted on a dried nitrocellulose filter and radioactivity is counted in the scintillation fluid.

The buffers used in the assay are as follows:

2 × binding buffer: 100 mM Tris, pH 7.5/1 mM EDTA/2 mM DTT/ 100 mM NaCl/10 mM MgCl$_2$/20 μM GTPγS/0.1% C$_{12}$E$_{10}$/~1000 cpm/pmol [^{35}S]GTPγS

Buffer I: 20 mM Tris, pH 7.5/1 mM EDTA/1 mM DTT/50 mM NaCl/ 0.1% C$_{12}$E$_{10}$/2 mM MgCl$_2$

Buffer II: 50 mM Tris, pH 7.5/1 mM EDTA/1 mM DTT/50 mM NaCl/10 mM MgCl$_2$/0.2% DDM

Washing buffer: 10 mM Tris, pH 7.5/100 mM NaCl/10 mM MgSO$_4$

Figure 2 shows the time course of GTPγS binding to RhoA in the absence or presence of 2 nM ΔPDZ-LARG and 30 nM Gα13. Diluted

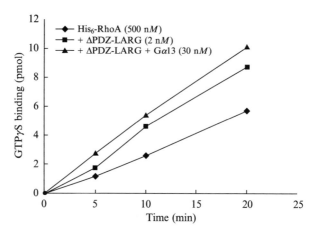

FIG. 2. The time course of GTPγS binding to RhoA. His$_6$-RhoA (500 nM) is incubated without (◆) or with ΔPDZ-LARG (2 nM; ■) and Gα13 (30 nM; ▲) at 30° for 5, 10, or 20 min.

RhoGEF protein or Rho dilution buffer (10 μl each) is added to reaction tubes (one for each condition). Next, AMF-activated Gα13 or 2× AMF (10 μl each) and 2× binding buffer (25 μl each) are added to each tube. His$_6$-RhoA protein (5 μl each) is added to initiate the reaction, and the mixture is incubated at 30°. At the indicated time, 50 μl of reaction mixture is transferred to a separate tube containing 3 ml of washing buffer to terminate the reaction. Figure 3 shows the dose–response curve of Gα13 for p115RhoGEF activation. p115RhoGEF (5 nM) is incubated with various concentrations of AMF-activated Gα13 for 5 min.

GEF Assay for Phosphorylated LARG

The GDP dissociation-based GEF assay using phosphorylated LARG is described. LARG is preincubated with Tec kinase in GEF assay buffer with 100 μM ATP at 20° for 40 min. Tec is prepared as follows. COS1 cells transfected with myc-tagged Tec are lysed in the lysis buffer (50 mM Tris, pH 7.5/150 mM NaCl/1% NP-40/1 mM EDTA/1 mM DTT/10 mM β-glycerophosphate/10 mM Na$_3$VO$_4$, and proteinase inhibitors). The lysate is centrifuged at 200,000g for 20 min. Tec is immunoprecipitated with anti-myc antibody 9E10 (Covance). Figure 4 shows the GDP dissociation assay of phosphorylated LARG. RhoA (100 nM) loaded with [^{3}H]GDP and AMF-activated Gα (200 nM) are added to the reaction mixture containing ΔPDZ-LARG (10 nM) and Tec. The reaction mixture is then incubated at 20° for 20 min. Gα12 stimulates the activity of phosphorylated but not

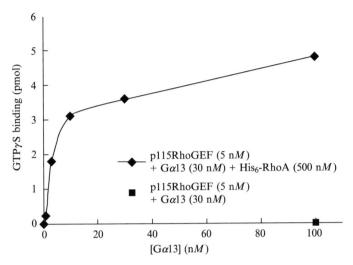

FIG. 3. The dose–response curve of Gα13 stimulation of p115RhoGEF activity. p115RhoGEF (5 nM) is incubated with His₆-RhoA (500 nM) at 30° for 5 min with 1, 3, 10, 30, or 100 nM AMF-activated Gα13 (◆). As a negative control, p115RhoGEF (5 nM) is incubated with 100 nM Gα13 without His₆-RhoA for 5 min (■).

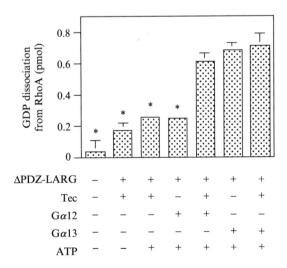

FIG. 4. GDP dissociation from RhoA by Gα12. Tec-phosphorylated LARG (10 nM) is incubated with AMF-activated Gα12 (200 nM) or Gα13 (200 nM) at 20° for 20 min. Gα12 activates phosphorylated but not nonphosphorylated LARG. From Suzuki et al. (2003), with permission.

nonphosphorylated LARG, whereas Gα13 stimulates both forms of LARG. These results indicate that the phosphorylation of LARG is responsible for Gα12-mediated RhoA activation (Suzuki et al., 2003).

Discussion

We have described two methods to detect RhoGEF activity in reconstitution experiments. From the results of time course and dose–response assays of GTPγS binding to RhoA, single point experiments can be performed at 5- or 10-min incubation times and a 30 nM Gα13 concentration to obtain an unsaturated response. The GTPγS binding assay using His$_6$-RhoA is convenient because the loading step of radioisotope is not required. Similar results are obtained from GTPγS binding or GDP dissociation assays. Compared with p115RhoGEF, LARG possesses higher basal activity in reconstitution assays. It is important to adjust the concentration of LARG carefully, as the reaction may saturate.

References

Fukuhara, S., Murga, C., Zohar, M., Igishi, T., and Gutkind, J. S. (1999). A novel PDZ domain containing guanine nucleotide exchange factor links heterotrimeric G proteins to Rho. J. Biol. Chem. 274, 5868–5879.

Hall, A. (1998). Rho GTPases and the actin cytoskeleton. Science 279, 509–514.

Hart, M. J., Jiang, X., Kozasa, T., Roscoe, W., Singer, W. D., Gilman, A. G., Sternweis, P. C., and Bollag, G. (1998). Direct stimulation of the guanine nucleotide exchange activity of p115 RhoGEF by Gα13. Science 280, 2112–2114.

Hart, M. J., Sharma, S., elMasry, N., Qiu, R.-G., McCabe, P., Polakis, P., and Bollag, G. (1996). Identification of a novel guanine nucleotide exchange factor for the Rho GTPase. J. Biol. Chem. 271, 25452–25458.

Kozasa, T. (1999). Purification of Recombinant G protein α and βγ subunits from Sf9 cells. In "Methods in Signal Transduction: G Protein Techniques of Analysis" (D. R. Manning, ed.), pp. 23–38. CRC Press, Boca Raton, FL.

Kozasa, T. (2003). Purification of G protein subunits from Sf9 insect cells using hexahistidine-tagged α and βγ subunits. In "Methods in Molecular Biology: G Protein Signaling" (A. V. Smrcka, ed.), Vol. 237, pp. 21–38. Humana Press, Totowa, NJ.

Kozasa, T., Jiang, X., Hart, M. J., Sternweis, P. M., Singer, W. D., Gilman, A. G., Bollag, G., and Sternweis, P. C. (1998). p115 RhoGEF, a GTPase activating protein for Gα12 and Gα13. Science 280, 2109–2111.

Suzuki, N., Nakamura, S., Mano, H., and Kozasa, T. (2003). Gα12 activates Rho GTPase through tyrosine-phosphorylated leukemia-associated RhoGEF. Proc. Natl. Acad. Sci. USA 100, 733–738.

Wells, C., Jiang, X., Gutowski, S., and Sternweis, P. C. (2002). Functional characterization of p115 RhoGEF. Methods Enzymol. 345, 371–382.

Subsection F

GRK Subfamily

[19] Analysis of G-Protein-Coupled Receptor Kinase RGS Homology Domains

By PETER W. DAY, PHILIP B. WEDEGAERTNER, and JEFFREY L. BENOVIC

Abstract

G-protein-coupled receptor kinases (GRKs) specifically phosphorylate agonist-occupied G-protein-coupled receptors (GPCRs). All seven mammalian GRKs contain an N-terminal domain that is homologous to the regulator of G-protein signaling (RGS) family of proteins. The RGS domain of GRK2 has been shown to interact specifically with $G\alpha_q$ family members. While the specificity and functional consequences of GRK/$G\alpha$ interaction remain somewhat poorly defined, GRK RGS homology (RH) domains likely function to provide specificity for GRK interaction with a particular $G\alpha$ subunit or GPCR/$G\alpha$ complex. Indeed, GRK2 binds $G\alpha_q$, α_{11}, and α_{14}, but not $G\alpha_{16}$, $G\alpha_s$, $G\alpha_i$, or $G\alpha_{12/13}$, while the RGS domains of GRK5 and GRK6 do not bind $G\alpha_{q/11}$, $G\alpha_s$, $G\alpha_i$, or $G\alpha_{12/13}$. In this chapter we describe various *in vitro* and intact cell strategies that can be used to elucidate the function of GRK RH domains.

Introduction

Many transmembrane signaling systems consist of specific G-protein-coupled receptors (GPCRs) that transduce the binding of a diverse array of extracellular stimuli into intracellular signaling events. In order to ensure that stimuli are translated into intracellular signals of appropriate magnitude, specificity, and duration, GPCR signaling cascades are tightly regulated. GPCRs are subject to three principal modes of regulation: (i) desensitization, the process whereby a receptor becomes refractory to continued stimuli; (ii) internalization, where receptors are physically removed from the cell surface by endocytosis; and (iii) downregulation, where total cellular receptor levels are decreased. GPCR desensitization is mediated primarily by second messenger-dependent kinases, such as protein kinase A (PKA) and protein kinase C (PKC), and by G-protein-coupled receptor kinases (GRKs). GRKs specifically phosphorylate activated GPCRs and initiate the recruitment of arrestins, which mediate receptor desensitization and internalization (Krupnick and Benovic, 1998; Pitcher *et al.*, 1998).

GRKs are found in metazoans and, in mammals, the seven GRKs can be divided into three subfamilies based on overall structural organization and homology: GRK1 (rhodopsin kinase) and GRK7; GRK2 (βARK1) and GRK3 (βARK2); and GRK4, GRK5 and GRK6. GRKs are serine/ threonine kinases with a tripartite modular structure. A central $\sim$330 amino acid catalytic domain, mostly related to those of other AGC kinases such as PKA, PKB, PKC, and PDK1, is flanked by an $\sim$180 residue N-terminal region and a divergent $\sim$60–160 amino acid C-terminal lipid-binding domain (Krupnick and Benovic, 1998; Pitcher *et al.*, 1998).

The N-terminal domains of all GRKs (residues 40–180) are homologous to the regulator of G-protein signaling (RGS) family of proteins (Siderovski *et al.*, 1996), multifunctional proteins that share an $\sim$130 amino acid RGS homology (RH) domain. RH domains can be grouped into six major subfamilies—R4, R7, R12, RZ, RL, and RA (axin)—with all but the RA family serving as negative regulators of G-protein signaling (Ross and Wilkie, 2000; Zhong and Neubig, 2001). GRK RH domains contain several features that distinguish them from other RH domains. First, the binding surface in GRKs is distinct from the binding interfaces used by RGS4 and RGS9 to bind Gα subunits and the surface used by axin to bind the APC helical peptide (Sterne-Marr *et al.*, 2003). Second, the GRK RH domain binds the GTPγS and GDP-AlF$_4^-$ forms of Gα_q comparably, in striking contrast to the preference of other RH domains for binding the GDP-AlF$_4^-$ form of Gα subunits (Carman *et al.*, 1999). Finally, although the GRK2 RH domain has the highest homology with the RH domain of axin (26% amino acid identity), GRK RH domains have been shown to interact specifically with Gα_q family members (Carman *et al.*, 1999; Sallese *et al.*, 2000; Usui *et al.*, 2000).

The recent X-ray crystal structure of GRK2 complexed with G$\beta_1\gamma_2$ provides important insight (Lodowski *et al.*, 2003). Interestingly, the crystal structure reveals that the RH, kinase, and PH domains of GRK2 form an equilateral triangle with the RH domain forming contacts with both kinase and PH domains. Moreover, the RH domain consists of two discontinuous regions with the standard nine helix bundle in the N-terminal region and two additional helices following the kinase domain, a structure similar to the RH domains of p115-RhoGEF and PDZ-RhoGEF (Chen *et al.*, 2001; Longenecker *et al.*, 2001). Docking analysis of the GRK2/G$\beta\gamma$ complex with Gα_q and the GPCR rhodopsin suggests that all three proteins should be able to bind to GRK2 simultaneously (Lodowski *et al.*, 2003). This would represent an effective way of turning off signaling by phosphorylating receptor and sequestering $G\alpha_q$ and $G\beta\gamma$, thereby preventing interaction with effector molecules.

While the specificity and functional consequences of GRK/$G\alpha$ interaction remain somewhat poorly defined, GRK RH domains likely function to provide specificity for the GRK interaction with a particular $G\alpha$ subunit or GPCR/$G\alpha$ complex. Indeed, GRK2 binds $G\alpha_q$, α_{11}, and α_{14} but not $G\alpha_{16}$, $G\alpha_s$, $G\alpha_i$, or $G\alpha_{12/13}$ (Carman et al., 1999; Day et al., 2003), whereas the RH domains of GRK5 and GRK6 do not bind $G\alpha_{q/11}$, $G\alpha_s$, $G\alpha_i$, or $G\alpha_{12/13}$ (Carman et al., 1999). GRK RH domains also function to attenuate signaling. Although the GRK2 RH domain does not enhance GTPase activity of $G\alpha_q$ in single turnover assays, it does function as a weak GAP in the receptor-stimulated activation of PLC-β (Carman et al., 1999). In addition, the GRK2 RH domain can effectively inhibit $G\alpha_q$ activation of PLC-β in vitro (Carman et al., 1999). GRK RH domains might also function to regulate GRK activity. GRKs are modular proteins and, as such, $G\alpha$ interaction with the RH domain may regulate the catalytic activity of the kinase. This would be analogous to the ability of $G\alpha_{13}$ interaction with the RGS domain of p115RhoGEF to activate GEF activity (Hart et al., 1998). Finally, GRK RH domains might also mediate additional protein interactions, as a number of RH domains appear to provide novel protein-binding surfaces. For example, the RH domain of axin binds the tumor suppressor protein APC (Kishida et al., 1998), a downstream target in the wnt signaling pathway. This article describes various in vitro and intact cell strategies that can be used to elucidate the function of GRK RH domains.

G-Protein Specificity of GRK RH Domains

GRK2 Interaction with $G\alpha_q$

The identification of an RH domain in the N terminus of GRK2 led to the discovery of the interaction between GRK2 and $G\alpha_q$ (Carman et al., 1999; Siderovski et al., 1996). The GRK2–$G\alpha_q$ interaction is activation dependent and specific for members of the $G\alpha_q$ family. Indeed, GRK2 binding to $G\alpha_q$ was first identified by passing a Triton X-100 solubilized bovine brain extract over a GRK2 affinity column in the presence of AlF_4^- (Fig. 1A) (Carman et al., 1999). In the same study, purified $G\alpha_q$ was shown to bind GRK2 with high affinity in an AlF_4^--dependent manner. Bovine brain $G\alpha_q$ also binds to the RH domain of GRK2 fused to GST [GST–GRK2-(1–178)] in an AlF_4^--dependent manner (Fig. 1B). GST-tagged versions of the GRK2 RH domain can also be used to study the GRK2–$G\alpha_q$ interaction in lysates from cells overexpressing $G\alpha_q$.

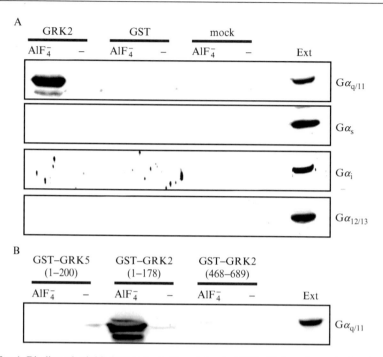

FIG. 1. Binding of soluble bovine brain G proteins to GRK affinity columns in the absence or presence of AlF_4^-. (A) Covalently bound GRK, GST-GRK, GST, and mock affinity columns were prepared and combined with a soluble bovine brain extract in the absence or presence of AlF_4^-. After washing the columns extensively, bound proteins were eluted by boiling with SDS sample buffer and were then subjected to SDS–PAGE. The gel was transferred to PVDF and subjected to immunoblotting with $G\alpha_{q/11}$-, $G\alpha_s$-, $G\alpha_i$-, and $G\alpha_{12/13}$-specific antibodies (indicated on right). (B) Experiments similar to those just described were performed using GST–GRK5-(1–200), GST–GRK2-(1–178), and GST–GRK2-(469–689) affinity columns and were immunoblotted with $G\alpha_{q/11}$-specific antibodies. Reproduced, with permission, from Carman *et al.* (1999).

Expression of GRK2 and GST-GRK2-RH Domains

Purified proteins allow examination of the GRK2–$G\alpha_q$ interaction in the absence of additional proteins that could enhance or inhibit the interaction. The purification of GRK2 has been described previously (Pronin *et al.*, 2002). Purified GRK2 can be used to create GRK2-coupled affinity resins (Carman *et al.*, 1999) and in steady-state GAP assays performed in reconstituted lipid vesicles in the presence of purified $G\alpha_q$, a $G\alpha_q$-coupled receptor and PLC-β (Biddlecome *et al.*, 1996; Carman *et al.*, 1999). GRK2 affinity columns have been used to estimate the affinity of the GRK2–Gα_q interaction. In steady-state assays, GRK2 displays a low level of GAP activity toward Gα_q (Carman *et al.*, 1999). Given the difficulty associated with purifying Gα_q and performing the aforementioned assays (Chidiac *et al.*, 2002), it may be more reasonable for most laboratories investigating the GRK2–Gα_q interaction to use purified GST–GRK2-(45–178) in combination with lysates from cells overexpressing Gα_q constructs. This approach has been used to identify residues in both the GRK2 RH domain and Gα_q that are involved in the GRK2–Gα_q interaction (Day *et al.*, 2003; Sterne-Marr *et al.*, 2003).

Purification of GST–GRK2-(45–178)

1. The nucleotide sequence encoding GRK2 residues 45–178 has been subcloned into the glutathione *S*-transferase fusion protein vector pGEX-2T, yielding GST–GRK2-(45–178). Use 4 ml of an overnight culture of BL21 cells transformed with GST–GRK2-(45–178) to inoculate a 200-ml culture in LB containing 5 μg/ml carbenicillin grown at 37° until the optical density (OD 600 nm) reaches 0.5. Expression of the GST fusion protein is induced with 0.5 mM isopropyl β-D-thiogalactoside and cells are grown for 3 h at 25°.

2. Cells are pelleted at 10,000 rpm for 10 min using an SS-34 (Sorvall) rotor, washed in STE buffer (20 mM Tris–HCl, 150 mM NaCl, 1 mM EDTA, pH 8.0), pelleted a second time, and then frozen at −70°.

3. Pellets are resuspended on ice in 10 ml STE buffer containing 100 μg/ml lysozyme and are incubated on ice for 15 min, followed by the addition of 5 mM dithiothreitol (DTT), Sarkosyl to 1.5% final concentration (from a 10% stock in STE), 100 μM phenylmethylsulfonyl fluoride (PMSF), 1 μg/ml leupeptin, and 1 μg/ml aprotinin.

4. Sonicate at least six times in 10-s bursts, followed by 15-s rest periods, until clarified. To remove insoluble protein, centrifuge for 10 min at 10,000 rpm using a GS-3 (Sorvall) rotor. Triton X-100 is then added to the supernatant to a 2% final concentration.

5. Fusion proteins are then immobilized on glutathione Sepharose 4B beads (from Amersham Life Sciences) by mixing for 1 h at 4° (0.3 ml of preswollen beads per 10 ml of extract). Wash once with 12 ml STE containing 1.5% Sarkosyl and 2% Triton X-100. Wash three times with STE followed by resuspension in STE containing 25% glycerol and 5 mM DTT. Adjust the volume with STE containing 25% glycerol and 5 mM DTT to yield a 25% slurry and store at −20°.

6. The amount of GST–GRK2-(45–178) bound to beads can be determined by eluting the purified protein in 50 mM Tris–HCl, pH 8.0,

containing 10 mM glutathione and 10 mM 2-mercaptoethanol at room temperature for 1 h, followed by Bradford assays.

Binding of GST-GRK2-(45–178) to Cellular Lysates

1. HEK-293 or COS cells are plated in 6-cm dishes so that they will be 60–70% confluent the next day. Transfect the cells with 2 μg of pcDNA3 containing Gα_q or the desired Gα subunit cDNA and 1 μg of empty pcDNA3 vector. The Gα_q subunits that we have used carry internal epitope tags to obviate difficulties associated with N- or C-terminally tagged Gα subunits and to allow the comparison of different Gα subunits by western blot analysis. The EE epitope replaces residues 171–176 of Gα_q with the residues EYMPTE. The HA epitope replaces residues 125–130 of Gα_q with DVPDYA.

2. The cells are lysed 24 h after transfection and prepared for incubation with the GST–GRK2-(45–178)-bound beads. Cells in 6-cm dishes are lysed in 300 μl of lysis buffer (20 mM Tris–HCl, pH 7.4, 1 mM EDTA, 1 mM DTT, 100 mM NaCl, 5 mM MgCl$_2$, 0.7% Triton X-100, 1 mM PMSF, 5 μg/ml leupeptin, 5 μg/ml aprotinin). Cells are scraped off the plate and collected in 1.5-ml centrifuge tubes and incubated on ice for 1 h with occasional vortexing.

3. The lysates are then centrifuged for 3 min at maximum speed in a microcentrifuge. Two hundred fifty microliters of the supernatant is removed and split into two tubes if AlF$_4^-$ will be used to activate Gα_q. To one of the tubes containing 125 μl of supernatant, add AlF$_4^-$ (25 μM AlCl$_3$ and 5 mM NaF). If the R183C or the Q209L constitutively active forms of Gα_q are being used, then all 250 μl of the supernatant can be used for the pull down. Twenty five microliters of the remaining supernatant should be saved to run alongside the immunoprecipitation samples.

4. The glutathione-Sepharose-bound GST–GRK2-(45–178) is washed three times in lysis buffer to remove any glycerol before being added to the lysates. To each tube of lysate add 8 μg of the GST–GRK2-(45–178). Rotate the tubes for 1–2 h at 4°.

5. After incubation of the lysates with the GST–GRK2-(45–178)-bound beads, samples are pelleted at low speed in a microcentrifuge for 3 min and beads are washed three times with lysis buffer.

6. Proteins are then eluted from the beads in 50 μl of SDS sample buffer and boiled for 5 min. Samples are subjected to 12% SDS–PAGE and then transferred to a PVDF membrane for western blot analysis. Gα subunits bound to GST–GRK2-(45–178) can be detected with the Gα_q-specific polyclonal antibody at 0.1 μg/ml or epitope-specific antibodies, EE monoclonal antibody (from Covance) at 2 μg/ml, or HA monoclonal antibody (12CA5 from Roche) at 1 μg/ml.

Analysis of GRK2 Interaction with Gα$_q$ in Cells

Although it is convenient to study the *in vitro* interaction of two purified proteins, it is also important to investigate the characteristics of the interaction in cells. The functional consequences of the GRK2–Gα$_q$ interaction can be assessed readily by the overexpression of both proteins in mammalian cells. Although immunoprecipitation does not necessarily replicate the conditions under which proteins interact in cells, it does allow examination of an interaction in the presence of other cellular proteins. Immunoprecipitation of GRK2 from cells overexpressing both GRK2 and Gα$_q$ confirms *in vitro* data that GRK2 binds Gα$_q$ in an activation-dependent manner (Fig. 2) (Carman *et al.*, 1999). Similarly, immunofluorescence experiments show that coexpression of constitutively active Gα$_q$, Gα$_q$-RC, or Gα$_q$-QL recruits GFP-GRK2-(45–178) to the plasma membrane where it colocalizes with the Gα$_q$ subunits (Sterne-Marr *et al.*, 2003). GRK2 binds to active Gα$_q$ and acts as an effector antagonist, and the effects of this interaction can be detected readily in cellular inositol phosphate assays. Cellular assays provide convenient means for the investigation of the GRK2–Gα$_q$ interaction.

Coimmunoprecipitation of Gα$_q$ with GRK2 and HA-GRK2-(45–178). The complex between GRK2 and Gα$_q$ can be detected by several different coimmunoprecipitation protocols. Wild-type GRK2 can be immunoprecipitated with either a polyclonal antibody specific for GRK2 (Carman *et al.*, 1999) or a monoclonal antibody that recognizes both GRK2 and GRK3 (Upstate Biotechnology). Alternatively, an HA-tagged version of the GRK2 RH domain can be immunoprecipitated with anti-HA antibodies.

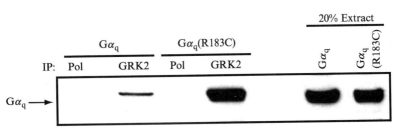

FIG. 2. Coimmunoprecipitation of activated Gα$_q$ and GRK2. COS-1 cells coexpressing GRK2 and either HA-tagged Gα$_q$ or HA-tagged Gα$_q$-R183C were harvested and lysed. Immunoprecipitation (IP) from cell extracts was performed by incubating with either GRK2-specific or nonspecific polyclonal (Pol) antibodies followed by incubation with protein A-agarose. Immunoprecipitated proteins were eluted with SDS sample buffer, and eluted proteins and initial cell extracts (20% of total used in IP) were subjected to SDS–PAGE and immunoblotting using an HA-specific monoclonal antibody. Reproduced, with permission, from Carman *et al.* (1999).

Finally, it is possible to immunoprecipitate $G\alpha_q$ using a polyclonal antibody that recognizes the extreme C terminus of $G\alpha_q$ ($G\alpha_q$ C-19 from Santa Cruz). However, it is not possible to detect the GRK2–$G\alpha_q$ interaction by immunoprecipitation of $G\alpha_q$ with antibodies that recognize internal HA or EE epitope tags, most likely because the binding of antibody to these epitopes, which are in close proximity to the GRK2-binding site, blocks or disrupts the $G\alpha_q$–GRK2 interaction.

1. HEK-293 or COS cells are plated on 6-cm dishes so that they will be roughly 60–70% confluent the next day. Cells are transfected with the desired constructs 24 h later. We use FuGENE 6 transfection reagent from Roche at a FuGENE to DNA ratio of 3:1 (for a 6-cm dish, 9 μl FuGENE and 3 μg of DNA). For HEK-293 cells, 2.7 μg of $G\alpha_q$ cDNA and 0.3 μg of GRK2 cDNA have yielded the most consistent results. Different ratios of DNA may be necessary in other cell types.

2. The cells are lysed 24 h after transfection and prepared for immunoprecipitation. Cells in 6-cm dishes are lysed in 300 μl of lysis buffer (20 mM Tris–HCl, pH 7.4, 1 mM EDTA, 1 mM DTT, 100 mM NaCl, 5 mM MgCl$_2$, 0.7% Triton X-100, 1 mM PMSF, 5 μg/ml leupeptin, 5 μg/ml aprotinin). Cells are scraped off the plate and collected in 1.5-ml centrifuge tubes and incubated on ice for 1 h with occasional vortexing.

3. The lysates are then centrifuged for 3 min at maximum speed in a microcentrifuge. Two hundred fifty microliters of the supernatant is removed and split into two tubes if AlF_4^- is used to activate $G\alpha_q$. To one of the tubes containing 125 μl of supernatant, add AlF_4^- (25 μM AlCl$_3$ and 5 mM NaF). If the R183C or the Q209L constitutively active forms of $G\alpha_q$ are being used, then all 250 μl of the supernatant can be used for immunoprecipitation. Twenty five microliters of the remaining supernatant should be saved to run alongside the immunoprecipitation samples.

4. GRK2 can be immunoprecipitated using 4 μg of a GRK2-specific polyclonal antibody per tube (Carman et al., 1999) or 1 μg of a GRK2/3-specific monoclonal antibody per tube (Upstate Biotechnology). For immunoprecipitation of $G\alpha_q$, 1 μg of the $G\alpha_q$ C-19 antibody from Santa Cruz works well. After addition of the appropriate antibody, 20 μl of protein A/G agarose-conjugated beads is added and the tubes are incubated for 2–3 h at 4°.

5. After incubation of the lysates with the antibody and beads, the samples are pelleted at low speed in a microcentrifuge for 3 min and the beads are washed three times with lysis buffer.

6. Proteins are then eluted from the agarose beads with 50 μl of SDS sample buffer and boiling for 5 min. Samples are subjected to 12% SDS–PAGE and transferred to a PVDF membrane. We use $G\alpha_q$ proteins that

carry internal HA or EE epitope tags and therefore probe the PVDF with either 1 μg/ml of 12CA5 or 2 μg/ml of EE-specific monoclonal antibody. For GRK2, 0.5 μg/ml of the GRK2/3-specific monoclonal antibody is used followed by horseradish peroxidase-conjugated secondary antibodies (1:10,000 dilution). Pierce ECL reagents are used for visualizing immunoblots.

Colocalization of Gα_q and GRK2 by Immunofluorescence. GRK2 is a cytoplasmic protein that is recruited to areas of active signaling at the plasma membrane (Krupnick and Benovic, 1998; Pitcher *et al.*, 1992). The GFP-tagged version of the RH domain of GRK2 is localized primarily in the cytoplasm and the nucleus with small amounts observable at the plasma membrane, similar to the localization of RGS2-GFP (Heximer *et al.*, 2001). Coexpression of constitutively active Gα_q is able to recruit GRK2-(45–178)-GFP to the plasma membrane (Fig. 3) (Sterne-Marr *et al.*, 2003). GRK2 RH domain mutants that do not bind Gα_q are not recruited to the plasma membrane by active Gα_q; conversely, Gα_q/Gα_{16} chimeras that do not bind the GRK2 RH domain fail to induce plasma membrane recruitment of GRK2-(45–178)-GFP (Day *et al.*, 2003; Sterne-Marr *et al.*, 2003). Thus, recruitment of the GFP-tagged RH domain of GRK2 to the plasma

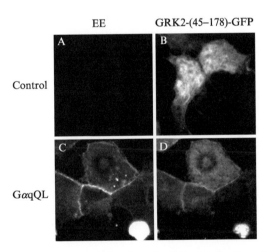

FIG. 3. Confocal microscopy comparing the subcellular localization of GRK2-(45–178)-GFP in cells cotransfected with activated Gα subunits. HEK-293 cells were transfected transiently with 0.02 μg of GRK2-(45–178)-GFP and either 0.7 μg of empty vector (control) or constitutively active Gα_q-QL. (A and C) Localization of the Gα subunits as visualized by immunostaining with EE-specific antibody. (B and D) Localization of GRK2-(45–178)-GFP. Cells were fixed and stained 48 h after transfection. More than 100 cells were examined in at least three independent experiments. Reproduced, with permission, from Day *et al.* (2003).

membrane by $G\alpha_q$ allows visualization of the $G\alpha_q$–GRK2 RH domain interaction in cells. Interestingly, the recruitment of full-length GRK2 to the plasma membrane by active $G\alpha_q$ is not detected easily.

1. HEK-293 cells are plated in six-well plates so that they will be roughly 70% confluent the next day. The cells are transfected with 0.02 μg of GRK2-(45–178)-GFP and either 1 μg of empty vector or 1 μg of constitutively active $G\alpha_q$.

2. After 24 h, transfected cells are replated onto glass coverslips, grown for an additional 24 h, and washed three times with PBS before fixing with 3.7% formaldehyde for 20 min. [If the EE antibody is used for visualization of the $G\alpha_q$ subunit, additional steps are required for the reduction of background associated with this antibody. Following formaldehyde fixation, cells are quenched with 50 mM ammonium chloride, washed three times with PBS, and followed by secondary fixation with methanol at $-20°$ for 4 min. This process reduces the background associated with this antibody greatly and retains GFP fluorescence, but also leads to a fair amount of damage to the cells, which decreases the quality of images that can be obtained. Fixing cells only in methanol, while possibly optimal for the EE antibody, destroys GFP fluorescence and thus cannot be used. Formaldehyde fixation alone is suitable for the $G\alpha_q$ polyclonal antibody or 12CA5 monoclonal antibody.]

3. Cells are washed three times with PBS and then incubated in blocking buffer consisting of TBS (50 mM Tris–HCl, pH 7.5, 150 mM NaCl) with 1% Triton X-100 and 2.5% nonfat milk. The coverslips are then incubated in blocking buffer containing a 1:100 dilution of the anti-$G\alpha_q$ polyclonal antibody (Santa Cruz) for 1 h. If the EE antibody is being used for the visualization of $G\alpha_q$, it should be diluted to 20 μg/ml in blocking buffer. [The HA-tagged version of $G\alpha_q$ is superior to the EE-tagged version and could be used; however, the GRK2-(45–178)-GFP construct also contains an HA tag.] Following the 1-h incubation, coverslips are washed five times with blocking buffer and then incubated in blocking buffer containing a 1:100 dilution of Alexa Fluor 594-conjugated goat antirabbit secondary antibody, for the $G\alpha_q$ polyclonal antibody, or Alexa Fluor 594-conjugated goat antimouse, for the EE monoclonal antibody.

4. The coverslips are washed and mounted on glass slides with Prolong antifade reagent. The slides are allowed to dry overnight before they are examined with a fluorescence microscope to visualize GRK2-(45–178)-GFP and $G\alpha_q$.

Measuring Inositol Phosphate Production as a Functional Assay for the $G\alpha_q$–GRK2 Interaction. Active $G\alpha_q$ stimulates the production of inositol phosphate and diacylglycerol by interacting with and activating the enzyme PLC-β. Coexpression of GRK2 or the RH domain of GRK2 inhibits inositol phosphate production stimulated by receptor-activated or constitutively active Gα_q (Carman *et al.*, 1999). RH domains can inhibit Gα signaling in two ways: by acting as GTPase-accelerating proteins (GAPs) and/or by acting as effector antagonists (Hepler *et al.*, 1997). *In vitro* assays show that GRK2 is not a GAP for Gα_q-RC; however, in the presence of the M$_1$ muscarinic acetylcholine receptor in lipid vesicles, GRK2 is a weak GAP for Gα_q-RC (Carman *et al.*, 1999). These results suggest that the majority of the GRK2-mediated inhibition of Gα_q-stimulated inositol phosphate production is via effector blockade; however, the presence of certain receptors may enhance the GAP activity of GRK2 toward Gα_q.

1. We have used both COS and HEK-293 cells in cellular inositol phosphate assays but find that HEK-293 cells generally yield lower background inositol phosphate counts. Plate HEK-293 cells in six-well dishes so that they will be 50% confluent the next day. Transfect the cells with a total of 1 μg of DNA. For assays in which stimulation of an exogenously expressed receptor is used to activate Gα_q, equal amounts of receptor and Gα_q DNA are transfected, usually 0.4 μg each. [Due to the promiscuous nature of Gα_q receptor coupling, it is possible to cotransfect Gα_q and the α_{2A}-adrenergic receptor (α_{2A}AR), normally a Gα_i-coupled receptor, and get functional α_{2A}AR-Gα_q coupling. This "forced" coupling allows the examination of signaling from transfected Gα_q subunits in the absence of background signaling from endogenous Gα_q, a requirement when examining mutant Gα_q (Day *et al.*, 2003). Alternatively, receptors that normally couple to Gα_q, such as the M$_1$ muscarinic acetylcholine receptor and the thromboxane A$_2$-α receptor, will couple functionally to endogenously expressed Gα_q when transfected into cells (Carman *et al.*, 1999).] Transfection of 0.2 μg of GRK2 or GRK2-K220R, a kinase inactive mutant, gives significant inhibition in this system (Day *et al.*, 2003). In assays using Gα_q-RC or Gα_q-QL to stimulate PLCβ, 0.1 μg of the Gα_q construct is transfected. The amount of GRK2 or GRK2 RH domain transfected differs depending on the Gα_q mutant being used. For Gα_q-RC, as little as 5 ng of GRK2 DNA will inhibit inositol phosphate production to some degree; however, a range from 10 to 200 ng GRK2 DNA gives the most consistent inhibitory profile. For Gα_q-QL, considerably more GRK2 is needed to inhibit inositol phosphate production, typically 100–500 ng.

2. After 24 h of transfection, trypsinize cells and resuspend in 5 ml of growth medium. Replate samples in triplicate, 500 μl/well, on 24-well plates and label for 16 h with 2 μCi/ml of [^{3}H]inositol. For each sample, plate an extra well of cells to confirm the expression of each protein by

western blotting. [In our system, we have found that coexpression of GRK2 or other RGS proteins can have a marked effect on the expression of $G\alpha_q$ or mutants of $G\alpha_q$. Therefore, checking the expression level of $G\alpha_q$ coexpressed with GRK2 or any other RGS protein is absolutely required.]

3. After labeling, aspirate the medium and wash with 0.5 ml assay medium (serum-free DMEM containing 20 mM HEPES, pH 7.4, and 5 mM LiCl) for 10 min at 37°. From this point forward, all conditions are nonsterile.

4. Aspirate the medium, add 0.5 ml assay media, and incubate for 1 h at 37°. If an agonist is being used, it should be added to the assay medium before addition to cells.

5. Aspirate the medium and add 0.75 ml of cold 20 mM formic acid (dilute from concentrated stock, approximately 23 M). Store for 30 min at 4°. The reactions can be stored overnight at 4° or frozen for longer term storage, but we find that results are more consistent if the samples are separated on columns the same day.

6. Equilibrate the columns (1 ml Dowex AG 1-X8 100–200 mesh columns) with 10 ml of 4 M ammonium formate and 0.2 M formic acid, followed by 5 ml of 0.18% ammonium hydroxide (6 ml of 30% concentrated ammonium hydroxide/liter).

7. After step 5, add 100 μl of 3% ammonium hydroxide to each sample well to neutralize the acid. The approximate pH is 8.0.

8. Put 6-ml scintillation vials under the Dowex columns and load the 850 μl of sample onto the columns. Let the sample drip through completely.

9. Add 1 ml of 0.18% ammonium hydroxide solution and let it drip through the columns.

10. The resulting 1.85-ml sample is the inositol fraction. Add 4 ml of Ultima Gold scintillation fluid (Packard Chemicals) to each scintillation vial. Cap, number, and shake the vials before counting in a scintillation counter.

11. Place the columns over a waste container and wash with 4 ml of 40 mM ammonium formate and 0.1 M formic acid. Let this drip through into the waste container. This wash contains radioactive material and should be disposed of appropriately.

12. Place the columns over new 6-ml scintillation vials and add 1 ml of 4 M ammonium formate and 0.2 M formic acid. Let this drip through the columns into the vials. This 1 ml is the inositol phosphate fraction. Add 4 ml of Ultima Flo AF scintillation fluid to each vial. Cap, number, and shake the vials before counting in a scintillation counter.

13. Immediately after each experiment is completed the columns should be washed with 10 ml of 4 M ammonium formate and 0.2 M formic

acid. The columns can be reused for long periods of time if they are washed properly. We have the best results if all of the solutions that are used on the columns are made with Milli-Q or similar water.

Specificity of the GRK–Gα Interaction

In the original identification of the interaction between $G\alpha_q$ and GRK2 it was demonstrated that GRK2 and GRK3 interacted with both $G\alpha_q$ and $G\alpha_{11}$, but neither GRK2 nor GRK3 interacted with $G\alpha_s$, $G\alpha_i$, or $G\alpha_{12/13}$ (Carman *et al.*, 1999). Subsequently, it has been demonstrated that GRK2 is able to interact with and inhibit all members of the $G\alpha_q$ family, $G\alpha_q$, $G\alpha_{11}$, and $G\alpha_{14}$ with the exception of $G\alpha_{16}$ (Fig. 4) (Day *et al.*, 2003). In contrast, the RH domain of GRK5 does not interact with $G\alpha_q$, and GRK5 and GRK6 do not inhibit inositol phosphate production stimulated by $G\alpha_q$ (Fig. 5) (Carman *et al.*, 1999). Nevertheless, it remains possible that the RH domain of other GRKs bind $G\alpha$ subunits other than $G\alpha_q$.

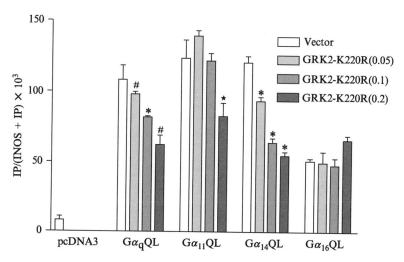

FIG. 4. Effect of GRK2 on $G\alpha_q$ family-stimulated inositol phosphate production. HEK-293 cells were cotransfected with EE-tagged constitutively active $G\alpha_qQL$, $G\alpha_{11}QL$, $G\alpha_{14}QL$, or $G\alpha_{16}QL$ and increasing amounts of GRK2-K220R, a kinase-deficient mutant. Twenty-four hours after transfection, cells were labeled with 2 μCi/ml of *myo*-[^{3}H]inositol. The next day, total inositol phosphate production was determined. Results shown are from a single experiment representative of at least three experiments, each performed in triplicate and displayed as the mean cpm $\pm$ SD of inositol phosphates over total inositol [IP/(INOS + IP) $\times$ 10^3]. Statistical significance of difference between the indicated bar and the control bar ($G\alpha$ in the absence of any GRK2-K220R) is denoted by a star ($p < 0.05$), a pound sign ($p < 0.01$), or an asterisk ($p < 0.005$). Reproduced, with permission, from Day *et al.* (2003).

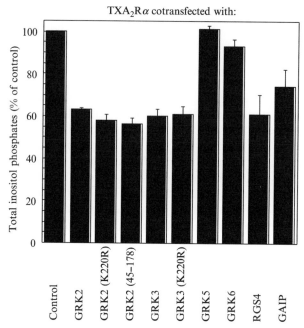

FIG. 5. Inhibition of $G\alpha_q$ signaling by RGS and GRK constructs. HEK-293 cells expressing the thromboxane A_2-α receptor ($TXA_2R\alpha$) along with vector (control) or the indicated GRK or RGS constructs were labeled metabolically with myo-[³H]inositol and then stimulated for 10 min with the $TXA_2R\alpha$ agonist U46619. Total [³H]inositol phosphates were quantitated by liquid scintillation counting, expressed as a percentage of control and plotted against the indicated experimental conditions (i.e., coexpressed constructs). Reproduced, with permission, from Carman *et al.* (1999).

Conclusions

GRK RH domains are specific $G\alpha$-binding domains that likely function in mediating GRK targeting to GPCR/$G\alpha$ protein complexes and enhancing signal turnoff. However, much remains to be learned about these domains. Do the RH domains in GRK1, GRK4, GRK5, GRK6, and GRK7 interact with $G\alpha$ subunits and, if so, what is the specificity of such interaction? Can GRKs function as effective GAPs in the presence of a particular GPCR/$G\alpha$ combination? Do $G\alpha$ subunits function to regulate the catalytic activity or substrate specificity of GRKs? In addition, do GRK RH domains provide a binding surface for additional protein/protein interactions? The techniques detailed in this chapter should prove useful in addressing some of these important questions.

References

Biddlecome, G. H., Berstein, G., and Ross, E. M. (1996). Regulation of phospholipase C-β1 by Gq and m1 muscarinic cholinergic receptor: Steady-state balance of receptor-mediated activation and GTPase-activating protein-promoted deactivation. *J. Biol. Chem.* **271**, 7999–8007.

Carman, C. V., Parent, J. L., Day, P. W., Pronin, A. N., Sternweis, P. M., Wedegaertner, P. B., Gilman, A. G., Benovic, J. L., and Kozasa, T. (1999). Selective regulation of Gα$_{q/11}$ by an RGS domain in the G protein-coupled receptor kinase, GRK2. *J. Biol. Chem.* **274**, 34483–34492.

Chen, Z., Wells, C. D., Sternweis, P. C., and Sprang, S. R. (2001). Structure of the rgRGS domain of p115RhoGEF. *Nature Struct Biol.* **8**, 805–809.

Chidiac, P., Gadd, M. E., and Hepler, J. R. (2002). Measuring RGS protein interactions with Gqα. *Methods Enzymol.* **344**, 686–702.

Day, P. W., Carman, C. V., Sterne-Marr, R., Benovic, J. L., and Wedegaertner, P. B. (2003). Differential interaction of GRK2 with members of the Gα$_q$ family. *Biochemistry* **42**, 9176–9184.

Hart, M. J., Jiang, X., Kozasa, T., Roscoe, W., Singer, W. D., Gilman, A. G., Sternweis, P. C., and Bollag, G. (1998). Direct stimulation of the guanine nucleotide exchange activity of p115 RhoGEF by Gα13. *Science* **280**, 2112–2114.

Hepler, J. R., Berman, D. M., Gilman, A. G., and Kozasa, T. (1997). RGS4 and GAIP are GTPase-activating proteins for Gqα and block activation of phospholipase C β by γ-thio-GTP-Gqα. *Proc. Natl. Acad. Sci. USA* **94**, 428–432.

Heximer, S. P., Lim, H., Bernard, J. L., and Blumer, K. J. (2001). Mechanisms governing subcellular localization and function of human RGS2. *J. Biol. Chem.* **276**, 14195–14203.

Kishida, S., Yamamoto, H., Ikeda, S., Kishida, M., Sakamoto, I., Koyama, S., and Kikuchi, A. (1998). Axin, a negative regulator of the wnt signaling pathway, directly interacts with adenomatous polyposis coli and regulates the stabilization of beta-catenin. *J. Biol. Chem.* **273**, 10823–10826.

Krupnick, J. G., and Benovic, J. L. (1998). The role of receptor kinases and arrestins in G protein-coupled receptor regulation. *Annu. Rev. Pharmacol. Toxicol.* **38**, 289–319.

Lodowski, D. T., Pitcher, J. A., Capel, W. D., Lefkowitz, R. J., and Tesmer, J. J. (2003). Keeping G proteins at bay: A complex between G protein-coupled receptor kinase 2 and Gβγ. *Science* **300**, 1256–1262.

Longenecker, K. L., Lewis, M. E., Chikumi, H., Gutkind, J. S., and Derewenda, Z. S. (2001). Structure of the RGS-like domain from PDZ-RhoGEF: Linking heterotrimeric G protein-coupled signaling to Rho GTPases. *Structure* **9**, 559–569.

Pitcher, J. A., Freedman, N. J., and Lefkowitz, R. J. (1998). G protein-coupled receptor kinases. *Annu. Rev. Biochem.* **67**, 653–692.

Pitcher, J. A., Inglese, J., Higgins, J. B., Arriza, J. L., Casey, P. J., Kim, C., Benovic, J. L., Kwatra, M. M., Caron, M. G., and Lefkowitz, R. J. (1992). Role of βγ subunits of G proteins in targeting the β-adrenergic receptor kinase to membrane-bound receptors. *Science* **257**, 1264–1267.

Pronin, A. N., Loudon, R. P., and Benovic, J. L. (2002). Characterization of G protein-coupled receptor kinases. *Methods Enzymol.* **343**, 547–559.

Ross, E. M., and Wilkie, T. M. (2000). GTPase-activating proteins for heterotrimeric G proteins: Regulators of G protein signaling (RGS) and RGS-like proteins. *Annu. Rev. Biochem.* **69**, 795–827.

Sallese, M., Mariggio, S., D'Urbano, E., Iacovelli, L., and De Blasi, A. (2000). Selective regulation of Gq signaling by G protein-coupled receptor kinase 2: Direct interaction of kinase N terminus with activated Gαq. *Mol. Pharmacol.* **57**, 826–831.

Siderovski, D. P., Hessel, A., Chung, S., Mak, T. W., and Tyers, M. (1996). A new family of regulators of G-protein-coupled receptors? *Curr. Biol.* **6**, 211–212.

Sterne-Marr, R., Tesmer, J. J., Day, P. W., Stracquatanio, R. P., Cilente, J. A., O'Connor, K. E., Pronin, A. N., Benovic, J. L., and Wedegaertner, P. B. (2003). GRK2:Gαq/11 interaction: A novel surface on an RGS homology domain for binding Gα subunits. *J. Biol. Chem.* **278**, 6050–6058.

Usui, H., Nishiyama, M., Moroi, K., Shibasaki, T., Zhou, J., Ishida, J., Fukamizu, A., Haga, T., Sekiya, S., and Kimura, S. (2000). RGS domain in the amino-terminus of G protein-coupled receptor kinase 2 inhibits Gq-mediated signaling. *Int. J. Mol. Med.* **5**, 335–340.

Zhong, H., and Neubig, R. R. (2001). Regulator of G protein signaling proteins: Novel multifunctional drug targets. *J. Pharmacol. Exp. Ther.* **297**, 837–845.

[20] Characterization of GRK2 RH Domain-Dependent Regulation of GPCR Coupling to Heterotrimeric G Proteins

By RACHEL STERNE-MARR, GURPREET K. DHAMI, JOHN J. G. TESMER, and STEPHEN S. G. FERGUSON

Abstract

Heterotrimeric guanine nucleotide (G)-coupled receptors (Gpcrs) form the largest family of integral membrane proteins. GPCR activation by an agonist promotes the exchange of GDP for GTP on the Gα subunit of the heterotrimeric G protein. The dissociated Gα and Gβγ subunits subsequently modulate the activity of a diverse assortment of effector systems. GPCR signaling via heterotrimeric G proteins is attenuated rapidly by the engagement of protein kinases. The canonical model for GPCR desensitization involves G protein-coupled receptor kinase (GRK)-dependent receptor phosphorylation to promote the binding of arrestin proteins that function to sterically block receptor:G-protein interactions. GRK2 and GRK3 have been shown to interact with Gαq via the regulator of G-protein signaling (RGS) homology (RH) domain localized within their amino-terminal domains. It now appears that the G-protein uncoupling of many GPCRs linked to Gαq, in particularly metabotropic glutamate receptors, may be mediated by the GRK2 RH domain via a phosphorylation-independent mechanism. This article reviews much of the background and methodology required for the characterization of the GRK2 phosphorylation-independent attenuation of GPCR signaling.

Introduction

Heterotrimeric guanine nucleotide (G) protein-coupled receptor (GPCR) signaling is a highly regulated process requiring a delicate balance among receptor activation, desensitization, and resensitization. At the receptor level, desensitization has evolved to protect against overstimulation, which can lead to cellular damage. GPCR desensitization is mediated by three families of regulatory molecules: second messenger-dependent protein kinases, G protein-coupled receptor kinases (GRKs), and arrestins (Ferguson, 2001, Lefkowitz, 1993). These proteins regulate the phosphorylation-dependent uncoupling of GPCRs from heterotrimeric G proteins. Two key mechanisms contribute to the phosphorylation-dependent desensitization of many GPCRs: phosphorylation by second messenger-dependent protein kinases and phosphorylation by GRKs. The second messenger-dependent protein kinases, protein kinase C (PKC) and cAMP-dependent protein kinase (PKA), are activated as a consequence of GPCR-stimulated second messenger responses. PKC and PKA phosphorylate GPCRs at serine and threonine residues within specific phosphorylation consensus motifs localized to the intracellular loop and carboxyl-terminal tail domains of GPCRs. In contrast, GRK-mediated phosphorylation of agonist-activated GPCRs promotes the membrane translocation and subsequent binding of arrestin proteins to the intracellular face of GPCRs (Barak et al., 1997; Lohse et al., 1990). Arrestin binding to the receptor sterically uncouples GPCRs from their cognate heterotrimeric G protein (Lohse et al., 1990). β-Arrestin proteins also target GPCRs for internalization in clathrin-coated vesicles (Ferguson et al., 1996; Goodman et al., 1996).

G Protein-Coupled Receptor Kinases

The GRK family of proteins is composed of seven members that can be divided into three subclasses based on sequence homology and function: (1) GRK1 (rhodopsin kinase) and GRK7, (2) GRK2 (βARK1) and GRK3 (βARK2), and (3) GRK4, GRK5, and GRK6. Each of the GRK isozymes specifically phosphorylates agonist-activated receptors at serine and threonine residues localized within the third intracellular loop and/or carboxyl-terminal tail domains of GPCRs (Ferguson, 2001; Lefkowitz, 1993). The GRK2 crystal structure confirms that GRK2 is composed of three functional domains: an amino-terminal regulator of G protein signaling (RGS) homology (RH) domain, a central protein kinase domain, and a carboxyl-terminal, Gβγ-binding pleckstrin homology (PH) domain (Lodowski et al., 2003). The plasma membrane targeting of GRKs involves a variety of mechanisms, but the membrane targeting of GRK2 and GRK3 is mediated by the association of their PH domain with the $G\beta\gamma$ subunit of hetero-trimeric G proteins and anionic phospholipids, such as phosphatidyl-inositides or phosphatidylserine (Pitcher et al., 1992, 1995). The loss of GRK2 binding to either phospholipids or $G\beta\gamma$ prevents both agonist- and GRK2-dependent phosphorylation of the β_2-adrenergic (β_2AR) (G_s-coupled) and the μ-opioid (G_i-coupled) receptors in intact cells, indicating that $G\beta\gamma$ and phospholipid engagement is essential for the phosphorylation function of GRK2 (Carman et al., 2000). GRK amino-terminal RH domains exhibit homology with the "RGS box" of RGS proteins (Siderovski et al., 1996).

Identification of GRK2 Binding to $G\alpha_{q/11}$

In general, the role of specific GRKs in regulating the phosphorylation and desensitization of GPCRs has been studied by the overexpression of wild-type and catalytically inactive GRKs (Freedman and Lefkowitz, 1996). Under these conditions, GRK2-dependent, but phosphorylation-independent, uncoupling of $G\alpha_q$-coupled GPCRs has been observed. Specifically, GRK2 phosphorylation-independent desensitization has been reported for the angiotensin II type 1A receptor (Oppermann et al., 1996), the endothelin A and B receptors (Freedman et al., 1997), the thromboxane A_2 receptor (TXA2R) (Carman et al., 1999), the parathyroid hormone receptor (Dicker et al., 1999), and metabotropic glutamate receptor 1 (mGluR1) (Dhami et al., 2002). For many of these receptors, expression of the amino-terminal domain of GRK2 is sufficient to attenuate signaling in the absence of phosphorylation (Carman et al., 1999; Dhami et al., 2002; Freedman et al., 1997; Sallese et al., 2000a; Usui et al., 2000). Carman et al. (1999) were the first to demonstrate biochemically that the GRK2 amino-terminal domain comprises a $G\alpha_{q/11}$-binding site. Constructs containing just the GRK2 RGS homology domain (residues 52–174) exhibit greater efficacy for attenuating both endothelin A and angiotensin II type 1A receptor second messenger responses than similar constructs that also contain the first 50 amino acid residues of the GRK2 amino terminus (Usui et al., 2000). This result is surprising considering that the first 30 amino acid residues of GRK1 are known to be important for the phosphorylation of rhodopsin by GRK1 (Palczewski et al., 1993). The ability of various segments of GRK2 to regulate second messenger production in transiently transfected cell lines correlates very well with the ability of the analogous fusion proteins to bind to $G\alpha_q$ in an AlF_4^--dependent manner in vitro (Sallese et al., 2000a; Usui et al., 2000). Thus, despite the observation that intact interactions between the GRK2 PH domain and both $G\beta\gamma$ and anionic phospholipids are required for the desensitization of G_i- and G_s-coupled receptors (Carman *et al.*, 2000), the GRK2 kinase and PH domains appear to be dispensable for the uncoupling of $G\alpha_{q/11}$ GPCR second messenger responses.

Sequences of the amino-terminal domains of all seven GRKs bear significant homology to RGS domains, as first revealed by *in silico* BLAST searches for sequence homology with RGS2 (Siderovski *et al.*, 1996). The GRK2 RGS box (residues 51–171) is related most closely to RGS12 (24% identity; 43% similarity) and RGS10 (22% identity; 44% similarity). However, the GRK2 RGS box is also equally closely related to axin (24% identity; 45% similarity), a protein that is not known to regulate heterotrimeric G proteins. Furthermore, the GRK2/3 RH domains are no more similar to the RGS boxes found in other GRKs (24–28% identity; 43–48% similarity) than they are to either the canonical RGS box or axin.

Typically, RGS proteins bind the GDP-aluminum tetrafluoride ($GDP \cdot AlF_4^-$)-bound form of $G\alpha_{q/11}$ with higher affinity than they bind to the GTPγS-bound "activated" form of $G\alpha_{q/11}$ and do not bind the inactive GDP-bound form of $G\alpha_{q/11}$ (Hepler *et al.*, 1997; Srinivasa *et al.*, 1998). In a biochemical search for GRK2-interacting proteins, Carman *et al.* (1999) identified $G\alpha_q$ and $G\alpha_{11}$ from bovine brain as AlF_4^--dependent GRK2- and GRK3-binding partners. This interaction is specific to $G\alpha_q$, $G\alpha_{11}$, and $G\alpha_{14}$, but not $G\alpha_{16}$ (Day *et al.*, 2003). Other $G\alpha$ proteins ($G\alpha_s$, $G\alpha_i$, $G\alpha_o$, and $G\alpha_{12/13}$) also do not bind to GRK2 (Carman *et al.*, 1999; Day *et al.*, 2003; Sallese *et al.*, 2000a). GST- or hexahistidine-tagged fusion proteins bearing segments of the amino terminus as small as amino acid residues 54–174 and as large as the entire N-terminal domain of GRK2 (amino acid residues 2–187) bind to $G\alpha_{q/11}$ in pull-down assays (Carman *et al.*, 1999). While all seven GRKs contain the 120 residue RGS homology domain, $G\alpha_{q/11}$ interactions are restricted to GRK2 and GRK3 (Carman *et al.*, 1999).

The investigation of GRK2 RH domain interactions with $G\alpha_{q/11}$ is facilitated by the use of a GTPase-defective, "constitutively activated" $G\alpha_q$ mutant (R183C) (Ingi *et al.*, 1998). $G\alpha_q$-R183C transfected into COS-1 cells stimulates endogenous phospholipase $C\beta$ (PLCβ) activity in the absence of GPCR coupling. This provides an assay for receptor-independent G protein-stimulated inositol phosphate formation to assess the GRK-dependent attenuation of PLCβ activity. In this assay, GRK2, GRK3, and catalytically inactive GRK2-K220R and GRK3-K220R mutants block PLCβ activation by $G\alpha_q$-R183C (Carman *et al.*, 1999). However, GRK5 expression does not prevent $G\alpha_q$-R183C-mediated stimulation of PLCβ, whereas GRK6 only modestly attenuates signaling but blocks PLCβ-mediated activation by AlF_4^- (Carman *et al.*, 1999; Willets *et al.*, 2001). When GRK2, GRK3, GRK5, and GRK6 are cotransfected with the TXA2Rα, only GRK2 and GRK3 block agonist-induced inositol phosphate formation (Carman *et al.*, 1999). Likewise, GRK2-K220R, but not GRK4-K216M/K217M, attenuates mGluR1a-stimulated PLCβ activity (Dale *et al.*, 2000; Dhami *et al.*, 2002; Sallese *et al.*, 2000b). Thus, it is possible that the regulation of mGluR1a activity by GRK4 and GRK5 is phosphorylation dependent (Dale *et al.*, 2000; Dhami *et al.*, 2002; Sallese *et al.*, 2000b). However, overexpression of GRK6 leads to mGluR1a phosphorylation in the absence of attenuated receptor signaling (Dale *et al.*, 2000). Together these results suggest that only GRK2 and GRK3 are involved in regulating the phosphorylation-independent desensitization of receptors that are specifically coupled to $G\alpha_q$, $G\alpha_{11}$, and $G\alpha_{14}$.

GRK2 Is Not a Classical RGS Protein

Although the role of GRK2 in regulating $G\alpha_q$ signaling remains to be fully elucidated, two features distinguish GRK2 from classical RGS proteins. RGS proteins accelerate $G\alpha$ GTPase activity by binding to the switch regions and thereby stabilizing the transition state for nucleotide hydrolysis (Berman *et al.*, 1996; Tesmer *et al.*, 1997). Accordingly, RGS proteins bind preferentially to $G\alpha$ transition state mimics (e.g., $G\alpha$-GDP $\cdot$ AlF$_4^-$) rather than binding to activated $G\alpha$ mimics (e.g., $G\alpha$-GTPγS) both *in vivo* and *in vitro* (Berman *et al.*, 1996; Ross and Wilkie, 2000; Srinivasa *et al.*, 1998). The exception to this rule is RGSZ1, which exhibits comparable affinities for $G\alpha_z$-GTPγS and $G\alpha_z$-GDP$\cdot$AlF$_4^-$ (Wang *et al.*, 1997). Like RGSZ1, GRK2 binds $G\alpha_q$-GTPγS nearly as well as it binds to $G\alpha_q$-GDP$\cdot$AlF$_4^-$ (Carman *et al.*, 1999). Thus, GRK2, like RGZ1, exhibits more avid binding to the activated state than typical RGS proteins. Second, GRK2 exhibits only weak activity as a $G\alpha_q$ GTPase-accelerating protein (GAP) (Carman *et al.*, 1999). For $G\alpha_q$, the rate of nucleotide dissociation in the absence of an activated GPCR is slow relative to GTP hydrolysis, making it impossible to assay using single nucleotide turnover approaches (Hepler *et al.*, 1993). However, the GTPase mutant $G\alpha_q$-R183C has been used in a single nucleotide turnover assay to demonstrate RGS2 GAP activity for $G\alpha_q$ (Ingi *et al.*, 1998). When full-length GRK2 is examined in the same assay, virtually no GAP activity for $G\alpha_q$-R183C is observed (Carman *et al.*, 1999). However, the reconstitution of $G\alpha_q$ with m1 muscarinic acetylcholine receptors in phospholipid membranes reveals a weak GAP activity for full-length GRK2 and GRK2(1–178) when compared to RGS4 (Carman *et al.*, 1999). Thus, the ability of GRK2 to function as a $G\alpha_q$ GAP *in vitro* appears to be receptor dependent.

In addition to acting as GAPs, RGS proteins can bind activated $G\alpha$ and preclude $G\alpha$:effector interactions. This mode of regulation has been dubbed "effector antagonism" (Hepler *et al.*, 1997; Yan *et al.*, 1997).

Effector antagonism is revealed using activated Gα subunits with bound nucleotide substrates that are resistant to GTP hydrolysis (e.g., GTPγS). Under these conditions, RGS4 prevents both Gα_q-GTPγS- and Gα_q-GDP·AlF$_4^-$-mediated activation of PLCβ *in vitro* and *in vivo*, respectively. The ability of GRK2(1–178) to prevent Gα_q-GDP·AlF$_4^-$-mediated stimulation of PLCβ1 activity *in vitro* has been compared to RGS4; GST-GRK2(1–178) is a more potent blocker of PLCβ1 than RGS4 in this assay (Carman *et al.*, 1999). Classical RGS proteins demonstrate variability in their capacity to function as effector antagonists. For example, although RGS2, RGS3, RGS4, RGS5, RGS10, and RGS16 are all equally effective Gα_q GAPs, RGS2, and RGS3 are more potent effector antagonists than RGS4, RGS5, and RGS16 (Anger *et al.*, 2004; Scheschonka *et al.*, 2000). Thus, given the apparently weak Gα_q GAP activity for GRK2, it is likely that the GRK2 RH domain may function to antagonize GPCR:effector coupling by antagonizing GPCR:G protein interactions and/or G protein:effector interactions.

Characterization of GRK2 RH Domain Residues Required for G$\alpha_{q/11}$ Binding

The first step required for the identification of GRK2 residues involved in G$\alpha_{q/11}$ interactions is to identify GRK2 domains that are competent in G$\alpha_{q/11}$ binding. Essential to this aim is the identification of GRK2 GST fusion proteins that, when expressed in *Escherichia coli*, are both soluble and stable for long-term storage. Thus, a diverse variety of amino-terminal GRK2 GST fusion proteins were each assessed for expression and stability (Fig. 1). Of these fusion proteins, GST-GRK2(70–178) binds to Gα_q, but is

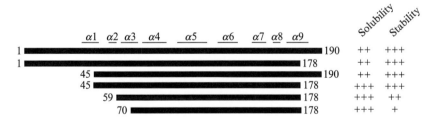

Fig. 1. Identification of optimal GRK2 RH domain fusion protein for mutagenesis studies. GST-GRK2 fusion protein constructs (indicated by bars) were generated and expressed in *E. coli*. The solubility of fusion proteins was assessed by the abundance of the appropriately sized product in the supernatant fraction following high-speed centrifugation and detection by Coomassie blue staining. Proteolysis to generate 27.5-kDa GST during storage at 4° was used as a measure of fusion protein instability. Thin lines at the top indicate the location of α helices 1–9 as determined from the structure of the GRK2:G$\beta\gamma$ complex (Lodowski *et al.*, 2003).

very unstable, as indicated by the proteolysis of the fusion protein following storage at −80°. In contrast, GST-GRK(45–178) exhibits increased solubility and stability when purified from *E. coli* and was used for mutagenesis studies examining $G\alpha_{q/11}$ interactions. However, even this fusion protein is prone to aggregation, as judged by size-exclusion chromatography. The explanation for this behavior is likely that two C-terminal α helices of the RH domain are omitted from this protein construct because they occur after the kinase domain in the primary sequence of GRK2 (Lodowski *et al.*, 2003). The typical assay used to identify $G\alpha_{q/11}$-binding mutants, the GST fusion protein pull-down assay, is outlined.

1. The GST fusion protein encoding residues 45–178 of bovine GRK2 is generated by polymerase chain reaction (PCR) amplification of bovine brain cDNA using primers that incorporated *Bam*HI and *Eco*RI restriction sites. The resulting PCR fragment is subcloned into *Bam*HI and *Eco*RI sites of the glutathione *S*-transferase fusion protein vector, pGEX-2T (Pharmacia), and is used to transform DH5α-competent bacteria.

2. The GST-GRK2(45–178) fusion protein is expressed in bacteria using standard methods (Sterne-Marr *et al.*, 2003). GST-GRK2(45–178) expresses at a concentration of 1 mg fusion protein per 40 ml of bacterial culture, which is sufficient for 120 pull-down assays (8 μg of fusion protein/ 1 ml assay volume). To induce GST-GRK2(45–178) expression, 40-ml cultures in Luria broth containing 5 μg/ml carbenicillin are grown at 37° to an optical density of 0.5 (600 nm), at which point the bacterial culture is treated with 0.5 mM isopropyl β-D-thlogalactoside for 3 h at 25°.

3. Cells are pelleted by centrifugation for 10 min at 7500g and washed with Tris buffer (20 mM Tris, pH 8.0, 150 mM NaCl, 1 mM EDTA) and frozen at −80° until required. Cell pellets are thawed on ice, resuspended, and incubated for 15 min in 0.4 ml of Tris buffer containing 100 μg/ml lysozyme, after which 10 mM 2-mercaptoethanol, 100 μM phenylmethylsulfonyl fluoride (PMSF; freshly prepared in isopropanol), leupeptin (1 μg/ml), benzamidine (20 μg/ml), 25 mM EDTA, and 1.5% N-laurylsarcosine is added to the lysate. To reduce viscosity, lysates are sonicated for four to six rounds of 10-s bursts separated by 15-s rest periods on ice.

4. Insoluble protein is then removed by centrifugation at 14,000g for 10 min at 4°.

5. The GST-GRK2(45–178) fusion protein is bound to glutathione-agarose beads at a ratio of 3 ml packed beads/mg lysate protein by mixing for 1 h at 4°. The beads are washed twice in Tris buffer, followed by two 5-min washes in Tris buffer containing 25% glycerol and 10 mM 2-mercaptoethanol to allow equilibration of the glycerol-containing

solution. Glutathione-agarose fusion protein beads can be stored at $-20°$ as a 50% slurry of beads in Tris buffer containing 25% glycerol and 10 mM 2-mercaptoethanol.

6. To assess Gα_q binding, equal amounts of normal and mutated GST-GRK2 fusion proteins must be present in each assay. To obtain accurate measurement of the amount of GRK2 associated with glutathione-agarose beads, fusion proteins are eluted from the beads in elution buffer (50 mM Tris–HCl, 10 mM glutathione, 10 mM 2-mercaptoethanol pH 8.0) at room temperature. Bradford protein assays are then performed on triplicate samples from each elution.

7. Bovine brain extract is used as a source of G$\alpha_{q/11}$ protein and is prepared as described by Carman et al. (1999). Fifty grams of brain tissue will yield $\sim$20 ml of brain extract at a protein concentration of 20 mg/ml. This is sufficient for more than 1000 pull-down binding assays under the standard conditions described later.

8. For GST-GRK2 pull-down assays, reactions are performed in 1-ml volumes containing 8 μg/ml of GST-GRK2 fusion protein, 200 μg/ml bovine brain protein extract in binding buffer (20 mM Tris–HCl, pH 8.0, 2 mM MgSO$_4$, 6 mM 2-mercaptoethanol, 100 mM NaCl, 0.05% C$_{12}$E$_{10}$, 5% glycerol, and 100 microMolar GDP) are incubated overnight (4-h incubations are sufficient) at 4° on an inversion wheel in the absence or presence of 30 μM aluminum chloride and 5 mM sodium fluoride (AlF$_4^-$) (Carman et al., 1999; Sterne-Marr et al., 2003).

9. Following the incubation, the glutathione-agarose beads are washed four times with binding buffer (in the absence or presence of AlF$_4^-$, as appropriate) and bound protein eluted in 50 μl of sodium dodecyl sulfate (SDS) sample buffer.

10. Twenty-microliter aliquots from each sample are subjected to SDS–polyacrylamide gel electrophoresis (SDS–PAGE) (12.5% gel). The proteins are transferred onto nitrocellulose membranes by semidry electroblotting. The membranes are blocked with 10% milk in wash buffer (TBS-T, 150 mM NaCl, 10 mM Tris–HCl, pH 7.0, 0.05% Tween 20) and probed with an anti-Gα_q-specific polyclonal antibody from Upstate Cell Signaling Solutions diluted 1:1000 in wash buffer containing 3% skim milk. The membranes are rinsed three times for 10 min with wash buffer and then incubated with secondary horseradish peroxidase-conjugated donkey antirabbit IgG (Amersham) diluted 1:2500 in wash buffer containing 3% skimmed milk. Membranes are rinsed twice with wash buffer and then are incubated with ECL western blotting detection reagents (Amersham) according to the manufacturer's specifications. SDS–PAGE is also used to verify that equal amounts of GST fusion protein are used in each assay, as well as to assess the extent of GST fusion protein degradation.

Note. Preliminary experiments suggest that the affinity GRK2 for $G\alpha_q$ lies in the low nanomolar range and thus the fusion protein is in excess in these reactions. To assess GRK2 mutants impaired greatly in $G\alpha_q$ binding, the concentration of fusion protein and bovine brain extract can be increased to 20 and 500 μg/ml, respectively.

Identification of Specific GRK2 RH Domain Residues Required for $G\alpha_{q/11}$ Binding

Ideally, it would have been useful to know the structure of the GRK2 RH domain prior to initiating mutagenesis studies. In the absence of a three-dimensional structure of GRK2, the GRK2 RH domain first identified *in silico* (Siderovski *et al.*, 1996) was aligned with RGS protein sequences of known structure (Sterne-Marr *et al.*, 2003) using a structure-based sequence alignment protocol implemented by Clustal W (Thompson *et al.*, 1994). At that time, the three-dimensional structures for RGS4, GAIP, RGS9, axin, PDZRhoGEF, and p115RhoGEF had been solved (Chen *et al.*, 2001; de Alba *et al.*, 1999; Longenecker *et al.*, 2001; Spink *et al.*, 2000; Tesmer *et al.*, 1997). The template for the alignment was the structure of the axin RH domain (Sterne-Marr *et al.*, 2003). Each of the RGS domains encoded by these proteins form all-helical protein domains composed of 9 α helices, with the exception of the PDZRhoGEF and p115RhoGEF RGS domains, which contain 11 α helices. The most conserved structural feature of this protein family was a four helix bundle formed by α helices 4–7, termed the bundle subdomain (Tesmer *et al.*, 1997). The remaining helices of the RH domain constitute the terminal subdomain. This model appeared to provide excellent predictions of the helices of the GRK2 RH domain, as well as most of the loops (Fig. 2A), particularly in the bundle subdomain. However, the structure of the $\alpha5:\alpha6$ loop, which was known to be a critical determinant for binding $G\alpha$ subunits in RGS proteins (Tesmer *et al.*, 1997), could not be modeled reliably. Because this loop had the most apparent similarity to the corresponding loop of axin, its structure was modeled based on the axin RH domain structure. However, this structural assignment proved to be incorrect (Lodowski *et al.*, 2003). One of the misleading features of this loop was L118, a hydrophobic residue whose side chain could be buried, if the $\alpha5:\alpha6$ loop were modeled as in axin. Instead, this residue protrudes from the surface of the GRK2 RH domain and is now believed to be involved in interactions with $G\alpha_q$ (Day *et al.*, unpublished results). The GRK2 RH domain model was built manually using the program O (Jones *et al.*, 1991) by choosing the most reasonable rotamers for nonidentical residues (Kleywegt and Jones, 1998). In the terminal subdomain (α helices 1, 2, 3,

8, and 9), which had higher sequence identity with other RH domains of known structure (usually GAIP), the GRK2 model was adjusted to reflect those models. The loop between the $\alpha 5$ and the $\alpha 6$ helices of the GRK2 RH domain had no obvious sequence homology to RH domains of known structure and was not included in the model presented in Fig. 2. The overall model was refined to idealize its stereochemistry.

Initially, R106 and D110 in the $\alpha 5$ helix of the GRK2 RH domain were identified as amino acid residues critical for GRK2:$G\alpha_{q/11}$ binding (Sterne-Marr et al., 2003). The mutation of these residues resulted in a loss of GDP·AlF$_4^-$-dependent binding of GRK2 to $G\alpha_{q/11}$ (Table I). Moreover, the plasma membrane recruitment of GRK2 by $G\alpha_q$-Q209L is prevented by the mutation of D110 to an alanine. Unlike what is observed for wild-type GRK2 and GRK3, GRK2 mutants defective in $G\alpha_{q/11}$ binding do not attenuate $G\alpha_q$-R183C-mediated inositol phosphate formation in COS-1 cells. The axin-based GRK2 RH homology model was subsequently used to predict amino acid residues that might be close to R106 and D110 in three-dimensional space and, therefore, might contribute to a $G\alpha_{q/11}$-binding surface. Five of the seven mutants predicted from the model exhibited perturbations in $G\alpha_{q/11}$ binding, which suggested that the model represented a reasonable approximation of the $G\alpha_{q/11}$-binding surface. Mutation of amino acid residues F109, M114, and E116 to alanine residues had the greatest effect on $G\alpha_{q/11}$ binding in the presence of GDP·AlF$_4^-$, the mutation of either L117 or V137 to alanine had a lesser effect on $G\alpha_{q/11}$ binding, and the mutation of Q133 and K139 had no reproducible effect on $G\alpha_{q/11}$ binding (Table I). Interestingly, a GRK2-K115A mutant exhibited only a modest decrease in GDP·AlF$_4^-$-dependent $G\alpha_{q/11}$ binding, but exhibited a dramatic increase in binding to GDP-bound $G\alpha_{q/11}$. Mutations within the $\alpha 4$ helix of the GRK2 RH domain do not effect GDP·AlF$_4^-$-dependent binding of GRK2 to $G\alpha_{q/11}$. Furthermore, mutations in the predicted $\alpha 3$:$\alpha 4$ and $\alpha 7$:$\alpha 8$ loops of the GRK2 RH domain (E78K, D160K, and K164A) had no effect on $G\alpha_{q/11}$ binding.

Determination of the GRK2:$G\beta\gamma$ crystal structure yielded unexpected results (Lodowski et al., 2003). First, GRK2 is a modular protein in which the three functional domains occupy vertices of a roughly equilateral triangle. The RH domain is located such that it is capable of receiving signals from $G\alpha_q$, $G\beta\gamma$, and the receptor, as well as transmitting those signals to the kinase domain. Second, the GRK2 RH domain is composed of 11 α helices. The first 9 helices are derived from amino acid residues 29–185 of GRK2 and form a two-lobed structure that resembles the RGS domains of RGS proteins, but the relative orientation of the bundle and terminal subdomains are different from other RGS domains (Fig. 2A, B). The $\alpha 10$ and $\alpha 11$ helices are derived from amino acid residues 513–547

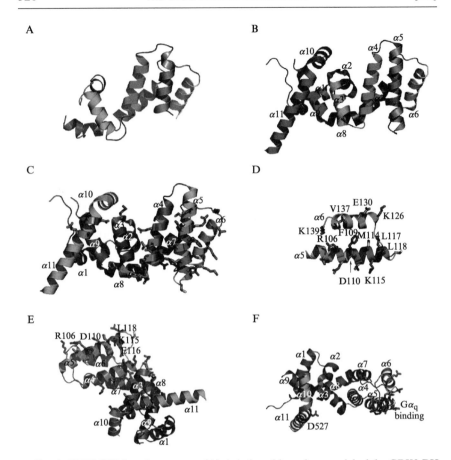

FIG. 2. GRK2 RH domain structure. (A) Axin-based homology model of the GRK2 RH domain. The RGS box of GRK2 (amino acid residues 51–171) was aligned with RH domains of known structure, and a homology model was built manually as described in the text. Note that two regions, the α1 helix and the α5:α6 loop, bear little sequence homology to analogous regions of other RH domains. In our original model (Sterne-Marr *et al.*, 2003), these regions were modeled after the axin RH domain but are omitted here. Helices are indicated by color: α2 is blue, α3 is turquoise, α4 is green, α5 is pink, α6 is yellow, α7 is orange, α8 is brown, and α9 is red. (B) The GRK2 RH domain determined from the GRK2:Gβγ crystal structure (Lodowski *et al.*, 2003). The helix color scheme is the same as in A, with the addition of α1 in gray, α10 in gold, and α11 in magenta. The axin-based homology model was a good predictor of the bundle subdomain (α helices 4–7), but a poor predictor of the terminal domain (α helices 1–3 and 8–11). The terminal subdomain was a region of variability among known RH domain structures, with the RhoGEF (Chen *et al.*, 2001; Longenecker *et al.*, 2001) and GRK2 RH domains containing extra α helices in this subdomain. In the GRK2 primary sequence, α helices 10 and 11 are separated from α helices 1–9 by the kinase domain. Differences between the axin-based homology model and the crystal structure are outlined in the text and can be visualized here. (C) Traditional view of the GRK2 RH domain indicating

located between the kinase and PH domains of GRK2 in the linear sequence. The length of α helices 1, 5, 8, and 9 is longer, whereas α helix 7 is shorter than predicted from both the structure of other RGS proteins and the axin-based GRK2 RGS homology model (Fig. 2A–C). The orientation of the α9 helix is also distinct, as it is packed against and is perpendicular to the α1 helix (Fig. 2C). Moreover, in the GRK2:G$\beta\gamma$ structure, the α10 helix of the RH domain contacts the kinase domain small lobe and the GRK2 RH α4–α5 loop makes contacts with the kinase large lobe. The RH domain also forms a hydrophobic interface with the GRK2 PH domain via the α1 helix. Finally, the α11 helix of the GRK2 RH domain is packed against α helices 1 and 9 on one face, but projects toward the membrane on the opposite face, suggesting that this region of the RH domain may contribute to membrane and/or receptor binding.

Initially, eight amino acid residues—R106, F109, D110, M114, K115, E116, L117, and V137—were predicted by mutagenesis to define the G$\alpha_{q/11}$-binding site for GRK2, and the majority of these residues are localized to the α5 helix in the axin-based model for the structure of the GRK2 RH domain. However, the GRK2:G$\beta\gamma$ crystal structure indicates that the α5 helix is longer than predicted in the axin-based homology model and suggests that additional residues localized to the C terminus of the α5 helix may also contribute to G$\alpha_{q/11}$ binding (Day et al., unpublished results). Thus, based on the predicted interface between GRK2 and Gα_q, additional α5 helix residues, such as L118, are predicted to project toward the solvent and to make critical contacts with the cleft between the helical and Ras domains of Gα_q (Lodowski et al., 2003;

residues that were mutagenized to test binding to either G$\alpha_{q/11}$ or mGluR1a. Mutated residues that perturbed G$\alpha_{q/11}$ binding are shown with magenta backbone, whereas mutated residues, which had no effect on G$\alpha_{q/11}$ binding, have green backbones. (D) G$\alpha_{q/11}$-binding site on GRK2. In the "traditional" RH domain view, the G$\alpha_{q/11}$-binding site faces the side of the structure. From the traditional view, the structure was rotated around the y axis $\sim$90° to bring the α5 and α6 helices proximal to the viewer and was then rotated $\sim$90° in the z axis to bring the α5 and α6 helices into a horizontal orientation. Color scheme is the same as in C, except that mutated residues that exhibit dramatic and mild defects in G$\alpha_{q/11}$ binding are shown in magenta and pink, respectively. (E) Membrane-proximal view of GRK2 RH domain showing location of the G$\alpha_{q/11}$-binding site with respect to the membrane. Because the structure of GRK2 was determined in complex with G$\beta\gamma$ subunits, the relative location of the plasma membrane can be predicted with a high degree of confidence. In this view, the GRK2 RH domain α11 helix is juxtaposed to the plasma membrane and the G$\alpha_{q/11}$-binding site points toward the cytoplasm. Color scheme is the same as in C, except α11 helix residues that were mutated to test binding to mGluR1a have an orange backbone if no effect was detected (T528A, A531E, E532A) or a blue backbone (D527A) if the mutation disrupted coimmunoprecipitation of GRK2 with mGluR1 (Dhami et al., 2004). (F) The structure in E was rotated so as to visualize the D527 side chain. Although solvent exposed, D527 points toward the α3 helix. Color scheme is the same as in C and E. (See color insert.)

TABLE I
SUMMARY OF GRK2 RH DOMAIN MUTATIONS AND THEIR EFFECTS ON $G\alpha_q$ BINDING, $G\alpha_q$-Q209L-MEDIATED MEMBRANE LOCALIZATION, AND MODULATION OF $G\alpha_q$-R183C ACTIVATION OF PHOSPHOLIPASE C (PLC)

GRK2 Mutant	Location (L = loop)	$G\alpha_q$			
		GDP·AlF$_4^-$ (pull-down assay)[a]	GTP (coimmuno-precipitation)[b]	Q209L (membrane localization)[c]	Inhibition of R183C-stimulated PLC activation[d]
WT		+++	+++	+	++
K62A	α2	+++			
H75A/L76A	α3/α4 L	+++			
E77A	α4	+++			
E78K	α4	+++			
K80A	α4	+++			
V83A	α4	+++			
E84A	α4	+++			
E87A	α4	+++			
K90A	α4	+++			
V103A/C104A	α5	+++			
R106A	α5	+/−	+		−
E107A/I108A	α5	+/−			
E107A	α5	+++			
F109I	α5	+/−			
D110A	α5	+/−	+	−	−
R106A/D110A	α5	−			
T111A	α5	+++			
M114A	α5	++	+		−
K115A	α5	+++		++	++
E116A	α5	+/−			
L117A	α5	+			
L118A	α5	+/−			
C120A	α5	+++			
K126A	α6	+++			
E130A	α6	++			
Q133A	α6	+++			
V137A	α6	++			+
K139A	α6	+++			
N156A	α7	+++			
D160K	α8	+++			
K164A	α8	+++			
D527A	α11		+++		
T528A	α11		+++		
A531E	α11		+++		
E532A	α11		+++		

[a]GST-GRK2(45–178) pull-down assays using bovine brain extract as the source of $G\alpha_{q/11}$ (Sterne-Marr et al., 2003).

Day *et al.*, unpublished results). Consistent with this prediction, the mutation of L118, but not T111 and C120, to alanine residues impairs $G\alpha_{q/11}$ binding (Table I). The GRK2:G$\beta\gamma$ crystal structure also predicts that the α6 helix also makes contacts with $G\alpha_{q/11}$, but the mutation of amino acid residues E130 and V137 affects $G\alpha_{q/11}$ binding only modestly and the mutation of amino acid residues K126 and K139 to alanine residues has no effect on $G\alpha_{q/11}$ binding (Day *et al.*, unpublished results). This discrepancy is likely related to subtle differences in the helical domains between $G\alpha_q$ and the $G\alpha_i$ structure upon which it is modeled (Lodowski *et al.*, 2003).

Mutagenesis suggests that implicated residues in the α5 (R106, F109, D110, M114, K115, E116, L117, and L118) and the α6 (E130 and V137) helices of the GRK2 RH domain form a continuous surface for $G\alpha_q$ binding (Fig. 2D, E). Amino acid residues R106, D110, M114, and L118 appear to provide the most important contact sites for $G\alpha_q$ binding. However, the orientation of other amino acid residues, such as K115, whose side chain is pointed toward helix α4, may modulate $G\alpha_q$ interactions, as mutation of these residues allows aberrant binding to GDP-bound $G\alpha_q$ (Sterne-Marr *et al.*, 2003). The $G\alpha_{q/11}$ binding surface of GRK2 is distinct from the interface used by other RGS proteins to bind $G\alpha$ and by axin to bind a peptide derived from APC (Spink *et al.*, 2000). Therefore, the $G\alpha_{q/11}$-binding site for GRK2 represents the third novel surface that RH domains have exploited to interact with other proteins.

Site-Directed Mutagenesis Tips

For GRK2, amino acid residues to be mutated are limited to those residues localized within the shortest stretch of the GRK2 RH domain that might be appropriately exposed, as predicted from the structural/homology model, to bind $G\alpha_{q/11}$. Furthermore, based on the alignment, only those residues that are predicted to be solvent exposed have been mutated to reduce the potential that the mutations might alter the GRK2 tertiary structure rather than $G\alpha_{q/11}$ interactions. In the early rounds of mutagenesis,

---

[b] Amino-terminally myc-tagged GRK2 (WT and mutants) were transfected into HEK 293 cells. GRK2 in the cell lysates was immunoprecipitated with the anti-myc monoclonal antibody, and immunoprecipitates were assayed for the presence of $G\alpha_{q/11}$ by immunoblotting with $G\alpha_{q/11}$-specific polyclonal antibodies (Dhami *et al.*, 2004).

[c] R106A, D110A, M114A, K115A, and V137A mutations were introduced into GRK2(45–178)-GFP and WT or mutant constructs were cotransfected with $G\alpha_{q/11}$-Q209L into HEK293 cells. GFP localization was visualized by confocal microscopy (Sterne-Marr *et al.*, 2003).

[d] WT or mutant GRK2 was cotransfected into COS-1 cells with $G\alpha_{q/11}$-R183C, and inositol phosphates were determined (Sterne-Marr *et al.*, 2003).

two to three adjacent amino acid residues are mutated to map sites of interaction. Subsequently, individual point mutations are created by site-directed mutagenesis.

1. Site-directed mutagenesis is performed using the Stratagene Quik-Change mutagenesis kit according to the manufacturer's specifications. The advantage of the kit is that it is simple to use, takes only 3 days to perform, and requires limited benchwork. The drawback of the Stratagene kit is that it is relatively expensive and occasionally introduces mutations at unspecified sites in the cDNA sequence. Thus, the sequence integrity of the entire cDNA clone must be confirmed by DNA sequencing.

2. In brief, complementary sense and antisense oligonucleotide primers are designed such that the codon nucleotides for the amino acid residue to be mutated are flanked on each side with a sufficient number of perfectly matching nucleotides to generate an optimal melting temperature of 78°. The equation for determining the melting temperature is provided by the manufacturer. Mutagenesis is most successful if, when calculating %GC content, only the annealing nonmutagenic region is considered.

3. A high-yield, proofreading thermostable DNA polymerase (e.g., Pfu-Turbo) is used to synthesize the entire template plasmid. At the extension temperature used (68°), the polymerase is unable to strand displace and therefore does not perform a "rolling" amplification of the input circular plasmid. Thus, although the procedure uses many rounds of denaturation, annealing, and extension steps, only the input template can be used as a template in the reaction and results in a linear and not exponential increase in the product.

4. The original template is removed from the reaction by restriction enzyme digestion such that nonmutagenized cDNA does not contribute to the transformation of competent *E. coli*. This is achieved by digestion of the plasmid cDNA with the restriction enzyme *Dpn*I. *Dpn*I only recognizes and cleaves methylated DNA. Thus, the cDNA template purified from *dam*$^+$ *E. coli* strains will be methylated and serve as a substrate for *Dpn*I, whereas the *in vitro*-synthesized cDNA will be unmethylated and resistant to *Dpn*I cleavage and thus remain competent to transform *E. coli*.

5. Minor modifications can be made to the QuikChange protocols to increase the cost effectiveness of using the Stratagene kit. Specifically, the protocols can be performed in reduced volumes without sacrificing successful mutagenesis. Stratagene suggests 50 μl of "supercompetent" *E. coli* and 1 μl of mutagenized cDNA product be utilized per transformation. The volume of cells can be cut to 25 μl and the amount of mutagenized cDNA product reduced to 0.5 μl. Thus, seven transformations instead of three transformations can be performed from 200 μl of supercompetent

cells. Furthermore, the volume of the mutagenesis reaction can be reduced from 50 to 25 μl. Under these modified conditions, only 8 μl of the reaction mixture is used to analyze the mutagenesis product by agarose gel electrophoresis, and the remaining 17 μl is treated with DpnI.

Mechanisms Underlying mGluR Desensitization

Metabotropic glutamate receptors (mGluRs) are members of the G-protein-coupled receptor superfamily and mediate slow synaptic glutamate responses by coupling to a range of second messenger cascades and ion channels via heterotrimeric G proteins (Conn and Pin, 1997; Dale et al., 2002; Nakanishi and Masu, 1994). This property allows mGluRs to translate relatively short neuronal activation into long-lasting changes in synaptic activity. The mGluR family includes eight distinct receptor subtypes that are subdivided into three groups based on sequence similarity and G-protein-coupling specificity: mGluR1 and mGluR5 (group I); mGluR2 and mGluR3 (group II); and mGluR4, 6, 7, and 8 (group III). Although mGluRs retain the seven transmembrane spanning helices characteristic of all GPCRs, these receptors bear absolutely no sequence homology to prototypic GPCRs, such as the β_2AR. Moreover, mGluRs are coupled preferentially to heterotrimeric G proteins via their second intracellular loop domain rather than the third intracellular loop domain utilized by classical GPCRs, such as the β_2AR (Gomeza et al., 1996). Group I mGluRs are coupled via $G\alpha_{q/11}$ to the activation of PLCβ, resulting in increased intracellular inositol-1,4,5-triphosphate (IP$_3$) and diacylglycerol concentrations, the release of calcium from intracellular stores, and the activation of PKC.

Receptor desensitization may be particularly relevant to the regulation of group I mGluR activity, as the overstimulation of these receptors results in excitotoxicity, leading to neuronal cell death associated with acute brain ischemia and neurotrauma (Alessandri et al., 1998; Nicoletti et al., 1996). The mechanisms underlying group 1 mGluR desensitization have been the subject of intensive investigation and have revealed that mGluR desensitization is mediated by both second messenger-dependent protein kinases and GRKs (Ciruela et al., 1999; Dale et al., 2000; Desai et al., 1996; Francesconi and Duvoisin, 2000; Gereau and Heinemann, 1998; Herrero et al., 1994; Sallese et al., 2000b). PKC phosphorylates several intracellular mGluR1 and mGluR5 serine and threonine residues, resulting in receptor desensitization (Francesconi and Duvoisin, 2000; Gereau and Heinemann, 1998). GRK2, GRK4, and GRK5 have been shown to contribute to the desensitization of mGluR1a in heterologous cell systems (Dale et al., 2000; Sallese et al., 2000b), and GRK4 regulates mGluR1 desensitization in cerebellar Purkinje cells (Sallese et al., 2000b).

Phosphorylation-Independent Attenuation of mGluR Signaling

Although mGluR signaling is attenuated following GRK2, GRK4, and GRK5 overexpression (Dale et al., 2000; Sallese et al., 2000b), unlike what is accepted for the majority of GPCRs, mGluR1 desensitization does not require GRK-mediated receptor phosphorylation and/or β-arrestin binding (Dale et al., 2001; Dhami et al., 2002, 2004). Evidence that mGluR1a desensitization by GRK2 is not dependent on GRK2-mediated receptor phosphorylation comes from studies examining the phosphorylation and desensitization of a truncated mGluR1a (1–886) mutant and the mGluR1 alternative splice variant mGluR1b (Dhami et al., 2002). Serial truncation of the mGluR1a carboxyl-terminal tail results in a loss of GRK2-mediated phosphorylation without a loss of GRK2 expression-dependent antagonism of both basal and agonist-stimulated PLCβ activity. Similarly, mGluR1b, a mGluR1 alternative splice variant lacking an extended carboxyl-terminal tail, is not phosphorylated by GRK2. Nonetheless, GRK2 overexpression effectively uncouples both basal and agonist-stimulated mGluR1b signaling. Maximal attenuation of mGluR1a and mGluR1b signaling requires only the expression of GRK2 (45–185) in HEK 293 cells. GRK2 binds to both mGluR1a and mGluR1b and is effectively coimmunoprecipitated with both receptors in the absence of chemical cross-linking (Dale et al., 2000; Dhami et al., 2002).

Dhami et al. (2004) examined whether GRK2 mutants impaired in $G\alpha_{q/11}$ binding are able to antagonize both basal and agonist-stimulated mGluR1a signaling. All of the GRK2 mutants impaired in $G\alpha_{q/11}$ binding interact normally with mGluR1a. However, we find that GRK2-R106A, GRK2-D110A, and GRK2-M114A exhibit a diminished capacity to reduce the V_{max} for agonist (quisqualate)-stimulated inositol phosphate (IP) formation when compared with wild-type GRK2. This is especially true at GRK2 expression levels that are physiologically relevant to mGluR1a signaling in vivo (Dhami et al., 2004). $G\alpha_{q/11}$ is also immunoprecipitated as a complex with mGluR1a in the absence of agonist; either agonist treatment or GRK2 overexpression promotes dissociation of the receptor/$G\alpha_{q/11}$ complex. However, receptor/$G\alpha_{q/11}$ complex formation is not disrupted by the expression of GRK2 mutants that do not bind $G\alpha_{q/11}$. Thus, intact GRK2:$G\alpha_{q/11}$ interactions appear to be required for GRK2 phosphorylation-independent attenuation of mGluR1a signaling (Dhami et al., 2004).

It has been suggested that GRK2 RH domain-dependent attenuation of mGluR1a signaling may occur as the consequence of the nonspecific sequestration of $G\alpha_{q/11}$ following GRK2 overexpression (Pao and Benovic, 2002). The inability of GRK2 mutants defective in $G\alpha_{q/11}$ binding to attenuate mGluR1a signaling does not rule out this possibility. However, the GRK2:G$\beta\gamma$ crystal structure suggests that the α11 helix of the second discontinuous segment of the GRK2 RH domain is oriented appropriately to make contact with the intracellular face of mGluR1a (Lodowski et al., 2003). Therefore, we examined whether this region of the GRK2 RH domain was involved in mGluR1 binding (Dhami et al., 2004). Mutation analysis of amino acid residues within the α11 helix of the GRK2 RH domain led to the identification of a GRK2-D527A mutant (Fig. 2F) that is (1) defective in binding mGluR1a, (2) unable to attenuate mGluR1a signaling, and (3) able to interact normally with G$\alpha_{q/11}$ in a GDP·AlF$_4^-$-dependent manner (Dhami et al., 2004). Taken together, these observations suggest that the phosphorylation-independent desensitization of mGluR1a is not the consequence of the nonspecific sequestration of G$\alpha_{q/11}$ following GRK2 overexpression. Rather, data provide a novel alternative mechanism to the widely accepted paradigm for GRK2 phosphorylation-dependent GPCR desensitization, whereby concomitant binding of GRK2 to both receptor and G protein regulates receptor:G-protein interactions in the absence of phosphorylation.

The remaining sections present experimental assays that are utilized to examine the contribution of GRKs to the phosphorylation-independent attenuation of GPCR responses. Specifically, we outline the methods used to examine the phosphorylation-independent regulation of mGluR1 signaling by GRK2.

Assaying Phosphorylation-Independent Receptor Desensitization

Two assays are performed to assess whether GRK2 contributes to the phosphorylation-independent desensitization of GPCRs: receptor-stimulated inositol phosphate formation and whole cell receptor phosphorylation. Because the sites for GRK2-mediated GPCR phosphorylation are often localized within the carboxyl-terminal tails of GPCRs (Ferguson, 2001), it is desirable to also test whether truncation of the carboxyl-terminal results in a loss of receptor phosphorylation. Receptor-mediated inositol phosphate formation in response to increasing concentrations of receptor agonist is tested in both the absence and the presence of both wild-type GRK2 and catalytically inactive GRK2-K220R mutant proteins. GRK2-mediated phosphorylation-independent receptor desensitization may be indicated if the expression of both GRK2 and GRK2-K220R results in attenuated inositol phosphate formation in response to the agonist in the absence of receptor phosphorylation. Additional experiments assaying the uncoupling of receptor responses following the overexpression of GRK2 mutants defective in $G\alpha_{q/11}$ binding and potentially impaired in receptor binding are also required.

Inositol Phosphate Formation Assay

To assess the effect of GRK2 on the phosphorylation-independent attenuation of receptor signaling, it is necessary to examine whether the overexpression of both wild-type GRK2 and the catalytically inactive GRK2-K220R mutants attenuate receptor-stimulated inositol phosphate formation in response to increasing concentrations of receptor agonist. The following section provides a simple protocol for the determination of inositol phosphate formation.

1. On day 1, HEK 293 cells seeded at 2×10^6 cells per 100 mm dish in 10 ml of Dulbecco's modified Eagle's medium (DMEM) containing 10% fetal bovine serum (FBS) and 100 μg/ml gentamicin are transfected in duplicate by calcium phosphate precipitation with receptor (10 μg in pcDNA3.1 expression vector) with and without either GRK2 or GRK2-K220R (5 μg in a pcDNA3.1 expression vector). Two dishes are transfected for each condition. For a detailed calcium phosphate transient transfection procedure, see Ferguson and Caron (2004).

2. Fifteen to 18 h after transfection, the culture medium is exchanged for 10 ml of fresh DMEM containing 10% FBS and 100 μg/ml gentamicin. Six to 8 h after changing the medium, the 100-mm cell culture dishes are washed twice with 5 ml of phosphate-buffered saline (PBS), treated with 2 ml of 0.05% trypsin (Gibco-Life Technologies) for 1 min, and then resuspended gently in 10 ml of DMEM containing 10% FBS and 100 μg/ml gentamicin. Cells transfected in duplicate are combined, and 1-ml aliquots of the cell suspensions are reseeded into duplicate wells of 24-well tissue culture dishes for each agonist concentration to be assayed. The 24-well tissue culture dishes are pretreated for 5 min with 1 ml of 10% collagen solution (Sigma) and are then washed twice with 2 ml of PBS. An additional 2.5-ml aliquot of the cell suspension for each transfection is seeded into another 6-well Falcon dish for either radioligand- or flow cytometric-based determination of cell surface receptor expression levels.

3. On the third day, the cell medium is aspirated and the cells are equilibrated overnight (16 h) in 0.5 ml of inositol-free DMEM containing 1 μCi/ml of myo-[^{3}H]inositol.

4. On the fourth day, the cells are washed three times with prewarmed ($37°$) Hank's buffered salt solution (HBSS; 116 mM NaCl, 20 mM HEPES, 11 mM glucose, 5 mM NaHCO$_3$, 4.7 mM KCl, 2.5 mM CaCl$_2$, 1.3 mM MgSO$_4$, 1.2 mM KH$_2$PO$_4$, pH 7.4) and are incubated 1 h in 0.5 ml of HBSS. Subsequently, the cells are washed once in HBSS

containing 10 mM LiCl and are incubated for 10 min in 0.5 ml of HBSS containing 10 mM LiCl.

5. The medium is aspirated, and 0.5 ml of HBSS (37°) is added to the cells for 30 min either lacking or containing increasing concentrations of receptor agonist. The reaction is stopped by the addition of 0.5 ml of 0.8 M perchloric acid. The samples are left on ice undisturbed for 30 min.

6. Eight hundred microliters of cell lysate is transfered into a test tube containing 400 μl of neutralizing buffer (0.72 M KOH, 0.6 M KHCO$_3$) and then the samples are stored, preferably overnight (at least 2 h), on ice to facilitate precipitation of the insoluble potassium perchlorate.

7. On the fifth day, the columns are prepared for affinity purification of inositol phosphates. The resin (Dowex AG1-X8 formate form resin 200–400 mesh) is rinsed with a large excess of distilled H$_2$O and is stirred gently with a pipette for 10–20 min. The resin is allowed to settle, and water is aspirated off until approximately a 50:50 solution of resin and water remains (20 g resin/50 columns). The resin is stirred gently with a pipette and 2 ml of resin slurry is added to each column. *Note.* Columns should be prepared just prior to use and *do not* reuse.

8. Samples are centrifuged for 15 min at 3000 rpm. A 50-μl aliquot ("total [3H]inositol") is removed from each sample and 5 ml of scintillation fluid is added to this aliquot.

9. Nine hundred microliters of the remaining sample is loaded onto a column and washed immediately with 5 ml of distilled H$_2$O. Subsequently, the columns are washed with 10 ml of distilled H$_2$O (2 × 5 ml) to remove free [3H]inositol. The free [3H]inositol must be removed completely because it is present in large excess relative to the levels of [3H]inositol phosphates.

10. The columns are washed with 10 ml (2 × 5 ml) of fresh 60 mM ammonium formate. Total [3H]inositol phosphates are eluted into scintillation vials with 4 ml of 0.1 M formic acid in 1 M ammonium formate and 15 ml of scintillation cocktail is added to each vial. [3H]Inositol phosphate formation is determined using a liquid scintillation counter. The percentage conversion of [3H]inositol to [3H]inositol phosphates is calculated as [((total DPMs eluted off the column) × (1.2/0.9) × 100)/((total [3H]inositol) × (1.2/0.05))].

Whole Cell Receptor Phosphorylation Protocol

To determine the role of GRK2 in the phosphorylation-independent attenuation of GPCR signaling, we examined the ability of GRK2 over-expression to increase the whole cell phosphorylation of mGluR1a, a carboxyl-terminal truncated mGluR1a (1–866) mutant, and the alternative mGluR1a splice variant mGluR1b that lacks an extended carboxyl-terminal tail (Dhami *et al.*, 2002). To facilitate immunoprecipitation, the mGluR1 receptor constructs are FLAG epitope tagged at the amino terminus (Dale *et al.*, 2000). Alternatively, mGluR1- and mGluR5-specific antibodies are available from Upstate Cell Signaling Solutions. A detailed protocol for whole cell GPCR phosphorylation is as follows.

1. On day 1, HEK 293 cells seeded at 2×10^6 cells per 100-mm dish in 10 ml of DMEM containing 10% FBS and 100 μg/ml gentamicin are transfected in duplicate by calcium phosphate precipitation (Ferguson and Caron, 2004) with receptor (10 μg) and with or without GRK2 (5 μg). Two dishes are transfected for each condition.

2. Fifteen to 18 h after transfection, the culture medium is exchanged for 10 ml of fresh DMEM containing 10% FBS and 100 μg/ml gentamicin. Six to 8 h after changing the medium, the 100-mm cell culture dishes are washed twice with 5 ml of PBS, treated with 2 ml of 0.05% trypsin for 1 min, and then resuspended gently in 10 ml of DMEM containing 10% FBS and 100 μg/ml gentamicin. Duplicate transfections are combined, and 2.5-ml aliquots of the cell suspension are reseeded into four wells of six-well tissue-culture dishes that are pretreated for 5 min with 1 ml of 10% collagen solution and then washed twice with 2 ml of PBS. An additional 2.5-ml aliquot of the cell suspension for each transfection condition is seeded in another six-well tissue culture dish for either radioligand- or flow cytometric-based determination of receptor expression.

3. On day 3, the HEK293 cells should be about 95% confluent; to avoid cells lifting off from the dishes, the following multiple wash steps are required for the phosphorylation assay. HEK 293 cells seeded in six-well tissue culture dishes are washed twice with phosphate-and serum-free DMEM and are then incubated for 1 h in 1 ml of the same medium at 37°. Prepare [^{32}P]orthophosphate labeling medium in 750 μl per well of phosphate- and serum-free DMEM (100 μCi/ml).

4. The medium is aspirated from the six-well dishes and 750 μl of [^{32}P]orthophosphate labeling medium is added to each well and incubated at 37° for 1–3 h. We aspirate the medium into a bottle trap to prevent contamination and for easy disposal. Duplicate wells are treated with 750 μl of phosphate- and serum-free DMEM either lacking or containing 2X concentration of receptor agonist and are incubated for 10 min at 37° (agonist is generally used at saturating concentrations). Cells are washed four times with 1 ml of PBS on ice and 400 μl of RIPA buffer (150 mM NaCl, 50 mM Tris, pH 7.4, 5 mM EDTA, 10 mM NaF, 10 mM Na$_2$pyrophosphate, 1% NP-40, 0.5% deoxycholate, 0.1% SDS, 0.1 mM PMSF, 10 μg/ml leupeptin, 5 μg/ml aprotinin, 1 μg/ml pepstatin A) is added to each well.

5. Cells are scraped on ice with a cell lifter, and duplicate lysates are combined in a 1-ml screw-capped microfuge tube and solubilized for 1 h at 4° on an inversion wheel.

6. Lysates are transferred to Beckmann polycarbonate centrifuge tubes and centrifuged for 20 min at 100,000 rpm in a TLA 100.2 rotor in a Beckman Optima TL ultracentrifuge.

7. Seven hundred fifty microliters of the lysate are removed (being careful not to disturb the pellet) and step 6 is repeated. The second centrifugation step is essential. Duplicate 15-μl aliquots are removed for protein determination and an additional 675 μl of the lysate is transferred (again taking care not to disturb the pellet) to 1-ml screw-capped microfuge tubes containing 100 μl of 20% (v/v) protein G-Sepharose beads resuspended in RIPA buffer plus 3% fraction V bovine serum albumin (BSA). The lysates are then precleared for 1 h at 4° on an inversion wheel.

8. Samples are centrifuged for 15 s at 13,000 rpm in a microcentrifuge, the lysate is removed to a second set of 1-ml screw-capped microfuge tubes containing 100 μl of 20% protein G-Sepharose beads, and 3 μl of monoclonal FLAG (M2) antibody (Eastman Kodak) is added. Lysates are incubated for 1–3 h at 4° on an inversion wheel.

9. Beads are washed three times by centrifugation with 1 ml of RIPA buffer containing 3% BSA. The lysate is removed from beads and beads are aspirated with a 29-gauge tuberculin syringe to dry the pelleted beads. Beads are resuspended in 50 μl of SDS sample buffer, and the immune complex is dissociated by incubation at 65° for 10 min. The volume of SDS sample buffer is then adjusted for differences in receptor protein content between sample, calculated as picomoles receptor per milligram lysate protein.

10. Forty-five-microliter aliquots from each sample are subjected to SDS–PAGE (12.5% gel). The gel is mounted, dried, and exposed to autoradiographic film in a sealed cassette at −80° for 24 h to 1 week. Alternatively, the gel can be exposed to a phosphoimager screen for 24 h (see Dhami et al., 2002).

Coimmunoprecipitation Assay

An important aspect for understanding the mechanism by which GRK2 contributes to the phosphorylation-independent attenuation of GPCR signaling is whether GRK2 overexpression is functioning to only sequester G$\alpha_{q/11}$ or whether GRK2 specifically regulates receptor/G-protein interactions. Therefore, it is necessary to assess the ability of both GRK2 and G$\alpha_{q/11}$ to associate with the GPCR of interest in the presence and absence of agonist.

1. Duplicate 100-mm dishes of HEK 293 cells (2.5×10^6 cells) are transfected with FLAG epitope-tagged receptor cDNA (10 μg), as outlined earlier, with and without GRK2 and GRK2 mutant constructs (5 μg) (see previous section).

2. Duplicate transfections are combined and reseeded into two new 100-mm dishes, collagen coated as described earlier. Cells are treated with and without agonist for 10 min, washed twice with ice-cold PBS, and lysed in 800 μl of cold lysis buffer (50 mM Tris, pH 8.0, 150 mM NaCl, 0.1% Triton X-100 containing protease inhibitors: 20 μg/ml leupeptin, 20 μg/ml aprotonin, and 20 μg/ml PMSF). For coimmunoprecipitation of GRK2 and Gα_q with mGluR1, no chemical cross-linker is required. However, a thio-cleavable chemical cross-linker may be required to stabilize receptor: GRK2 and/or receptor:Gα_q complexes for other GPCRs (Freedman *et al.*, 1997). Receptor:Gα_q complexes may also be regulated by the addition of 100 μM GDP, 2 mM MgSO$_4$, 30 μM aluminum chloride, and 5 mM sodium fluoride to solubilized cells (i.e., induction of the GDP·AlF$_4^-$-bound form of the G-protein α subunit).

3. Cells are solubilized for 1 h at $4°$ on an inversion wheel following which the particulate fraction is removed by centrifugation for 20 min at 100,000 rpm in a TLA 100.2 rotor in a Beckman Optima TL ultracentrifuge. The protein concentration of the soluble fraction is determined, and 500 μg of supernatant protein is transferred to 1 ml screw-capped microfuge tubes containing 100 μl of 20% (v/v) protein G-Sepharose beads resuspended in lysis buffer plus 3% fraction V BSA. The lysates are then precleared for 1 h at $4°$ on an inversion wheel.

4. Samples are centrifuged for 15 s at 13,000 rpm in a microcentrifuge, and the clarified lysate is transfered to a second set of 1-ml screw-capped microfuge tubes containing 100 μl of 20% protein G-Sepharose beads and 3 μl of monoclonal FLAG (M2) antibody (Eastman Kodak) is added. Lysates are incubated for 1–3 h at $4°$ on an inversion wheel.

5. The beads are washed three times by centrifugation with 1 ml of lysis buffer containing 3% BSA. The lysate is removed from beads and beads are aspirated with a tuberculin syringe to dry the pelleted beads. Beads are resuspended in 50 μl of SDS sample buffer, and the immune complex is dissociated by incubation at $65°$ for 10 min.

6. Forty-five-microliter aliquots from each sample are subjected to SDS–PAGE (12.5% gel). The proteins are transferred onto nitrocellulose membranes by semidry electroblotting. The membranes are blocked with 10% milk in wash buffer (TBS-T: 150 mM NaCl, 10 mM Tris–HCl, pH 7.0, 0.05% Tween 20) and then incubated with either rabbit GRK2- or Gα_q-specific antibodies (Upstate Cell Signaling Solutions) diluted 1:1000 in wash buffer containing 3% skim milk. The membranes are rinsed three

times for 10 min with wash buffer and are then incubated with secondary horseradish peroxidase-conjugated donkey antirabbit IgG (Amersham) diluted 1:2500 in wash buffer containing 3% skim milk. Membranes are rinsed twice with wash buffer and are then incubated with ECL western blotting detection reagents (Amersham) according to the manufacturer's specifications.

Acknowledgments

G.K.D. is the recipient of a Canadian Institutes of Health Research (CIHR) doctoral training award. S.S.G.F. holds a Canada Research Chair in Molecular Neuroscience, Premier's Research Excellence Award and is a Career Investigator of the Heart and Stroke Foundation of Ontario. This work was supported by CIHR Grant MA-15506 to S.S.G.F., National Science Foundation Grants MCB9728179 and MCB0315888 to R.S.M., and American Heart Association Scientist Development Grant 0235273N and a Research Corporation Cottrell Scholar grant to J.J.G.T.

References

Alessandri, B., and Bullock, R. (1998). Glutamate and its receptors in the pathophysiology of brain and spinal cord injuries. *Prog. Brain Res.* **116**, 303–330.

Anger, T., Zhang, W., and Mende, U. (2004). Differential contribution of GTPase activation and effector antagonism to the inhibitory effect of RGS proteins on G_q-mediated signaling in vivo. *J. Biol. Chem.* **279**, 3906–3915.

Barak, L. S., Ferguson, S. S. G., Zhang, J., and Caron, M. G. (1997). A β-arrestin/green fluorescent protein biosensor for detecting G protein-coupled receptor activation. *J. Biol. Chem.* **272**, 27497–27500.

Berman, D. M., Kozasa, T., and Gilman, A. G. (1996). The GTPase-activating protein RGS4 stabilizes the transition state for nucleotide hydrolysis. *J. Biol. Chem.* **271**, 27209–27212.

Carman, C. V., Parent, J. L., Day, P. W., Pronin, A. N., Sternweis, P. M., Wedegaertner, P. B., Gilman, A. G., Benovic, J. L., and Kozasa, T. (1999). Selective regulation of $G\alpha_{q/11}$ by an RGS domain in the G protein-coupled receptor kinase, GRK2. *J. Biol. Chem.* **274**, 34483–34492.

Carman, C. V., Barak, L. S., Chen, C., Liu-Chen, L. Y., Onorato, J. J., Kennedy, S. P., Caron, M. G., and Benovic, J. L. (2000). Mutational analysis of $G\beta\gamma$ and phospholipid interaction with G protein-coupled receptor kinase 2. *J. Biol. Chem.* **275**, 10443–10452.

Chen, Z., Wells, C. D., Sternweis, P. C., and Sprang, S. R. (2001). Structure of the RGS domain of p115RhoGEF. *Nature Struct. Biol.* **8**, 805–809.

Ciruela, F., Giacometti, A., and McIlhinney, R. A. (1999). Functional regulation of metabotropic glutamate receptor type 1c: A role for phosphorylation in the desensitization of the receptor. *FEBS Lett.* **462**, 278–282.

Conn, P. J., and Pin, J.-P. (1997). Pharmacology and functions of metabotropic glutamate receptors. *Annu. Rev. Pharmacol. Toxicol.* **37**, 205–237.

Dale, L. B., Babwah, A. V., and Ferguson, S. S. G. (2002). Mechanisms of metabotropic glutamate receptor desensitization: Role in the patterning of effector enzyme activation. *Neurochem. Int.* **41**, 319–326.

Dale, L. B., Bhattacharya, M., Anborgh, P. H., Murdoch, B., Bhatia, M., Nakanishi, S., and Ferguson, S. S. G. (2000). G protein-coupled receptor kinase mediated desensitization of metabotropic glutamate receptor 1α protects against cell death. *J. Biol. Chem.* **275**, 38213–38220.

Dale, L. B., Bhattacharya, M., Seachrist, J. L., Anborgh, P. H., and Ferguson, S. S. G. (2001). Agonist-dependent and -independent internalization of metabotropic glutamate receptor 1a: β-Arrestin isoform-specific endocytosis. *Mol. Pharmacol.* **60**, 1243–1253.

Day, P. W., Carman, C. V., Sterne-Marr, R., Benovic, J. L., and Wedegaertner, P. B. (2003). Differential interaction of GRK2 with members of the Gα$_q$ family. *Biochemistry* **42**, 9176–9184.

de Alba, E., De Vries, L., Farquhar, M. G., and Tjandra, N. (1999). Solution structure of human GAIP (Galpha interacting protein): A regulator of G protein signaling. *J. Mol. Biol.* **291**, 927–939.

Desai, M. A., Burnett, J. P., Mayne, N. G., and Schoepp, D. D. (1996). Pharmacological characterization of desensitization in a human mGlu1 alpha-expressing non-neuronal cell line co-transfected with a glutamate transporter. *Bri. J. Pharmacol.* **118**, 1558–15564.

Dhami, G. K., Anborgh, P. H., Dale, L. B., Sterne-Marr, R., and Ferguson, S. S. G. (2002). Phosphorylation-independent regulation of metabotropic glutamate receptor signaling by G protein-coupled receptor kinase 2. *J. Biol. Chem.* **277**, 25266–25272.

Dhami, G. K., Dale, L. B., Anborgh, P. H., O'Connor-Halligan, K. E., Sterne-Marr, R., and Ferguson, S. S. G. (2004). G protein-coupled receptor kinase 2 RGS homology domain binds to both metabotropic glutamate receptor 1a and Gαq to attenuate signaling. *J. Biol. Chem.* **279**, 16614–16620.

Dicker, F., Quitterer, U., Winstel, R., Honold, K., and Lohse, M. J. (1999). Phosphorylation-independent inhibition of parathyroid hormone receptor signaling by G protein-coupled receptor kinases. *Proc. Natl. Acad. Sci. USA* **96**, 5476–5481.

Ferguson, S. S. G. (2001). Evolving concepts in G protein-coupled receptor endocytosis: The role in receptor desensitization and signaling. *Pharmacol. Rev.* **53**, 1–24.

Ferguson, S. S. G., and Caron, M. G. (2004). Green fluorescent protein tagged β-arrestin translocation as a measure of G protein-coupled receptor activation. *Methods Mol. Biol.* **237**, 121–126.

Ferguson, S. S. G., Downey, W. E., III, Colapietro, A.-M., Barak, L. S., Ménard, L., and Caron, M. G. (1996). Role of β-arrestin in mediating agonist-promoted G protein-coupled receptor internalization. *Science* **271**, 363–366.

Freedman, N. J., Ament, A. S., Oppermann, M., Stoffel, R. H., Exum, S. T., and Lefkowitz, R. J. (1997). Phosphorylation and desensitization of human endothelin A and B receptors. Evidence for G protein-coupled receptor kinase specificity. *J. Biol. Chem.* **272**, 17734–17743.

Freedman, N. J., and Lefkowitz, R. J. (1996). Desensitization of G protein-coupled receptors. *Recent Prog. Horm. Res.* **51**, 319–351.

Francesconi, A., and Duvoisin, R. M. (2000). Opposing effects of protein kinase C and protein kinase A on metabotropic glutamate receptor signaling: Selective desensitization of the inositol triphosphate/Ca^{2+} pathway by phosphorylation of the receptor-G protein-coupling domain. *Proc. Natl. Acad. Sci. USA* **97**, 6185–6190.

Gereau, R. W., and Heinemann, S. F. (1998). Role of protein kinase C phosphorylation in rapid desensitization of metabotropic glutamate receptor 5. *Neuron* **20**, 143–151.

Gomeza, J., Joly, C., Kuhn, R., Knopfel, T., Bockaert, J., and Pin, J.-P. (1996). The second intracellular loop of metabotropic glutamate receptor 1 cooperates with the other intracellular domains to control coupling to G-proteins. *J. Biol. Chem.* **271**, 2199–2205.

Goodman, O. B., Jr., Krupnick, J. G., Santini, F., Gurevich, V. V., Penn, R. B., Gagnon, A. W., Keen, J. H., and Benovic, J. L. (1996). β-Arrestin acts as a clathrin adaptor in endocytosis of the β_2-adrenergic receptor. *Nature* **383**, 447–450.

Hepler, J. R., Berman, D. M., Gilman, A. G., and Kozasa, T. (1997). RGS4 and GAIP are GTPase-activating proteins for $G_{q\alpha}$ and block activation of phospholipase Cβ by γ-thio-GTP-$G_{q}\alpha$. *Proc. Natl. Acad. Sci. USA* **94**, 428–432.

Hepler, J. R., Kozasa, T., Smrcka, A. V., Simon, M. I., Rhee, S. G., Sternweis, P. C., and Gilman, A. G. (1993). Purification from Sf9 cells and characterization of recombinant Gq alpha and G11 alpha: Activation of purified phospholipase C isozymes by G alpha subunits. *J. Biol. Chem.* **268**, 14367–14375.

Herrero, I., Miras-Portugal, T., and Sanchez-Prieto, J. (1994). Rapid desensitization of the metabotropic glutamate receptor that facilitates glutamate release in rat cerebrocortical nerve terminals. *Eur. J. Neurosci.* **6**, 115–120.

Ingi, T., Krumins, A. M., Chidiac, P., Brothers, G. M., Chung, S., Snow, B. E., Barnes, C. A., Lanahan, A. A., Siderovski, D. P., Ross, E. M., Gilman, A. G., and Worley, P. F. (1998). Dynamic regulation of RGS2 suggests a novel mechanism in G-protein signaling and neuronal plasticity. *J. Neurosci.* **18**, 7178–7188.

Jones, T. A., Zou, J. Y., Cowan, S. W., and Kjeldgaard, M. (1991). Improved methods for building protein models in electron density maps and the location of errors in these models. *Acta Crystallogr.* **A47**, 110–119.

Kleywegt, G., and Jones, T. (1998). Databases in protein crystallography. *Acta Crystallogr. D Biol. Crystallogr.* **54**, 1119–1131.

Lefkowitz, R. J. (1993). G protein-coupled receptor kinases. *Cell* **74**, 409–412.

Lodowski, D. T., Pitcher, J. A., Capel, W. D., Lefkowitz, R. J., and Tesmer, J. J. G. (2003). *Science* **300**, 1256–1262.

Lohse, M. J., Benovic, J. L., Codina, J., Caron, M. G., and Lefkowitz, R. J. (1990). β-Arrestin: A protein that regulates β-adrenergic receptor function. *Science* **248**, 1547–1550.

Longenecker, K. L., Lewis, M. E., Chikumi, H., Gutkind, J. S., and Derewenda, Z. S. (2001). Structure of the RGS-like domain from PDZ-RhoGEF: Linking heterotrimeric G protein-coupled signaling to Rho GTPases. *Structure* **9**, 559–569.

Nakanishi, S., and Masu, M. (1994). Molecular diversity and functions of glutamate receptors. *Annu. Rev. Biophys. Biomol. Struct.* **23**, 319–348.

Nicoletti, F., Brunom, V., Copanim, A., Casabona, G., and Knopfel, T. (1996). Metabotropic glutamate receptors: A new target for the therapy of neurodegenerative disorders? *Trends Neurosci.* **19**, 267–271.

Oppermann, M., Freedman, N. J., Alexander, R. W., and Lefkowitz, R. J. (1996). Phosphorylation of the type 1A angiotensin II receptor by G protein-coupled receptor kinases and protein kinase C. *J. Biol. Chem.* **271**, 13266–13272.

Palczewski, K., Buczylko, J., Lebioda, L., Crabb, J. W., and Polans, A. S. (1993). Identification of the N-terminal region in rhodopsin kinase involved in its interaction with rhodopsin. *J. Biol. Chem.* **268**, 6004–6013.

Pao, C. S., and Benovic, J. L. (2002). Phosphorylation-independent desensitization of G protein-coupled receptors? *Sci. STKE.* **153**, PE42.

Pitcher, J. A., Inglese, J., Higgins, J. B., Arriza, J. L., Casey, P. J., Kim, C., Benovic, J. L., Kwatra, M. M., Caron, M. G., and Lefkowitz, R. J. (1992). Role of $\beta\gamma$ subunits of G proteins in targeting the β-adrenergic receptor kinase to membrane-bound receptors. *Science* **257**, 1264–1267.

Pitcher, J. A., Touhara, K., Payne, E. S., and Lefkowitz, R. J. (1995). Pleckstrin homology domain-mediated membrane association and activation of the β-adrenergic receptor kinase requires coordinate interaction with $G_{\beta\gamma}$ subunits and lipid. *J. Biol. Chem.* **270**, 11707–11710.

Ross, E. M., and Wilkie, T. M. (2000). GTPase-activating proteins for heterotrimeric G proteins: Regulators of G protein signaling (RGS) and RGS-like proteins. *Annu. Rev. Biochem.* **69**, 795–827.

Sallese, M., Mariggio, S., D'Urbano, E., Iacovelli, L., and De Blasi, A. (2000a). Selective regulation of Gq signaling by G protein-coupled receptor kinase 2: Direct interaction of kinase N terminus with activated $G\alpha q$. *Mol. Pharmacol.* **57**, 826–831.

Sallese, M., Salvatore, L., D'Urbano, E., Sala, G., Storto, M., Launey, T., Nicoletti, F., Knopfel, T., and De Blasi, A. (2000b). The G-protein-coupled receptor kinase GRK4 mediates homologous desensitization of metabotropic glutamate receptor 1. *FASEB J.* **14**, 2569–2580.

Scheschonka, A., Dessauer, C., Sinnaraja, S., Chidiac, P., Shi, C.-S., and Kehrl, J. H. (2000). RGS3 is a GTPase-activating protein for $G_i\alpha$ and $G_q\alpha$ and a potent inhibitor of signaling by GTPase-deficient forms of $G_q\alpha$ and $G_{11}\alpha$. *Mol. Pharmacol.* **58**, 719–728.

Siderovski, D. P., Hessel, A., Chung, S., Mak, T. W., and Tyers, M. (1996). A new family of regulators of G-protein-coupled receptors? *Curr. Biol.* **6**, 211–212.

Spink, K. E., Polakis, P., and Weis, W. I. (2000). Structural basis of the Axin-adenomatous polyposis coli interaction. *EMBO J.* **19**, 2270–2279.

Srinivasa, S. P., Watson, N., Overton, M. C., and Blumer, K. J. (1998). Mechanisms of RGS4, a GTPase-activating protein for G protein alpha subunits. *J. Biol. Chem.* **273**, 1529–1533.

Sterne-Marr, R., Tesmer, J. J. G., Day, P., Stracquatanio, R. P., Cilente, J.-A., O'Connor, K. E., Pronin, A., Benovic, J. L., and Wedegaertner, P. B. (2003). G protein-coupled receptor kinase 2/$G\alpha_{q/11}$ interaction: A novel surface on a regulator of G protein signaling homology domain for binding $G\alpha$ subunits. *J. Biol. Chem.* **278**, 6050–6058.

Tesmer, J. J. G., Berman, D. M., Gilman, A. G., and Sprang, S. R. (1997). Structure of RGS4 bound to AlF$_4^-$-activated G(i alpha1): Stabilization of the transition state for GTP hydrolysis. *Cell* **89**, 251–261.

Thompson, J. D., Higgins, D. G., and Gibson, T. J. (1994). CLUSTAL W: Improving the sensitivity of progressive multiple sequence alignment through sequence weighting, position-specific gap penalties and weight matrix choice. *Nucleic Acids Res.* **22**, 4673–4680.

Usui, H., Nishiyama, M., Moroi, K., Shibasaki, T., Zhou, J., Ishida, J., Fukamizu, A., Haga, T., Sekiya, S., and Kimura, S. (2000). RGS domain in the amino-terminus of G protein-coupled receptor kinase 2 inhibits Gq-mediated signaling. *Int. J. Mol. Med.* **5**, 335–340.

Wang, J., Tu, Y., Woodson, J., Song, X., and Ross, E. M. (1997). A GTPase-activating protein for the G protein Galphaz: Identification, purification, and mechanism of action. *J. Biol. Chem.* **272**, 5732–5740.

Willets, J. M., Challiss, R. A., Kelly, E., and Nahorski, S. R. (2001). G protein-coupled receptor kinases 3 and 6 use different pathways to desensitize the endogenous M3 muscarinic acetylcholine receptor in human SH-SY5Y cells. *Mol. Pharmacol.* **60**, 321–330.

Yan, Y., Chi, P. P., and Bourne, H. R. (1997). RGS4 inhibits Gq-mediated activation of mitogen-activated protein kinase and phosphoinositide synthesis. *J. Biol. Chem.* **272**, 11924–11927.

[21] Analysis of Differential Modulatory Activities of GRK2 and GRK4 on $G\alpha_q$-Coupled Receptor Signaling

By ANTONIETTA PICASCIA, LOREDANA CAPOBIANCO,
LUISA IACOVELLI, and ANTONIO DE BLASI

Abstract

G-protein-coupled receptor kinases (GRK) contain a regulator of G-protein signaling (RGS)-like domain located at the N terminus (GRK-Nter) of their sequence. This domain is present in all the GRK subtypes, but the RGS-like domain of GRK2 was documented to be functionally active, as it is able to interact selectively with $G\alpha_q$ (both *in vitro* and in cells) and to inhibit $G\alpha_q$-dependent signaling. In contrast GRK4, GRK5, and GRK6 are unable to interact with $G\alpha_q$. This article describes the methodology to investigate the modulatory activity of GRK2 and GRK4 on GPCR-stimulated $G\alpha_q$ signaling. This analysis is essentially based on three types of experiments: (a) study of the effect of the GRK-Nter on GPCR-dependent signaling; (b) analysis of the binding of GRK-Nter to $G\alpha_q$ *in vitro*; and (c) analysis of the interaction of GRK with $G\alpha_q$ in cells.

Introduction

The continuous communication among cells is fundamental in maintaining the coherent function and specialization of the system in which they are a part. Central for the maintainance of this homeostasis are cell surface receptors that receive and transduce inside the cells signals coming from outside. G-protein-coupled receptors (GPCRs) form the largest known family of cell surface receptors and share the characteristic structure of seven transmembrane domains. These receptors bind to a variety of agonists, such as photons, neurotransmitters, hormones, and chemokines, thus regulating very diverse functions as vision, olfaction, neurotransmission, proliferation, and immune processes. The signal transduction process triggered by agonist binding to a GPCR consists of discrete steps, and one of the earliest is the coupling of this complex to the heterotrimeric G protein that dissociates in $G\alpha$ and $G\beta\gamma$ subunits (Hamm, 1998). These subunits activate or inhibit a number of effector enzymes, such as adenylyl cyclase, phospholipase C, ion channels, or, as identified more recently, c-Src and mitogen-activated protein kinases (MAPK), resulting in a variety of cellular functions.

GPCR-mediated signal transduction is strictly regulated by multiple mechanisms acting at different levels of signal propagation. After agonist stimulation, the receptor is desensitized by G-protein-coupled receptor kinase (GRK) phosphorylation and subsequent binding of arrestin proteins (homologous desensitization) (Lefkowitz, 1998; Palczewski, 1997). The activated α subunit of the G protein (Gα) can, in turn, be inhibited by regulator of G-protein signaling (RGS) proteins (Berman and Gilman, 1998; De Vries and Farquhar, 1999; Hepler, 1999; Tesmer et al., 1997). These RGS proteins are functionally homologous to the well-known GTPase-activating proteins (GAPs) acting on small G proteins such as Ras. In fact, RGS proteins interact directly with Gα subunits, thus increasing their intrinsic GTPase activity.

The existence of multiple mechanisms that regulate GPCR signaling may have relevant functional consequences. For example, GRK, which acts at the receptor level, should desensitize all the signaling pathways activated by one receptor, whereas RGS proteins only desensitize the signaling activated by a given Gα protein. In addition to desensitizing the receptor by phosphorylation, GRK2 is able to regulate Gα_q selectively. The GRK2 N-terminal domain interacts directly with the activated Gα_q and regulates its signaling in a phosphorylation-independent manner (Sallese et al., 2000b). GRK2 can regulate different GPCR-mediated responses by multiple mechanisms that depend on the coupling to different G proteins.

GPCR Homologous Desensitization

One of the earlier regulatory events of GPCR signaling is homologous desensitization (Lefkowitz, 1998). The term "homologous" means that the desensitization of the receptor is initiated by activation of the same receptor by the agonist. When the agonist binds to the receptor, it acquires the suitable conformation to be phosphorylated by one GRK. Phosphorylation can occur at the C terminus of the receptor, as for rhodopsin and β_2-adrenergic receptors, or at the third intracellular loop, as for M$_2$-muscarinic and α_2-adrenergic receptors (Eason et al., 1995). GRK-phosphorylated receptors are desensitized only minimally, but this phosphorylation increases the affinity for arrestin. Binding of arrestin to the receptor then induces maximal homologous desensitization. The next event is receptor sequestration in endosomal vesicles, where it will be degraded (downregulation) or dephosphorylated by specific phosphatases and then recycled to the membrane.

The GRK Gene Family

The GRK family is composed of seven cloned members named GRK1 to GRK7, according to the order of their identification (Fig. 1). GRK1, GRK2, and GRK3 were previously named rhodopsin kinase (RK), βARK1, and βARK2, respectively (Lefkowitz, 1998; Palczewski, 1997). Importantly, GRK2, 3, 5, and 6 are widely distributed, whereas GRK1, GRK7, and GRK4 have selected sites of expression in that GRK1 and GRK7 are localized exclusively in the retina, whereas GRK4 has been found in sperm and, with lower levels of expression, in brain and kidney (Sallese *et al.*, 1997).

Based on sequence homology, GRKs are classified in three subfamilies: GRK1 and GRK7 are in the first (RK subfamily), GRK2 and 3 form the second (βARK subfamily), and GRK4, 5, and 6 constitute the third (GRK4 subfamily) (Fig. 1). From an evolutionary point of view, RK and GRK4 subfamilies are more strictly related.

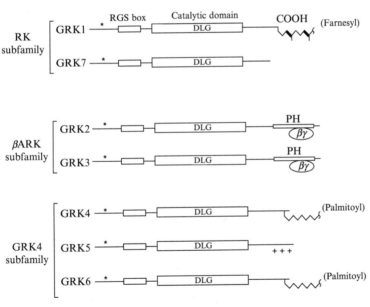

FIG. 1. GRK gene family. Schematic representation of the seven GRK variants showing the common central catalytic domain [with the asparate–leucine–glycine (DLG) sequence that constitutes the ATP-binding pocket], the PH domain in GRK2 and GRK3, the CaM-interacting region (asterisk), and the putative "RGS box." The principal mechanisms of membrane anchoring are indicated: Farnesylation for GRK1, Gβγ binding for GRK2 and GRK3, palmitoylation for GRK4 and 6, and positive charges for GRK5 (+++).

The overall structure of GRK consists of a central core, which is the catalytic domain, flanked by the amino-terminal region and the C-terminal region (Fig. 1). For example, GRK2 consists of a central catalytic domain (F191-G443), a C-terminal domain (G443-L689) involved in membrane targeting, and an N-terminal domain (M1-F191).

The catalytic domain that contains the ATP-binding site is the most conserved region among all GRK subtypes. Sites of alternative splicing have been identified for some GRK subtypes. No splice variants have been identified in human GRK2 and 3 and only one was found in GRK1. For human GRK4, two sites of alternative splicing have been identified, one at the N-terminal domain (exon 2) and the other at the C-terminal domain (exon 15), thus resulting in four splice variants (Premont et al., 1996; Sallese et al., 1997). Alternative splicing could be predicted for GRK5 and 6, as they are closely related to GRK4. So far, splice variants in rat GRK6, which are different from those of GRK4, have been demonstrated (Firsov and Elalouf et al., 1997).

The membrane-targeting domains of GRKs are located mainly at their C termini, but the anchoring mechanisms are somewhat different. GRK1 has a farnesylation site, while a palmitoylation site is present in both GRK4 and GRK6 (Palczewski, 1997). GRK2 and 3 have an extended C terminus in which a pleckstrin homology (PH) domain is located and binding to $G\beta\gamma$ targets them to the membrane (Fig. 1). The $G\beta\gamma$-binding domain partially overlaps with the PH domain (Aragay et al., 1998; Inglese et al., 1994; Koch et al., 1993). GRK5 likely binds to membrane phospholipids by positively charged amino acid clusters located at the C terminus.

Phosphorylation of GPCR by GRKs

Studies performed on GRK-mediated phosphorylation of receptors or peptides show several hallmark characteristics. First of all, GRKs do not demonstrate clear consensus sequences in their receptor substrates. Initial studies with model peptide substrates showed some substrate specificity. GRK1 preferentially phosphorylates serine or threonine residues followed by several acidic residues, whereas GRK2 phosphorylates substrates in which acidic residues precede the target serine or threonine (Onorato et al., 1991). GRK5 and GRK6 most actively phosphorylate peptides containing basic residues amino-terminal to the serine target residues (Eason et al., 1995; Fredericks et al., 1996; Ohguro et al., 1993; Prossnitz et al., 1995). Further investigations, based on studies with GPCRs, came to the conclusion that GRKs recognize activated (agonist-bound) conformations of GPCRs rather then linear sequences. Moreover, interaction of GRKs with activated receptors potently activates these enzymes. The ability of laboratories (Carman *et al.*, 1999; Diker *et al.*, 1999; Iacovelli *et al.*, 2003; Sallese *et al.*, 2000b) have used the GRK2 kinase-dead mutant GRK2-K220R, which was transfected transiently (usually in HEK293 cells) in parallel with the wild-type kinase. The level of expression of both wild-type and kinase-dead GRK2 was similar. Unexpectedly, it was found that, in some cases, the GRK2-K220R was able to inhibit GPCR-mediated signaling as well as the wild-type kinase. For example, agonist-stimulated inositol phosphate (IP) production mediated by a variety of G_q-coupled GPCR (including the parathyroid hormone, the α_1-adrenergic, the $5HT_{2C}$, the TSH, the mGluR1 receptors) was inhibited in cells transfected with the GRK2-K220R mutant (Carman *et al.*, 1999; Diker *et al.*, 1999; Diviani *et al.*, 1996; Iacovelli *et al.*, 2003; Sallese *et al.*, 2000b). Subsequent studies demonstrated that this effect was mediated by the RGS-like domain located within the N terminus of GRK2 (see later).

The GRK2 kinase-dead mutant was also able to inhibit the GPCR-stimulated signaling mediated by $G\beta\gamma$. For example, it was shown that agonist stimulation of the mGluR4 receptor, which is coupled to G_i, inhibits adenylyl cyclase (mediated by $G\alpha_i$) and promotes MAPK stimulation (mediated by $G\beta\gamma$). Cotransfection of GRK2 wild type or GRK2-K220R each resulted in the selective inhibition of receptor-mediated MAPK activation (Iacovelli *et al.*, 2004). This regulatory effect, which is phosphorylation independent, is mediated by the PH domain present within the C terminus of GRK2. This domain binds to free $G\beta\gamma$, thus promoting the translocation of GRK2 to plasma membranes and the activation of the kinase for receptor phosphorylation and desensitization. In addition, this interaction results in the inhibition of the signaling mediated by free $G\beta\gamma$.

Based on these results, GRK2 appears as a multidomain kinase able to modulate GPCR signaling by multiple mechanisms that are either phosphorylation dependent or phosphorylation independent (Sallese *et al.*, 2000a). The kinase activity is important for agonist-dependent receptor phosphorylation and (homologous) desensitization, whereas the RGS-like domain (the N terminus) and the PH domain (at the C terminus) are able to interact with the activated G-protein subunits to regulate selectively the signaling pathways downstream of the receptor (Fig. 2).

RGS Domain of GRKs

Sequence analysis in comparison with RGS proteins revealed that GRK2 contains an RGS homology domain located within the N terminus of this kinase (amino acids 54 to 175) (Siderovski *et al.*, 1996; reviewed in Bünemann, 1999). It was also observed that the RGS-like domain is present in the sequence of all the GRK subtypes. Further studies demonstrated that

activated receptors to enhance the activity of GRKs towar
substrates suggests the hypothesis that, in a cellular setting,
phosphorylate nonreceptor substrates following receptor activa

GRK recognition of the activated receptor is probably con
receptor conformational modifications following agonist binding.
ent mechanistic models have been proposed in order to clarify th
GPCR–GRK interaction: the stable complex model and the hyst
(Palczewski, 1997). In the first model, a stable ternary comple
occupied receptor and GRK is formed in a stoichiometric r
complex involves geometrical complementarity of the N ter
GRK and different domains of the receptor. Several lines of evic
that GRKs interact with receptors also at sites distinct from
phosphorylation. Truncated receptors that are not substrates
phorylation retain their ability to bind the GRKs and to activ:
toward peptides substrates (Kameyama et al., 1994; Kim et al.,
et al., 1994). In complex with the receptor, the catalytic activity
activated toward the C terminus of the receptor with whi
complex or toward exogenous peptides (Chen et al., 1993;
Benovic, 1991). Dissociation of the complex occurs when eith
the receptor is phosphorylated (Pulvermüller et al., 1993).

The hysteresis model provides that, after interacting v
bound receptor, the GRK remains in the active conforma'
time, even after dissociation from the receptor. The activa'
capable of phosphorylating both activated and nonactivat(
model is particularly suitable for explaining the high-gair
observed on rhodopsin consequent to exceedingly lov
particular, activated GRK1 is able to phosphorylate bo'
nonactivated rhodopsin (Aton, 1989; Binder et al., 1'
models account for some features of GRK activation.
be learned about the mechanisms of GRK regulation.

It seems that the vast majority of G-protein-coupl(
sent substrates of GRKs due to the fact that the proc
desensitization is a general characteristic of this f:
Consistently for a large number of GPCRs, a role c
desensitization has been documented.

Regulation of GPCR Signaling by Phosphorylation
   Mechanisms

Kinase-dead mutants of GRKs, in which kinase
by site-directed mutagenesis, were developed tc
GRK-dependent phosphorylation in receptor des

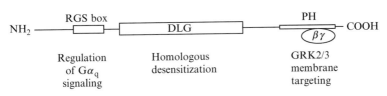

FIG. 2. Schematic representation of the three main functional domains of GRK2/3. The N-terminal domain, which contains an RGS homology region, interacts with $G\alpha_q$. The catalytic central domain interacts and phosphorylates the GPCR, thus leading to homologous desensitization. The C terminus, which contains a PH domain, interacts with $G\beta\gamma$.

the N terminus domain of GRK2 and GRK3 (GRK2/3-Nter), which contain the RGS box, was able to interact with activated $G\alpha_q$ and to inhibit $G\alpha_q$-dependent signaling (Carman et al., 1999; Sallese et al., 2000b). This interaction was selective, as GRK2-Nter did not interact with $G\alpha_s$, $G\alpha_o$, $G\alpha_i$, and $G\alpha_{12}$. In contrast, the GRK4, GRK5, and GRK6 N terminus domains were not able to bind to $G\alpha_q$ nor modulate $G\alpha_q$-dependent signaling, indicating that a functional RGS-like domain is present only on GRK2 and GRK3 (Carman et al., 1999; Iacovelli et al., 2003; Sallese et al., 2000b). Unlike the classical RGS proteins, the RGS-like domain of GRK2/3 possesses a weak GAP activity toward $G\alpha_q$ (Carman et al., 1999). It is therefore likely that the ability of the GRK2-Nter domain to inhibit $G\alpha_q$-mediated signaling is mainly due to sequestration of the activated $G\alpha_q$ by the GRK2-Nter.

Experimental Procedures

Analysis of the Regulatory Activity of the RGS-like Domain of GRKs

This analysis is essentially based on three types of experiments: (a) study of the effect of the GRK-Nter on GPCR-dependent signaling; (b) analysis of the binding of GRK-Nter to $G\alpha_q$ *in vitro*; and (c) analysis of the interaction of GRK with $G\alpha_q$ in cells.

To study the modulatory activity of the RGS-like domain on $G\alpha_q$-coupled GPCR signaling, we cotransfect the receptor of interest with the wild-type GRK, the kinase-dead GRK mutant, and the recombinant N terminus domain of the kinase (Figures 3 and 4). The agonist-stimulated IP production is measured. When the GRK regulates the signaling by a phosphorylation-independent mechanism, the agonist-stimulated IP production is inhibited (as compared to cells transfected with the receptor alone) in cells cotransfected with the kinase-dead GRK as well as with the wild-type kinase. The inhibition by the expression of the GRK-Nter

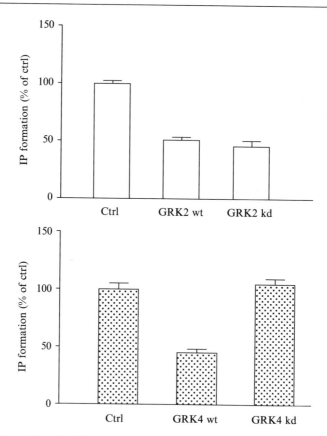

FIG. 3. Regulation of mGlu$_{1a}$ receptor signaling by GRK2 and GRK4. HEK293 cells were transfected with the mGlu$_{1a}$ receptor and cotransfected with mock (ctrl), GRK2, or GRK4 wild-type (wt) or kinase-dead mutant (kd) and IP formation stimulated by quisqualate (100 μM) was measured. Agonist-stimulated IP formation was inhibited by both wt- and kinase-dead GRK2, suggesting a phosphorylation-independent mechanism. In contrast, only the wt GRK4 (but not the kinase-dead GRK4 mutant) inhibited IP formation, indicating that the kinase activity of GRK4 is necessary for mGlu$_{1a}$ receptor regulation. This indicates that GRK2 (but not GRK4) contains a functional RGS domain.

indicates that the RGS-like domain of the kinase is mediating this effect. When the RGS homology domain is not active (such as in the case of GRK4 regulating the mGluR1 receptor), only the wild-type GRK inhibits agonist-stimulated IP production.

The interaction between the RGS-like domain of GRKs and the Gα subunits *in vitro* can be determined by direct binding experiments (Fig. 5). In these experiments, the GST-Nter fusion protein is incubated with the

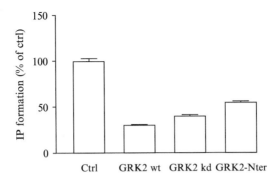

FIG. 4. Regulation of 5-HT$_{2C}$ receptor-mediated signaling. 5-HT$_{2C}$ receptor-stimulated IP formation in HEK293 cells transfected with 5-HT$_{2C}$ receptor and cotransfected with the vector (ctrl) or with GRK2 wt, GRK2 kinase dead mutant (GRK2kd), or GRK2-Nter expression constructs. Data are expressed as the percentage of maximal stimulation by serotonin (5-HT; 10 μM for 30 min) in control cells. This experiment shows that agonist-stimulated IP formation is inhibited by the recombinant GRK2-Nter containing the RGS-like domain.

$G\alpha$ subunit in the presence or absence of AlF$_4^-$ to activate the $G\alpha$. The $G\alpha$ should bind to the GST-Nter only when in the active form (i.e., in the presence of AlF$_4^-$). In typical experiments, only $G\alpha_q$ (but not $G\alpha_s$) binds GRK2-Nter in the presence of AlF$_4^-$. The binding of $G\alpha_q$ to the GRK4-Nter is negligible, confirming that GRK4 does not have a functionally active RGS-like domain. The use of a constitutively active $G\alpha_q$ mutant ($G\alpha_q$-Q209L), which also binds to the GRK2-Nter in the absence of AlF$_4^-$, provides a useful control.

The interaction between the RGS-like domain of GRKs and the $G\alpha$ subunits in cells can be investigated by coimmunoprecipitation (Fig. 6). HEK293 cells transfected with the appropriate $G\alpha$ target of the GRK of interest, plus the receptor coupled to the G-protein investigated, are stimulated with the receptor agonist before immunoprecipitation (with the anti-$G\alpha$ antibody). When the RGS-like domain interacts with the $G\alpha$ subunit, the GRK is found in the immunoprecipitate (revealed by immunoblot) in an agonist-dependent manner. In our experiments, GRK2 (but not GRK4) was coimmunoprecipitated with $G\alpha_q$ in the presence of quisqualate (the agonist of the mGluR1 receptor). In cells transfected with the constitutively active $G\alpha_q$-Q209L mutant, GRK2 was coimmunoprecipitated with $G\alpha_q$-Q209L even in the absence of the agonist, although quisqualate treatment was able to further increase the level of GRK2 coimmunoprecipitated.

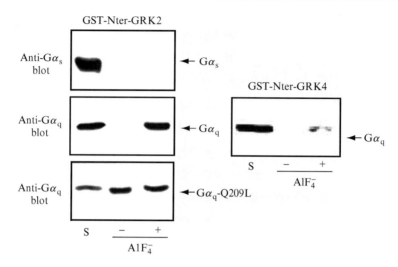

FIG. 5. Binding of GRK2-Nter and GRK4-Nter to activated $G\alpha_q$. Recombinant purified GST-GRK-Nter proteins conjugated to glutathione agarose beads were incubated with cytosolic proteins (150 μg) from $G\alpha_q$-transfected HEK293 cells in the absence (−) or presence (+) of 30 μM $AlCl_3$ and 20 mM NaF to activate the $G\alpha_q$ (AlF_4^-). Incubation (1 h at 4°) was stopped by centrifugation (300,000g). After three extensive washings, $G\alpha_q$ bound to the column was detected by an immunoblot using the anti-$G\alpha_q$ antibody. Starting material (S) (30–40 μg of cytosolic preparation, about one-fourth of total cytosol used for binding) is also included in the immunoblot. A similar procedure was used to determine binding of the GRK2-Nter to $G\alpha_s$ or to the constitutively active $G\alpha_q$-Q209L mutant. In the presence of AlF_4^-, GRK2-Nter binds to the activated $G\alpha_q$ but not to $G\alpha_s$ (nor to $G\alpha_o$ and $G\alpha_i$, data not shown). Interaction with the constitutively active $G\alpha_q$-Q209L mutant is also observed in the absence of AlF_4^-. The interaction of GRK4-Nter with $G\alpha_q$ is poor.

Study of the Modulatory Effect of the GRK-Nter on GPCR-Dependent Signaling

To investigate the regulation of a G_q-coupled receptor by GRK2 or GRK4, GPCR-stimulated IP production is measured in HEK293 cells transiently transfected to express GPCR alone or in combination with GRKs (wild-type, kinase-dead mutant, or recombinant GRK4-Nter domain).

Clones and Mutants. The GRK2 kinase-dead mutant GRK2-K220R was generated by site-directed mutagenesis of a single amino acid in the GRK2 sequence. Similarly, the GRK4 kinase-dead mutant GRK4-(K216M, K217M) was generated by mutation of two adjacent lysine residues to methionines. In both mutants, the kinase activity was completely disrupted by the mutagenesis.

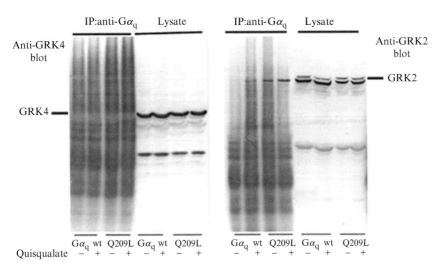

FIG. 6. Coimmunoprecipitation of Gα_q and GRKs from HEK293 cells. HEK293 cells coexpressing Gα_q or Gα_q (Q209L) (as indicated) and either GRK4 (left blot) or GRK2 (right blot) were untreated or exposed to 100 μM quisqualate for 5 min. Cells were then harvested and lysed. Immunoprecipitation (IP, lanes 1–4 of each blot) was performed using an anti-Gα_q antibody as described in the text. The initial cell extract (15%) was included for comparison (lysate, lanes 5–8 of each blot). GRK2 was immunoprecipitated with Gα_q in an agonist-dependent manner in cells expressing wild-type Gα_q (wt) and in an agonist-independent manner in cells expressing the constitutively active Gα_q-Q209L mutant. GRK4 was not co-immunoprecipitated with Gα_q nor with Gα_q-Q209L, indicating that this kinase does not interact with Gα_q.

The GRK2 N-terminal region (Ala2-Thr187) cDNA was cloned by polymerase chain reaction (PCR). This fragment was subcloned in the eukaryotic expression vector pcDNA3HisC (Invitrogen, Carlsbad, CA) to be used for transfection. A similar procedure can be used to generate the recombinant N terminus of GRK4 or other GRKs.

Cell Culture and Transfection. HEK293 cells are maintained in Dulbecco's modified Eagle's medium (DMEM) supplemented with 10% (v/v) fetal bovine serum at 37° in a humidified 7% CO_2 atmosphere. Transfections are performed on 70% confluent monolayers in 100-mm tissue culture dishes. The cells are transfected with the lipofectamine method in which a solution containing DNA–liposome complexes is layered directly onto the cells. Transfection is performed using the commercially available LipofectAMINE PLUS reagent (Invitrogen). The DNA–liposome complexes are formed by mixing the GPCR expression vector (8 μg cDNA/2 × 10^6 cells) with GRK2 (8 μg cDNA/2 × 10^6 cells), K220R (8 μg cDNA/2 × 10^6 cells), N-ter

(8 μg cDNA/2 $\times$ 10^6 cells), or C-ter (8 μg cDNA/2 $\times$ 10^6 cells) and the lipofectamine reagent (30 μl/100-mm dishes) in the serum-free transfection medium OPTIMEM (GIBCO) (5 ml/100-mm dishes). When needed, the empty pcDNA3 vector is added to keep the total amount of DNA added per dish constant. Cells are incubated with the transfection medium containing DNA–liposome complexes for 3–5 h and then this medium is replaced with culture medium (incubated overnight).

IP Measurement. One day after transfection, transfected HEK293 cells are reseeded in 24-well plates at the density of 20 $\times$ 10^4 cells per well. Forty-eight hours after transfection, cells are labeled for 24 h with 5 μCi/ml of *myo*-[2-^{3}H]inositol (Amersham) in the inositol-free medium M199 (GIBCO). Seventy-two hours after transfection, labeled cells are washed twice with prewarmed HBSS$^+$, incubated at 37° for 15 min in an incubation buffer (HBSS$^+$ is Hank's buffered salt solution plus 350 mg/liter $NaHCO_3$, 10 mM HEPES, 10 mM LiCl, pH 7.3), and then incubated with the indicated agonists at 37° for 30 min. The reaction is stopped aspirating the incubation buffer and adding ice-cold 6% perchloric acid/1.5 mM EDTA. Cells were scraped and transferred into a microfuge tube, and the pH is brought to pH 7 with 10 mM KOH. Cell debris is removed by centrifugation (5 min at room temperature in a microfuge). Supernatants, containing the intracellular inositol phosphates, are applied to ion-exchange chromatography columns of Dowex AG1-X8 (formate form) (200–400 mesh, 350-μl bed volume). Each column is washed with 2 column volumes of unlabeled inositol (5 mM), 2 column volumes of 5 mM sodium tetraborate/60 mM ammonium formate, and eluted by 3 ml of 0.1 M formic acid/1.5 M ammonium formate as described previously (Iacovelli *et al.*, 2003). Total IPs are quantified by counting β emissions. Eluates are mixed with scintillation solution and subjected to scintillation counting.

Analysis of Binding of GRK-Nter to $G\alpha_q$ In Vitro

Preparation of Bacterial Recombinant Proteins. The GRK2 N-terminal region (Ala2-Thr187) cDNA is amplified using the primers 5'-CG*GG-ATCC*GCGGACCTGGAGGC-3' (forward) and 5'-TCA*GGTCAGGT-GGATGTTGAGC*-3' (reverse) (native sequences are underlined; the *Bam*HI restriction site is in italics) in 100 μl of PCR buffer (10 mM Tris–HCl, pH 8.3, 50 mM KCl, 2 mM $MgCl_2$), 0.4 μg of each primer, 200 μM of dNTPs, 2.5 U of *Thermus acquaticus* DNA polymerase (Amplitaq Perkin Elmer). Amplification is carried out for 30 cycles, 94° 1 min (denaturation), 56° 1 min (annealing), and 72° 1 min (extension). Amplification ends with 10 min at 72° to complete the polymerization of any incomplete chains. The PCR product is restriction digested with *Bam*HI and ligated into pGEX-4T1 (Pharmacia Biotech) digested with *Bam*HI and *Sma*I. The ligase product is used to transform competent cells and for minipreps and large-scale plasmid purification. A similar procedure is used to generate the GRK4-Nter fusion protein.

Recombinant GST-Nter-GRK2 and GST-Nter-GRK4 fusion proteins are expressed in *Escherichia coli* BL21 by induction for 3 h with 1 m*M* isopropyl-D-thiogalactoside. The fusion proteins are purified essentially according to Frangioni and Neel (1993). Ten milliliters of bacteria is grown overnight in LB plus 100 μg/ml ampicillin. The day after, 100 ml of sterile LB medium is inoculated with 2 ml of the overnight culture, and the bacterial suspension bacteria is grown at 37° for 1 h with constant shaking at 225 rpm. At this point, the polypeptide is induced for 3 h at 37° with 1 m*M* isopropyl-D-thiogalactoside. The bacterial suspension is then collected in 250-ml centrifuge bottles and centrifuged for 10 min at 7000 rpm at 4° in a JA 14-type rotor (Beckman Inc). Bacteria are lysed by sonication in STE buffer [10 m*M* Tris–HCl, pH 7.5, 150 m*M* NaCl, 5 m*M* dithiothreitol (DTT), 1 m*M* EDTA] containing 100 μg/ml lysozyme, 1.5% *N*-laurylsarcosine, and protease inhibitor cocktail. After clarification by a centrifugation step (10,000g for 5 min), the lysate is incubated with glutathione agarose beads for 1 h at 4°, washed eight times with ice-cold phosphate-buffered saline (PBS), and placed in storage buffer I (75 m*M* HEPES, 150 m*M* NaCl, 5 m*M* DTT, 10% glycerol) until used.

Purification of $G\alpha_q$ Protein. The purification of $G\alpha_q$ is performed according to Kozasa and Gilman (1995). Sf9 cells are cultured on a monolayer using TNM-FH (Life Technology) medium containing 10% fetal bovine serum and antibiotics (2.5 μg/ml fungizone, 50 μg/ml streptomicin, 50 μg/ml penicillin). To produce recombinant $G\alpha_q$, Sf9 cells are infected with baculoviruses encoding the $G\alpha_q$ subunit, the β1 subunit, and the His6-γ2 subunit (one plaque-forming unit per cell for each virus). Seventy-two hours later, the cells are harvested and lysed in ice-cold lysis buffer (50 m*M* Na HEPES, pH 8, 100 m*M* NaCl, 0.1 m*M* EDTA, 3 m*M* MgCl$_2$, 10 m*M* 2-mercaptoethanol, 50 μ*M* GDP, plus a mixture of protease inhibitors and 0.03 mg/ml lima bean trypsin inhibitor) using a polytron tissue disrupter (Janke and Kundel) at low speed for 40 s on ice. Unbroken cells and cell nuclei are pelleted by centrifugation (800g for 5 min) and discarded. The supernatant is then centrifuged at 100,000g for 30 min at 4°. The resultant pellet is suspended in wash buffer (50 m*M* Na HEPES, pH 8, 50 m*M* NaCl, 3 m*M* MgCl$_2$, 10 m*M* 2-mercaptoethanol, 50 μ*M* GDP, plus a mixture of protease inhibitors) and centrifuged at 100,000g for 30 min at 4°. The washed pellet (membranes) is frozen in liquid nitrogen and stored at −80°. Membranes are thawed and resuspended to 5 mg/ml in the ice-cold wash buffer. Sodium cholate is added to a final concentration of 1%

(w/v), and the mixture is stirred on ice for 1 h prior to centrifugation at 100,000g at 4° for 40 min. The remaining procedures are carried out at 4° unless otherwise specified. The supernatants (membrane extract) are collected, diluted five-fold with buffer A (20 mM Na HEPES, pH 8, 100 mM NaCl, 1 mM MgCl$_2$, 10 mM 2-mercaptoethanol, 50 μM GDP, 0.5% BRIJ 58, plus a mixture of protease inhibitors) and loaded onto a Ni-NTA column (Qiagen, Chatsworth, CA), which has been equilibrated with buffer A. The column is washed with 25 column volumes of buffer B (20 mM Na HEPES, pH 8, 300 mM NaCl, 1 mM MgCl$_2$, 10 mM 2-mercaptoethanol, 50 μM GDP, 0.5% BRIJ 58, 5 mM imidazole, plus a mixture of protease inhibitors). The column is then incubated at room temperature with buffer C (20 mM Na HEPES, pH 8, 100 mM NaCl, 0.2 mM MgCl$_2$, 10 mM 2-mercaptoethanol, 5 μM GTPγS, 0.2% sodium cholate, 5 mM imidazole, plus a mixture of protease inhibitors) and washed with the same buffer C. This step is utilized to elute an endogenous (Sf9 cell-derived) Gα_i-like protein. The Gα_q is eluted by an AlF$_4^-$-containing buffer (20 mM Na HEPES, pH 8, 50 mM NaCl, 50 mM MgCl$_2$, 10 mM 2-mercaptoethanol, 50 μM GDP, 1% sodium cholate, 5 mM imidazole, 50 μM AlCl$_3$, 10 mM NaF, plus a mixture of protease inhibitors). The eluted Gα_q is inactivated by exchanging the elution buffer with storage buffer (20 mM HEPES, pH 8.0, 100 mM NaCl, 1 mM EDTA, 3 mM MgCl$_2$, 3 mM DDT, 0.7% 3-[(3-cholamidopropyl)dimethylamino]propanesulfonate, 0.5 μM GDP) and used for protein–protein interaction experiments with GST-N-ter or GST-RGS4 fusion proteins.

Binding of Gα to the GRK N-Terminal. Cytosolic proteins (150 μg) from HEK293 cells transfected with the indicated Gα subunits or 100 ng of purified Gα_q are mixed with 40 μl of slurry containing GST-GRK-Nter fusion proteins bound to glutathione agarose beads in a final volume of 400 μl of binding buffer (20 mM Tris–HCl, pH 7.5, 1 mM DTT, 100 mM NaCl, 0.1% lubrol, 10 μM GDP, 3 mM MgCl$_2$), in the presence or absence of 47 mM MgCl$_2$, 30 μM AlCl$_3$, and 20 mM NaF. After 1 h at 4°, the beads are washed three times with 1 ml of ice-cold binding buffer, and resins containing the eventual bound proteins are analyzed by immunoblotting.

Analysis of the Interaction of GRK with Gα_q in Cells

Immunoprecipitation. After treatments, cells are washed rapidly in ice-cold PBS and solubilized in Triton X-100 lysis buffer (10 mM Tris–HCl, pH 7.4, 150 mM NaCl, 1% Triton-X 100, 1 mM EDTA, 10% glycerol, 1 mM phenylmethysulfonyl fluoride, 10 μg/ml aprotinin, 1 mM sodium orthovanadate, 50 mM sodium fluoride, and 10 mM β-glycerophosphate) for 15 min. The lysates are clarified by centrifugation (10,000g for 10 min).

The protein concentration of supernatants is estimated by the modified Bradford assay. For immunoprecipitation, 600 μg of total protein is incubated with 5 μg of anti-$G\alpha_q$ antibodies for 2 h at 4° followed by the addition of 50 μl protein A-Sepharose preequilibrated in HNTG buffer (20 mM HEPES, 150 mM NaCl, 0.1% Triton X-100, 5% glycerol) and an additional 1-h incubation at 4°. Immunoprecipitates are washed four times in HNTG buffer, and the pellets are boiled in Laemmli's buffer for 5 min before electrophoresis. Immunoprecipitates and starting materials are subjected to 10% sodium dodecyl sulfate–polyacrylamide gel electrophoresis under reducing condition. After electrophoresis, proteins are transferred to a PVDF membrane, and immunoblotting is performed using anti-GRK2 (Upstate Biotechnology) or anti-GRK4 (Santa Cruz) antibodies.

References

Aragay, A. M., Ruiz-Gomez, A., Penela, P., Sarnago, S., Elorza, A., Jimenez-Sainz, M. C., and Mayor, F., Jr. (1998). G protein-coupled receptor kinase 2 (GRK2): Mechanisms of regulation and physiological functions. *FEBS Lett.* **430,** 37–40.

Aton, B. R. (1989). Illumination of bovine photoreceptor membranes causes phosphorylation of both bleached and unbleached rhodopsin molecules. *Biochemistry* **25,** 677–680.

Berman, D. M., and Gilman, A. G. (1998). Mammalian RGS proteins: Barbarians at the gate. *J. Biol. Chem.* **273,** 1269–1272.

Binder, B. M., Biernbaum, M. S., and Bownds, M. D. (1990). Light activation of one rhodopsin molecule causes the phosphorylation of hundreds of others: A reaction observed in electropermeabilized frog rod outer segments exposed to dim illumination. *J. Biol. Chem.* **265,** 15333–15340.

Carman, C. V., Parent, J. L., Day, P. W., Pronin, A. N., Sternweis, P. M., Wedegaertner, P. B., Gilman, A. G., Benovic, J. L., and Kozasa, T. (1999). Selective regulation of $G\alpha_{q/11}$ by an RGS domain in the G protein-coupled receptor kinase, GRK2. *J. Biol. Chem.* **274,** 34483–34492.

Chen, C. Y., Dion, S. B., Kim, C. M., and Benovic, J. L. (1993). β-adrenergic receptor kinase: Agonist-dependent receptor binding promotes kinase activation. *J. Biol. Chem.* **268,** 7825–7831.

De Vries, L., and Farquhar, M. G. (1999). RGS proteins: More than just GAPs for heterotrimeric G proteins. *Trends Cell Biol.* **9,** 138–143.

Dicker, F., Quitterer, U., Winstel, R., Honold, K., and Lohse, M. J. (1999). Phosphorylation-independent inhibition of parathyroid hormone receptor signaling by G protein-coupled receptor kinases. *Proc. Natl. Acad. Sci. USA* **96,** 5476–5481.

Diviani, D., Lattion, A.-L., Larbi, N., Kanapuli, P., Pronin, A., Benovic, J. L., and Cotecchia, S. (1996). Effect of different G protein-coupled receptor kinases on phosphorylation and desensitization of the alpha1B-adrenergic receptor. *J. Biol. Chem.* **271,** 5049–5059.

Eason, M. G., Moreira, S. P., and Liggett, S. B. (1995). Four consecutive serines in the third intracellular loop are the sites for β-adrenergic receptor kinase-mediated phosphorylation and desensitization of the α2A-adrenergic receptor. *J. Biol. Chem.* **270,** 4681–4688.

Firsov, D., and Elalouf, J. M. (1997). Molecular cloning of two rat GRK6 splice variants. *Am. J. Physiol.* **273,** 953–961.

Frangioni, J. V., and Neel, B. J. (1993). Solubilization and purification of enzymatically active gluthatione S-transferase (pGEX) fusion proteins. *Anal. Biochem.* **210,** 179–187.

Fredericks, Z. L., Pitcher, J. A., and Lefkowitz, R. J. (1996). Identification of the G protein-coupled receptor kinase phosphorylation sites in the human β_2-adrenergic receptor. *J. Biol. Chem.* **271,** 13796–13803.

Hamm, H. E. (1998). The many faces of G protein signaling. *J. Biol. Chem.* **273,** 669–672.

Hepler, J. R. (1999). Emerging roles for RGS proteins in cell signalling. *Trends Pharmacol. Sci.* **20,** 376–382.

Iacovelli, L., Salvatore, L., Capobianco, L., Picascia, A., Barletta, E., Storto, M., Mariggiò, S., Gallese, M., Porcellini, A., Nicoletti, F., and De Blasi, A. (2003). Role of G protein-coupled receptor kinase 4 and β-arrestin 1 in agonist-stimulated metabotropic glutamate receptor 1 internalization and activation of mitogen-activated protein kinases. *J. Biol. Chem.* **278,** 12433–12442.

Iacovelli, L., Capobianco, L., Lula, M., Di Giorgi Gerevini, V., Picascia, Balhos, J., Melchiorri, D., Nicoletti, F., and De Blasi, A. (2004). Regulation of mGlu4 metabotropic glutamate receptor signaling by type-2 G-protein coupled receptor kinase (GRK2). *Mol. Pharmacol.* **65,** 1103–1110.

Inglese, J., Luttrell, L. M., Iniguez-Lluhi, J. A., Touhara, K., Koch, W. J., and Lefkowitz, R. J. (1994). Functionally active targeting domain of the beta-adrenergic receptor kinase: An inhibitor of G beta gamma-mediated stimulation of type II adenylyl cyclase. *Proc. Natl. Acad. Sci. USA* **91,** 3637–3641.

Kameyama, K., Haga, K., Haga, T., Moro, O., and Sadee, W. (1994). Activation of a GTP-binding protein and a GTP-binding-protein-coupled receptor kinase (beta-adrenergic-receptor kinase-1) by a muscarinic receptor m2 mutant lacking phosphorylation sites. *Eur. J. Biochem.* **226,** 267–276.

Kim, C. M., Dion, S. B., and Benovic, J. L. (1993). Mechanism of beta-adrenergic receptor kinase activation by G proteins. *J. Biol. Chem.* **268,** 15412–15418.

Koch, W. J., Inglese, J., Stone, W. C., and Lefkowitz, R. J. (1993). The binding site for the beta gamma subunits of heterotrimeric G proteins on the beta-adrenergic receptor kinase. *J. Biol. Chem.* **11,** 8256–8260.

Kozasa, T., and Gilman, A. G. (1995). Purification of recombinant G proteins from Sf9 cells by hexahistidine tagging of associated subunits: Characterization of alpha 12 and inhibition of adenylyl cyclase by alpha z. *J. Biol. Chem.* **270,** 1734–1741.

Lefkowitz, R. J. (1998). G-protein coupled receptors. *J. Biol. Chem.* **273,** 18667–18680.

Ohguro, H., Palczewski, K., Ericsson, L. H., Walsh, K. A., and Johnson, R. S. (1993). Sequential phosphorylation of rhodopsin at multiple sites. *Biochemistry* **32,** 5718–5724.

Onorato, J. J., Palczewski, K., Regan, J. W., Caron, M. G., Lefkowitz, R. J., and Benovic, J. L. (1991). Role of acidic amino acids in peptide substrates of the beta-adrenergic receptor kinase and rhodopsin kinase. *Biochemistry* **30,** 5118–5125.

Palczewski, K., and Benovic, J. L. (1991). G protein-coupled receptor kinases. *Trends Biochem. Sci.* **16,** 387.

Palczewski, K. (1997). GTP-binding-protein-coupled receptor kinases. *Eur. J. Biochem.* **248,** 261–269.

Premont, R. T., Koch, W. J., Inglese, J., and Lefkowitz, R. J. (1994). Identification, purification, and characterization of GRK5, a member of the family of G protein-coupled receptor kinases. *J. Biol. Chem.* **269,** 6832–6841.

Premont, R. T., Macrae, A. D., Stoffel, R. H., Chung, N., Pitcher, J. A., Ambrose, C., Inglese, J., MacDonald, M. E., and Lefkowitz, R. J. (1996). Characterization of the G protein-coupled receptor kinase GRK4. Identification of four splice variants. *J. Biol. Chem.* **271,** 6403–6410.

Prossnitz, E. R., Kim, C. M., Benovic, J. L., and Ye, R. D. (1995). Phosphorylation of the N-formyl peptide receptor carboxyl terminus by the G protein-coupled receptor kinase, GRK2. *J. Biol. Chem.* **270,** 1130–1137.

Pulvermuller, A., Palczewski, K., and Hofmann, K. P. (1993). Interaction between photoactivated rhodopsin and its kinase: Stability and kinetics of complex formation. *Biochemistry* **32,** 14082–14088.

Sallese, M., Mariggiò, S., Collodel, G., Moretti, E., Piomboni, P., Baccetti, B., and De Blasi, A. (1997). G protein-coupled receptor kinase GRK4: Molecular analysis of the four isoforms and ultrastructural localization in spermatozoa and germinal cells. *J. Biol. Chem.* **272,** 10188–10195.

Sallese, M., Iacovelli, L., Cumashi, A., Capobianco, L., Cuomo, L., and De Blasi, A. (2000a). Regulation of G protein-coupled receptor kinase subtypes by calcium sensor proteins. *Biochim. Biophys. Acta* **1498,** 112–121.

Sallese, M., Mariggio, S., D'Urbano, E., Iacovelli, L., and De Blasi, A. (2000b). Selective regulation of Gq signaling by G protein-coupled receptor kinase 2: Direct interaction of kinase N terminus with activated Galphaq. *Mol. Pharmacol.* **57,** 826–831.

Siderovski, D. P., Hessel, A., Chung, S., Mak, T. W., and Tyers, M. (1996). A new family of regulators of G protein-coupled receptors? *Curr. Biol.* **6,** 211–212.

Tesmer, J. J. G., Berman, D. M., Gilman, A. G., and Sprang, S. R. (1997). Structure of RGS4 bound to AlF_4^--activated $Gi\alpha 1$: Stabilisation of the transition state for GTP hydrolysis. *Cell* **89,** 251–261.

Subsection G

Other RGS Proteins

[22] Identification and Functional Analysis of Dual-Specific A Kinase-Anchoring Protein-2

By L. L. Burns-Hamuro, D. M. Barraclough, and S. S. Taylor

Abstract

Since the cloning of dual-specificity A kinase-anchoring protein 2 (D-AKAP2), there has been considerable progress in understanding the structural features of this AKAP and its interaction with protein kinase A (PKA). The domain organization of D-AKAP2 is quite unique, containing two tandem, putative RGS domains, a PKA-binding motif, and a PDZ (PSD95/Dlg/ZO1)-binding motif. Although the function of D-AKAP2 has remained elusive, several reports suggest that D-AKAP2 is targeted to cotransporters in the kidney and that it may play a role in regulating transporter activity. In addition, the finding that a single nucleotide polymorphism in the PKA-binding region of D-AKAP2 may contribute to increased morbidity and mortality emphasizes the potential importance of this protein in pathogenesis. The first part of this article focuses on initial efforts to identify and clone D-AKAP2, followed by tissue localization and expression profiles. The latter half of the article focuses on the domain organization of D-AKAP2 and its interaction with PKA. Finally, a comprehensive analysis of the PKA binding motif is described, which has led to the development of novel peptides derived from D-AKAP2 that can be useful tools in probing the function of this AKAP in cellular and animal models.

Introduction

Dual-specific A kinase-anchoring protein 2 (D-AKAP2) is unique among mammalian regulator of G-protein signaling (RGS) proteins in that it contains two RGS domains and a protein kinase A (PKA)-binding motif (Fig. 1A). It was initially identified in a yeast two hybrid assay designed to screen for interacting proteins of the type I regulatory subunit (RI) of protein kinase A (Huang *et al.*, 1997a). The C-terminal 40 residues of the protein identified in the screen interacted specifically and with high affinity to three of the four regulatory subunit isoforms of PKA, defining it as a

A

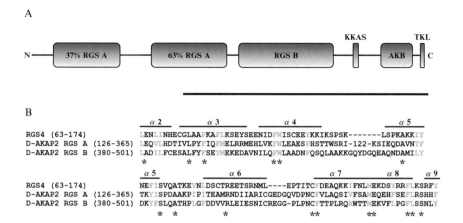

B

		α 2 α 3 α 4 α 5
RGS4	(63-174)	LENLINHECGLAAFKAFLKSEYSEENIDFWISCEEYKKIKSPSK-------LSPKAKKIY
D-AKAP2 RGS A	(126-365)	LEQVLHDTIVLPYFIQFMELRRMEHLVKFWLEAESFHSTTWSRI-122-KSIEQDAVNTF
D-AKAP2 RGS B	(380-501)	LADIILFCESALFYFSEYMEKEDAVNILQFWLAADNFQSQLAAKKGQYDGQEAQNDAMILY
		* * * ** *

		α 5 α 6 α 7 α 8 α 9
RGS4	(63-174)	NEFISVQATKEVNLDSCTREETSRNML----EPTITCFDEAQKKIFNLMEKDSYRRFLKSRFY
D-AKAP2 RGS A	(126-365)	TKYISPDAAKPIPITEAMRNDIIARICGEDGQVDPNCFVLAQSIVFSAMEQEHFSEFLRSHHF
D-AKAP2 RGS B	(380-501)	DKYFSLQATHPLGFDDVVRLEIESNICREGG-PLPNCFTTPLRQAWTTMEKVFLPGFLSSNLY
		* * * ** ** ** *

FIG. 1. (A) Domain organization of human D-AKAP2. RGS, regulator of G-protein signaling domain; KKAS, phosphorylation consensus site for PKA; AKB, A kinase-binding domain; TKL, PDZ-binding motif. (B) Sequence alignment of the two RGS domains of D-AKAP2 (RGS A and RGS B) with RGS4 using Clustal W alignment software. The asterisk (*) denotes identical residues, and residues in gray denote conserved hydrophobic residues observed for this fold (Tesmer et al., 1997). Reprinted from Hamuro et al. (2002), with permission.

"dual-specific" A kinase-anchoring protein (D-AKAP). Subsequent cloning of mouse and human genes revealed two regions of unmistakable homology to RGS domains (Hamuro et al., 2002a; Huang et al., 1997a; Wang et al., 2001) (Fig. 1B).

In general, AKAPs act as scaffold proteins, coordinating the activity of PKA by binding and localizing PKA and other signaling proteins to form localized transduction units, which serve to regulate the activity of the kinase and to increase the fidelity of receptor-specific cAMP-mediated signaling (Colledge and Scott, 1999; Edwards and Scott, 2000). As a scaffold protein, D-AKAP2 has the potential to coordinate a signaling complex that links cAMP signaling with G-protein-coupled receptor (GPCR) signaling. Although there have been no Gα-binding partners identified for D-AKAP2, there are some interesting leads as to the functional implications of this protein.

D-AKAP2 has been linked indirectly to a Na^+-dependent phosphate cotransporter (NaPi-IIa) in the brush border of the kidney via its interaction with PDZ proteins, PDZK1 and NHERF-1 (Gisler et al., 2003a,b). Both PDZK1 and NHERF-1 contain multiple PDZ domains and interact directly with the C terminus of NaPi-IIa (Gisler et al., 2001; Hernando et al., 2002). PDZK1 contains four tandem PDZ domains and is a major component in the brush border cells of the kidney (Gisler *et al.*, 2003a,b). The third PDZ domain of PDZK1 binds to the NaPi-IIa cotransporter, while the fourth domain binds to the C-terminal PDZ-binding motif of D-AKAP2 (Gisler *et al.*, 2003a). NaPi-IIa is primarily responsible for inorganic phosphate reabsorption in the proximal tubule of the kidney, which is regulated by diet and the parathyroid hormone (PTH) (Murer *et al.*, 2003). Regulated endocytosis of the cotransporter is mediated by PTH and involves a network of protein kinases, which is believed to be responsible for regulating the level of the receptor at the brush border membrane (Bacic *et al.*, 2003; Murer *et al.*, 2003). Interestingly, NHERF-1 also interacts with the Na^+/H^+ exchanger (NHE-3) and recruits PKA via the ezrin AKAP to downregulate channel activity (Weinman *et al.*, 2000, 2001). The detailed mechanism of D-AKAP2-mediated PKA regulation of the NaPi-IIa cotransporter is not known. However, the implication is that D-AKAP2 is recruiting PKA to NaPi-IIa indirectly through the PDZ proteins and is potentially involved in regulating the level of the cotransporter at the membrane by receptor endocytosis and/or vesicular trafficking (Gisler *et al.*, 2003a).

In addition to the association of D-AKAP2 with cotransporters in the kidney, genomic efforts have identified several, single nucleotide polymorphisms (SNPs) associated with the human D-AKAP2 gene, two of which are located in exons and code for amino acid sequence changes in the protein (Kammerer *et al.*, 2003). One of the SNPs is located in the N-terminal RGS domain of D-AKAP2 and codes for an arginine-to-histidine change at position 249 (R249H). The functional consequences of this residue change are not known, as no interacting G proteins have been identified. The other coding SNP is located in the PKA-binding region of the protein and codes for an amino acid change at position 646 (I646V). This sequence change alters binding to only the type I regulatory subunit (RIα) of PKA. In a search for SNPs associated with morbidity and mortality, it was found that an older population of healthy volunteers had a reduced frequency of Val at position 646 compared with a younger population. This suggests that this particular sequence change, which enhances the affinity of PKA by a factor of three, increases the mortality and/or morbidity of the aging population, removing this segment from the healthy donating pool. Although a particular disease has not been associated with the SNP, preliminary evidence suggests an association with cardiac pathology. Patients that are homozygous for the valine allele (Val/Val) have statistically distinct parameters on echocardiogram measurements (Kammerer *et al.*, 2003). Although this is not a direct indication of a disease process, this finding underscores the potential importance of this protein and continued efforts are underway to investigate the biology of D-AKAP2 further.

The first half of this article begins by describing (1) the identification of D-AKAP2 using a yeast two hybrid screen, (2) cloning and characterization of mouse and human proteins, (3) *in Vitro* and *in vivo* association of the protein with PKA, and (4) protein expression levels in mouse tissue and evidence for mitochondrial localization. The second half of the article presents a detailed characterization of the domain structure of this protein, including a model of the RGS domain of D-AKAP2, and describes how we have been able to optimize the selectivity of the R-binding motif of D-AKAP2 using peptide arrays. This has enabled the design of isoform-selective peptide disruptors of AKAP-mediated PKA localization that can be used to elucidate the function of this dual-specificity AKAP in cell and animal studies.

Identification and Cloning of D-AKAP2

D-AKAP2 was initially identified in a yeast two hybrid screen using Ret/ptc2 as the bait (Huang *et al.*, 1997a). Ret/ptc2 is an oncoprotein with the C-terminal domain of the Ret receptor tyrosine kinase fused to the N terminus of the RIα subunit of PKA that contains the dimerization/docking (D/D) domain and is critical for AKAP binding. After screening a mouse embryonic cDNA library, several interacting clones were identified, eight of which coded for novel protein fragments designated R potential binding protein (RPP): RPP7, RPP8, and RPP9. Both RPP7 and RPP8 were identified as novel binding proteins of PKA and were characterized further. RPP7 was identified as D-AKAP1 and characterization of this protein can be found elsewhere (Huang *et al.*, 1997b, 1999; Ma and Taylor, 2002).

To obtain the full-length clone of RPP8, an adult mouse testis cDNA library expressed in λgt10 was screened using ^{32}P-labeled RPP8 cDNA as described in Huang *et al.* (1997a). Positive phage cDNAs were isolated and cloned into pBluescript KS(+). The cDNAs were sequenced and used to search the BLAST database at the National Library of Medicine, National Institutes of Health. From the mouse testis cDNA library, the full-length cloned appeared to code for a 372 amino acid protein (Fig. 1A, black bar). The cDNA contained a stop codon just upstream of what was believed at the time to be the initiator methionine, followed by a stop codon and a poly(A) tail. There was no overall nucleotide sequence homology to known proteins when the sequence was compared with sequences from the GenBank database. However, amino acid residues 70–125 showed homology to RGS domains from several RGS proteins.

After the cloning of mouse D-AKAP2, it became apparent from western blots of mouse tissue extracts using several polyclonal antibodies directed against different regions of the protein that at least some expressed protein contained a higher apparent molecular mass than the coding sequence predicted. Estimates from the western blot were 70–100 kDa, whereas the molecular mass of the coding sequence was ~43 kDa. Verification of a larger D-AKAP2 gene came from Fisher and Chatterjee when they submitted the gene sequence for human D-AKAP2 (GenBank ID AF037439) to the National Center for Biotechnology Information (NCBI), which suggested a larger coding region. To help resolve this discrepancy, human D-AKAP2 was cloned.

Human D-AKAP2 was cloned from a human brain cDNA library (Clontech) using 5′ and 3′ primers to the coding region of the human gene as described in Wang et al. (2001). After amplifying the cDNA by polymerase chain reaction (PCR), a 2.5-kb insert was ligated into a plasmid (pSP73) and sequenced. The cDNA of human D-AKAP2 codes for a 74-kDa, 662 amino acid protein. Using the BLAST human genome database, the gene was localized to chromosome 17q and contained 15 exons. The previously cloned mouse gene, which coded for 372 amino acids, was 94% identical to the C terminus of the human protein. This region included the C-terminal RGS domain and the PKA-binding site. A longer mouse sequence for D-AKAP2 (GenBank ID AK049399) has been submitted to GenBank by the Fantom Consortium and the Riken Genome Exploration Research Group (Okazaki et al., 2002). This sequence was identified from a 7-day mouse embryonic, whole body cDNA library and codes for a 662 amino acid protein that is 93% identical to human D-AKAP2, verifying that cDNAs for both mouse and human code for identical length, highly homologous proteins. However, protein expression patterns from mouse tissue suggest that alternative length proteins can be expressed in a tissue-specific manner, as described later.

D-AKAP2 Interacts with Regulatory Subunit Isoforms of PKA

The interaction of D-AKAP2 with PKA was initially verified by expressing the C-terminal 40 residues of D-AKAP2 (RPP8) as a GST fusion protein using the plasmid pGEX-KG as described in Huang et al. (1997a). The RPP8 fusion protein, designated GST-RPP8, is expressed in *Escherichia coli* BL21(DE3). Bacteria are grown at 37° for approximately 8 h until an optical density of 0.8 absorbance units at 600 nm is reached. After lowering the temperature to 24°, protein expression is induced with 0.5 mM isopropyl β-D-thiogalactoside for 10–12 h. The cell pellet is frozen at −20°. Cell pellets are lysed by sonication in PBS (10 mM potassium phosphate, 150 mM NaCl, pH 7.4) with 0.1% Triton X-100, 1 mM EDTA, 5 mM 2-mercaptoethanol, and the protease inhibitors 1 mM phenylmethylsulfonyl fluoride and 5 mM

benzamidine. After centrifugation, the soluble fraction containing GST-RPP8 is incubated with glutathione resin for 2 h at 4°. Resin-bound GST-RPP8 is rinsed extensively with the same buffer used for cell lysis, and 100–200 μg of purified PKA regulatory subunit (RIα, RIβ, RIIα, or RIIβ) is added and incubated for 2 h at 4°. After extensive washing of the resin with PBS, the bound regulatory subunit is eluted by boiling in SDS gel-loading buffer and is analyzed by SDS–PAGE. Three of the four regulatory subunit isoforms of PKA (RIα, RIIα, and RIIβ) interact specifically with GST-RPP8 and not with a GST alone control (Huang *et al.*, 1997a).

Making use of the fact that the regulatory subunits of PKA bind cAMP with high affinity, *in vivo* association of D-AKAP2 with regulatory subunit isoforms of PKA in mouse brain extracts was investigated using a cAMP resin (Wang *et al.*, 2001). Mouse brain extracts are prepared by homogenizing the tissue in cold homogenization buffer [20 mM HEPES, pH 7.4, 20 mM NaCl, 5 mM EDTA, 5 mM EGTA, 0.5% Triton X-100, 1 mM dithiothreitol (DTT)] and a protease inhibitor cocktail (Calbiochem). After centrifuging at 10,000g for 30 min, the supernatants are incubated with cAMP agarose (Sigma) overnight. As a negative control, a similar incubation is carried out in the presence of 50 mM cAMP, which serves as a competitive inhibitor of regulatory subunit binding to the cAMP affinity resin. The resin is washed several times with a high salt wash buffer (10 mM HEPES, pH 7.4, 1.5 mM MgCl$_2$, 10 mM KCl, 0.5 M NaCl, 0.1% Nonidet P-40, 1 mM DTT, and protease inhibitors) and a low salt buffer (high salt buffer without NaCl). Any associated protein is eluted with 25 mM cAMP by rotating for 1 h at room temperature. The eluted protein is precipitated (10% TFA), and interacting proteins are detected by western blot using commercially available antibodies for RIα, RIIα (Transduction Laboratories, Lexington, KY), RIIβ (Biomol, Plymouth Meeting, PA), and anti-D-AKAP2. D-AKAP2 associates with both RIIα and RIIβ, but not RIα in this assay (Wang *et al.*, 2001). The inability to detect an interaction with RIα may be due to the reduced binding affinity to this isoform as described later, or to the fact that the RIα isoform is not detected at appreciable levels in brain extracts.

RNA and Protein Expression Levels of D-AKAP2 and Mitochondrial Localization

The mRNA and protein expression levels for D-AKAP2 are evaluated from various mouse tissues (Huang *et al.*, 1997a; Wang *et al.*, 2001). To investigate the tissue expression level of D-AKAP2 mRNA, Northern blots containing 2 μg of poly(A)$^+$ RNA from adult mouse tissues are probed with ^{32}P-labeled RPP8 cDNA. A 5-kb mRNA is detected in all mouse tissues examined (Huang *et al.*, 1997a). A weak 10-kb signal is detected in the brain, skeletal muscle, kidney, and testis. Furthermore, the testes contain additional smaller mRNA fragments of 3.4, 2.8, and 1.8 kb, suggesting alternative processing in the testes.

D-AKAP2 is expressed ubiquitously in mouse tissue. Western blots using polyclonal antibodies against D-AKAP2 reveal that the mouse tissue examined (white adipose tissue, brown adipose tissue, skeletal muscle, tongue, small intestine, kidney, lung, brain, pancreas, heart, spleen, and liver) express some level of the protein (Wang *et al.*, 2001). The protein is most abundant in heart, brain, pancreas, and brown adipose tissue with variations in molecular weight seen throughout different tissues. For example, two prominent bands are detected in brain extracts that have very similar molecular weights. In addition, brown adipose tissue, skeletal muscle, small intestine, kidney, and heart all have high molecular weight and low molecular species, suggesting alternative mRNA splicing or degradation of the protein in these tissues (Wang *et al.*, 2001). Gisler *et al.* (2003a) have cloned D-AKAP2 cDNA from a mouse kidney cDNA library and found that the cDNA coded a protein 372 amino acids, similar to the initial clone identified from an adult mouse testes library (Huang *et al.*, 1997a). The molecular mass of endogenously expressed D-AKAP2 from adult mouse kidney was slightly larger than 40 kDa, consistent with the cDNA clone from kidney (Gisler *et al.*, 2003a). Therefore, there appears to be heterogeneity in the length of the protein expressed in mouse and this heterogeneity seems to be tissue specific and suggestive of alternative splicing.

The cellular distribution of D-AKAP2 was evaluated using fractionated mouse brown adipose tissue and by staining intact cells (Wang *et al.*, 2001). Mitochondria are isolated from cell lysates using centrifugation and a sucrose gradient. Homogenized brown adipose tissue in 250 mM sucrose is spun at 8500g to obtain the cytosolic fraction. The pellet is resuspended and recentrifuged at 800g to remove the nuclear pellet. The supernatant is centrifuged for an additional 8500g to obtain the mitochondrial pellet. The mitochondrial pellet is washed in 100 mM KCl with 1% bovine serum albumin (BSA) and the quality of the mitochondria is verified by electron microscopy (Wang *et al.*, 2001). Aliquots from the cytosolic, nuclear, and mitochondrial fractions are resolved by SDS–PAGE and the protein is detected by a western blot. D-AKAP2 is found in the cytosolic, nuclear, and mitochondrial fractions from brown adipose tissue with a greater concentration in the mitochondria and nuclear fractions. Immunostaining of intact cells revealed that D-AKAP2 can associate with the mitochondria in addition to being cytoplasmic. Mitochondrial localization was observed in both a mouse cell line (C2 mouse myocytes) and primary rat cardiomyocytes

(Wang *et al.*, 2001). The low level of cytoplasmic and nuclear staining, in addition to the concentrated mitochondrial staining pattern for this protein, suggests that there may be multiple intracellular pools of D-AKAP2 in the cell. Immunohistochemistry and confocal microscopy of rat tissue revealed that at the neuromuscular junction, D-AKAP2 is rather diffuse as well and is present in both presynaptic and postsynaptic borders (Perkins *et al.*, 2001). Interestingly, in surrounding muscle, D-AKAP2 is present near the actin region and does not colocalize with mitochondria (Perkins *et al.*, 2001). This suggests that D-AKAP2 localization with the mitochondria is dependent on the cell type. This, in addition to the ubiquitous expression patterns, suggests that D-AKAP2 may have multiple cellular roles and/or is involved in a basal cellular function.

Defining the Domain Organization of D-AKAP2: Modeling the RGS Domain

AKAPs in general are large proteins that contain multiple recognition domains for many different interacting proteins. Several approaches were taken to delineate structural features of D-AKAP2. Initial sequence homology modeling revealed two regions of D-AKAP2 that contained homology to RGS domains. The domain boundaries of these RGS domains were defined further and a structural homology model of the C-terminal RGS domain was constructed using the known structure of RGS4 (Hamuro *et al.*, 2002a). In addition, the domain boundaries of the protein were evaluated using limited trypsin proteolysis and isotopic amide hydrogen exchange, which measures the exchange rate of backbone amide hydrogens in a protein (Hamuro *et al.*, 2002a).

Human D-AKAP2 was used to search the NCBI CDD database using RPS-BLAST. Two domains matching SMART 000315 and Pfam PF00615 were identified as RGS domains (Fig. 1A). The N-terminal RGS domain (RGS A) was split into two segments representing 37% (residues 125–168) and 63% (residues 292–368) of the domain. Interestingly, the split segments were separated by an ~122 amino acid insert that maps between helix 4 and 5 of the predicted RGS domain based on homology modeling to RGS4 (Fig. 1B). The C-terminal RGS domain (RGS B) includes residues 380–505. The presence of two RGS domains and the split nature of RGS A are features recognized to be present only in lower eukaryotic RGS domain-containing proteins (De Vries *et al.*, 2000). In a phylogenetic classification of RGS domain-containing proteins, the RGS B domain of D-AKAP2 was found to be distinct from other mammalian proteins (Zheng *et al.*, 1999). This may explain the inability to detect an interaction with traditional Gα proteins.

The PKA-binding site of D-AKAP2 [A kinase-binding (AKB) domain] is located at the C terminus of the protein (Fig. 1A). The AKB domain is characteristic of other AKAP proteins in that there are repeating units of hydrophobic (particularly branched chain residues) and hydrophillic residues, which form an amphipathic helical docking motif. This motif is described in detail in the next section. The last C-terminal residues of the protein contain a PDZ (PSD95/Dlg/ZO1)-binding motif. As described earlier, this motif has been implicated in tethering D-AKAP2 to a NaPi-IIa cotransporter in the brush border of the kidney via PDZ proteins (Gisler et al., 2003a).

A structural homology model of the RGS B domain of D-AKAP2 was constructed using the SwissModel server as described in Hamuro et al. (2002a). The X-ray structure coordinates for chain E of AlF_4^--activated $G_{i\alpha1}$ with RGS4 from rat (PDB ID: 1AGR_E) used as a template. RGS4 is a helical protein containing nine alpha helices connected by turns and loops (Tesmer et al., 1997). The amino acid sequence of RGS B was threaded onto the crystal structure of RGS4, including helices 2 through 9. Helix 1 did not show significant homology to RGS4 and was not used. Once the sequence was threaded onto the structure using the first approach mode of the SwissModel server, the final structural model was subjected to additional energy minimization using the GROMOS 43b1 force field as implemented in the DeepBlue SwissPDB viewer (Guex and Peitsch, 1997) using 5000 cycles of steepest descent followed by 5000 cycles of conjugate gradients. All computations were done in vacuo, without reaction field. Harmonic constraints of 50 C factors and 2500 C factors were used for each method. The final model was validated using the Biotech Validation Suite for Protein structures (http://biotech.ebi.ac.uk:8400). The aligned structures of RGS4 and the modeled RGS B domain of D-AKAP2 are shown in Fig. 2A. The alignment of the backbone α carbons of the RGS4 template and RGS B were within 0.7 Å with a Z-score of 6.0, indicating a high level of structural similarity.

In addition to sequence and structural homology modeling of D-AKAP2 to predict the domain organization of the protein, the folded regions of the protein were probed experimentally using hydrogen/deuterium exchange mass spectrometry (DXMS). Isotopic exchange of protein backbone amide hydrogens has been around for many decades, reviewed in Englander et al. (1997). The idea behind this approach is that if a protein is exposed to deuterium (^{2}H), labile hydrogens will exchange at some rate with the isotope, resulting in a heavy atom derivative at the site of exchange. The heavy atom can then be followed by nuclear magnetic resonance or mass spectrometry. Amide hydrogens from the protein backbone exchange with deuterium over a range of timescales that correlate with the

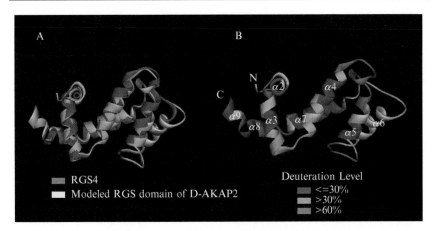

FIG. 2. (A) A backbone structural alignment of RGS4 and modeled RGS B domain of D-AKAP2. (B) Deuterium exchange data after 3000 s on exchange mapped onto the structure of the modeled RGS B domain. Reprinted from Hamuro *et al.* (2002), with permission. (See color insert.)

degree of solvent exposure of the hydrogen and the structural context (i.e., hydrogen-bonding network) of the backbone amide hydrogen (Engen and Smith, 2001). This enables the technique to be used to map folded regions of proteins (Hamuro *et al.*, 2002b; Resing *et al.*, 1999; Yan *et al.*, 2002; Zhang and Smith, 1993), protein–protein interactions (Anand *et al.*, 2002; Ehring, 1999; Mandell *et al.*, 2001), and small ligand-binding sites (Andersen *et al.*, 2001; Engen *et al.*, 1999; Hamuro *et al.*, 2002c). An overview of this technique to evaluate the conformational properties of a protein is illustrated in Fig. 3. The protein is incubated in deuterated buffer for various lengths of time ranging from 10 s to several days. The deuterated protein is then quenched in acidic conditions to prevent back-exchange of the deuterium and the sample is cleaved with an acid-insensitive protease (pepsin), which cleaves the protein into numerous peptide fragments of varying lengths. The peptides are resolved over a C18 column in line with an electrospray mass spectrometry and the resulting mass of the peptides is determined. Peptides from regions of the protein that are highly solvent exposed will contain a greater number of deuterons (higher MW) than peptides from regions that are in solvent-inaccessible regions or folded regions of the protein. A detailed description of this technique can be found in other reviews (Engen and Smith, 2001; Englander *et al.*, 1997; Hamuro *et al.*, 2003; Hoofnagle *et al.*, 2003). The following is a summary of the technique used for the analysis of D-AKAP2 (Hamuro *et al.*, 2002b).

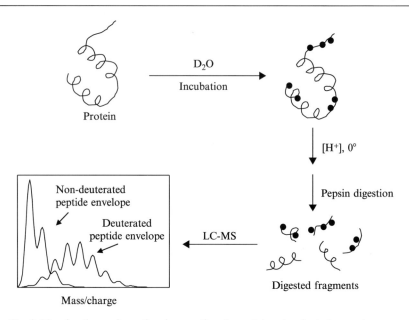

FIG. 3. Mapping the conformational properties of a protein using deuterium exchange mass spectrometry (DXMs). The protein is incubated in a deuterated buffer solution for various times ranging from 10 s to several days. To determine the location of the exchanged hydrogens, the protein is quenched under acidic conditions near $0°$ and is digested into numerous peptide fragments using an acid-insensitive protease (i.e., pepsin). The peptide fragments are separated by high-performance liquid chromatography (HPLC), and the extent of deuteration is determined by measuring the mass-to-charge ratio of each peptide isotopic envelope by mass spectrometry.

Full-length human D-AKAP2 (662 amino acids) is not expressed as a soluble protein in bacteria. We use a soluble form of mouse D-AKAP2 (375 amino acids), which includes the C-terminal half of RGS A, the full RGS B domain, the PKA-binding site, and the PDZ-binding motif for DXMS experiments. To determine the folded regions of D-AKAP2, the pepsin-cleaved peptide fragments are first identified using LC-MS/MS and Sequest software (ThermoFinnigan, Inc). Peptides representing >90% of the D-AKAP2 sequence are selected for deuterium exchange experiments. For a typical experiment, deuterated samples of D-AKAP2 are prepared by diluting 1 μl of a D-AKAP2 stock solution (50 μM) with 19 μl of deuterated buffer (10 mM HEPES, pH 7.4, 150 mM NaCl) following on-exchange incubation times of 10–3000 s. The samples are then quenched in 30 μl of 0.8% formic acid, 3.2 M GuHCl, on ice to prevent back exchange. The sample is injected into a pepsin column (66-μl bed volume, Upchurch

Scientific) up front of a C18 column that are both in line with an electro-spray mass spectrometry (Micromass LC-Q), as reviewed in Woods and Hamuro (2001). Nondeuterated (ND) and fully deuterated (FD) (in 95% acidic deuterated buffer for 48 h at room temperature) control samples are processed similarly. The molecular weights of the peptide fragments are determined by measuring the centroid of the peptide isotopic envelope using Magtran software. The percentage deuteration level for each peptide is calculated as

$$\% \text{ deuteration level} = [(m(P)-m(N))/(m(F)-m(N))] \times 100$$

where $m(P)$, $m(N)$, and $m(F)$ are the centroid values of partially deuter-ated, nondeuterated, and fully deuterated peptides, respectively.

The percentage deuteration levels for D-AKAP2 at room temperature between 10 and 3000 s are plotted in Fig. 4. The blue colors represent regions that have a low percentage deuteration and represent regions of the protein that are either folded and/or solvent inaccessible. The red regions of the protein are areas that are more solvent accessible and/or less structured regions of the protein. The RGS domains, AKB domain, and the PDZ motif are indicated with a box. The slower exchanging regions of the protein are located in or near these domains. Sites that are suscepti-ble to trypsin or GluC protease cleavage are indicated with arrows and map primarily within fast exchanging regions. A highly disordered region connects RGS B with the PKA-binding site. Interestingly, this region contains a consensus site for PKA phosphorylation (highlighted with an asterisk in Fig. 4) and is phosphorylated by PKA *in vitro* (data not shown). However, the structural or functional consequences of this phosphorylation are not known. If deuterium exchange data are plotted onto the modeled structure of RGS B from D-AKAP2, it is clear that the slowly exchange regions are located in the α helices and the faster exchanging regions map to the turns and loops (Fig. 2B). Exchange data support the structural model of this domain. The turn between helix 5 and 6 is the most sol-vent-accessible region of the RGS domain. This region in RGS4 is impor-tant for stabilizing the switch region of Gα and is implicated in enhancing GTPase activity.

The AKB domain is heavily protected from exchange in the intact protein, suggesting that this region is highly ordered in the absence of PKA (Fig. 4). In contrast, the PDZ-binding motif appears to be more disordered given the faster exchange rate for this region. This may be important for the access of interacting PDZ domains. It will be important to establish if the PKA-binding site is altered conformationally by PDZ domain binding.

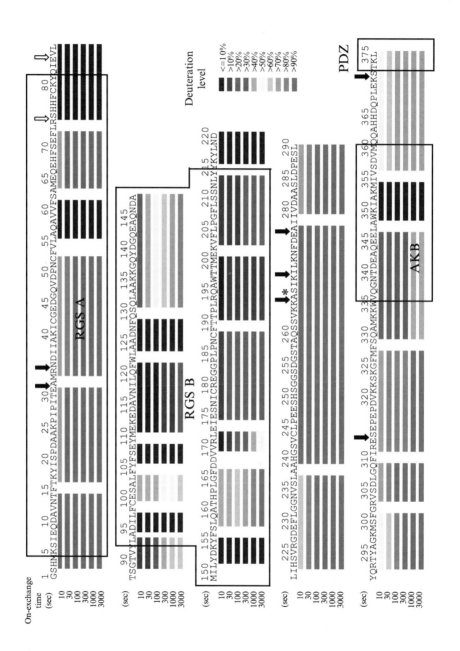

Designing Peptide Disruptors of AKAP-Mediated PKA Localization Using the AKB Domain of D-AKAP2

D-AKAP2 interacts with both type I and type II regulatory subunits of PKA. The regulatory subunits of PKA are not functionally redundant in the cell, which raises an interesting possibility that D-AKAP2 can tether two different types of PKA responsiveness. Type I isoforms are generally diffuse in the cytoplasm with some specific examples of cellular localization, whereas type II isoforms usually have distinct localization patterns in cells (Skalhegg and Tasken, 2000). Type I holoenzymes are also activated by lower concentrations of cAMP relative to the type II holoenzyme, suggesting that the type I enzyme can be a more sensitive trigger for cAMP-dependent kinase activation (Feliciello et al., 2001; Skalhegg and Tasken, 2000). The type I isoform (RIα) has been implicated in cancer (Cho-Chung et al., 1995), immune cell regulation (Torgersen et al., 2002), and the Carney complex (Casey et al., 2000; Kirschner et al., 2000), a syndrome characterized by benign tumor formation, especially in heart tissue. More recent evidence has shown that a single nucleotide polymorphism identified in human D-AKAP2 is associated with morbidity and/or mortality (as described earlier) and affects binding to only the type I isoform (RIα). These unique features attributed to RIα have prompted us to design an AKAP peptide that can bind tightly and specifically to the dimerization/docking (D/D) domain of both RIα and RIIα (Burns-Hamuro et al., 2003). By binding to the D/D of the regulatory subunit, the peptide can displace the localization of the holoenzyme (R_2C_2) in an isoform-dependent manner and shed light on PKA isoform-specific signaling processes.

In collaboration with Sequenom, Inc. (San Diego, CA) and Jerini AG (Berlin, Germany), we constructed peptide substitution arrays consisting of 27 amino acids from the AKB domain of D-AKAP2 to map the sequence

---

FIG. 4. Domain organization of D-AKAP2 probed by deuterium exchange mass spectrometry. A soluble form of mouse D-AKAP2 (372 amino acids), which included the C-terminal half of RGS A, the full RGS B domain, the PKA-binding site, and the PDZ-binding motif, was used for DXMS experiments. The deuteration level of several D-AKAP2 peptides after on exchange for 10–3000 s is presented. The peptides are denoted under the sequence of D-AKAP2 as color bars, representing the extent of deuteration (see legend). The on-exchange time is listed to the left of the sequence. Blue represents heavily protected areas, suggesting an ordered region and red represents highly exchanging regions, suggesting a disordered region. Residues corresponding to the partial RGS A domain, the full RGS B domain, the AKB domain, and the PDZ-binding motif are indicated. Limited proteolysis cleavage sites by either trypsin or Glu-C are illustrated after 1 (black arrows) and 24 (white arrows) h of digestion. The asterisk above serine 267 is highlighting an in vitro PKA phosphorylation site. Reprinted from Hamuro et al. (2002), with permission. (See color insert.)

requirements for D-AKAP2 binding to type I and type II regulatory subunits (Burns-Hamuro *et al.*, 2003). Each position in the 27 residue peptide is replaced by the other 19 L-amino acids using standard SPOT-synthesis protocols (Frank, 2002). The peptides are synthesized on an amino-functionalized cellulose membrane as distinct spots using a SPOT synthesizer with a β-alanine dipeptide spacer included between the C terminus of the peptide and the membrane support. The peptides are extended by using standard fluorenylmethoxycarbonyl (FMOC) solid-phase peptide synthesis (Wellings and Atherton, 1997), followed by cleavage of the side chain protection groups using trifluoroacetic acid (Guy and Fields, 1997). Selected peptides on the membrane are synthesized in duplicate and cleaved from the membrane using ammonia vapor in the dry state (Wenschuh *et al.*, 2000) and the sequence is verified by mass spectrometry.

To screen for regulatory subunit binding to the peptide array membranes, the membranes are treated similarly to western blots. The peptide arrays are preincubated with T-TBS blocking buffer (TBS, pH 8.0/0.05% Tween 20 in the presence of a blocking reagent; Roche Diagnostics chemiluminescence detection kit 1500694). The arrays are then incubated with GFP fusion proteins of each regulatory subunit D/D domain; GFP-RIαD/D and GFP-RIIαD/D at a final concentration of 1.0 μg/ml for 2 h in T-TBS blocking buffer. After three 10-min washes, anti-GFP antibody 3E6 (Quantum Biotechnologies, Montreal) is added at 1 μg/ml for 1 h. After extensive washing, antimouse IgG peroxidase-labeled antibody (Sigma) is applied for 1 h at 1 μg/ml. Bound protein is detected using a chemiluminescence substrate and the LumiImager (Roche Diagnostics).

Results from the substitution array verified that the AKB-binding motif was α helical. There is a distinct repeating pattern of substituted residues spanning residue 12–25 that dramatically affect binding to the regulatory subunits (Fig. 5). These critical binding epitopes map to one face of an α helix. It is also apparent that the binding interaction for RII is more localized, whereas the RI interaction is dispersed over more residues (Fig. 5, highlighted by horizontal bar). Several key differences at particular substitutions are also evident (Fig. 5, circled residues). For example, substituting valine at position 21 with a tryptophan maintains binding to RI but abolishes binding to RII. The substitution arrays provided initial leads as to which substitutions could be engineered into the D-AKAP2 AKB motif to give desired isoform-selective binding.

Individual peptides are synthesized that contain single or multiple mutations that were suggested from the peptide array as having selective binding to either RI or RII. A C-terminal cysteine is engineered into each peptide during synthesis to facilitate conjugation of a fluorescent probe.

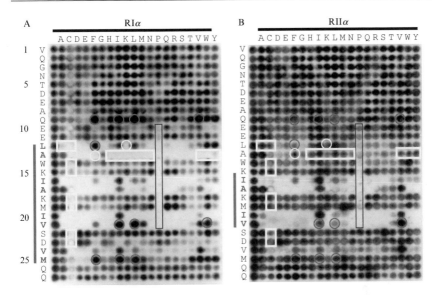

FIG. 5. A peptide substitution array of the 27 residue AKB domain of D-AKAP2 probed with either RIα (A) or RIIα (B) using an antibody sandwich detection system as described in the text. Each spot represents an immobilized peptide prepared by SPOT synthesis. All 20 amino acids (top) are substituted along each position of the AKB domain (left). Residues most sensitive to substitution are indicated in red. The bar to the left of each array highlights the region sensitive to substitution. Residues that are circled green represent peptides that show preferential binding to RIα. Residues that are boxed yellow represent peptides that show preferential binding to RIIα. Proline substitutions disrupt binding to both RI and RII along a defined sequence length (blue box). Reprinted from Burns-Hamuro et al. (2003), with permission. (See color insert.)

The peptide is labeled fluorescently with tetramethylrhodamine-5-maleimide (TMRM) (Molecular Probes), which is dissolved in dimethyl sulfoxide at 25 mM. The peptides are labeled by incubating with a threefold molar excess of TMRM for 16 h at 4° in 20 mM Tris, pH 7.0, and 1 mM Tris-(2-carboxyethyl) phosphine, hydrochloride (TCEP) (a nonthiol-reducing agent). The labeled peptides are resolved using a reverse-phase C18 column with a water/acetonitrile gradient containing 0.1% trifluoroacetic acid. The concentration of each peptide is determined at 541 nm using the extinction coefficient for the fluorescent probe (91,000 M^{-1} cm^{-1}).

Fluorescence polarization is used to quantitate the binding of each peptide to the regulatory subunit isoforms. This technique is ideal for measuring the binding of small molecular weight species (such as peptides) to large molecular weight proteins, as the property being measured is mobility of the labeled peptide. When the peptide becomes bound to a large protein, the mobility of the fluorophore-labeled peptide can be sufficiently reduced so as to result in a change in fluorescence polarization. This technique allows for solution determination of affinities and requires no separation of bound species. RIα and RIIα are diluted serially with either 1 or 10 nM of fluorescently labeled peptide in 10 mM HEPES, pH 7.4, 0.15 M NaCl, 3 mM EDTA, and 0.005% Surfactant P20. The samples are incubated for at least 1 h at room temperature, and fluorescence polarization is monitored using a Fluoromax-2 (JY Horiba, Edison, NJ) equipped with Glan-Thompson polarizers. The sample is excited at 541 nm (5- to 10-nm bandpass) and emission is monitored at 575 nm (5- to 10-nm bandpass). Three separate binding experiments are averaged and fit to a 1:1 binding model using the nonlinear application in GRAPHPAD PRISM version 3.00 (GraphPad, San Diego, CA). Table I summarizes the dissociation constants obtained for several of the peptides examined.

Both RI and RII selective binding peptides were designed using the peptide array approach. Experiments are underway to test whether these peptides, when added exogenously to a cell system, can selectively disrupt PKA-RI- vs PKA-RII-mediated anchoring. Preliminary evidence suggests that these peptide sequences will have the desired selectively *in vivo* as well. The peptide sequence that binds selectively to RI *in vitro* also only colocalizes with cotransfected RI and not RII as demonstrated in an artificial targeting system described previously (Burns-Hamuro *et al.*, 2003). In addition, the RII selective binding peptide also colocalizes with RII and not RI using this system (Burns-Hamuro *et al.*, 2003).

TABLE I

SUMMARY OF THE DISSOCIATION CONSTANTS (K_D) FOR PKA ISOFORM-SELECTIVE BINDING PEPTIDES

Peptide[a]	RIα (nM)	RIIα (nM)
WT (dual)	48 ± 4	2.2 ± 0.2
VQGNTDEAQEELAWKIAKMIVSDVMQQ		
AKB (PKA-RI)	5.2 ± 0.5	456 ± 33
FEELAWKIAKMIWSDVFQQ		
AKB (PKA-RII)	2493 ± 409	2.7 ± 0.1
VQGNTDEAQEELLWKIAKMIVSDVMQQ		
AKB (PKA-Null)	998 ± 66	>10,000
VQGNTDEAQEELAWKIEKMIWSDVMQQ		

[a] Substituted residues are underlined in bold. Reprinted from Burns-Hamuro *et al.* (2003), with permission.

Conclusions

D-AKAP2 is an intriguing protein and evidence suggests its functional importance as an AKAP, as a PDZ-binding protein, and as a potential GPCR regulator. The implication that D-AKAP2 may be imported for recruiting PKA for vesicular trafficking and/or receptor endocytosis of kidney transporters is the first step toward identifying the functional significance of this protein. However, ubiquitous expression of D-AKAP2 in many different mouse tissues suggests that D-AKAP2 function is not restricted to the kidney. D-AKAP2 may also interact with other PDZ domains involved in organizing signaling networks at membranes. Interestingly, two additional RGS proteins, Gα interacting protein (GAIP) and RGS12 also contain PDZ-binding motifs (De Vries et al., 1998; Snow et al., 1998). GAIP has been implicated in vesicular trafficking associated with clathrin-coated vesicles and trans-Golgi-derived vesicles (De Vries et al., 1998; Wylie et al., 1999, 2003). Given the preliminary association with a cardiac phenotype from the SNP analysis (Kammerer et al., 2003), D-AKAP2 may be important for cell signaling in cardiac cells. Future work will clarify the functional role of D-AKAP2.

References

Anand, G. S., Hughes, C. A., Jones, J. M., Taylor, S. S., and Komives, E. A. (2002). Amide H/2H exchange reveals communication between the cAMP and catalytic subunit-binding sites in the RIa subunit of protein kinase A. J. Mol. Biol. 323, 377–386.

Andersen, M. D., Shaffer, J., Jennings, P. A., and Adams, J. A. (2001). Structural characterization of protein kinase A as a function of nucleotide binding: Hydrogen-deuterium exchange studies using matrix-assisted laser desorption ionization-time of flight mass spectrometry detection. J. Biol. Chem. 276, 14204–14211.

Bacic, D., Capuano, P., Gisler, S. M., Pribanic, S., Christensen, E. I., Biber, J., Loffing, J., Kaissling, B., Wagner, C. A., and Murer, H. (2003). Impaired PTH-induced endocytotic down-regulation of the renal type IIa Na+/Pi-cotransporter in RAP-deficient mice with reduced megalin expression. Pflug. Arch. 446, 475–484.

Burns-Hamuro, L. L., Ma, Y., Kammerer, S., Reineke, U., Self, C., Cook, C., Olson, G. L., Cantor, C. R., Braun, A., and Taylor, S. S. (2003). Designing isoform-specific peptide disruptors of protein kinase A localization. Proc. Natl. Acad. Sci. USA 100, 4072–4077.

Casey, M., Vaughan, C. J., He, J., Hatcher, C. J., Winter, J. M., Weremowicz, S., Montgomery, K., Kucherlapati, R., Morton, C. C., and Basson, C. T. (2000). Mutations in the protein kinase A R1alpha regulatory subunit cause familial cardiac myxomas and Carney complex. J. Clin. Invest. 106, R31–R38.

Cho-Chung, Y. S., Pepe, S., Clair, T., Budillon, A., and Nesterova, M. (1995). cAMP-dependent protein kinase: Role in normal and malignant growth. Crit. Rev. Oncol. Hematol. 21, 33–61.

Colledge, M., and Scott, J. D. (1999). AKAPs: From structure to function. Trends Cell Biol. 9, 216–221.

De Vries, L., Lou, X., Zhao, G., Zheng, B., and Farquhar, M. G. (1998). GIPC, a PDZ domain containing protein, interacts specifically with the C terminus of RGS-GAIP. *Proc. Natl. Acad. Sci. USA* **95**, 12340–12345.

De Vries, L., Zheng, B., Fischer, T., Elenko, E., and Farquhar, M. G. (2000). The regulator of G protein signaling family. *Annu. Rev. Pharmacol. Toxicol.* **40**, 235–271.

Edwards, A. S., and Scott, J. D. (2000). A-kinase anchoring proteins: Protein kinase A and beyond. *Curr. Opin. Cell Biol.* **12**, 217–221.

Ehring, H. (1999). Hydrogen exchange/electrospray ionization mass spectrometry studies of structural features of proteins and protein/protein interactions. *Anal. Biochem.* **267**, 252–259.

Engen, J. R., Gmeiner, W. H., Smithgall, T. E., and Smith, D. L. (1999). Hydrogen exchange shows peptide binding stabilizes motions in Hck SH2. *Biochemistry* **38**, 8926–8935.

Engen, J. R., and Smith, D. L. (2001). Investigating protein structure and dynamics by hydrogen exchange MS. *Anal. Chem.* **73**, 256A–265A.

Englander, S. W., Mayne, L., Bai, Y., and Sosnick, T. R. (1997). Hydrogen exchange: The modern legacy of Linderstrom-Lang. *Prot. Sci.* **6**, 1101–1109.

Feliciello, A., Gottesman, M. E., and Avvedimento, E. V. (2001). The biological functions of A-kinase anchor proteins. *J. Mol. Biol.* **308**, 99–114.

Frank, R. (2002). The SPOT-synthesis technique: Synthetic peptide arrays on membrane supports—principles and applications. *J. Immunol. Methods* **267**, 13–26.

Gisler, S. M., Madjdpour, C., Bacic, D., Pribanic, S., Taylor, S. S., Biber, J., and Murer, H. (2003a). PDZK1. II. An anchoring site for the PKA-binding protein D-AKAP2 in renal proximal tubular cells. *Kidney Int.* **64**, 1746–1754.

Gisler, S. M., Pribanic, S., Bacic, D., Forrer, P., Gantenbein, A., Sabourin, L. A., Tsuji, A., Zhao, Z. S., Manser, E., Biber, J., and Murer, H. (2003b). PDZK1. I. A major scaffolder in brush borders of proximal tubular cells. *Kidney Int.* **64**, 1733–1745.

Gisler, S. M., Stagljar, I., Traebert, M., Bacic, D., Biber, J., and Murer, H. (2001). Interaction of the type IIa Na/Pi cotransporter with PDZ proteins. *J. Biol. Chem.* **276**, 9206–9213.

Guex, N., and Peitsch, M. C. (1997). SWISS-MODEL and the Swiss-PdbViewer: An environment for comparative protein modeling. *Electrophoresis* **18**, 2714–2723.

Guy, C. A., and Fields, G. B. (1997). Trifluoroacetic acid cleavage and deprotection of resin-bound peptides following synthesis by Fmoc chemistry. *Methods Enzymol.* **289**, 67–83.

Hamuro, Y., Burns, L., Canaves, J., Hoffman, R., Taylor, S., and Woods, V. (2002a). Domain organization of D-AKAP2 revealed by enhanced deuterium exchange-mass spectrometry (DXMS). *J. Mol. Biol.* **321**, 703.

Hamuro, Y., Burns, L. L., Canaves, J. M., Hoffman, R. C., Taylor, S. S., and Woods, V. L., Jr. (2002b). Domain organization of D-AKAP2 revealed by enhanced deuterium exchange-mass spectrometry (DXMS). *J. Mol. Biol.* **321**, 703–714.

Hamuro, Y., Coales, S. J., Southern, M. R., Nemeth-Cawley, J. F., Stranz, D. D., and Griffin, P. R. (2003). Rapid analysis of protein structure and dynamics by hydrogen/deuterium exchange mass spectrometry. *J. Biomol. Tech.* **14**, 171–182.

Hamuro, Y., Wong, L., Shaffer, J., Kim, J. S., Stranz, D. D., Jennings, P. A., Woods, V. L., Jr., and Adams, J. A. (2002c). Phosphorylation-driven motions in the COOH-terminal Src kinase, Csk, revealed through enhanced hydrogen-deuterium exchange and mass spectrometric studies (DXMS). *J. Mol. Biol.* **323**, 871–881.

Hernando, N., Deliot, N., Gisler, S. M., Lederer, E., Weinman, E. J., Biber, J., and Murer, H. (2002). PDZ-domain interactions and apical expression of type IIa Na/P(i) cotransporters. *Proc. Natl. Acad. Sci. USA* **99**, 11957–11962.

Hoofnagle, A. N., Resing, K. A., and Ahn, N. G. (2003). Protein analysis by hydrogen exchange mass spectrometry. *Annu. Rev. Biophys. Biomol. Struct.* **32**, 1–25.

Huang, L. J., Durick, K., Weiner, J. A., Chun, J., and Taylor, S. S. (1997a). D-AKAP2, a novel protein kinase A anchoring protein with a putative RGS domain. *Proc. Natl. Acad. Sci. USA* **94**, 11184–11189.

Huang, L. J., Durick, K., Weiner, J. A., Chun, J., and Taylor, S. S. (1997b). Identification of a novel protein kinase A anchoring protein that binds both type I and type II regulatory subunits. *J. Biol. Chem.* **272**, 8057–8064.

Huang, L. J., Wang, L., Ma, Y., Durick, K., Perkins, G., Deerinck, T. J., Ellisman, M. H., and Taylor, S. S. (1999). NH$_2$-terminal targeting motifs direct dual specificity A-kinase-anchoring protein 1 (D-AKAP1) to either mitochondria or endoplasmic reticulum. *J. Cell Biol.* **145**, 951–959.

Kammerer, S., Burns-Hamuro, L. L., Ma, Y., Hamon, S. C., Canaves, J. M., Shi, M. M., Nelson, M. R., Sing, C. F., Cantor, C. R., Taylor, S. S., and Braun, A. (2003). Amino acid variant in the kinase binding domain of dual-specific A kinase-anchoring protein 2: A disease susceptibility polymorphism. *Proc. Natl. Acad. Sci. USA* **100**, 4066–4071.

Kirschner, L. S., Carney, J. A., Pack, S. D., Taymans, S. E., Giatzakis, C., Cho, Y. S., Cho-Chung, Y. S., and Stratakis, C. A. (2000). Mutations of the gene encoding the protein kinase A type I-alpha regulatory subunit in patients with the Carney complex. *Nature Genet.* **26**, 89–92.

Ma, Y., and Taylor, S. (2002). A 15-residue bifunctional element in D-AKAP1 is required for both endoplasmic reticulum and mitochondrial targeting. *J. Biol. Chem.* **277**, 27328–27336.

Mandell, J. G., Baerga-Ortiz, A., Akashi, S., Takio, K., and Komives, E. A. (2001). Solvent accessibility of the thrombin-thrombomodulin interface. *J. Mol. Biol.* **306**, 575–589.

Murer, H., Hernando, N., Forsten, I., and Biber, J. (2003). Regulation of Na/Pi transporter in the proximal tubule. *Annu. Rev. Physiol.* **65**, 531–542.

Okazaki, Y., Furuno, M., Kasukawa, T., Adachi, J., Bono, H., Kondo, S., Nikaido, I., Osato, N., Saito, R., Suzuki, H., Yamanaka, I., Kiyosawa, H., Yagi, K., Tomaru, Y., Hasegawa, Y., Nogami, A., Schonbach, C., Gojobori, T., Baldarelli, R., Hill, D. P., Bult, C., Hume, D. A., Quackenbush, J., Schriml, L. M., Kanapin, A., Matsuda, H., Batalov, S., Beisel, K. W., Blake, J. A., Bradt, D., Brusic, V., Chothia, C., Corbani, L. E., Cousins, S., Dalla, E., Dragani, T. A., Fletcher, C. F., Forrest, A., Frazer, K. S., Gaasterland, T., Gariboldi, M., Gissi, C., Godzik, A., Gough, J., Grimmond, S., Gustincich, S., Hirokawa, N., Jackson, I. J., Jarvis, E. D., Kanai, A., Kawaji, H., Kawasawa, Y., Kedzierski, R. M., King, B. L., Konagaya, A., Kurochkin, I. V., Lee, Y., Lenhard, B., Lyons, P. A., Maglott, D. R., Maltais, L., Marchionni, L., McKenzie, L., Miki, H., Nagashima, T., Numata, K., Okido, T., Pavan, W. J., Pertea, G., Pesole, G., Petrovsky, N., Pillai, R., Pontius, J. U., Qi, D., Ramachandran, S., Ravasi, T., Reed, J. C., Reed, D. J., Reid, J., Ring, B. Z., Ringwald, M., Sandelin, A., Schneider, C., Semple, C. A., Setou, M., Shimada, K., Sultana, R., Takenaka, Y., Taylor, M. S., Teasdale, R. D., Tomita, M., Verardo, R., Wagner, L., Wahlestedt, C., Wang, Y., Watanabe, Y., Wells, C., Wilming, L. G., Wynshaw-Boris, A., Yanagisawa, M. *et al.* (2002). Analysis of the mouse transcriptome based on functional annotation of 60,770 full-length cDNAs. *Nature* **420**, 563–573.

Perkins, G. A., Wang, L., Huang, L. J., Humphries, K., Yao, V. J., Martone, M., Deerinck, T. J., Barraclough, D. M., Violin, J. D., Smith, D., Newton, A., Scott, J. D., Taylor, S. S., and Ellisman, M. H. (2001). PKA, PKC, and AKAP localization in and around the neuromuscular junction. *BMC Neurosci.* **2**, 17.

Resing, K. A., Hoofnagle, A. N., and Ahn, N. G. (1999). Modeling deuterium exchange behavior of ERK2 using pepsin mapping to probe secondary structure. *J. Am. Soc. Mass Spectrom.* **10**, 685–702.

Skalhegg, B. S., and Tasken, K. (2000). Specificity in the cAMP/PKA signaling pathway. Differential expression, regulation, and subcellular localization of subunits of PKA. *Front. Biosci.* **5**, D678–D693.

Snow, B. E., Hall, R. A., Krumins, A. M., Brothers, G. M., Bouchard, D., Brothers, C. A., Chung, S., Mangion, J., Gilman, A. G., Lefkowitz, R. J., and Siderovski, D. P. (1998). GTPase activating specificity of RGS12 and binding specificity of an alternatively spliced PDZ (PSD-95/Dlg/ZO-1) domain. *J. Biol. Chem.* **273,** 17749–17755.

Tesmer, J. J., Berman, D. M., Gilman, A. G., and Sprang, S. R. (1997). Structure of RGS4 bound to AlF4-activated G(i alpha1): Stabilization of the transition state for GTP hydrolysis. *Cell* **89,** 251–261.

Torgersen, K. M., Vang, T., Abrahamsen, H., Yaqub, S., and Tasken, K. (2002). Molecular mechanisms for protein kinase A-mediated modulation of immune function. *Cell Signal.* **14,** 1–9.

Wang, L., Sunahara, R. K., Krumins, A., Perkins, G., Crochiere, M. L., Mackey, M., Bell, S., Ellisman, M. H., and Taylor, S. S. (2001). Cloning and mitochondrial localization of full-length D-AKAP2, a protein kinase A anchoring protein. *Proc. Natl. Acad. Sci. USA* **98,** 3220–3225.

Weinman, E. J., Steplock, D., Donowitz, M., and Shenolikar, S. (2000). NHERF associations with sodium-hydrogen exchanger isoform 3 (NHE3) and ezrin are essential for cAMP-mediated phosphorylation and inhibition of NHE3. *Biochemistry* **39,** 6123–6129.

Weinman, E. J., Steplock, D., and Shenolikar, S. (2001). Acute regulation of NHE3 by protein kinase A requires a multiprotein signal complex. *Kidney Int.* **60,** 450–454.

Wellings, D. A., and Atherton, E. (1997). Standard Fmoc protocols. *Methods Enzymol.* **289,** 44–67.

Wenschuh, H., Volkmer-Engert, R., Schmidt, M., Schulz, M., Schneider-Mergener, J., and Reineke, U. (2000). Coherent membrane supports for parallel microsynthesis and screening of bioactive peptides. *Biopolymers* **55,** 188–206.

Woods, V. L., and Hamuro, Y. (2001). High resolution, high-throughput amide deuterium exchange-mass spectrometry (DXMS) determination of protein binding site structure and dynamics: Utility in pharmaceutical design. *J. Cell. Biochem.* **S37,** 89–98.

Wylie, F., Heimann, K., Le, T. L., Brown, D., Rabnott, G., and Stow, J. L. (1999). GAIP, a Galphai-3-binding protein, is associated with Golgi-derived vesicles and protein trafficking. *Am. J. Physiol.* **276,** C497–C506.

Wylie, F. G., Lock, J. G., Jamriska, L., Khromykh, T., L., Brown, D., and Stow, J. L. (2003). GAIP participates in budding of membrane carriers at the trans-Golgi network. *Traffic* **4,** 175–189.

Yan, X., Zhang, H., Watson, J., Schimerlik, M. I., and Deinzer, M. L. (2002). Hydrogen/deuterium exchange and mass spectrometric analysis of a protein containing multiple disulfide bonds: Solution structure of recombinant macrophage colony stimulating factor-beta (rhM-CSFb). *Pro. Sci.* **11,** 2113–2124.

Zhang, Z., and Smith, D. L. (1993). Determination of amide hydrogen exchange by mass spectrometry: A new tool for protein structure elucidation. *Prot. Sci.* **2,** 522–531.

Zheng, B., De Vries, L., and Gist-Farquhar, M. (1999). Divergence of RGS proteins: Evidence for the existence of six mammalian RGS subfamilies. *Trends Biochem. Sci.* **24,** 411–414.

Section II

Other Heterotrimeric G-Protein Signaling Regulators

Subsection A

Activators

[23] Purification and Functional Analysis of Ric-8A: A Guanine Nucleotide Exchange Factor for G-Protein α Subunits

By GREGORY G. TALL and ALFRED G. GILMAN

Abstract

Ric-8A (synembryn) has been shown to accelerate the *in vitro* guanine nucleotide exchange activities of most G-protein α subunits (with the exception of Gα_s). Methods are presented in this article for the purification of Ric-8A and functional analysis of the effects Ric-8A has on G-protein α subunit guanine nucleotide-binding activities. The use of Ric-8A to prepare GTPγS-Gα and nucleotide-free Gα (in complex with Ric-8A) is described.

Introduction

Activation of heterotrimeric G proteins results from the guanine nucleotide exchange activity of heptahelical G-protein-coupled receptors (GPCRs) (Gilman, 1987). Nonreceptor guanine nucleotide exchange factors (GEFs), or activators, have been discovered with structures and subcellular localizations that are distinct from those of classical GPCRs (Cismowski *et al.*, 2001; Takesono *et al.*, 1999; Tall *et al.*, 2003). These novel activators present exciting opportunities to uncover unappreciated mechanisms for the regulation of G proteins *in vivo*. In addition, the unique properties of these proteins make them useful for study of the biochemical properties of several G proteins *in vitro*.

Ric-8A is a predominantly soluble non-GPCR GEF that acts on a broad subset of Gα proteins *in vitro* (Tall *et al.*, 2003). Ric-8 (synembryn) was first identified in *Caenorhabditis elegans*, where mutation of the single *C. elegans* Ric-8 gene affected synaptic transmission and spindle pole movements in embryogenesis in ways that suggested interactions between Ric-8 and Gα_q or Gα_o, respectively (Miller and Rand, 2000; Miller *et al.*, 1996, 2000). Two mammalian homologs of Ric-8, termed Ric-8A and Ric-8B, were identified and found to bind directly to distinct subsets of

METHODS IN ENZYMOLOGY, VOL. 390   0076-6879/04 $35.00

$G\alpha$ proteins. Ric-8A catalyzes exchange of GDP for GTP using $G\alpha_i$, $G\alpha_o$, $G\alpha_q$, $G\alpha_{12}$, (Tu, Y. and Ross, E. M., personal communication) and $G\alpha_{13}$ (but not $G\alpha_s$) (Singer, W. D. and Sternweis, P. C., personal communication) as substrates (Tall *et al.*, 2003). Ric-8B interacts with $G\alpha_q$ and $G\alpha_s$, but attempts to purify functional Ric-8B and study its potential guanine nucleotide exchange-stimulating activities have not yet been successful (Klattenhoff *et al.*, 2003; Tall *et al.*, 2003).

This article presents methods used for the expression, purification, and functional analysis of Ric-8A. The latter includes demonstrations of the unique effects of purified Ric-8A on the GTP-binding and steady-state GTPase activities of $G\alpha$ proteins and the use of purified Ric-8A to prepare GTPγS-bound $G\alpha$ proteins and nucleotide-free $G\alpha$:Ric-8A complexes.

Purification of Ric-8A

Production of Baculovirus and Expression of GST-Ric-8A in Sf9 Cells

Recombinant rat Ric-8A (GenBank AY177754) is cloned into the baculovirus transfer vector pFastBacGSTTev (Tall *et al.*, 2003). Recombinant baculoviruses encoding GST-Ric-8A are generated following the instructions in the Bac-to-Bac manual (Invitrogen, Inc., Carlsbad, CA). *Sf*9 cells are grown and maintained in IPL41 medium containing 10% (v/v) fetal bovine serum, 0.1% (v/v) surfactant Pluronic F-68, and 10 μg/ml gentamicin. Transfected viral supernatants are amplified further in 15-ml cultures (2×10^6 cells/ml) for 5 days at $27°$ on a rotating shaker (150 rpm). The primary viral supernatant is collected, and 2.5 ml of this supernatant is amplified further for 72 h in a 50-ml *Sf*9 cell culture (2×10^6 cells/ml). A tertiary viral supernatant is generated by amplifying 10 ml of the secondary viral supernatant for 60 h in a 200-ml *Sf*9 cell culture. To express GST-Ric-8A, the medium is changed to IPL41, 1% (v/v) fetal bovine serum, 1% (v/v) lipid concentrate (Invitrogen, Inc.), 0.1% (v/v) surfactant Pluronic F-68, and 10 μg/ml gentamicin. Eight 1-liter cultures of *Sf*9 cells ($3-4 \times 10^6$ cells/ml) are infected with 10–30 ml of the tertiary viral supernatant and shaken for 48 h. Cells are pelleted and the pellets are flash frozen in liquid nitrogen and stored at $-80°$ until use.

Lysis and Glutathione-Sepharose Chromatography

Cell pellets are suspended in 1.2 liters of lysis buffer [20 m*M* Na HEPES, pH 8.0, 150 m*M* NaCl, 5 m*M* EDTA, 1 m*M* dithiothreitol (DTT), and the protease inhibitors 3.3 μg/ml leupeptin, 3.3 μg/ml lima bean trypsin inhibitor, 2.3 μg/ml phenylmethylsulfonyl fluoride

(PMSF), 2.1 μg/ml 1-chloro-3-tosylamido-7-amino-2-heptanone hydrochloride (TLCK), and 2.1 μg/ml L-1-p-tosylamino-2-phenylethyl chloromethyl ketone (TPCK)]. All steps are conducted at 4°. The cell suspension is lysed by nitrogen cavitation in a Parr bomb (Parr Instrument Company, Moline, IL). The lysate is centrifuged for 10 min at 2000g, and the supernatant is saved. The pellet is suspended in 500 ml of lysis buffer, homogenized with a glass dounce homogenizer (pestle B; Kontes, Co. Vineland, NJ), and centrifuged for 10 min at 2000g. This supernatant is combined with the lysate supernatant and centrifuged for 45 min at 100,000g.

A 10-ml bed volume of glutathione-Sepharose CL-4B (Amersham Biosciences, Inc., Piscataway, NJ) is packed in a 2.6-cm-diameter Econocolumn (Bio-Rad Laboratories, Hercules, CA) and equilibrated with lysis buffer. The 100,000g supernatant is charged with fresh protease inhibitors from 1000× stocks and applied to the column at 0.5–1.0 ml/min with a peristaltic pump. The column is washed with lysis buffer (250 ml), lysis buffer containing 300 mM NaCl (250 ml), and again with lysis buffer (100 ml). The drained column is removed from the pump, the bottom is plugged, and 5 ml of lysis buffer containing 2000 U of tobacco etch virus (Tev) protease (Invitrogen, Inc.) is added to the glutathione-Sepharose. The top is plugged, and an even slurry is created by slowly inverting the column several times. The column is stored upright, and digestion with Tev protease proceeds overnight at 4°. The column is unplugged, and the gravity-driven flow through is collected (Tev eluate). Two 10-ml aliquots of fresh lysis buffer are applied consecutively to the column and collected. To trap GST or undigested GST-Ric-8A that has bled from the column, the entire Tev eluate is reapplied to the drained column and collected. After draining the column completely, lysis buffer is applied and the flow through is collected until no observable protein is eluted, as determined by periodically monitoring aliquots of the eluate with the Bio-Rad protein assay detection reagent (Bio-Rad Laboratories).

Alternatively, to prepare undigested GST-Ric-8A, the washed glutathione-Sepharose is eluted with 50 ml of lysis buffer containing 20 mM reduced glutathione, pH 8.0. A substantial amount of GST protein (perhaps the result of incomplete translation of the GST-Ric-8A product) is present in reduced glutathione eluates. GST-Ric-8A can be resolved from GST using anion-exchange chromatography, as described later.

Mono-Q Chromatography

Anion-exchange chromatography is used to purify the Tev protease-digested Ric-8A or GST-Ric-8A. The Tev eluate (or reduced glutathione eluate) is filtered through a 0.22-μm PVDF Durapore filter (Millipore, Inc.,

Billerica, MA) and is loaded onto a Mono-Q HR10/10 column equilibrated with lysis buffer (Amersham Biosciences, Inc.) at 1 ml/min. The column is washed with 15 ml of lysis buffer and eluted with a 90-ml linear gradient to lysis buffer containing 500 mM NaCl. Tev-digested Ric-8A elutes as two peaks at approximately 260 and 290 mM NaCl (Fig. 1). GST-Ric-8A elutes as a single peak at approximately 350 mM NaCl. Mass spectrometry reveals that the Ric-8A peak eluting at 260 mM NaCl is a mixture of lesser phosphorylated forms of the protein and that the peak eluting at 290 mM contains highly phosphorylated Ric-8A (Tall, T. C. and Thomas, C., unpublished observations). The GEF activities of these modified forms of Ric-8A (measured *in vitro*) are indistinguishable. The variably phosphorylated species of Ric-8A are routinely pooled and exchanged into lysis buffer containing 1 mM EDTA using a 50,000 MWCO Amicon ultra 15 concentrator (Millipore, Inc.). As a more economical option to the Mono-Q HR 10/10 column, an Amersham Hi-Trap Q column (5 ml) can be used with a similar protocol. However, this column does not resolve the phosphorylated forms of Tev-digested Ric-8A.

Assays of the GEF Activity of Ric-8A

Inactive heterotrimeric G proteins contain GDP-bound α subunits associated with a dimer of β and γ subunits. Classically, activation is initiated when an agonist-bound GPCR interacts with the heterotrimer

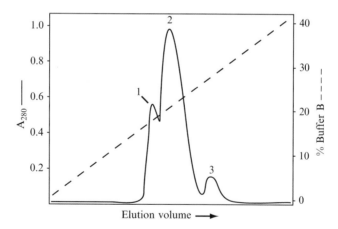

Fig. 1. Mono-Q purification of Tev-digested Ric-8A. The Tev protease eluate from the glutathione-Sepharose column was purified over a Mono-Q HR 10/10 column with a linear NaCl gradient. Peak 1 (260 mM NaCl) is a mixture of lesser phosphorylated forms of Tev-digested Ric-8A, peak 2 (290 mM NaCl) is highly phosphorylated Tev-digested Ric-8A, and peak 3 (350 mM NaCl) is undigested GST-Ric-8A.

and facilitates the release of GDP from the $G\alpha$ protein. GDP release is followed rapidly by binding of GTP and dissociation of $G\beta\gamma$ and the receptor from GTP-$G\alpha$. G-protein inactivation follows hydrolysis of GTP by $G\alpha$ and reassociation of GDP-$G\alpha$ with $\beta\gamma$. Purified recombinant Ric-8A is a GEF for a broad subset of $G\alpha$ proteins (Tall *et al.*, 2003). However, Ric-8A does not measurably promote nucleotide exchange using trimeric G proteins as substrates. Ric-8A-stimulated nucleotide exchange with $G\alpha$ proteins as substrates can be assayed using a number of methods that measure the rates of specific steps of the G-protein cycle (GTPγS binding or GDP release) or the entire cycle (steady-state GTPase activity).

Guanine Nucleotide-Binding Assays

The observed rate of binding of GTP or GTPγS to a $G\alpha$ protein in solution is limited by the slower rate at which $G\alpha$ proteins release GDP. To measure the activation of G proteins by GPCRs, it is necessary to reconstitute G-protein heterotrimers and GPCRs into phospholipid vesicles. The kinetics of nucleotide release or binding or steady-state GTPase activity is then measured using the reconstituted vesicles (Berstein *et al.*, 1992; Cerione and Ross, 1991). Uniquely, Ric-8A is a soluble GEF, and the kinetics of Ric-8A-stimulated nucleotide exchange can be examined directly in detergent-free solution using $G\alpha$ proteins that are not modified covalently with lipids or in detergent-containing solutions using lipidated $G\alpha$ subunits.

Ric-8A-mediated stimulation of GTPγS binding to $G\alpha$ proteins is measured using a nitrocellulose filter-binding method (Krumins and Gilman, 2002; Sternweis and Robishaw, 1984). Purified Ric-8A and the $G\alpha$ to be assayed are thawed rapidly at $30°$ from $-80°$ cryostorage and placed on ice. The GTPγS-binding reaction buffers are formulated as follows: 20 mM Na HEPES, pH 8.0, 100 mM NaCl, 10 mM MgSO$_4$, 1 mM EDTA, and 1 mM DTT. For lipidated $G\alpha_i$, $G\alpha_o$, $G\alpha_{12}$, or $G\alpha_{13}$, 0.05% (v/v) polyoxyethylene 10 laurel ether (C12E10) is included and, for myristoylated $G\alpha_q$, 0.05% (v/v) Genapol C100 detergent (Calbiochem, San Diego, CA) is included in the reaction buffer. No detergents are required if unmodified $G\alpha$ proteins prepared from *Escherichia coli* are used. To initiate the $G\alpha$ GTPγS-binding reaction, a $2\times$ reaction mixture consisting of the appropriate reaction buffer, Ric-8A (100 to 400 nM), and 20 μM [^{35}S]GTPγS (10,000 cpm/pmol) is added to an equal volume of $G\alpha$ protein (400 nM) diluted in reaction buffer at $20°$ ($30°$ for $G\alpha_i$). At appropriate time points (typically 1- to 2-min intervals for the first 10 min, followed by longer intervals to 30 min for $G\alpha_o$, $G\alpha_q$, $G\alpha_{12}$, and $G\alpha_{13}$ or to 2 h for $G\alpha_i$), duplicate 20-μl aliquots are removed and added to 100 μl of chilled stop solution [20 mM Tris–HCl, pH 7.7, 100 mM NaCl, 2 mM MgSO$_4$,

0.05% (v/v) C12E10, and 1 mM GTP]. The quenched reactions are diluted with 6 ml of wash buffer (50 mM Tris–HCl pH 7.7, 100 mM NaCl, and 2 mM MgSO$_4$) and are passed through BA-85 nitrocellulose filters (Schleicher and Schuell, Inc., Keene, NH) by vacuum filtration. These filters are washed four times with 2 ml of wash buffer and dried under a heat lamp. The dried filters are placed in 4 ml of scintillation fluid (ICN, Santa Mesa, CA) and counted. Data are expressed as moles of GTPγS bound per mole of Gα over the time course. The Ric-8A-stimulated rate of GTPγS binding can be calculated from the linear portion of this curve (typically the first 3–5 min). Figure 2 shows the intrinsic and Ric-8A-stimulated GTPγS-binding rates for Gα_q and Gα_{i1}.

Soluble Steady-State GTPase Assays

In the absence of a GEF, measurement of steady-state GTPase activity for a G-protein α subunit is indicative of the rate of GDP dissociation. As a Gα GEF, Ric-8A stimulates the steady-state GTPase activities of several Gα proteins dramatically (Tall et al., 2003). Ric-8A-stimulated Gα steady-state GTPase activities are calculated by quantifying the amount of ^{32}P$_i$ hydrolyzed from [γ-^{32}P]GTP by Gα over a given time period. The basic Gα steady-state GTPase assay has been described in detail elsewhere and makes use of the fact that all intact nucleotides in the reaction, but not free P$_i$, can be bound to activated acidic charcoal (Krumins and Gilman, 2002; Wang et al., 1999). Methods for the preparation of the 5% (w/v) charcoal stop suspension and HPLC purification of [γ-^{32}P]GTP for this assay have been described (Krumins and Gilman, 2002).

Tev-digested Ric-8A (in concentrations ranging from 0 to 2.5 μM, Fig. 3) is incubated with 1 μM [γ-^{32}P]GTP (10,000–40,000 cpm/pmol) and the identical Gα-specific buffer/detergent reaction mixtures described earlier. To initiate the steady-state reaction, an equal volume of the Gα protein (20–100 nM) in the appropriate reaction buffer is added. At specific time points after this addition, typically 1- to 2-min intervals for 8–10 min, duplicate 20-μl aliquots are removed and added to 1 ml of ice-cold 5% (w/v) charcoal stop solution. The charcoal is pelleted at 4000g for 10 min, and 510 μl of the clear supernatant is removed carefully and added to a scintillation vial containing 4 ml of scintillation fluid. Data are expressed as moles of P$_i$ released per mole of Gα, generated over time courses for each concentration of Ric-8A tested. A rate is then calculated for each concentration of Ric-8A tested. The dose–response curve plotted in Fig. 3 shows that Ric-8A activates Gα_q in detergent solution at a maximal steady-state GTPase rate of $\sim$0.15 min^{-1} with an EC$_{50}$ of $\sim$250 nM. When native heterotrimeric or recombinant Gα_q was co-reconstituted into lipid

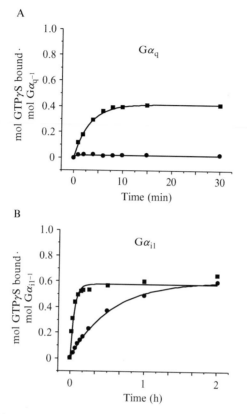

FIG. 2. Purified Ric-8A is a guanine nucleotide exchange factor for several Gα subunits. (A) Gα$_q$ (200 nM, 20°) or (B) Gα$_{i1}$ (200 nM, 30°) was incubated with Ric-8A (200 nM) in GTPγS-binding reactions. At the specified time points, duplicate aliquots were removed from each reaction, quenched, adsorbed to nitrocellulose filters, washed, and dried. The dried filters were subjected to scintillation counting, and the amount of GTPγS bound per mole of Gα protein was plotted versus time. The fraction of active Gα$_q$ and Gα$_{i1}$ in each preparation was estimated to be 45 and 60%, respectively.

vesicles with the m1 muscarinic receptor and stimulated with carbachol, a Gα$_q$ single turnover GTPase rate of 0.8 min^{-1} was established (Berstein *et al.*, 1992; Mukhopadhyay and Ross, 1999). Given the relative similarity of the Ric-8A-stimulated- and carbachol/m1-stimulated rates and considering the obvious differences in these assays (solution based versus membrane reconstituted and the absence or presence of βγ dimers), we speculate that the maximal observed rate of Ric-8A-stimulated Gα$_q$ steady-state GTPase activity in solution approaches the rate of Gα$_q$ GTP hydrolysis.

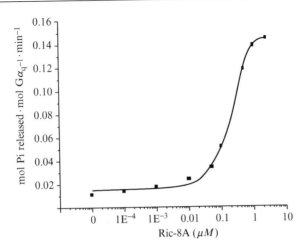

FIG. 3. Dose–response curve of $G\alpha_q$ steady-state GTPase rate stimulated by Ric-8A GEF activity (*i.e.*, acceleration of GDP release—the rate-limiting step in the guanine nucleotide cycle). $G\alpha_q$ (45 n*M*) was incubated with the indicated concentrations of Ric-8A, and steady-state GTPase activity was measured. Duplicate aliquots were removed 2, 4, 6, and 8 min after the start of each reaction and were quenched with activated charcoal. The charcoal was pelleted, and the free $^{32}P_i$ present in the supernatant was quantified by scintillation counting. A reaction rate was calculated for each concentration of Ric-8A tested and the dose–response plotted versus Ric-8A concentration (logarithmic scale).

The influence of other regulators is conveniently assessed using the steady-state GTPase assay. For example, G-protein $\beta\gamma$ subunits inhibit Ric-8A-stimulated nucleotide exchange (Tall *et al.*, 2003). This is observed readily by the addition of $G\beta\gamma$ subunits of interest to Ric-8A-stimulated steady-state GTPase assays.

Use of Ric-8A as a Catalyst to Prepare the Nucleotide-Free $G\alpha$–Ric-8A Complex and GTPγS-Bound $G\alpha$ Proteins

Ric-8A-mediated guanine nucleotide exchange proceeds directionally in the presence of millimolar concentrations of Mg^{2+} *in vitro*, i.e., Ric-8A measurably promotes the exchange of GDP for GTP. Mechanistically, Ric-8A acts as a $G\alpha$ GDP release factor, forming a stable complex with nucleotide-free $G\alpha$. GTP then binds to the $G\alpha$ protein and dissociates the Ric-8A:$G\alpha$ complex. Nucleotide-selective binding of GTP to $G\alpha$ associated with Ric-8A occurs because of the higher affinity of $G\alpha$ proteins for Mg^{2+}-GTP compared to GDP and, possibly, because Ric-8A acts like a chaperone that stabilizes $G\alpha$ in the open state. Only the $G\alpha$ conformational change that occurs upon GTP, but not GDP, binding releases Ric-8A from $G\alpha$, thus ensuring directional exchange. Nucleotide-free complexes of Ric-8A and $G\alpha_{i1}$ are stable in the presence of micromolar concentrations of GDP, but not GTP (Tall et al., 2003).

The molecular weight of the Ric-8A:$G\alpha$ complex is ~100,000, whereas those of $G\alpha$ proteins are between 40,000 and 45,000. Thus, the Ric-8A:$G\alpha$ complex can be separated readily from free monomeric $G\alpha\cdot$GTPγS by size-exclusion (gel filtration) chromatography; however, monomeric Ric-8A (60.1 kDa) is not separated from the complex as easily. If the desired outcome is the preparation of a nucleotide-free Ric-8A–$G\alpha$ complex, a high concentration of EDTA is included in the reaction mixture and gel filtration buffers and excess $G\alpha$ is used to drive Ric-8A toward complex formation. Conversely, if the desired outcome is the preparation of $G\alpha$-GTPγS, millimolar concentrations of Mg^{2+} are included in the reaction with excess GTPγS and Ric-8A.

Preparation of Ric-8A:$G\alpha$ Nucleotide-Free Complex

To prepare a large quantity of the nucleotide-free complex (e.g., for crystallization), 5 mg of Ric-8A (166 μM) is incubated with 6.5 mg of $G\alpha_{i1}$ (322 μM) in a 500-μl reaction containing complex gel filtration buffer (20 mM Na HEPES, pH 8.0, 150 mM NaCl, 5 mM EDTA, and 1 mM DTT) for 20 min at 22°. The reaction mixture is gel filtered over tandem Superdex-75 and -200 HR10/30 size-exclusion columns (Amersham Biosciences) in the same buffer at 0.15 ml/min (Fig. 4). Fractions (300 μl) are collected and those containing complex are pooled and concentrated using an ultrafree 15-ml 30000 MWCO ultrafiltration device (Millipore, Inc.). Two predominant peaks are observed: the first is the nucleotide-free complex and the second is excess monomeric $G\alpha_{i1}$. If millimolar concentrations of Mg^{2+} are included in the reaction mixture and gel filtration buffers, some monomeric Ric-8A elutes as a shoulder just after the peak of complex (not shown, but similar to the profile seen in Fig. 5A).

Preparation of $G\alpha$-GTPγS

Preparations of GTPγS-bound $G\alpha$ proteins are valuable for various biochemical assays. Examples include the examination of $G\alpha$-stimulated or -inhibited effector activities in vitro and characterization of the $G\alpha\cdot$GTP- or $G\alpha\cdot$GDP-binding preference(s) of other interacting proteins. For $G\alpha$ proteins where GTPγS binding in solution is extremely slow, such as $G\alpha_q$ and $G\alpha12/13$ family members, Ric-8A can be used to improve the yield of $G\alpha\cdot$GTPγS dramatically; $G\alpha_{i1}$ is used as an example here because of the ease with which it can be obtained (Lee et al., 1994). A 500-μl

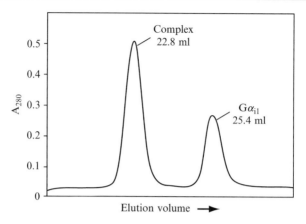

FIG. 4. Purification of a Ric-8A:Gα nucleotide-free complex. Ric-8A (5 mg) and $G\alpha_{i1}$ (6.5 mg) were incubated for 20 min in complex gel filtration buffer and gel filtered over tandem Superdex-75 and -200 HR 10/30 columns. The first peak eluted at 22.8 ml and is the nucleotide-free $G\alpha_{i1}$: Ric-8A complex. The second peak eluted at 25.4 ml and is excess monomeric $G\alpha_{i1}$.

reaction containing 1 mg of Tev-digested Ric-8A (33.3 μM) and 400 μg of $G\alpha_{i1}$ (20 μM) in 20 mM Na HEPES, pH 8.0, 150 mM NaCl, 10 mM MgSO$_4$, 1 mM EDTA, 1 mM DTT, and 100 μM GTPγS is incubated for 1 h at 22°. [^{35}S]GTPγS (500–1000 cpm/pmol) can be included to monitor the progress of the reaction before and after gel filtration. The reaction mixture is gel filtered as described earlier, with Gα·GTPγS gel filtration buffer consisting of 20 mM Na HEPES, pH 8.0, 150 mM NaCl, 2 mM MgSO$_4$, 1 mM EDTA, and 1 mM DTT. GTPγS-$G\alpha_{i1}$ elutes as a distinct peak at 25.6 ml, whereas monomeric Ric-8A elutes at 23.8 ml and residual unreacted $G\alpha_i$ in complex with Ric-8A elutes at 22.6 ml (Fig. 5A). The $G\alpha_{i1}$ in this example bound nearly 1 mol of GTPγS per mole of $G\alpha_{i1}$.

As a test to show the utility of $G\alpha_i$·GTPγS in unambiguously demonstrating a nucleotide-specific interaction between Gα and a putative modulator or effector, 400 μg of $G\alpha_i$·GTPγS prepared by the method shown in Fig. 5A is mixed with 1 mg of fresh Ric-8A in 20 mM HEPES, pH 8.0, 150 mM NaCl, 2 mM MgSO$_4$, 1 mM EDTA, 1 mM DTT, and 100 μM GTPγS for 1 h at 22° and then gel filtered as before. No Ric–8A:Gα complex was formed (Fig. 5B). This contrasts with the experiment shown in Fig. 4 where virtually all of the Ric-8A formed a complex when $G\alpha_i$·GDP was included. Together, these experiments confirm the nucleotide-binding preference of Ric-8A for $G\alpha_i$·GDP.

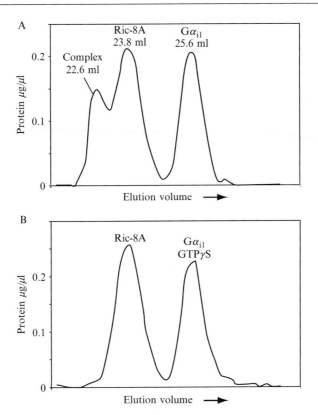

FIG. 5. Production of $G\alpha\cdot GTP\gamma S$ using Ric-8A GEF activity. (A) Ric-8A (1 mg) and $G\alpha_{i1}$ (400 μg) were incubated in $G\alpha\cdot GTP\gamma S$ gel filtration buffer (containing 100 μM GTPγS and 10 mM MgSO$_4$) and then gel filtered over tandem Superdex-75 and -200 HR 10/30 columns. The first peak (a shoulder of the second peak) eluted at 22.6 ml and is the $G\alpha_{i1}$:Ric-8A complex. The second peak eluted at 23.8 ml and is monomeric Ric-8A. The third peak eluted at 25.6 ml and is monomeric $G\alpha_{i1}\cdot GTP\gamma S$ that had 0.9 mol of GTPγS bound per mole of $G\alpha_{i1}$. (B) Ric-8A (1 mg) and $G\alpha_{i1}\cdot GTP\gamma S$ (400 μg) prepared by the method shown in A were incubated in $G\alpha\cdot GTP\gamma S$ gel filtration buffer (containing 100 μM GTPγS and 10 mM MgSO$_4$) for 1 h and gel filtered over tandem Superdex-75 and -200 HR 10/30 columns. No peak of complex was detected. The first peak eluted at 23.8 ml and is monomeric Ric-8A. The second peak eluted at 25.7 ml and is monomeric $G\alpha_{i1}\cdot GTP\gamma S$.

References

Berstein, G., Blank, J. L., Smrcka, A. V., Higashijima, T., Sternweis, P. C., Exton, J. H., and Ross, E. M. (1992). Reconstitution of agonist-stimulated phosphatidylinositol 4,5-bisphosphate hydrolysis using purified m1 muscarinic receptor, Gq/11, and phospholipase C-β1. *J. Biol. Chem.* **267**, 8081–8088.

Cerione, R. A., and Ross, E. M. (1991). Reconstitution of receptors and G proteins in phospholipid vesicles. *Methods Enzymol.* **195**, 329–342.

Cismowski, M. J., Takesono, A., Bernard, M. L., Duzic, E., and Lanier, S. M. (2001). Receptor-independent activators of heterotrimeric G-proteins. *Life Sci.* **68**, 2301–2308.

Gilman, A. G. (1987). G proteins: Transducers of receptor-generated signals. *Annu. Rev. Biochem.* **56**, 615–649.

Klattenhoff, C., Martin, M., Soto, X., Guzman, L., Romo, X., Garcia, M. D., Mellstrom, B., Naranjo, J. R., Hinrichs, M. V., and Olate, J. (2003). Human brain Synembryn interacts with Gsα and Gqα and is translocated to the plasma membranes in response to isoproterenol and carbachol. *J. Cell. Physiol.* **195**, 151–157.

Krumins, A. M., and Gilman, A. G. (2002). Assay of RGS protein activity *in vitro* using purified components. *Methods Enzymol.* **344**, 673–685.

Lee, E., Linder, M. E., and Gilman, A. G. (1994). Expression of G-protein α subunits in *Escherichia coli*. *Methods Enzymol.* **237**, 146–163.

Miller, K. G., Alfonso, A., Nguyen, M., Crowell, J. A., Johnson, C. D., and Rand, J. B. (1996). A genetic selection for *Caenorhabditis elegans* synaptic transmission mutants. *Proc. Natl. Acad. Sci. USA* **93**, 12593–12598.

Miller, K. G., Emerson, M. D., McManus, J. R., and Rand, J. B. (2000). RIC-8 (Synembryn): A novel conserved protein that is required for Gqα signaling in the *C. elegans* nervous system. *Neuron* **27**, 289–299.

Miller, K. G., and Rand, J. B. (2000). A role for RIC-8 (Synembryn) and GOA-1 (Goα) in regulating a subset of centrosome movements during early embryogenesis in *Caenorhabditis elegans*. *Genetics* **156**, 1649–1660.

Mukhopadhyay, S., and Ross, E. M. (1999). Rapid GTP binding and hydrolysis by Gq promoted receptor and GTPase-activating proteins. *Proc. Natl. Acad. Sci. USA* **96**, 9539–9544.

Sternweis, P. C., and Robishaw, J. D. (1984). Isolation of two proteins with high affinity for guanine nucleotides from membranes of bovine brain. *J. Biol. Chem.* **259**, 13806–13813.

Takesono, A., Cismowski, M. J., Ribas, C., Bernard, M., Chung, P., Hazard, S., III, Duzic, E., and Lanier, S. M. (1999). Receptor-independent activators of heterotrimeric G-protein signaling pathways. *J. Biol. Chem.* **274**, 33202–33205.

Tall, G. G., Krumins, A. M., and Gilman, A. G. (2003). Mammalian Ric-8A (Synembryn) is a heterotrimeric Gα protein guanine nucleotide exchange factor. *J. Biol. Chem.* **278**, 8356–8362.

Wang, J., Tu, Y., Mukhopadhyay, S., Chidiac, P., Biddlecome, G. H., and Ross, E. M. (1999). GTPase-activating proteins (GAPs) for heterotrimeric G proteins. *In* "G Proteins: Techniques of Analysis" (D. R., Manning, ed.), pp. 123–151. CRC Press, Boca Raton, FL.

[24]   Tubulin as a Regulator of G-Protein Signaling

By Mark M. Rasenick, Robert J. Donati,
Juliana S. Popova, and Jiang-Zhou Yu

Abstract

Tubulin is known to form high-affinity complexes with certain G proteins. The formation of such complexes allows tubulin to activate Gα and
fosters a system whereby elements of the cytoskeleton can influence G-
protein signaling. This article describes the interaction between tubulin and
G proteins and discusses methods for examining this interaction.

The Cytoskeleton and G-Protein Signaling

Initial studies suggesting an interaction between the cytoskeleton and
G-protein signaling were carried out in leukocytes, where it was seen that
disruption of microtubules by drugs such as colchicine or vinblastine
increased adenylyl cyclase activity in the presence of β-adrenergic agonists
(Kennedy and Insel, 1979; Rudolph *et al.*, 1979). This was followed by a
study that demonstrated that treatment of rat cerebral cortex synaptic
membranes with microtubule-disrupting drugs increased Gs-stimulated
adenylyl cyclase activity due to an apparent release of Gs from a cytoskeletal "tether" that prevented it from a facile interaction with adenylyl
cyclase (Rasenick *et al.*, 1981). Later experiments revealed that microtubules were not involved in these phenomena, but tubulin dimers were
essential (Rasenick and Wang, 1988).

Tubulin binds to the α subunits of Gi1, Gs, and Gq with a dissociation
constant (K_d) of about 130 nM (Wang *et al.*, 1990). *In vitro* studies show
that the intrinsic GTPase of tubulin is activated by Giα or Gsα in the
GDP-bound form. This allows removal of the GTP "cap" on the growing
end of a microtubule and facilitates dynamic behavior of that microtubule
(Roychowdhury *et al.*, 1999). This article focuses on the regulation of
G-protein activation by tubulin rather than the modulation of the
cytoskeleton by G-protein signaling.

Binding of Tubulin by Gα Chimeric Proteins *In Vitro*

We have demonstrated previously that tubulin binds to Gsα, Giα1, and
Gqα, whereas transducin (Gtα) consistently failed to bind tubulin (Wang
et al., 1990). In order to determine the regions of these closely related

G proteins responsible for binding tubulin, reciprocal chimeras containing the region 237–270 from Giα1 or Gtα (chimera 1, chimera 2, and chimera 3) were tested. This region spans the area between the switch II and switch III domains and contains both α3 and β5 regions. Tubulin is purified from ovine brains through two cycles of polymerization–depolymerization as described (Shelanski et al., 1973). Isolated microtubule proteins are passed through a phosphocellulose column to separate tubulin from microtubule-associated proteins (Roychowdhury et al., 1993). In order to prepare ^{125}I-tubulin, 100 μg of phosphocellulose-purified tubulin (see Wang et al., 1990) in 100 μl of PIPES buffer (100 mM PIPES, pH 6.9, 2 mM EGTA, 1 mM MgCl$_2$) is applied to a 12 × 75-mm glass tube precoated with 100 μg IODO-GEN (Pierce) dissolved in 100 μl of triethanolamine and dried in a ventilated hood. Two microcuries of Na [125]I (Amersham) is then added and the reaction is allowed to proceed with gentle agitation for 15 min. The reaction is terminated by the addition of 100 μl of PIPES buffer containing 4 mM dithiothreitol (DTT). The free iodide is removed by ultrafiltration by loading the tubulin suspension on a 3-ml Bio-Gel P-6DG (Bio-Rad, Richmond, CA) desalting column and this step is repeated over a second column. The desalted ^{125}I-labeled tubulin is first centrifuged at 16,000g for 10 min to remove denatured protein, and the supernatant containing the iodinated tubulin is used in protein-binding experiments. ^{125}I-labeled tubulin prepared in this manner was validated for retention of biological activity by polymerization in the presence of glycerol (Wang et al., 1990).

G-protein constructs were made and purified as described (Chen et al., 2003). While the G-protein constructs described herein contain an amino-terminal hexahistidine tag and were purified by chromatography over a nickel agarose column, their affinity for tubulin is comparable to native G proteins purified from mammalian brain (Wang et al., 1990). Purified Giα1, chimera 1, chimera 2, chimera 3 (Giα 1–237-Gtα 237–270-Giα 271–355), and bovine transducin (Gtα) and ovalbumin (Sigma) are applied to nitrocellulose membrane (Midwest Scientific) in spots containing 200, 100, 50, 25, or 10 ng. The nitrocellulose is then incubated with 10% bovine serum albumin in 100 mM PIPES buffer at room temperature for 2 h to block nonspecific binding and is subsequently incubated with 100 nM iodinated tubulin in 100 mM PIPES buffer at room temperature for 2 h. The membrane is then washed with PIPES buffer three times followed by the autoradiographic detection of ^{125}I-labeled tubulin bound to Giα (Fig. 1A). The affinity of tubulin for Giα and chimera 3 is estimated by immobilizing these proteins (100 ng) on nitrocellulose and hybridizing with various concentrations of ^{125}I-labeled tubulin (25, 50, 100, 200, 300, and 500 nmol) and quantified in a γ counter (Beckman Gamma 9000). Scatchard analysis is performed using Graph Pad Prism software (San Diego, CA) (Fig. 1B).

Evidence for Tubulin–G-Protein Interactions

Brief incubation of tubulin-Gpp(NH)p with synaptic membranes induced an inhibition of adenylyl cyclase, which persisted after membrane washing (Rasenick and Wang, 1988). If tubulin did not have a hydrolysis-resistant GTP analog bound, it was inactive with respect to adenylyl cyclase inhibition. Investigations of the mechanism for this process revealed that, subsequent to the formation of a complex between tubulin and $Gi\alpha$, tubulin transfers GTP directly to that G protein, thereby activating the $G\alpha$ subunit and thus bypassing the requirement for a Gi-coupled receptor. This process was studied with pure proteins, with tubulin added to membrane preparations and in permeable cells.

Transactivation of $G\alpha$ by Tubulin In Vitro

The use of photoaffinity GTP analogs as a tool for the study of GTP binding proteins has been described in detail by Rasenick et al. (1994) and is discussed herein only as it pertains to the analysis of tubulin as a regulator of G-protein signaling.

To prepare tubulin with the radiolabeled photoaffinity GTP isomer ^{32}P-AAGTP (tubulin-^{32}P-p3-1,4-azidoanilido-p1-5′-GTP) bound the guanine nucleotide on tubulin (GDP or GTP) is removed with charcoal pretreatment, followed by incubation with ^{32}P-AAGTP as described previously (Rasenick et al., 1994). ^{32}P-AAGTP is also available commercially from ALT Corp. (Lexington, KY). This procedure allows for the incorporation of 0.82–0.84 mol AAGTP/mol tubulin.

Tubulin (1 μM) is incubated with 100 μM of (α-^{32}P) p3-1,4-azidoanili-do-p1-5′-GTP [AAGTP, prepared as described by Rasenick et al. (1994)] on ice for 30 min in buffer (10 mM HEPES, pH 7.4, 5 mM MgCl$_2$, 150 mM NaCl, 1 mM 2-mercaptoethanol). Free AAGTP is removed with two passes through a Bio-Gel P6-DG column (Bio-Rad). All tubulin preparations are stored in aliquots under liquid nitrogen until use (Roychowdhury et al., 1993).

Tubulin-AAGTP is incubated with equimolar concentrations of $Gi\alpha 1$, chimera 3, or $Gt\alpha$ at 30° for 30 min followed by photoactivation with UV irradiation on ice for 3 min with a 9-W Mineralight at a distance of 3 cm. Ten microliters of the samples is mixed with an equal volume of 2× SDS–PAGE sample buffer, heated for 5 min at 65°, and then loaded on a 10% SDS gel for electrophoresis. The gel is stained with Coomassie brilliant

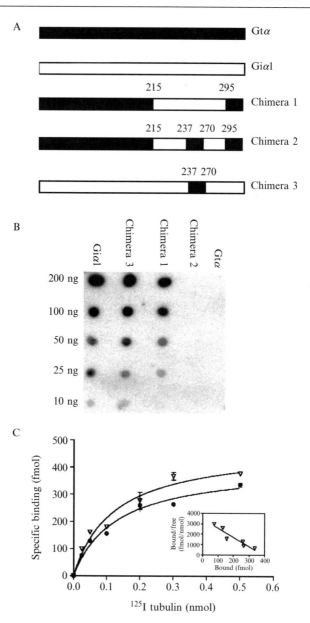

FIG. 1. Specificity and affinity of ^{125}I-labeled tubulin binding to Gα. (A) Schematic model of Giα/Gtα chimera constructs. (B) ^{125}I-labeled tubulin binding to G-protein α subunits. Giα1, chimera 1, chimera 2, chimera 3, and Gtα were applied to a nitrocellulose sheet in the amounts indicated and air dried at room temperature. Following this, tubulin binding was assessed by overlay with ^{125}I-labeled tubulin and autoradiography as described. One of three

blue and then dried. AAGTP transferred from tubulin to G protein is quantified in a phosphorimager (Storm 840; Molecular Dynamics). Figure 2 shows the results of a typical experiment.

In the presence of equimolar concentrations of tubulin-AAGTP and Giα1, AAGTP becomes approximately equal in its distribution between the two proteins. The addition of GTP in concentrations in excess of 1000-fold over either G protein does not alter the transactivation process (Roychowdhury and Rasenick, 1994). Should the recipient G protein be occupied with a hydrolysis-resistant GTP analog (cold AAGTP, GTPγS, or GppNHp), AAGTP is not transferred from tubulin to Gα (Rasenick and Wang, 1988). Furthermore, if AAGTP binding to tubulin is made covalent prior to incubation with G protein, no transactivation occurs (Rasenick and Wang, 1988).

Figure 2 also illustrates that a Giα–Gtα chimera (where regions 237–270 of Giα are replaced by transducin) is not a substrate for transactivation but is capable of inhibiting that process (Chen et al., 2003). A Giα/Gtα chimera (chimera 2) that does not bind to tubulin (similar to transducin) has no effect on the act of transactivation. Chimera 3, however, binds tubulin with an affinity comparable to that of Giα1 (Fig. 1). Thus, chimera 3 functions as a "dominant negative" to block the transactivation process. Expression of this chimera in cells has profound effects on the microtubule cytoskeleton (Chen et al., 2003).

Transactivation of Gsα by Tubulin in C6 Glioma Membranes

Direct effects on adenylyl cyclase (inhibition) were seen when tubulin was incubated with synaptic membranes and, under these same conditions, tubulin-AAGTP, incubated with those membranes, showed significant transactivation of Giα1 (Rasenick and Wang, 1988). Because tubulin had equal affinities for both Gsα and Giα1, it seemed odd that Giα1 was the primary substrate for added tubulin and the effect of adding tubulin to rat cerebral cortex synaptic membranes was the inhibition of adenylyl cyclase.

---

similar experiments is shown. (C) Saturation isotherm and Scatchard plot for tubulin binding to Giα1 and chimera 3. Data were derived from dot blotting performed with a method similar to that for B. One hundred nanograms of Giα1 or chimera 3 was applied to each spot. Triplicate nitrocellulose spots corresponding to the total binding and nonspecific binding (determined in the presence of 100-fold excess unlabeled tubulin) were cut out and counted in an LKB Rack γ counter. The saturation isotherm for specific binding of ^{125}I-labeled tubulin to chimera 3 (∇) or Giα1 ($\bullet$) is plotted. (*Inset*) A Scatchard plot derived from these data for chimera 3. The K_d and B_{max} for Giα1 were 121 nM and 386 fmol/ng tubulin, respectively. For chimera 3, these data were 123 nM and 473 fmol/ng, respectively. Data were calculated from two similar experiments.

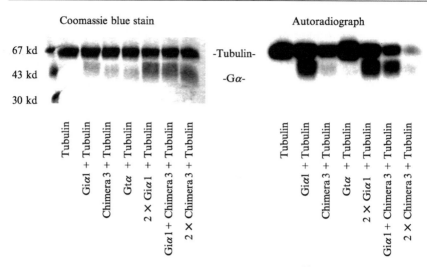

FIG. 2. Transactivation of Gα constructs by tubulin. Tubulin-[^{32}P]AAGTP was incubated with equimolar Giα1, chimera 3, Gtα, Giα1 plus chimera 3, or twice molar Giα1, or chimera 3 as indicated. AAGTP binding to tubulin or Gα was made covalent by UV irradiation, and samples were prepared for SDS–PAGE. [^{32}P]AAGTP bound to tubulin or Gα (as a result of transactivation) was quantified by phosphorimaging analysis (Molecular Dynamics Storm 860). Images (protein stain on the left; phosphorimage on the right) shown are from one of three similar experiments.

This issue became somewhat resolved when it was observed, in coimmunoprecipitation studies, that about 90% of the Gsα in synaptic membranes was complexed with tubulin (Yan *et al.*, 1996). Thus, tubulin added to those membranes was more likely to target and transactivate Giα1, resulting in the inhibition of adenylyl cyclase. We have theorized, however, that tubulin participates in the stimulation of adenylyl cyclase in response to a signal that elevates Ca^{2+} (Yan *et al.*, 1996, 2001). As such, it was important to determine whether tubulin could transactivate Gsα.

The pellet containing the membrane fraction isolated from C6 glioma cells as described by Rasenick and Kaplan (1986) is used to study tubulin transactivation of endogenous membrane Gsα. Isolated membranes are stored in appropriate aliquots under liquid nitrogen until use. Transactivation reactions are performed by incubating C6 membranes (40 μg of protein) with 2.4 μM of either ^{32}P-AAGTP or tubulin-^{32}P-AAGTP in the presence or absence of 100 μM isoproterenol in 10 mM HEPES buffer, pH 7.4, containing 5 mM MgCl$_2$ and 1 mM 2-mercaptoethanol at a total volume of 100 μl for 10 min in a shaking water bath at 37°. DTT should not be substituted for 2-mercaptoethanol, as the former compound will prevent nitrene-free radicals from binding to their target, whereas the latter will simply block nonspecific association of those free radicals. Ovalbumin at the same molar concentration as tubulin is added to the samples incubated with free ^{32}P-AAGTP in order to equalize total protein. Because AAGTP is photosensitive, all experiments are carried out under yellow fluorescent lights (Phillips F40-GO). At the end of the incubation, samples are UV irradiated for 3 min on ice to covalently link ^{32}P-AAGTP to tubulin, and the reactions are quenched with ice-cold 2 mM HEPES buffer containing 5 mM MgCl$_2$ and 4 mM DTT, pH 7.4. After centrifugation at 20,000g for 10 min at 4°, membrane pellets are separated from the supernatants, washed with HEPES buffer, and dissolved in SDS sample buffer (Laemmli, 1970) containing 50 mM DTT. Supernatant samples are also dissolved in SDS sample buffer to determine the amount of tubulin-AAGTP that becomes membrane associated. Gel electrophoresis of the samples is followed by autoradiography or phosphor image analysis (Storm 840, Molecular Dynamics) of the dried gels. Purified Gsα is run as a standard for the identification and quantification of Gs.

C6 glioma membranes are devoid of Giα1 (Fig. 4, inset) and, as such, the primary substrate for transactivation by tubulin-AAGTP is Gsα. The level of activation of Gsα observed upon treatment with isoproterenol correlates well with the level of transactivation of Gsα mediated by tubulin in C6 membranes (Fig. 3) (Yan et al., 2001).

Activation of Gsα in Permeable Cells

Previous studies have been done using purified membranes from various cell lines and tissue types, but important cellular components can be lost in an in vitro system. Using a permeable cell system, while still not a "pure" in vivo system, allows the investigator to study in vivo cellular processes more accurately. Measurement of adenylyl cyclase activity in C6 glioma cells made permeable with saponin was first described by Rasenick and Kaplan (1986). In that study, the authors were able to demonstrate that the properties of permeable cells resembled those of intact cells more so than the use of purified membrane preparations. GTP and GTPγS activated adenylyl cyclase in membranes from both vehicle and isoproteronol-treated C6 cells. However, in saponin-permeable C6 cells, adenylyl cyclase was activated by GTP and GTPγS only in the presence of isoproterenol, suggesting that some component present in permeable cells, but lost in membrane preparations, may be important for adenylyl cyclase regulation by cytoskeletal components.

C6 glioma cells are subcultured into 24-well sterile plates and used at 80–90% confluence. The cell monolayers are washed three times with

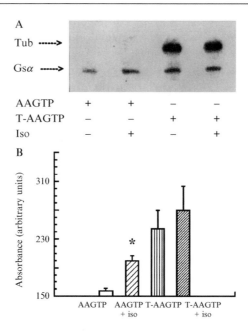

FIG. 3. Transctivation of Gsα by tubulin-AAGTP vs free AAGTP in C6 glioma membranes. Comparison with isoproterenol activation. C6 glioma membranes were incubated with ^{32}P-AAGTP (AAGTP) or tubulin-^{32}P-AAGTP (T-AAGTP), with or without isoproterenol (iso), as indicated. Each sample was subjected to SDS–PAGE (70 μg of protein in each lane) and autoradiography. (A) A representative autoradiogram from one of three identical experiments with similar results is shown. Tubulin (Tub) and Gsα are identified. (B) Mean absorbance of the identical Gsα-^{32}P-AAGTP bands, labeled under the conditions described and estimated by densitometry ($\pm$ SEM, $n = 3$), is presented. Isoproterenol potentiated Gsα binding of free ^{32}P-AAGTP (*$P < 0.05$, Student's t test), but isoproterenol treatment does not increase significantly the level of Gsα transactivation by tubulin-^{32}P-AAGTP.

200 μl of Locke's solution (154 mM NaCl, 2.6 mM KCl, 2.15 mM K_2 HPO_4, 0.85 mM KH_2PO_4, 10 mM glucose, 2 mM $CaCl_2$, 1 mM $MgCl_2$, pH 7.4) for 5 min at 37°. Next, 200 μl of saponin solution in KG buffer (140 mM potassium glutamate, pH 6.8, 2 mM ATP plus 100 μg/ml saponin) is added for 2.5 min at room temperature. The plates are washed three times with 200 μl of KG buffer (without saponin). Various concentrations of isoproterenol, tubulin-GppNHp, or GppNHp are added to triplicate wells and are incubated for 3 min at room temperature. Following this, 150 μl of reaction buffer [α^{32}P-ATP (totaling 10^6 cpm), 0.5 mM ATP, 1 mM $MgCl_2$, 0.5 mM isobutylmethylxanthine in a modified Hank's buffer containing 50 mM HEPES, 80 mM $CaCl_2\cdot2H_2O$, and 0.1% glucose] is added to each

well in order to initiate the reaction. The reaction is allowed to take place for 10 min at 37°. The reaction is stopped with 300 μl of ice-cold 15 mM HEPES buffer, pH 7.4, and the entire plate is placed on dry ice until frozen. Plates are removed from the dry ice and thawed, and cells are scraped into 1.5-ml microfuge tubes. The wells are rinsed with an additional 100 μl of 15 mM HEPES buffer and combined with scraped cells. The tubes are boiled for 8 min and centrifuged for 8 min at 15,000g. The supernatants are transferred to 12 × 75-mm glass tubes, and the ^{32}P-cAMP generated is separated by sequential chromatography on dowex and alumina columns according to Salomon (1979).

Figure 4 shows that while GTP analogs cannot activate Gsα and adenylyl cyclase in permeable cells in the absence of a Gs-linked receptor agonist, tubulin with a hydrolysis-resistant GTP analog bound is capable of bypassing a tightly coupled receptor to activate the Gsα directly. It is hypothesized that tubulin transactivated Gsα in these permeable cells, leading to the activation of adenylyl cyclase.

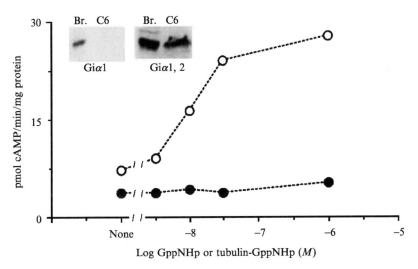

Fig. 4. Activation of adenylyl cyclase by tubulin-GppNHp versus GppNHp in permeable C6 glioma cells. C6 cells were grown in 24-well plates and were made permeable with saponin. Indicated concentrations of tubulin-GppNHp (○) or GppNHp (●) were added to triplicate wells and incubated for 3 min at room temperature in modified Hank's buffer containing 50 mM HEPES, 80 mM CaCl$_2$·2H$_2$O, and 0.1% glucose. Following this, cells were assayed for adenylyl cyclase as described in the text. Means of triplicate determinations (± SEM) for one of four similar experiments are shown. (*Inset*) C6 cells contain Giα2 (right panel) and Giα3 but not Giα1 (left panel). Western blots of C6 cells are shown. Giα1, the form of Giα that interacts with tubulin, is absent. Br, cerebral cortex membranes; C6, C6 glioma cell membranes.

Measurement of ^{32}*P-AAGTP Transfer from Tubulin to Gsα*

This assay is exactly parallel to the adenylyl cyclase assay, except that tubulin-AAGTP is added to cells and transactivation of Gsα is measured. This allows a direct comparison of the role of tubulin in activating Gsα with the activation of adenylyl cyclase. Tubulin-^{32}P-AAGTP (100 nM) is added to permeable C6 cells for 10 min at 37° in a total volume of 200 μl modified Hank's buffer containing 0.5 mM ATP. Following this incubation, the plates are placed on ice and subjected to UV light (9 W, 254 nm) for 1 min at a distance of about 4 cm. This cross-linking reaction is stopped by adding 1.5 ml of 2 mM HEPES, pH 7.4, containing 5 mM MgCl$_2$ and 4 mM DTT. The cells are scraped, transferred into glass tubes, and sonicated for 30 s. The lysed cell preparations are then centrifuged for 10 min at 600g at 4°. The supernatants are decanted and used to prepare crude membrane fractions by means of high-speed centrifugation (100,000g for 30 min at 4°). The high-speed supernatants are concentrated by centrifugation dialysis using Centriprep-10 ultrafiltration devices (Amicon, Beverly, MA). All samples are dissolved in SDS–PAGE sample buffer containing 50 mM DTT for electrophoresis on 10% polyacrylamide gels. Gels are stained with Coomassie brilliant blue, dried, and subjected to phosphorimage analysis.

Consistent with the suggestion from aforementioned data that tubulin functions as a regulator of G-protein signaling, we observed transactivation of Gsα by tubulin-AAGTP (Fig. 5) under agonist-free conditions where only Gsα was capable of activating adenylyl cyclase. Free AAGTP was able to bind to Gsα in permeable cells, but only if an agonist was present (Fig. 3). Thus, tubulin is capable of bypassing a Gs-coupled receptor in order to activate Gsα and, subsequently, adenylyl cyclase. This is consistent with the scheme outlined by Yan *et al.* (2002) where a spike in Ca^{2+} is suggested to facilitate tubulin interaction with Gsα, allowing a neurotransmitter that elevates Ca^{2+}, such as glutamate, to participate, indirectly, in the activation of adenylyl cyclase.

Gqα Transactivation by Tubulin: Bidirectional Regulation of
  Phospholipase Cβ1

Unlike Gsα and adenylyl cyclase, Gqα and PLCβ1 are regulated in an agonist-dependent manner by tubulin. This was determined using a system reduced to the principal components of the PLCβ1 cascade (Popova *et al.*, 1997). In these experiments, Sf9 cells are infected in different combinations with baculoviruses expressing the m1 muscarinic receptor, Gqα, or PLCβ1 cDNAs at a ratio of 1:1:1 and a multiplicity of infection of 5. Construction of the recombinant baculoviruses has been described previously (Boguslavsky *et al.*, 1994; Graber *et al.*, 1992; Parker *et al.*, 1991).

Gsα immunoblot   Autoradiogram

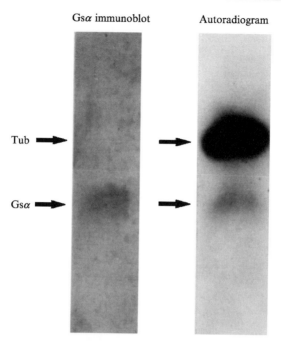

FIG. 5. Transfer of AAGTP from tubulin to Gsα in permeable C6 glioma cells. C6 glioma cells were made permeable with saponin and incubated with 100 nM tubulin-^{32}P-AAGTP as described in the text. Following incubation, cells were UV irradiated for 2 min and membrane and soluble fractions were prepared. Proteins released from the membranes were concentrated and 70 μg of protein was subjected to SDS–PAGE, western blotting, and autoradiography. The shorter (45 kDa) form of Gsα predominates in C6 cells and is identified with an arrow. Tubulin (Tub) is also indicated with an arrow. The anti-Gs immunoblot and its respective autoradiogram from one of four similar experiments are shown. Similar experiments in which free ^{32}P-AAGTP instead of tubulin-^{32}P-AAGTP is added to permeable cells show no labeling of Gsα. Thus, tubulin is capable of transactivating Gsα even when added to permeable cells.

Cells are harvested after 65 h and sonicated in ice-cold 20 mM HEPES, pH 7.4, 1 mM MgCl$_2$, 100 mM NaCl, 1 mM DTT, 0.3 mM phenylmethylsulfonyl fluoride. After centrifugation at 500g, the supernatant is collected and centrifuged at 100,000g for 30 min at 4°. The membrane pellet is washed, resuspended in the same buffer, and frozen in aliquots in liquid nitrogen. The expression of recombinant proteins is verified by immunoblotting with antisera specific for the m$_1$ muscarinic receptor (# 71, from Dr. Luthin, Philadelphia, PA), Gαq/11 (# 0945, from Dr. David Manning, Philadelphia, PA), or PLCβ1 (antiholoenzyme from Dr. S-G Rhee, Bethesda, MD), each at a dilution of 1:500. Biotinylated goat antirabbit IgG and streptavidin–alkaline phosphatase conjugate or ECL is utilized for detection and quantitation. Expression levels usually vary by no more than 10% for a given recombinant polypeptide. Receptor-binding studies using [³H]QNB (0.02–4.00 nM) as a ligand are also performed to assess the number and affinity of the m_1 muscarinic receptors (Luthin and Wolfe, 1984). Nonspecific binding, measured in the presence of 1 μM atropine, is usually less than 10%. Saturation isotherms are analyzed using GraphPad Prism software.

Sf9 cell membranes, expressing recombinant proteins, are incubated with 1 μM tubulin-[³²P]AAGTP in the presence or absence of 10 μM carbachol in 100 mM PIPES buffer, pH 6.9, 2 mM EGTA, 1 mM MgCl₂ (buffer A) for 10 min at 23° with constant shaking. Incorporation of AAGTP into Gqα is determined as for C6 membranes described earlier.

To test tubulin-GppNHp-potentiated activation of PLCβ1 via Gqα transactivation, 20 μg of isolated Sf9 cell membranes containing recombinant proteins and 40 μl of the PIP₂ substrate mixture are incubated in a Branson water bath sonicator for 15 min at 4°. Omitting certain baculoviruses (m1AChR, Gqα, or PLCβ1) allows one to determine that all three proteins are required to observe tubulin activation of PLCβ1 or effective transactivation of Gqα (Popova *et al.*, 1997). The substrate mixture consists of PIP₂ at a final concentration of 30 μM (1 μCi of [³H]PIP₂) evaporated under a stream of nitrogen and subsequently sonicated on ice for 5 min in 20 mM Tris–maleate buffer, pH 6.8, containing 6 mM MgCl₂, 4 mM ATP, EGTA, and CaCl₂ to give a final concentration of 50 nM Ca²⁺ (Fabiato, 1988) and 1 mM deoxycholate. The PIP₂ substrate mixture is used immediately upon preparation. Ten microliters of tubulin-GppNHp (30 nM final concentration) and 10 μl of carbachol (10 μM final concentration) are subsequently added to reaction mixtures at the final volume of 80 μl. The reaction tubes are incubated for 15 min at 37° with constant shaking. Aqueous and lipid phases are separated as described previously (Popova and Dubocovich, 1995), and [³H]inositol trisphosphate ([³H]IP₃) in the aqueous phase is quantified by liquid scintillation counting.

When tubulin was added to these Sf9 membranes, there was a rapid activation of PLCβ1, providing that the tubulin was "charged" with hydrolysis-resistant GTP analog (Popova *et al.*, 1997). Activation of mAChR appeared to be required for this process, and the activation of those receptors actually recruited tubulin to the membrane along the same time course as tubulin transactivated Gq and activated PLCβ1 (Fig. 6) (Popova and Rasenick, 2000; Popova *et al.*, 1997). Further experiments suggested that the inhibition of PLCβ1 activation seen at higher tubulin concentrations was due to tubulin binding PIP₂, the substrate for PLCβ1 (Popova *et al.*, 2002). This rendered the substrate unavailable to the enzyme and suggested that there was a "feedback inhibition" in the regulation of PLCβ1 by tubulin.

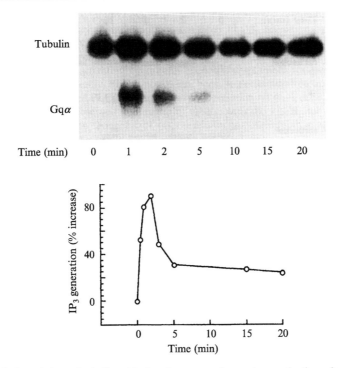

FIG. 6. Association of tubulin with the plasma membrane, transactivation of Gqα, and PLCβ1 activation follow a similar time course. (*Top*) Time course of carbachol-evoked tubulin-[^{32}P]AAGTP association with the membrane and transactivation of Gqα. Membranes were incubated for the indicated times with tubulin-[^{32}P]AAGTP and carbachol at 23°, followed by UV irradiation, SDS–PAGE (70 μg of membrane protein in each lane), and autoradiography as described. The experiment shown is representative of three with similar results. In the absence of carbachol, no increase in tubulin membrane association or nucleotide transfer to Gqα was detected at any time point tested (data not shown). All carbachol-induced effects are blocked by atropine (100 nM) (Popova and Rasenick, 2000). (*Bottom*) Carbachol-evoked activation of PLCβ1 by tubulin-GppNHp. Sf9 cell membranes containing the indicated recombinant proteins were incubated with the PIP$_2$ substrate as described. Tubulin-GppNHp and carbachol were subsequently added and incubated for the indicated times at 37°C as described. The experiment shown is representative of three with similar results. Control activity (0 time point) was 0.61 + 0.09 nmoles IP$_3$/min/mg protein. When GppNHp was tested under these experimental conditions, a linear increase in PLCβ1 activation was observed and no downturn was seen (Popova *et al.*, 1997).

Conclusions

Data accumulated over the past two decades have allowed us to construct a scheme whereby tubulin participates in the regulation of G-protein signaling and G proteins act to modify the microtubule cytoskeleton. The recent development of a dominant-negative Gα (i.e., chimera 3) that blocks the transactivation process is likely to increase exponentially our understanding of the role of tubulin in regulating G-protein signaling.

References

Boguslavsky, V., Rebecchi, M., Morris, A. J., Jhon, D.-Y., Rhee, S. G., and McLaughlin, S. (1994). Effect of monolayer surface pressure on the activities of phosphoinositide-specific phospholipase C-β1, -γ1, and δ1. *Biochemistry* **33**, 3032–3041.

Chen, N. F., Yu, J. Z., Skiba, N. P., Hamm, H. E., and Rasenick, M. M. (2003). A specific domain of Giα required for the transactivation of Giα by tubulin is implicated in the organization of cellular microtubules. *J. Biol. Chem.* **278**, 15285–15290.

Fabiato, A. (1988). Computer programs for calculating total from specified free or free from specified total ionic concentrations in aqueous solutions containing multiple metals and ligands. *Methods Enzymol.* **157**, 378–417.

Graber, S. G., Figler, R. A., and Garrison, J. C. (1992). Expression and purification of functional G protein α subunits using a baculovirus expression system. *J. Biol. Chem.* **267**, 1271–1280.

Kennedy, M., and Insel, P. (1979). Inhibitor of microtubule assembly enhance beta-adrenergic and prostaglandin El-stimulated cyclic AMP accumulation in S49 lymphoma cells. *Mol. Pharmacol.* **16**, 215.

Laemmli, U. K. (1970). Cleavage of structural proteins during the assembly of the head of bacteriophage T4. *Nature (Lond)* **227**, 680.

Luthin, G. R., and Wolfe, B. B. (1984). [3H]Pirenzepine and [3H]quinuclidinyl benzilate binding to brain muscarinic cholinergic receptors: Differences in measured receptor density are not explained by differences in receptor isomerization. *Mol. Pharmacol.* **26**, 164–169.

Parker, E. M., Kameyama, K., Higashijima, T., and Ross, E. M. (1991). Reconstitutively active G protein-coupled receptors purified from baculovirus-infected insect cells. *J. Biol. Chem.* **266**, 519–527.

Popova, J. S., and Dubocovich, M. L. (1995). Melatonin receptor-mediated stimulation of phosphoinositide breakdown in chick brain slices. *J. Neurochem.* **64**, 130–138.

Popova, J. S., Garrison, J. C., Rhee, S. G., and Rasenick, M. M. (1997). Tubulin, Gq, and phosphatidylinositol 4,5-bisphosphate interact to regulate phospholipase Cbeta1 signaling. *J. Biol. Chem.* **272**, 6760–6765.

Popova, J. S., and Rasenick, M. M. (2000). Muscarinic receptor activation promotes the membrane association of tubulin for the regulation of Gq-mediated phospholipase Cbeta1 signaling. *J. Neurosci.* **20**, 2774–2782.

Popova, J. S., Greene, A. K., Wang, J., and Rasenick, M. M. (2002). Phosphatidylinositol 4,5-bisphosphate modifies tubulin participation in phospholipase Cbeta1 signaling. *J. Neurosci.* **22**, 1668–1678.

Rasenick, M. M., and Kaplan, R. S. (1986). Guanine nucleotide activation of adenylate cyclase in saponin permeable glioma cells. *FEBS Lett.* 296–301.

Rasenick, M. M., Talluri, M., and Dunn, W. J., III (1994). Photoaffinity guanosine 5'triphosphate analogs as a tool for the study of GTP-binding proteins. *Methods Enzymol.* **237**, 100–110.

Rasenick, M. M., and Wang, N. (1988). Exchange of guanine nucleotides between tubulin and GTP-binding proteins that regulate adenylate cyclase: Cytoskeletal modification of neuronal signal transduction. *J. Neurochem.* **51**, 300–311.

Rasenick, M. S., Stein, P. J., and Bitensky, M. W. (1981). The regulatory subunit of adenylate cyclase interacts with cytoskeletal components. *Nature* **294,** 560–562.

Roychowdhury, S., Panda, D., Wilson, L., and Rasenick, M. M. (1999). G protein alpha subunits activate tubulin GTPase and modulate microtubule polymerization dynamics. *J. Biol. Chem.* **274,** 13485–13490.

Roychowdhury, S., and Rasenick, M. M. (1994). Tubulin-G protein association stabilizes GTP binding and activates GTPase: Cytoskeletal participation in neuronal signal transduction. *Biochemistry* **33,** 9800–9805.

Roychowdhury, S., Wang, N., and Rasenick, M. M. (1993). G protein binding and G protein activation by nucleotide transfer involve distinct domains on tubulin: Regulation of signal transduction by cytoskeletal elements. *Biochemistry.* **32,** 4955–4961.

Rudolph, S. A., Hegstrand, L. R., Greengard, P., and Malawista, S. (1979). The interaction of colchicine with hormone-sensitive adenylate cyclase in human leukocytes. *Mol. Pharmacol.* **16,** 805.

Salomon, Y. (1979). Adenylate cyclase assay. *Adv. Cyclic. Nucleotide Res.* **10,** 35–55.

Shelanski, M., Gaskin, F., and Cantor, C. (1973). Microtubule assembly in the absence of added nucleotides. *Proc. Natl. Acad. Sci. USA* **70,** 765.

Wang, N., Yan, K., and Rasenick, M. M. (1990). Tubulin binds specifically to the signal-transducing proteins, Gs alpha and Gi alpha 1. *J. Biol. Chem.* **265,** 1239–1242.

Yan, K., Greene, E., Belga, F., and Rasenick, M. M. (1996). Synaptic membrane G proteins are complexed with tubulin *in situ. J. Neurochem.* **66,** 1489–1495.

Yan, K., Popova, J. S., Moss, A., Shah, B., and Rasenick, M. M. (2001). Tubulin stimulates adenylyl cyclase activity in C6 glioma cells by bypassing the β-adrenergic receptor: A potential mechanism of G protein activation. *J. Neurochem.* **76,** 182–190.

[25] Nucleoside Diphosphate Kinase–Mediated Activation of Heterotrimeric G Proteins

By Susanne Lutz, Hans-Jörg Hippe,
Feraydoon Niroomand, and Thomas Wieland

Abstract

Formation of GTP by nucleoside diphosphate kinase (NDPK) can contribute to receptor independent G protein activation. Apparently, the NDPK B isoform forms complexes with $G\beta\gamma$ dimers and thereby phosphorylates His266 in $G\beta1$ subunits. Phosphorylated His266 mediates G protein activation by a transfer of the high energetic phosphate onto GDP, thus leading to *de novo* synthesis of GTP. Moreover, it has been demonstrated that the sarcolemmal content of NDPK isoforms is increased in hearts with terminal congestive heart failure leading to enhanced G protein activation. Similar data were reported in a rat model for β-adreno-ceptor-induced cardiac hypertrophy. We therefore describe in this chapter several methods which can be used for analysis of NDPK mediated G protein activation: (1) The quantification of NDPK isoforms in highly

METHODS IN ENZYMOLOGY, VOL. 390   0076-6879/04 $35.00

purified cardiac sarcolemmal membranes, (2) the enrichment of the NDPK B/G$\beta\gamma$-complex from preparations of the retinal G protein transducin, (3) the analysis of the enhanced NDPK activated and high energy phosphate transfer in a neonatal rat cardiac myocyte derived cell line stably over-expressing NDPK (H10 cells), and (4) the increased activation of adenylyl cyclase by the enhanced receptor-independent activation of the stimulatory G protein α subunit in these cells.

Introduction

Nucleoside diphosphate kinase (NDPK; EC 2.7.4.6) catalyzes the transfer of terminal phosphate groups from $5'$-triphosphate to $5'$-diphosphate nucleotides. Four isoforms that display the enzymatic activity via a phosphohistidine intermediate have been identified. One of the major reactions catalyzed by NDPK is the phosphate transfer from ATP to GDP to maintain levels of GTP. Only a small fraction of cellular NDPK binds to the plasma membrane, where it may serve the synthesis of GTP, required for the activation of G proteins. However, the binding affinity of GTP to G proteins is an order of magnitude below intracellular concentrations. Hence, a more direct contribution of NDPK to G-protein activation has been proposed for a long time (Kimura, 1993). While numerous studies *in vitro* (for review, see Piacentini and Niroomand, 1996) have revealed G-protein activation through the enzymatic activity of NDPK, i.e., synthesis of GTP from a nucleoside triphosphate and GDP, the significance of this phenomenon for signal transduction in living cells has been questioned (Otero, 2000). The most direct putative mechanism for an NDPK-induced G-protein activation would be the formation of GTP from GDP still bound to the G-protein α subunit. Despite many attempts, such a direct phosphate transfer mechanism on the G protein has never been proven and is sterically unlikely. A complex formation of NDPK B with G$\beta\gamma$ in different tissues has been detected (Cuello *et al.*, 2003). In this complex, NDPK B mediates the phosphorylation of Gβ at His266, which has been shown before to induce G-protein activation by phosphate transfer (Wieland *et al.*, 1992, 1993). Coexpression of NDPK isoforms and the α subunit of the G$_s$ protein (G$_s\alpha$), followed by the measurement of cAMP synthesis and phosphorylation of Gβ, revealed that the NDPK B-G$\beta\gamma$ complex can induce G$_s\alpha$ activation in a receptor-independent manner in living cells (Cuello *et al.*, 2003; Hippe *et al.*, 2003). These provide evidence that the increase of sarcolemmal NDPK observed in hearts from patients with severe congestive heart failure or in a rat model for cardiac hypertrophy might be of pathophysiological importance (Lutz *et al.*, 2001, 2003).

This article therefore describes the methods used to analyze an increase in sarcolemmal NDPK, the purification of the G$\beta\gamma$-NDPK B complex, and the stable overexpression of NDPKs in H10 cells, which can be used for the analysis of enhanced G-protein and effector activation.

Detection of Sarcolemma-Bound NDPK Isoforms in Samples from Human Heart

A higher amount of NDPK isoforms bound to the sarcolemma in failing human hearts compared to nonfailing controls argues for a translocation of NDPK during the pathogenesis of the disease. To analyze this translocation, tissue homogenates, cytosolic fractions, particulate fractions, and highly purified sarcolemmal membranes have to be prepared from samples of human hearts. The amount and composition of NDPK isoforms are subsequently analyzed by immunoblotting and determination of the enzymatic activity.

Preparation of Highly Purified Sarcolemma from Human Heart Tissue

Buffers. Buffer I contains 10 mM HEPES, pH 7.5, 750 mM NaCl, and 2 mM NaN$_3$. Buffer II consists of 10 mM HEPES, pH 7.5, 10 mM NaHCO$_3$, and 2 mM NaN$_3$. Buffer III consists of 100 mM Tris–HCl, pH 7.5, 50 mM Na$_4$P$_2$O$_7$, 300 mM NaCl, and 2 M sucrose. Buffer IV contains 100 mM Tris–HCl, pH 7.5, 50 mM Na$_4$P$_2$O$_7$, 300 mM NaCl, and 600 mM sucrose. Buffer V consists of 10 mM HEPES, pH 7.5, and 250 mM sucrose.

Directly after extraction the human heart is placed on ice and separated in septum, left, and right ventricular tissue. The heart compartments are further cut in 1 × 1-cm^3 pieces, immediately frozen in liquid N$_2$, and stored at −80° until further use. Depending on the laboratory equipment, up to six different tissues can be handled in parallel. In general, all subsequent steps are carried out in a cold room, with precooled buffers and plasticware. All centrifugation steps are accomplished in prechilled centrifuges.

The human heart tissues are allowed to thaw slowly in the cold room and are then minced into small pieces with a pair of scissors or a blender (Jones *et al.*, 1988). Ice-cold buffer I (100 ml) is added. The tissue is homogenized three times for 15 s with a Polytron homogenizer at 20,000 rpm followed by a 30-s cooling period on ice. If not otherwise indicated, all following homogenization steps are performed at this speed with an intermittent 30-s cooling period. An aliquot of the homogenate (H) is filtered through a thrice-folded gauze, frozen in liquid nitrogen, and stored at −80° to serve as a control for subsequent analyses. Thereafter, the homogenate is centrifuged at 21,500g in a fixed angle rotor (Beckman, JA10) for 30 min. The supernatant is discarded, and the pellet is resuspended in 100 ml buffer I, homogenized three more times for 15 s, and centrifuged as described earlier. The supernatant is discarded and the pellet is resuspended in 150 ml buffer II. Homogenization for 15 s is repeated three times and followed by another centrifugation. The pellet is resuspended thoroughly in 26 ml buffer II, homogenized three times for 30 s, and centrifuged again. The membrane-containing supernatant is filtered through a thrice-folded gauze to remove remaining tissue particles. To increase the yield of sarcolemma, the pellet is again resuspended in 26 ml buffer II and homogenized three more times for 30 s. After centrifugation, the supernatant is also filtered through the thrice-folded gauze. The supernatants are combined and centrifuged at 55,000g for 40 min (Beckman Ti60 fixed angle rotor). The supernatant is discarded and the pellet is resuspended in 5 ml ice-cold double distilled water by pipetting up and down, homogenized with 20 strokes of a glass–glass Dounce homogenizer, and transferred in a 25-ml ultracentrifugation tube. Buffer III (5 ml) is added and mixed thoroughly, but not vigorously by inverting the tube several times. A discontinuous sucrose gradient is formed by carefully layering 7 ml buffer IV onto the resuspended protein solution. Finally, 7 ml buffer V is layered on top of the gradient. The sucrose gradients are centrifuged for 70 min at 370,000g (Beckman Ti60 fixed angle rotor). After centrifugation, a cloudy white interphase between the upper and the middle phase forms. This sarcolemma-containing interphase is collected, transferred into a new centrifugation tube, and filled up with ice-cold double distilled water. The sarcolemmal membranes (SM) are pelleted at 170,000g for 30 min, resuspended in an appropriate volume (0.5–2 ml, to achieve a protein concentration of about 1 μg/μl) buffer V and homogenized with 20 strokes of a glass–glass Dounce homogenizer. Small aliquots are frozen in liquid N_2 and stored at $-80°$.

Preparation of Cytosolic and Total Particulate Fractions

A part of the homogenate (H, see earlier discussion) obtained during the membrane preparation procedure is centrifuged at 100,000g for 60 min (Beckman Ti60 fixed angle rotor). The supernatant (cytosolic fraction, CF) is frozen in liquid N_2 and stored at $-80°$ in aliquots until use. The pellet is resuspended in a 1000-fold excess of buffer I, homogenized three times for 30 s at 20,000 rpm with a Polytron homogenizer, and centrifuged once again at 100,000g for 60 min. The supernatant is discarded and the pellet (particulate fraction, PF) is resuspended in buffer I, homogenized for 30 s at 20,000 rpm, frozen in liquid N_2, and stored at $-80°$.

Assay for NDPK Activity

For determination of enzymatic NDPK activity, 0.3–1 μg protein of fractionated heart tissue is used. The samples are incubated on ice with 1 μg alamethicin per microgram protein in 50 mM triethanolamine, pH

7.5, for 20 min in a volume of 50 μl. Alamethicin is a pore-forming iono-phore, which permeabilizes membrane vesicles and therefore allows mea-surement of the NDPK activity not only in inside-out, but also in right side-out, vesicles. The reaction is started by adding 50 μl of the reaction mix, consisting of 0.2 mM cAMP, 0.2 mM ATP, 6 mM MgCl$_2$, 2 mM EDTA, 2 mM dithiothreitol (DTT), 0.2 mM GDP, and 1 μCi of [8-^{3}H]GDP. The samples are incubated for 10 min at 37° in a water bath, and the reaction is terminated by adding 10 μl of a 10% (w/v) SDS solution. Guanine nucleo-tides are separated by thin-layer chromatography. First, for each sample, 2 μl of a GMP-GDP-GTP solution (each 3 mM) is spotted as a carrier near the bottom of a PEI cellulose plate and then 10 μl of the sample is spotted in four steps (2.5 μl each) onto the carrier. The plate is developed in a reservoir with a freshly prepared soluble phase consisting of equal parts of 1 M LiCl and 2 M formic acid. The separated guanine nucleotides are visualized with UV light at 254 nm, the areas containing GTP, GDP, and GMP are cut off, and 4 ml scintillation liquid is added. To elute the nucleotides, the samples are shaken vigorously for 2 h. Radioactivity is determined in a liquid scintil-lation spectrometer. As an example, the distribution of NDPK activity in fractions from failing and nonfailing human hearts is shown in Fig. 1A.

Immunoblot Analysis

Ten micrograms of sarcolemmal membrane protein from different left ventricular tissues is separated by discontinuous sodium dodecyl sul-fate–polyacrylamide gel electrophoresis (SDS–PAGE) (Laemmli, 1970). The resolving gel consists of 15% acrylamide/bisacrylamide (37.5:1), 0.375 M Tris–HCl, pH 8.8, 0.1% (w/v) SDS, 0.1% (w/v) ammonium persul-fate, and 0.04% (v/v) tetramethylethylenediamine (TEMED). A stacking gel containing 5% acrylamide/bisacrylamide (37.5:1), 0.125 M Tris–HCl, pH 6.8, 0.1% (w/v) SDS, 0.1% (w/v) ammonium persulfate, and 0.04% (v/v) TEMED is layered on top of the resolving gel. Samples are boiled in SDS sample buffer, with a final concentration of 0.05 M Tris–HCl, pH 6.8, 2% (w/v) SDS, 10% (v/v) glycerol, 25 mM DTT, and bromphenol blue at 95°C for 5 min. Electro-phoresis is carried out in an electrophoresis buffer containing 25 mM Tris, pH 8.2, 0.25 M glycine, and 0.1% (v/v) SDS with constant voltage (200 V) for 50 min. The proteins are then transferred onto a nitrocellulose membrane in a cooled wet blot system for 1 h at 100 V. The transfer buffer contains 25 mM Tris–HCl, pH 8.3, 192 mM glycine, and 20% (v/v) methanol. Proteins on the membrane are stained in 0.1% (w/v) Ponceau S/5% (v/v) acetic acid and are documented by video-based imaging to verify equal loading. Thereafter, the membrane is destained with TBST [10 mM Tris–HCl, pH 7.4, 150 mM NaCl, 0.1% (v/v) Tween 20] and blocked in 5% (w/v) bovine serum albumin fraction V in TBST for 1 h at room temperature. After three washing steps, 5 min each

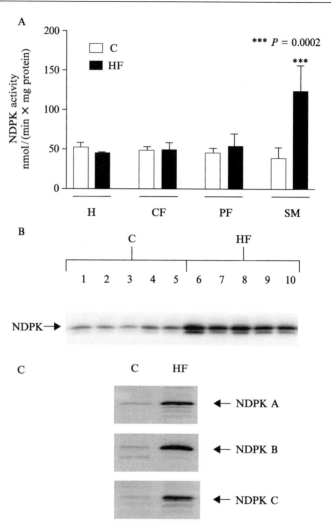

FIG. 1. Translocation of NDPK A, B, and C to the sarcolemmal membrane in failing human hearts. (A) NDPK activities in different fractions of the membrane preparation from control (C) and failing hearts (HF). One microgram protein from homogenates (H), cytosolic (CF), and particulate fractions (PF) was used to determine NDPK activity by measuring the conversion of $[^3H]GDP$ to $[^3H]GTP$ as described in the text. To determine NDPK activity in the sarcolemmal membranes (SM), 0.3 μg protein was sufficient. (B) Immunoblot showing levels of NDPK in sarcolemmal membranes from five human control hearts (C) and five failing human (HF) hearts. Ten micrograms membrane protein from left ventricular tissues was separated by SDS–PAGE and transferred onto a nitrocellulose membrane. NDPK was detected with a nonisoform-specific, polyclonal nm23 antibody. (C) Immunoblots showing levels of the NDPK isoforms A, B, and C in one control (C) and one failing human heart (HF). NDPK A, B, and C are detected with isoform-specific, polyclonal antibodies.

with TBST, the nitrocellulose membrane is incubated with the different NDPK antibodies. To detect NDPK A and B, a 1:200 dilution of an nm23H1/H2 polyclonal antibody from Santa Cruz (sc-343) in TBST is used. For specific detection of the NDPK isoforms A, B, and C, custom-made polyclonal antibodies are used (kindly provided by Dr. Ion Lascu, University of Bordeaux, France). These antibodies are diluted 1:1000 (NDPK A and B) and 1:3000 (NDPK C) in TBST. Blots are incubated for 1 h at room temperature with the primary antibodies, followed by three washing steps with TBST for 10 min. As a secondary antibody, an antirabbit horseradish peroxidase-coupled antibody is used according to the manufacturer's recommendations. After three final washing steps, immunoreactivity is visualized with an enhanced chemiluminescence system. The detection of total NDPK and the enrichment of the isoforms NDPK A, B, and C in the SM fraction of failing human hearts are shown in Figs. 1B and C.

5'-Nucleotidase Activity Assay

To compare the purity of the membrane preparations, the activity of the 5'-nucleotidase is determined as a specific sarcolemmal marker (Emmelot and Bos, 1966). Three micrograms sarcolemmal membrane protein is incubated for 20 min on ice with 1 μg alamethicin per microgram membrane protein in 180 μl of 50 mM Tris–HCl, pH 7.2. The reaction is initiated by adding 20 μl of a reaction mixture containing 1 M KCl, 50 mM Tris–HCl, pH 7.2, and 100 mM AMP as substrate. Samples are incubated for 20 min at 37° in a water bath. The reaction is stopped by adding 50 μl of 25% (w/v) trichloroacetic acid. To detect free phosphate, 250 μl of a 5% (w/v) SDS solution, 250 μl 2,4-diaminophenol dihydrochloride (1 mg/ml in 1% sodium sulfite), and 250 μl of 1.22% (w/v) ammonium molybdate in 1 M sulfuric acid are added. The samples are incubated for 20 min at room temperature. The color of the samples changes from colorless to violet/blue. The absorbance at 660 nm is determined. To create a standard curve, absorbance induced by 0, 20, 40, 60, 80, and 100 μl 2 mM monopotassium phosphate is measured. The phosphate content of the test samples is calculated by linear regression. The 5'-nucleotidase activity of human heart sarcolemmal membranes is given in Table I.

TABLE I
5'-NUCLEOTIDASE ACTIVITY IN HUMAN SARCOLEMMAL MEMBRANES

	Control	Heart failure
5'-Nucleotidase (μmol $\times$ h^{-1} $\times$ mg^{-1})	20 $\pm$ 2.3 ($n = 5$)	22 $\pm$ 3.5 ($n = 13$)

Purification of the NDPK B/Transducin $\beta\gamma$ Complex

We have been able to demonstrate a specific complex formation of the NDPK B with heterotrimeric G-protein $\beta\gamma$ dimers (G$\beta\gamma$) (Cuello *et al.*, 2003). Also, the retinal rod heterotrimeric G-protein transducin (G$_t$) is able to form a complex with NDPK B through its specific $\beta\gamma$ dimer (G$_t\beta\gamma$, i.e., G$\beta_1\gamma_1$). Due to the unique properties of G$_t$, i.e., the elution of large amounts of nearly pure G$_t$ from rod outer segment (ROS) membranes in a hypotonic, GTP-containing but detergent-free buffer, the NDPK/G$_t\beta\gamma$ complex can be separated from noncomplexed G$_t$ by a two-step chromatographic procedure.

Buffers

ROS storage buffer contains 0.1 M NaH$_2$PO$_4$/NaOH, pH 6.5, 1 mM MgCl$_2$, and 0.1 mM EDTA. Isotonic buffer consists of 10 mM Tris–HCl, pH 7.5, 100 mM NaCl, 5 mM MgCl$_2$, 0.1 mM EDTA, 1 mM DTT, and 1 mM phenylmethylsulfonyl fluoride (PMSF). Hypotonic buffer contains 10 mM Tris–HCl, pH 7.5, 0.1 mM EDTA, 1 mM DTT, and 1 mM PMSF. Blue Sepharose buffer A consists of 10 mM Tris–HCl, pH 7.5, 6 mM MgCl$_2$, 1 mM DTT, and 20% (v/v) glycerol. Blue Sepharose buffer B additionally contains 500 mM KCl. Hydroxyapatite buffer A consists of 10 mM Tris–HCl, pH 6.5, 6 mM MgCl$_2$, 1 mM DTT, and 20% (v/v) glycerol. Hydroxyapatite buffer B additionally contains 400 mM KH$_2$PO$_4$.

Preparation of Rod Outer Segment Membranes and Elution of G$_t$

Retinae are isolated from eyes obtained from freshly slaughtered animals, and rod outer segment membranes are prepared essentially as described (Papermaster and Dreyer, 1974). Bleached ROS membranes obtained from about 200 eyes are suspended in 10 ml of ROS storage buffer, frozen in liquid N$_2$, and stored at $-80°$ until further use. After thawing, membranes obtained from about 600 eyes are washed six times with 20 ml isotonic buffer and six more times with 20 ml hypotonic buffer. For G$_t$ elution, washed ROS membranes (about 80 mg protein) are resuspended with a 21-gauge needle in 5 ml hypotonic buffer containing 100 μM of the stable GTP analog guanosine-5'-O-(β,γ-imino)triphosphate (Gpp[NH]p). After pelleting of the membranes by ultracentrifugation (40 min at 225,000g, 4°, Ti70 fixed angle rotor, Beckman), the G$_t$-containing supernatant is removed and stored on ice. Elution of the membranes is repeated twice, and the transducin-containing supernatants are combined and concentrated to a final volume of 5 ml in an ultrafiltration cell (Amicon, PM 10 membrane). For removal of unbound Gpp[NH]p and exchange of buffer, the eluate (2.5 ml each) is applied to two PD 10 gel filtration columns (Amersham Biosciences) equilibrated with Blue Sepharose buffer A. G_t is eluted from each column with 3.5 ml Blue Sepharose buffer A.

Resolution of $G_t\beta\gamma$ Dimers from $G_t\alpha$ Subunits and the NDPK/$G_t\beta\gamma$ Complex

For all protein chromatography described herein, the use of a protein chromatography system, e.g., FPLC, is recommended. First, 5 g of Blue Sepharose Cl-6B (Amersham Biosciences) is resuspended in 200 ml water overnight at 4°, and the resulting 20-ml gel matrix is filled into an XK16/20 column (Amersham Biosciences). The column is equilibrated with 100 ml of Blue Sepharose buffer A at a flow rate of 1 ml/min. Thereafter, the 7-ml transducin-containing eluate from the PD 10 columns (about 2 mg of protein, see earlier) is applied to the column at a flow rate of 0.5 ml/min. The column is washed with 100 ml of Blue Sepharose buffer A at 1 ml/min, and protein elution is monitored by absorption at 280 nm with a UV monitor. Fractions of 4 ml are collected. As shown in Fig. 2A, free $G_t\beta\gamma$ dimers exhibit only a weak interaction with the matrix and elute from the column within the first 30 ml. Thereafter, 100 ml of Blue Sepharose buffer B is applied to the column. The increased ionic strength releases $G_t\alpha$ together with the NDPK/$G_t\beta\gamma$ complex in a single peak from the column (Fig. 2). The $G_t\beta\gamma$-containing fractions are pooled, frozen in liquid N_2, and stored at −80° for further use. The $G_t\alpha$-containing fractions are pooled and concentrated in an ultrafiltration cell (Amicon, PM 10 membrane) until a protein concentration of 1 mg/ml is reached.

Separation of NDPK/$G_t\beta\gamma$ Complex from $G_t\alpha$

A 5-ml hydroxyapatite column (Econo Cartridge CHTII, BioRad Laboratories) is equilibrated with 20 ml hydroxyapatite buffer A. The concentrated $G_t\alpha$-containing fractions from the Blue Sepharose column are applied to the column at a flow rate 0.5 ml/min. Protein elution is monitored by absorption at 280 nm with a UV monitor. Fractions of 1 ml are collected. The column is washed with 20 ml hydroxyapatite buffer A. Then, 10 mM KH_2PO_4 is applied to the column by gradient mixing of hydroxyapatite buffers A and B and this concentration is held for 10 ml of flow. As shown in Figs. 2B and C, $G_t\alpha$ elutes from the column already at this low phosphate concentration. Thereafter, the KH_2PO_4 (10–400 mM) concentration is increased linearly in a 25-ml gradient. The NDPK/$G_t\beta\gamma$ complex elutes from the column at about 250 mM of KH_2PO_4. Due to its low abundance the complex is not visible as a protein peak in the absorption at 280 nm. Fractions of the gradient have therefore to be screened by the more sensitive NDPK activity assay (see earlier). Alternatively, the

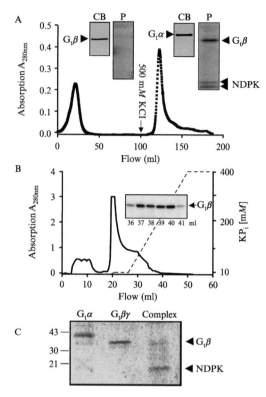

FIG. 2. Resolution of free G_t subunits from the NDPK/$G_t\beta\gamma$ complex. (A) Chromatography on Blue Sepharose CL6B. Two milligrams of G_t eluted from ROS membranes with GppNHp was applied to a Blue Sepharose CL6B FPLC column. The protein content of the fractions was monitored by absorbance at 280 nm. The addition of 500 mM KCl to the eluent is indicated. See text for details. (Insets) The two peak fractions in each gradient were assayed for protein content by SDS–PAGE and staining with Coomassie brillant blue (CB) and for $G\beta$ and NDPK phosphorylating activity using $[\gamma\text{-}^{32}\text{P}]\text{GTP}$ (P). (B) Chromatography on hydroxyapatite. One milligram of the $G_t\alpha$ pool was applied to an Econo cartridge CHTII column. Bound proteins were eluted with the indicated potassium phosphate gradient (dotted line). Protein content was monitored by absorbance at 280 nm. Fifteen microliters of each fraction was tested for $G\beta$ phosphorylating activity using $[\gamma\text{-}^{32}\text{P}]\text{GTP}$. (Inset) An autoradiograph of phosphorylated proteins after SDS–PAGE. (C) $G_t\beta\gamma$ dimers (0.3 μg) (first peak of the Blue Sepharose column), $G_t\alpha$ (0.3 μg) (main peak of the hydroxyapatite column), and proteins in the peak fraction (1 ml) of the $G\beta$ phosphorylating activity (precipitated with trichloroacetic acid) were resolved by SDS–PAGE. Proteins were stained with silver.

$G\beta$/NDPK phosphorylation assay (see Fig. 2B) can be used. Note, however, that high concentrations of phosphate inhibit this assay, and therefore the sensitivity of the phosphorylation assay is increased by dialysis of the fractions against hydroxyapatite buffer A. Separation of $G_t\alpha$ and uncomplexed $G_t\beta\gamma$ from the NDPK/$G_t\beta\gamma$ complex is shown in Fig. 2C.

*Determination of Transiently Phosphorylated Gβ and NDPK by
Phosphorylation with [γ-³²P]GTP or [³⁵S]GTPγS*

Both NDPK and Gβ can be phosphorylated at histidine residues. Whereas the phosphorylation of NDPK represents the transfer of an NTP γ-phosphate to His118 as the first step in the enzymatic activity catalyzed by NDPK (Morera *et al.*, 1995), the phosphorylation of Gβ at His266 apparently requires phosphorylated NDPK as a phosphate donor (Cuello *et al.*, 2003). The best NTP substrate for the phosphorylation of both proteins is GTP or its synthetic analog guanosine 5′-*O*-(3-thio)triphosphate (GTPγS). Both nucleotides additionally offer the advantage that they are poor substrates for protein kinases using ATP. Phosphorylation with [γ-³²P]GTP (6000 Ci/mmol) or [³⁵S]GTPγS (2000 Ci/mmol) therefore offers a simple assay to visualize phospho-NDPK and phospho-Gβ with low background phosphorylation, e.g., in fractions obtained during purification procedures (see Fig. 2B) or even in crude membrane preparation (see Fig. 3).

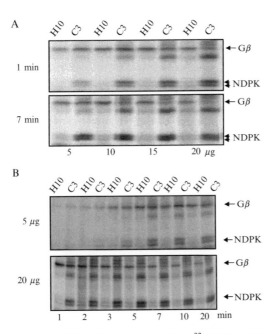

FIG. 3. Phosphorylation of H10 cell membranes with [γ-³²P]GTP. (A) Increasing amounts of H10 cell membranes obtained from control cells H10 or cells threefold overexpressing the human NDPK B (C3) were phosphorylated with 10 n*M* [γ-³²P]GTP for 1 min (top) and 7 min (bottom) (B) Five (top) or 20 (bottom) μg of membrane protein was phosphorylated for increasing periods of time. Autoradiographs after SDS–PAGE are shown. Phosphorylated Gβ and phosphorylated NDPK are indicated.

The assay is usually performed in Eppendorf tubes in a total volume of 20 μl in a reaction mixture containing 50 mM triethanolamine–HCl, pH 7.4, 150 mM NaCl, 2 mM MgCl$_2$, 1 mM EDTA, 1 mM dithiothreitol, 10 nM [γ-^{32}P]GTP, or 20 nM [^{35}S]GTPγS (Amersham Biosciences). Usually, 5 μl of the four-fold concentrated reaction mixture is given into a tube on ice. Before incubation the tubes are warmed for 5 min at 30°, and the protein, e.g., 15 μl of a membrane suspension in 10 mM triethanolamine–HCl, pH 7.4, is added and incubated for the desired periods of time (see Fig. 2). The reaction is terminated by the addition of 10 μl of threefold concentrated sample buffer (Laemmli 1970), followed by a 1-h incubation at room temperature. As the phosphoamidate bond of a phosphohistidine is heat sensitive, boiling of the samples should be avoided. Proteins are separated by discontinuous SDS–PAGE on gels containing 10–12% (w/v) acrylamide. The dye front containing free nucleotides is cut off from the gel. The gel is rinsed for 15 min in a 10% (v/v) glycerol/water mixture, dried, and autoradiographed. As phosphoamidate bonds are sensitive to acid, staining of the proteins by methods using an acidic fixation, e.g., Coomassie brilliant blue staining, should be avoided. The use of prestained molecular weight markers is recommended.

Due to the high energetic nature of the phosphohistidine in Gβ as well as in the NDPK, phosphates can be retransferred onto GDP, and thus the phosphorylation is transient. Note that especially in membranes, GTPases create a constant pool of GDP and thus, depending on the amount of the used membrane protein, the phosphorylation is extremely short lived. For example, when 20 μg of H10 cell membranes has been subjected to phosphorylation with [γ-^{32}P]GTP, the maximal phosphorylation is already reached after 1–2 min of incubation. Any prolonged incubation results in a loss of incorporated radioactivity due the aforementioned mechanism (Fig. 3B, bottom). Therefore, optimal conditions for obtaining good phosphorylation results have to be adjusted in each system by varying incubation times and protein amounts. In Fig. 3, an example for such an adjustment for membranes of H10 cells is shown. It further demonstrates the alterations occurring in the kinetics of Gβ phosphorylation by a threefold overexpression of NDPK B in H10 cells (see later).

Analysis of NDPK-Induced G$_s\alpha$ Activation in H10 Cells

To demonstrate the activation of G proteins by NDPK in intact cells, the approach of coexpressing NDPK and G$_s\alpha$ protein in a cell line derived from neonatal rat cardiac myocytes (H10 cells; Jahn et al., 1996) is used. First, cell clones stably overexpressing the human NDPK isoforms A (nm23-H1) and B (nm23-H2) and the catalytically inactive mutant of NDPK B, nm23-H2H118N (M118), are generated. In a second step, increasing amounts of

$G_s\alpha$ are coexpressed by means of adenoviral gene transfer. An increase in cAMP accumulation due to the stimulation of adenylyl cyclase in these cells is used as readout for enhanced $G_s\alpha$ activation.

Generation of H10 Cell Clones Stably Expressing NDPK Isoforms and Mutants

H10 cells are cultured at 33° in Dulbecco's modified Eagle medium (DMEM) supplemented with 10% fetal calf serum (FCS), 2 mM L-glutamine, and 10 μg/ml gentamycin. For transfection of H10 cells, SuperFect (Qiagen), a transfection reagent based on specifically designed activated dendrimers, is used. Coding sequences of nm23-H1, nm23-H2, and M118 are subcloned into the pIRESpuro vector (Clontech), a bicistronic expression vector, which allows translation of the nm23 genes and the puromycin resistence gene from one messenger RNA. Twenty hours before transfection, H10 cells are seeded at a density of 7×10^5 per 60-mm dish in 5 ml supplemented DMEM. On the day of transfection, cells should have grown to 80% confluence. For transfection, 8 μg of the plasmid is diluted (pIRESpuro-EGFP or pIRESpuro-NDPK constructs) in 150 μl DMEM without serum and antibiotics at room temperature. Fifty microliters SuperFect transfection reagent (Qiagen) is added to the DNA solution and mixed by pipetting up and down five times (no vortexing). The samples are incubated for 10 min at room temperature to allow for transfection complex formation. One milliliter of supplemented DMEM is added to the samples and mixed by pipetting up and down twice. Meanwhile, the cells are washed once with serum-free DMEM and the medium is replaced with the 1.2-ml transfection sample per 60-mm dish. After 4 h, the transfection medium is removed, the cells are washed three times with 4 ml phosphate buffered saline and 5 ml of supplemented DMEM is added. The transfection efficiency is approximately 20% as judged by the expression of enhanced green fluorescent protein (EGFP). Cells are passaged onto 10-cm culture dishes at 1:15 dilution into the selective DMEM containing 2 μg/ml puromycin 48 h after transfection. Puromycin-resistent clones are isolated after 2 weeks in culture using cloning cylinders and 100 μl accutase. Cell clones are propagated from 12-well plates to 10-cm dishes. To identify positive clones, NDPK expression is screened by immunoblot analysis and NDPK activity assays (see earlier discussion). Once positive cell clones are identified, continuous culture under selective pressure with 2 μg/ml puromycin is recommended.

Adenoviral Overexpression of $G_s\alpha$ in H10 Cell Clones

Efficient overexpression of $G_s\alpha$ is performed by adenoviral gene transfer. Adenoviruses (serotype 5) encoding either rat $G_s\alpha$ (Ad5$G_s\alpha$) or EGFP

(Ad5GFP) are a kind gift of Dr. Thomas Eschenhagen, Hamburg, Germany. The EGFP encoding virus serves as a control virus and monitors infection efficiency. The day before infection cells are seeded at a density of 10^5 cells per well in 12-well tissue culture dishes. Cells were washed once with serum-free DMEM. Viral stocks (about 10^9 biological active viral particles per μl) are diluted in 300 μl serum-free DMEM per well to obtain the desired multiplicities of infection (MOI, ratio of biological active viruses to cells, e.g., 50, 100, and 200). Cells are incubated with the virus-containing medium for 20 min at 33°. Then 700 μl DMEM containing 4% FCS is added per well. Cells are harvested after 24–48 h in culture and are used in the assays.

Preparation of H10 Cell Homogenates and Membranes

Analysis of NDPK and $G_s\alpha$ expression in H10 cell is performed by immunoblotting with specific antibodies (see earlier discussion) in the crude membrane and cell homogenate, respectively. For preparation of the homogenate, H10 cell clones are grown in monolayer on 10-cm dishes until confluence is reached. Cells are washed twice with ice-cold phosphate-buffered saline. Thereafter, cells are scraped off in 1 ml per dish of a homogenization buffer containing 10 mM Tris–HCl, pH 7.4, 0.1 mM EDTA, and 1 mM PMSF and transferred to a 12-ml tube. Cells are homogenized twice for 10 s with a Polytron homogenizer at 20,000 rpm followed by a 30-s cooling period on ice. To obtain a crude membrane fraction, the homogenate is centrifuged at 100,000g for 30 min at 4°. The pellet is resuspended in 1 ml of homogenization buffer. Centrifugation and resuspension are repeated two more times. The final pellet is resuspended in 250 μl homogenization buffer per 10-cm dish. The homogenates and the membranes are frozen in liquid N_2 and stored in aliquots at $-80°$ until further use.

Determination of cAMP Accumulation in Intact H10 Cells

To quantify $G_s\alpha$ activation by NDPK, we measured adenylyl cyclase-mediated cAMP accumulation in intact H10 cell clones infected with the recombinant adenoviruses using a competitive enzyme immunoassay. Cells are cultured and infected with the appropriate adenovirus as described earlier. After a 36-h culture in DMEM containing 4% FCS, cells are serum starved 4 h before starting the assay. The accumulation of cAMP is determined in 400 μl of an assay buffer containing 20 mM HEPES, pH 7.4, 100 μM isobutylmethylxanthine, and 100 μM propranolol, an inverse agonist of the β-receptor that negates spontaneously or constitutive receptor activity. Cells are allowed to equilibrate in the assay buffer for 30 min at 33°. Thereafter, the assay medium is removed, cells are broken by the addition of 400 μl of 0.1 M hydrochloric acid, scraped off, and transferred into an Eppendorf tube. After centrifugation at 20,000g for 15 min at 4°, 100 μl of the supernatant is used for the competitive enzyme immunoassay for cAMP (R&D System, Wiesbaden, Germany) strictly following the manufacturer's instructions. The acid-insoluble cell pellets are neutralizied with 0.1 M NaOH, and the protein content is determined according to Bradford (1976). The production of cAMP is normalized to the protein content. The overexpression of $G_s\alpha$ by adenoviral gene transfer and the increase in $G_s\alpha$-mediated cAMP accumulation, as well as membrane-associated NDPK activity in NDPK B-overexpressing cell clones, are shown in Fig. 4.

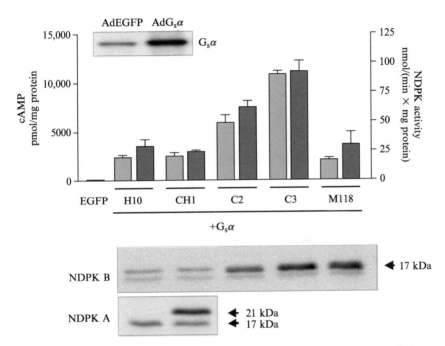

FIG. 4. cAMP accumulation and membrane-associated NDPK activity in H10 cell clones with combined overexpression of NDPK and $G_s\alpha$. cAMP content (gray bars, left y axis) was determined in H10 cells infected with AdEGFP (EGFP) or AdG$_s\alpha$ (H10 + G$_s\alpha$, MOI = 100) and in a cell clone with ~6-fold overexpression of NDPK A (CH1), three cell clones overexpressing NDPK B ~2-fold (C2), ~3-fold (C3), or ~3-fold the catalytically inactive NDPK B mutant (M118) infected with AdG$_s\alpha$. Adenoviral infection resulted in a ~10-fold overexpression of G$_s\alpha$ compared to AdEGFP-infected cells. (Inset) A representative immunoblot is shown. NDPK activity was quantified as formation of [³H]GTP in a 1-μg cell membrane of the indicated cell clones (black bars, right y axis). Representative immunoblots with specific antibodies against NDPK B (top) and NDPK A (bottom) are shown below.

References

Bradford, M. M. (1976). A rapid and sensitive method for the quantitation of microgram quantities of protein utilizing the principle of protein-dye binding. *Anal. Biochem.* **72,** 248–254.

Cuello, F., Schulze, R. A., Heemeyer, F., Meyer, H. E., Lutz, S., Jakobs, K. H., Niroomand, F., and Wieland, T. (2003). Activation of heterotrimeric G proteins by a high energy phosphate transfer via nucleoside diphosphate kinase (NDPK) B and Gβ subunits: Complex formation of NDPK B with G$\beta\gamma$ dimers and phosphorylation of His-266 in Gβ. *J. Biol. Chem.* **278,** 7220–7226.

Emmelot, P., and Bos, C. J. (1966). Studies on plasma membranes. 3. Mg^{2+}-ATPase, (Na^{+}-K^{+}-Mg^{2+})-ATPase and 5'-nucleotidase activity of plasma membranes isolated from rat liver. *Biochim. Biophys. Acta* **120,** 369–382.

Hippe, H. J., Lutz, S., Knorr, K., Vogt, A., Jakobs, K. H., Wieland, T., and Niroomand, F. (2003). Activation of heterotrimeric G proteins by a high energy phosphate transfer via nucleoside diphosphate kinase (NDPK) B and Gβ subunits: Specific activation of $G_s\alpha$ by a NDPK B-G$\beta\gamma$ complex in H10 cells. *J. Biol. Chem.* **278,** 7227–7233.

Jahn, L., Sadoshima, J., Green, A., Parker, C., Morgan, K. G., and Izumo, S. (1996). Conditional differentiation of heart- and smooth muscle-derived cells transformed by a temperature-sensitive mutant of SV40 T antigen. *J. Cell Sci.* **109,** 397–407.

Jones, L. R. (1988). Rapid preparation of canine cardiac sarcolemmal vesicles by sucrose flotation. *Methods Enzymol.* **157,** 85–91.

Kimura, N. (1993). Role of nucleoside diphosphate kinase in G-protein action. *In* "Handbook of Experimental Pharmacology" (B. F., Dickey and L, Birnbaumer, eds.), Vol. 108II, pp. 485–498. Springer-Verlag, Berlin.

Laemmli, U. K. (1970). Cleavage of structural proteins during the assembly of the head of bacteriophage T4. *Nature* **227,** 680–685.

Lutz, S., Mura, R., Baltus, D., Movsesian, M. A., Kübler, W., and Niroomand, F. (2001). Increased activity of membrane-associated nucleoside diphosphate kinase and inhibition of cAMP synthesis in failing human myocardium. *Cardiovasc. Res.* **49,** 48–55.

Lutz, S., Mura, R. A., Hippe, H. J., Tiefenbacher, C., and Niroomand, F. (2003). Plasma membrane-associated nucleoside diphosphate kinase (nm23) in the heart is regulated by β-adrenergic signaling. *Br. J. Pharmacol.* **140,** 1019–1026.

Morera, S., Chiadmi, M., LeBras, G., Lascu, I., and Janin, J. (1995). Mechanism of phosphate transfer by nucleoside diphosphate kinase: X-ray structures of the phosphohistidine intermediate of the enzymes from *Drosophila* and *Dictyostelium*. *Biochemistry* **34,** 11062–11070.

Otero, A. S. (2000). NM23/nucleoside diphosphate kinase and signal transduction. *J. Bioenerg. Biomembr.* **32,** 269–275.

Papermaster, D. S., and Dreyer, W. J. (1974). Rhodopsin content in the outer segment membranes of bovine and frog retinal rods. *Biochemistry* **13,** 2438–2444.

Piacentini, L., and Niroomand, F. (1996). Phosphotransfer reactions as a means of G protein activation. *Mol. Cell. Biochem.* **157,** 59–63.

Wieland, T., Nürnberg, B., Ulibarri, I., Kaldenberg-Stasch, S., Schultz, G., and Jakobs, K. H. (1993). Guanine nucleotide-specific phosphate transfer by guanine nucleotide-binding regulatory protein β-subunits: Characterization of the phosphorylated amino acid. *J. Biol. Chem.* **268,** 18111–18118.

Wieland, T., Ronzani, M., and Jakobs, K. H. (1992). Stimulation and inhibition of human platelet adenylyl cyclase by thiophosphorylated transducin $\beta\gamma$-subunits. *J. Biol. Chem.* **267,** 20791–20797.

Subsection B
Inhibitors

[26] Purification and *In Vitro* Functional Analyses of
RGS12 and RGS14 GoLoco Motif Peptides

By RANDALL J. KIMPLE,
FRANCIS S. WILLARD, and DAVID P. SIDEROVSKI

Abstract

The GoLoco motif is a short polypeptide sequence that binds to hetero-trimeric G-protein α subunits of the adenylyl cyclase-inhibitory ($G\alpha_{i/o}$) subclass in a nucleotide-dependent manner (i.e., solely to the GDP-bound ground state). This article describes methods used for the expression, purification, and *in vitro* evaluation of membrane-permeant tag fusion peptides derived from the GoLoco motif regions of "regulator of G-protein signaling" proteins type 12 (RGS12) and 14 (RGS14) and a consensus GoLoco sequence from the multiple GoLoco motif protein AGS3. Three different fluorescence-based assays are described for evaluating the *in vitro* function of these GoLoco peptides as guanine nucleotide dissociation inhibitors, including measurements of GTPγS binding and Gα subunit activation by the planar ion aluminum tetrafluoride.

Introduction

Heterotrimeric guanine nucleotide binding or "G" proteins, composed of Gα, Gβ, and Gγ subunits, activate a variety of downstream signaling pathways (reviewed in Hamm, 1998). The standard model of G-protein-coupled receptor (GPCR) signaling assumes that the duration of signaling from both the Gα subunit and the G$\beta\gamma$ dimer is controlled by the lifetime of GTP-bound Gα. This assumption has been challenged by our discovery (Siderovski *et al.*, 1999) of an additional class of Gα-regulatory proteins (Fig. 1), each containing characteristic "GoLoco" motifs that bind specifically to Gα·GDP (reviewed in Kimple *et al.*, 2002b). We and others have demonstrated that GoLoco motif-containing proteins exhibit unique specificity for their Gα partners. For example, the GoLoco motif of Purkinje cell protein-2 (Pcp2) acts as a guanine nucleotide dissociation inhibitor (GDI) for both Gα_i and Gα_o subunits (Kimple *et al.*, 2002a; Natochin *et al.*, 2001),

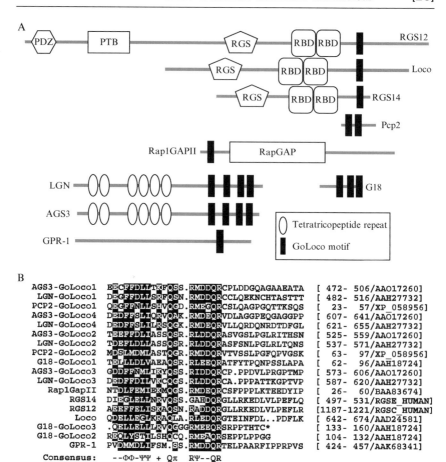

FIG. 1. GoLoco motifs are found in a diverse set of G-protein-signaling regulatory proteins. (A) Domain organization of single- and multi-GoLoco motif-containing proteins as described by the Simple Modular Architecture Research Tool (SMART website: http:// smart.embl-heidelberg.de). Domain abbreviations used are as follows: PDZ, PSD-95, Dlg, ZO-1 homology domain; PTB, phosphotyrosine-binding domain; RGS, regulator of G-protein signaling box; RBD, Ras-binding domain; GoLoco, $G\alpha_{i/o}$-Loco interacting motif; RapGAP, Rap GTPase-activating protein domain. Note that the N terminus of the *C. elegans* protein GPR-1 (Colombo et al., 2003) is predicted, by protein fold recognition algorithms (e.g., 3D-PSSM: http://www.sbg.bio.ic.ac.uk/~3dpssm), to form a stable all-α-helical fold composed of tetratricopeptide repeats (TPRs); however, individual TPRs are not currently identifiable by SMART in the GPR-1 sequence and are thus not depicted in the GPR-1 protein schematic. (B) Multiple sequence alignment of GoLoco motifs. For each representative protein, SwissProt ID (or GenBank accession number) and location of sequence range within protein are shown in square brackets. Amino acid abbreviations used in the consensus are (-), acidic; Φ, hydrophobic; Ψ, long aliphatic; π, small side chain.

whereas the GoLoco motifs of RGS12, RGS14, and G18 are GDIs only for $G\alpha_i$ subunits (Kimple *et al.*, 2001, 2004). The association of a GoLoco motif with a GDP-bound $G\alpha$ subunit is thought to prevent both the activation of $G\alpha$ effectors and the reassociation of the $G\alpha\beta\gamma$ heterotrimer. GoLoco motif-containing proteins have roles in many intracellular processes, including asymmetric cell division, mitotic spindle organization, and chromosomal segregation (e.g., Colombo *et al.*, 2003; reviewed in Willard *et al.*, 2004a). While a clearly defined biochemical function for GoLoco-$G\alpha$·GDP protein complexes has yet to be established, we have used GoLoco peptides to delineate the $G\alpha$ coupling of dopamine receptors in AtT20 pituitary lactotroph cells (Webb *et al.*, 2004). Thus, GoLoco peptides can be used as "perturbagens" (Kimple *et al.*, 2002b) of heterotrimeric G-protein signaling to aid in our understanding of the G-protein coupling profile of GPCRs.

This article describes the design, cloning, expression, and purification of membrane-permeable GoLoco peptides, as well as the validation of their activity by an *in vitro* assay of GDI activity. It also presents additional *in vitro* assays useful for validating the functional competence of GoLoco peptides. GoLoco peptides are purified by fusion to the membrane-permeable sequence (MPS) of the Kaposi's sarcoma virus fibroblast growth factor modified with a cyanogen bromide (CNBr)-cleavable N-terminal ketosteroid isomerase (KSI) fusion protein and a C-terminal hexahistidine (His_6) tag. Kuliopulos and Walsh (1994) have previously described the use of an N-terminal KSI tag to target fusion proteins to inclusion bodies upon expression in *Escherichia coli*. Isolation of the insoluble fraction allows the MPS-GoLoco peptide to be solubilized by guanidine denaturation. The KSI-MPS-GoLoco-His_6 fusion protein can then be purified by immobilized metal affinity chromatography after which the KSI and His_6 tags are removed easily by CNBr cleavage. It is important to note that CNBr cleavage occurs after methionine residues and results in the presence of a C-terminal homoserine lactone. This reactive group can be used to attach various moieties to the peptide, such as biotin or fluorescein, via a primary amine coupling. However, any peptide produced using this system must not contain internal methionine residues, to avoid internal cleavage. Kuliopulos and Walsh (1994) have successfully mutated internal methionines to isoleucine without affecting the biological activity of the expressed polypeptide. MPS-GoLoco peptides can then be purified further via additional chromatographic steps. The ability to purify significant quantities of MPS-GoLoco peptides using this protocol should facilitate cellular studies on the effects of GoLoco peptides on heterotrimeric G-protein signaling and cell division processes.

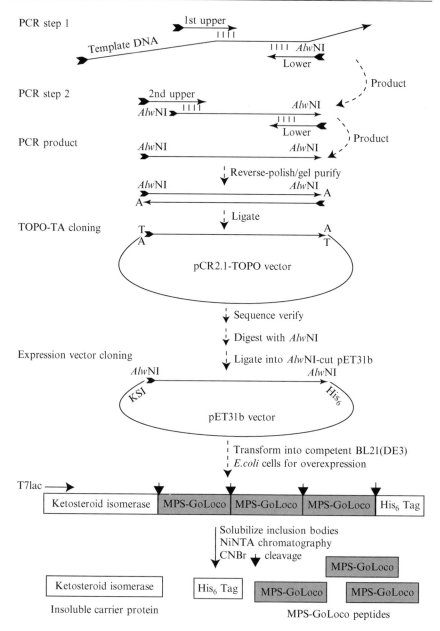

FIG. 2. Schematic of MPS-GoLoco construction, expression, and purification. The open reading frame that encodes the GoLoco peptide tagged with an N-terminal membrane-permeable sequence (MPS) is initially created by a two-step PCR protocol, followed by

Purification of GoLoco Peptides

Design of Membrane-Permeable GoLoco Peptides

To facilitate the intracellular delivery of GoLoco peptides, we chose to couple the GoLoco motif of rat RGS12 (EAEEFFELISKAQSNRA-DDQRGLLRKEDLVLPEFLR; amino acids 1186 to 1221 of SwissProt accession number O08774) and rat RGS14 (DIEGLVELLNRVQSS-GAHDQRGLLRKEDLVLPEFLQ; amino acids 496 to 531 of SwissProt accession number O08773) to the MPS of Kaposi's sarcoma virus fibroblast growth factor (Lin *et al.*, 1995). In addition to this MPS polypeptide, several other reported moieties can be used for the intracellular delivery of peptides and proteins, as reviewed by Hawiger (1999) and Schwarze *et al.* (2000). These include the cell permeation sequence from the HIV-1 Tat protein (Frankel and Pabo, 1988), the *Drosophila* Antennapedia (Antp) homeotic transcription factor (Joliot *et al.*, 1991), the herpes simplex virus 1 DNA-binding protein VP22 (Elliott and O'Hare, 1997), or attachment of a palmitoyl (Covic *et al.*, 2002) or myristate (Eichholtz *et al.*, 1993) lipid moiety. To allow us to encode these peptides genetically and express them in bacteria, we choose to use a proteinaceous membrane-permeable tag. In particular, we preferred the neutrally charged MPS sequence (LAAVAL-LPAVLLALLAK), as opposed to the basic HIV-1 Tat (GGGYGRK-KRRQRRRG) or Antp (RQIKIWFQNRRMKWKK) sequences, given the possibility of unfavorable intramolecular electrostatic interactions with the acidic GoLoco peptide sequence. The length of a single MPS-GoLoco polypeptide ($\sim$53 amino acids) makes this fusion difficult to synthesize in high yields without failure sequences using traditional [N-(9-fluorenyl)-methoxycarbonyl] blocking group (Fmoc) chemistry; therefore, we chose to use the KSI fusion system of Kuliopulos and Walsh (1994) for the expression of MPS-GoLoco polypeptides in bacteria instead of conjugating separately purified synthetic GoLoco and MPS polypeptide sequences (e.g., Zhang *et al.*, 1998).

In order to add an N-terminal MPS sequence to the GoLoco motif of RGS12 or RGS14, two sequential polymerase chain reaction (PCR) amplifications are performed. In the first reaction, the GoLoco motif is amplified and the last nine amino acids (AVLLALLAK) of the MPS sequence are added by the first sense primer ("1st upper"; Fig. 2). In addition, a 3' *Alw*NI

cloning into a carrier vector such as pCR2.1-TOPO, digestion with *Alw*NI and subcloning into a KSI fusion expression vector such as pET31b, overexpression of the KSI-MPS-GoLoco-His$_6$ fusion polypeptide, cleavage of the desired MPS-GoLoco peptide with cyanogen bromide (CNBr), and purification of MPS-GoLoco peptides by additional chromatography.

cut site (CAGatg'CTG) is engineered into each antisense primer ("lower"; Fig. 2). To break any secondary structure in the MPS fusion peptide, a proline codon is inserted between the MPS sequence and the GoLoco sequence. The second PCR reaction uses the same antisense primer, but substitutes a single, nested sense primer ("2nd upper"; Fig. 2) for both RGS12 and RGS14 (collectively abbreviated "R1x"). The second sense primer encodes a 5' *Alw*NI restriction site and the remainder of the MPS sequence (LAAVALLP). The primers used for PCR are detailed in Fig. 3.

Cloning of MPS-GoLoco PCR Products

To generate MPS-GoLoco sequences for cloning, the open reading frames (ORF) of both RGS12 and RGS14, described previously (Snow *et al.*, 1997), are used as DNA templates for PCR amplification. The PCR reactions are performed in a PTC-150 programmable thermal cycler (MJ Research Inc.) using a "touch-down/touch-up" protocol described previously (Snow *et al.*, 2002). The optimal annealing temperature (T_{an}) for sense/antisense primer pairs is calculated by the program Oligo (Molecular Biology Insights; http://www.oligo.net/): for the indicated primer pairs in Fig. 3, an optimal T_{an} of ~62° is predicted. However, PCR amplifications using the touch-down/touch-up protocol with an optimal annealing

MPS-R14GL 1st upper:
    5'-gccgtgctgctagccttgttagccaaacccgacattgaaggcctagtgga-3'
    AlaValLeuLeuAlaLeuLeuAlaLys*Pro*AspIleGluGluLeuVal

MPS-R14GL lower:
    5'-cca**CAGcatCTG**cagaaattcaggaaggaccaggtcctctttgc-3'

MPS-R12GL 1st upper:
    5'-ccgtgctgctagccttgttagccaagcccgaagcagaggagtttttgag-3'
    ValLeuLeuAlaLeuLeuAlaLys*Pro*GluAlaGluGluPhePheGlu

MPS-R12GL lower:
    5'-gg**CAGcatCTG**gcgaagaaactcaggcagtaccaggtcttcctttc-3'

MPS-R1xGL 2nd upper:
    5'-aa**CAGatgCTG**gctgcggttgctctccttcctgccgtgctgctagccttgttag-3'
    GlnMetLeuAlaAlaValAlaLeuLeuProAlaValLeuLeuAlaLeuLeu

FIG. 3. PCR amplification primers used for cloning of MPS-GoLoco open reading frames. Shown below each sense primer is the corresponding open reading frame translation using three-letter amino acid abbreviations. The *Alw*NI restriction site (CAGnnnCTG) is highlighted in bold. The proline residue introduced between MPS and GoLoco polypeptides to reduce secondary structure is indicated in italics.

temperature of $62°$ were unsuccessful. Therefore, a $T_{an} = 58°$ is used for the touch-down/touch-up protocol as follows:

Step 1	95.0°	00:40 s	(initial denaturation phase)
Step 2	94.0°	00:20 s	
Step 3	64.0°	00:20 s − 2.0°/cycle	(using "increment" option)
Step 4	72.0°	00:45 s + 1 s/cycle	(using "extend" option)
Step 5	4 times to step 2		(using GOTO command)
Step 6	94.0°	00:25 s	
Step 7	56.0°	00:20 s + 1.0°/cycle	
Step 8	72.0°	00:50 s + 1 s/cycle	
Step 9	4 times to step 6		
Step 10	94.0°	00:25 s	
Step 11	60.0°	00:20 s	
Step 12	72.0°	00:55 s + 1 s/cycle	
Step 13	17 times to step 10		
Step 14	72.0°	05:00 s	(final extension phase)
Step 15	4.0°	00:00	(i.e., continuous incubation)
Step 16	end		

The first round of PCR reactions is set up in 0.25-ml thin-wall reaction tubes (MJ Research, Inc.) using 200 ng of template DNA, 1.0 μl of 1st upper primer (25 μM stock), 1.0 μl of lower primer (25 μM stock), 5 μl of Pfu-Turbo buffer, 1.0 μl of dNTPs (10 mM stock), 1.0 μl of Pfu-Turbo proofreading, thermostable DNA polymerase (Stratagene), and 40 μl of sterile H_2O. Following completion of the thermal cycling, 1 μl of the PCR amplification mixture is used to set up the second amplification reaction [i.e., with 1.0 μl of the 2nd upper primer (25 μM stock), 1.0 μl of lower primer (25 μM stock), 5 μl of Pfu-Turbo buffer, 1.0 μl of dNTPs (10 mM stock), 1.0 μl of Pfu-Turbo polymerase, and 40 μl of sterile H_2O]. After completion of the second PCR amplification, 0.5 μl of *Taq* polymerase is added to each reaction tube and the PCR products are "reversed polished" for 10 min at $72°$ to add overhanging adenosines compatible with "TA-cloning" vectors. The entire reaction mixture is subsequently resolved by electrophoresis through 2.0% agarose (buffered with 40 mM Tris–acetate, pH 8.2, 1 mM EDTA), and DNA of the correct molecular weight (as visualized by ethidium bromide staining of coelectrophoresed DNA markers) is purified from gel slices using chaotropic salt solubilization and binding to silica gel spin column membranes (QIAquick gel extraction system; Qiagen Inc., Valencia, CA). DNA is eluted in 30 μl of sterile solution containing 10 mM Tris, pH 8.0, and 1 mM EDTA.

Five microliters of gel-purified PCR product is added to 1 μl of topoisomerase I-activated, linearized, $3'$ deoxythymidine-tailed vector (pCR2.1-TOPO; Invitrogen, San Diego, CA) and is incubated for 5 min at room temperature. The ligation product is then transformed into *E. coli*

TOP10F' cells, and clones with inserts are identified by a β-galactosidase chromogenic assay (i.e., lacZ complementation "blue/white" selection) according to the manufacturer's instructions. Multiple white colonies are selected and plasmid DNA is isolated by alkaline lysis of 5-ml overnight liquid cultures and subsequent anion-exchange chromatography (Qiagen plasmid minikit, Qiagen Inc.). Correct-sized inserts ($\sim$180 bp) are confirmed by plasmid DNA digestion with EcoRI restriction endonuclease and 3% agarose gel electrophoresis [2% NuSieve 3:1 agarose (Cambrex Bio Science)/1% low EEO agarose (Fisher Scientific)], prior to fluorescent dideoxynucleotide sequencing (ABI/Perkin Elmer, Foster City, CA) using the T7 primer (5′-TAATACGACTCACTATAGGG-3′).

Subcloning of MPS-GoLoco Expression Constructs

pCR2.1-TOPO clones containing sequence-validated MPS-GoLoco cDNA inserts are next subjected to restriction digestion with AlwNI (New England Biolabs, Beverly, MA) and liberated inserts are separated from pCR2.1-TOPO vector DNA on a 3% NuSieve/agarose gel. Correct-sized DNA inserts ($\sim$180 bp) are cut out of the gel and purified from gel slices using chaotropic salt solubilization as described earlier. Following gel purification, inserts are ligated into the pET-31b(+) vector, digested previously with AlwNI and dephosphorylated (Novagen), using the Roche rapid DNA ligation kit (Roche Applied Science, Indianapolis, IN). Briefly, 1 μl of pET31b vector is added to 5 μl of MPS-GoLoco insert, 2 μl of DNA dilution buffer, and 2 μl of sterile H_2O. The 10-μl DNA mixture is incubated at 55° for 5 min and allowed to cool to room temperature for 10 min to allow for vector and insert overhangs to anneal. Ten microliters of T4 DNA ligation buffer and 1 μl of T4 DNA ligase are added to the mixture, and the ligation reaction is allowed to proceed for 15 min at room temperature. Four microliters of ligated vector is then added to 40 μl of XL10 Gold supercompetent cells (Stratagene) and the mixture is allowed to incubate on ice for 20 min. Cells are heat shocked by a 45-s incubation at 42° followed immediately by cooling on ice for 2 min. Fifty microliters of Luria broth (LB) is added to the competent cell/DNA mixture, and the entire mixture is placed at 37° for 30 min. Fifty microliters of the transformation reaction is then streaked onto LB agar plates containing 50 μg/ml carbenicillin and allowed to incubate overnight at 37°.

After at least 16 h, individual colonies are chosen to inoculate 5-ml overnight cultures of LB broth containing 50 μg/ml carbenicillin. Following at least 16 h of shaking at 220 rpm in a 37° incubator, pET31b-MPS-GoLoco plasmids are isolated from the E. coli liquid culture by alkaline lysis and anion-exchange chromatography as described previously. To screen for proper insertion of MPS-GoLoco DNA segments, diagnostic restriction endonuclease digests are performed prior to sequencing using the T7 primer. Sequence-verified plasmids are transformed into BL21(DE3) cells for overexpression of the KSI-MPS-GoLoco-His$_6$ fusion protein. While this method allows for the insertion of tandem repeats of a given template (as shown in Fig. 2), the use of tandem repeats is recommended only for peptides less than 25 amino acids in length. Therefore, we selected only those plasmids containing a single MPS-GoLoco insertion for subsequent protein expression and purification.

Overexpression and Purification of KSI-MPS-GoLoco-His$_6$ Protein

For overexpression of fusion protein, bacteria are grown to an OD$_{600\ nm}$ of 0.6–0.8 at 37° before induction with 1 mM isopropyl β-D-thiogalactoside. After an additional 4-h incubation at 37°, cell pellets are snap frozen at −80° (Fig. 4A, lane i). Prior to purification, E. coli cell pellets are resuspended in 25 ml (per 2 liters of original culture volume) of buffer N1 (5 mM imidazole, 20 mM Tris, pH 8.0, 500 mM NaCl), and bacteria are ruptured by lysis on an AMINCO French press (SLM Instruments Inc., Urbana, IL). Cellular lysates are centrifuged at 100,000g for 30 min at 4°, after which the supernatant is removed (Fig. 4A, lane s). The remaining pellet is washed with 25 ml of buffer N1 and centrifuged again at 100,000g for 30 min at 4° (Fig. 4A, lane r). Next, the insoluble pellet is resuspended in 25 ml of buffer N2 (5 mM imidazole, 20 mM Tris, pH 8.0, 500 mM NaCl, 6 M guanidine HCl) using a Potter–Elvehjem tissue grinder (Corning Life Sciences, Acton, MA) to extract the water-insoluble proteins. The remaining insoluble material is removed by further centrifugation at 100,000g for 30 min at 4°.

The guanidine-soluble fraction (Fig. 4A, lane p) is loaded onto a 5-ml nickel-nitrilotriacetic acid (NiNTA) column (Amersham Pharmacia) equilibrated previously with 50 ml of buffer N2 (Fig. 4B). After protein loading, the column is washed sequentially with 30 ml of buffer N2 and 30 ml of buffer N3 (15 mM imidazole, 20 mM Tris, pH 8.0, 500 mM NaCl, 6 M guanidine HCl). The KSI-MPS-GoLoco-His$_6$ protein is then eluted with 30 ml of buffer N4 (300 mM imidazole, 20 mM Tris, pH 8.0, 500 mM NaCl, 6 M guanidine HCl). The remaining protein is eluted with buffer N5 (500 mM imidazole, 20 mM Tris, pH 8.0, 500 mM NaCl, 6 M guanidine HCl).

CNBr Cleavage of KSI-MPS-GoLoco-His$_6$ Fusion Protein

The KSI-MPS-GoLoco-His$_6$ fusion protein recovered from NiNTA chromatography is dialyzed at 4° overnight into 2 liters of MilliQ-purified sterile deionized H$_2$O using Spectra/Por dialysis membrane (6000–8000

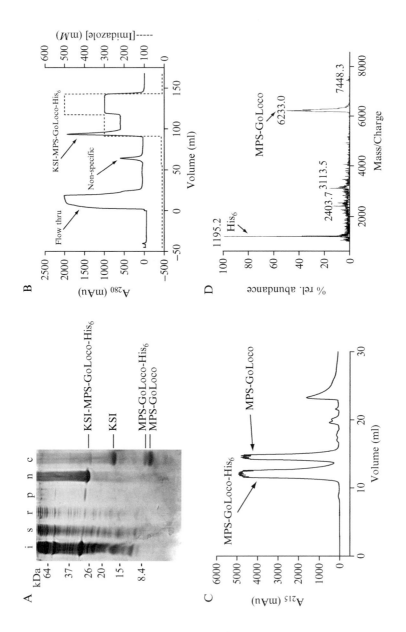

molecular weight cutoff; Spectrum Laboratories, Inc., Rancho Dominguez, CA). To ensure maximum removal of the 6 M guanidine, the H_2O is exchanged four times. The white precipitate containing the KSI-MPS-GoLoco-His$_6$ protein that forms during dialysis is then transferred into a sterile 50-ml polypropylene conical tube, and the dialysis bag is rinsed with 25 ml of H_2O to collect as much precipitate as possible. The precipitate (Fig. 4A, lane n) is then centrifuged at 2750g at 4° for 30 min in a Centra CL3R clinical centrifuge (ThermoIEC, Needham Heights, MA). After removal of the supernatant, the remaining pellet is dissolved in 6 ml of 80% formic acid and transferred to a 50-ml round-bottom flask containing 0.2 g of CNBr. Nitrogen gas is bubbled in for 5 min, and the flask is covered with parafilm and wrapped in aluminum foil. The peptide/formic acid/CNBr solution is stirred slowly at room temperature for 24 h, after which the solution is transferred to several 1.5-ml microcentrifuge tubes (Fig. 4A, lane c). The solution is allowed to evaporate to dryness in a Speed-Vac rotary evaporator (Svc100, ThermoSavant, Holbrook, NY). Each pellet is resuspended in 100 μl of 30% acetonitrile and 0.1% trifluoroacetic acid and is allowed to rotate overnight. The remaining insoluble KSI protein is removed by centrifugation at 10,000g for 10 min, after which the supernatant, containing MPS-GoLoco and His$_6$ peptides, is pooled by transfer to a new 1.5-ml microcentrifuge tube.

To complete purification of the CNBr-released MPS-GoLoco peptides, the peptide mixture is separated by size-exclusion chromatography (buffer: 30% acetonitrile, 0.1% trifluoroacetic acid) over a Superdex peptide column (Amersham Pharmacia; Fig. 4C). Peak fractions (measured by

---

FIG. 4. Purification of MPS-GoLoco peptides. (A) KSI-MPS-GoLoco-His$_6$ protein samples resolved by 12% SDS–PAGE and Coomassie staining during purification. Lane abbreviations are as follows: i, induced sample; s, soluble fraction; r, first rinse; p, insoluble pellet; n, after NiNTA purification and precipitation; c, after CNBr cleavage. Positions of full-length KSI-MPS-GoLoco-His$_6$ protein and cleaved ketosteroid isomerase (KSI) protein are indicated at right. (B) Absorbance at 280 nm (in milliabsorbance units or "mAu") and imidazole concentration (mM) tracings of KSI-MPS-GoLoco-His$_6$ purification during nickel-NTA chromatography. The NiNTA column was equilibrated for 50 ml (−50 to 0 ml on x axis) prior to loading of protein (0–30 ml on x axis). A wash of 15 mM imidazole (60–90 ml) results in the elution of nonspecifically-bound protein. KSI-MPS-GoLoco-His$_6$ was eluted with 30 ml of 300 mM imidazole (90–120 ml). (C) Following CNBr cleavage and KSI precipitation, 100 μl of MPS-GoLoco peptide mixture was injected onto a Superdex peptide column and resolved by size-exclusion chromatography. Peptide elution was monitored by absorbance at 215 nm (milliabsorbance units). (D) MALDI mass spectrometry was used to evaluate the purity of the final peptide product. The MPS-GoLoco peptide has a predicted molecular mass of 6232 Da, whereas the His$_6$ tag (LLEHHHHHH) has a predicted molecular mass of 1195 Da.

absorbance at 215 nm) are pooled and lyophilized via freeze drying in a Virtis Freezemobile 6 (SP Industries Inc., Gardiner, NY) prior to storage at $-80°$. The final purity of MPS-GoLoco peptides is assessed by mass spectrometry (Fig. 4D).

Fluorescence-Based Assays of GoLoco Peptide Activity

Real-Time Measurement of Gα Subunit Binding to Fluor-Tagged GTPγS

Instead of using conventional, radioactive GTPγ[^{35}S] filter-binding assays, we monitor the spontaneous exchange of GDP for GTP by Gα subunits using a real-time BODIPY FL GTPγS binding assay (first described by McEwen et al., 2001) to test for GoLoco peptide-mediated GDI activity (Kimple et al., 2003). The fluorescent BODIPY moiety is linked to GTPγS via the γ-phosphate such that, when in solution, the nucleoside folds back upon the BODIPY fluor, resulting in self-quenching and a low quantum yield. Upon binding to Gα subunits, the fluorescent moiety exhibits a greatly increased quantum yield that can be monitored in real time.

We perform all assays of GDI activity in a Perkin Elmer LS55 spectrofluorimeter with a multiturret cuvette adapter. Using this system, up to four independent BODIPY GTPγS binding experiments can be performed simultaneously. Each of the four cuvettes is loaded with 1 ml of TEM buffer (10 mM Tris, pH 8.0, 1 mM EDTA, 10 mM MgCl$_2$) plus 1 μM BODIPY FL GTPγS (Molecular Probes Inc., Eugene, OR). Recombinant, GDP-bound Gα$_{i1}$ protein is diluted in TEM buffer to a volume of 40 μl, which provides a final concentration of 200 nM protein upon addition to the 1 ml TEM + 1 μM BODIPY FL GTPγS cuvette. Enough MPS-GoLoco peptide is added to the Gα$_{i1}$ protein, or buffer control, to give a final concentration of 1 μM peptide and is allowed to incubate for 15 min at 25° prior to addition to the 1-ml cuvette (at 500 s; Fig. 5A). The solution in each cuvette is excited at a wavelength of 485 nm and is monitored for emission at 530 nm (slit widths set at 3.0 nm). As illustrated in the example assay in Fig. 5A, addition of the MPS-tagged RGS14 GoLoco peptide (MPS-R14GL) to GDP-bound Gα$_{i1}$ slows the exchange of GDP for BODIPY FL GTPγS dramatically. This finding is consistent with our observations that MPS-GoLoco peptides, despite their N-terminal, cell permeation tag extensions, act as GDIs for Gα$_i$ subunits with activities equivalent to their nontagged GoLoco peptide counterparts.

Activation by AlF$_4^-$: Real-Time Measurement of Gα Subunit Intrinsic Tryptophan Fluorescence

We and others have reported that mutation of the invariant arginine within the GoLoco motif cripples the ability of GoLoco proteins and peptides to act as GDIs (Kimple *et al.*, 2002a; Peterson *et al.*, 2002). Such loss-of-function GoLoco mutants should serve as valuable negative controls in experiments employing cell-permeant GoLoco peptides. For example, we have used the peptide synthesis facility of Tufts University (www.tucf.org) to synthesize two N-terminally palmitoylated ("palm") GoLoco peptides derived from the AGS3 consensus sequence (Kimple *et al.*, 2001; Peterson *et al.*, 2000), one of which contains an arginine-to-phenylalanine substitution (underlined) at this important residue:

palmAGS3:   palmitoyl-TMGEEDFFDLLAKSQSKRMDDQRVDLAG
palmAGS3^{R23F}:   palmitoyl-TMGEEDFFDLLAKSQSKRMDDQ<u>F</u>VDLAG

We have used these peptides in an *in vitro* assay of Gα activation by the planar ion aluminum tetrafluoride (AlF$_4^-$) (Higashijima *et al.*, 1987; Phillips and Cerione, 1988), an assay employed previously in characterizing the GoLoco motifs of AGS3, RGS12, and RGS14 (De Vries *et al.*, 2000; Kimple *et al.*, 2001, 2002a). In the presence of activating ligands such as AlF$_4^-$, G-protein α subunits exhibit an ~60% increase in fluorescence emission at 340 nm ($\lambda_{ex} = 290$ nm) (Higashijima *et al.*, 1987) caused by the rearrangement of tryptophan-211 within the conformationally flexible switch II region (Faurobert *et al.*, 1993; Lan *et al.*, 1998). Measurements of intrinsic Gα tryptophan fluorescence are performed on the LS55 spectrofluorimeter with excitation at 292 nm and emission at 342 nm (slit widths 2.5 and 5.0 nm, respectively). Recombinant Gα_{i1} is diluted to 200 nM in preactivation buffer (100 mM NaCl, 100 μM EDTA, 2 mM MgCl$_2$, 20 μM GDP, 20 mM Tris–HCl, pH 8.0) and is incubated at 30°. Lyophilized palmitoylated peptides are resuspended to 1 mM in 1% dimethyl sulfoxide (DMSO) (v/v in H$_2$O) and subsequently diluted in preactivation buffer to a final concentration of 2 μM in cuvette also containing 200 nM Gα_{i1} protein. Peptides and Gα_{i1} protein are allowed to incubate in the cuvette for 150 s prior to the addition of 20 mM NaF and 30 μM AlCl$_3$ (final concentrations). As illustrated in Fig. 5B, only the wild-type peptide slows the activation rate of Gα_{i1} upon the addition of aluminum tetrafluoride; the arginine-to-phenylalanine mutant peptide exhibits no inhibitory activity.

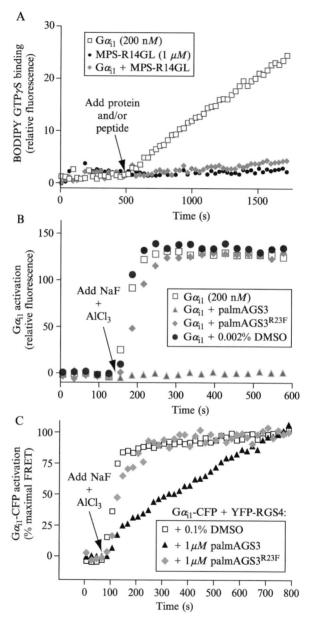

FIG. 5. Fluorescence-based biochemical assays of GoLoco peptide activity. GoLoco peptides act as guanine nucleotide dissociation inhibitors (GDIs) and slow the activation of $G\alpha_i$ subunits by aluminum tetrafluoride (AlF_4^-). (A) Time course of 1 μM BODIPY FL GTPγS binding to 200 nM $G\alpha_{i1}\cdot$GDP preincubated with 1 μM MPS-R14GL peptide.

Activation by AlF$_4^-$: Real-Time Measurement of Gα/RGS Box
Fluorescence Resonance Energy Transfer

The inhibitory effect of GoLoco motif peptides on Gα activation by AlF$_4^-$ can also be measured in real time using a novel assay employing fluorescence resonance energy transfer (FRET) between Gα$_{i1}$-CFP and YFP-RGS4 fusion proteins (described in detail in Willard *et al.*, 2004b). This assay uses the exquisite sensitivity of the RGS box of RGS4 for the AlF$_4^-$-induced transition state of Gα$_{i1}$ over GDP- or GTPγS-bound forms of Gα$_{i1}$ (Berman *et al.*, 1996). We have found that Gα$_{i1}$-CFP and YFP-RGS4 fusion proteins provide an AlF$_4^-$-sensitive FRET pair and thus can facilitate real-time monitoring of GoLoco peptide-mediated inhibition of Gα$_{i1}$ activation.

FRET is measured at 20° in a 1-ml cuvette containing 140 n*M* each of Gα$_{i1}$-CFP and YFP-RGS4 fusion proteins in a buffer consisting of 10 m*M* Tris–HCl, pH 8.0, 1 m*M* EDTA, 10 m*M* MgCl$_2$, 200 m*M* NaCl, and 2 μ*M* GDP. Test samples contain either the GoLoco peptide or a vehicle control (e.g., 0.1% DMSO to match peptide solubilization conditions). Peptide/ FRET pair samples are incubated in the cuvettes for 5 min to induce complex formation between Gα$_{i1}$-CFP and GoLoco motif peptides. Fluorescence measurements are initiated during this preincubation: excitation energy is given at a wavelength of 433 nm (2.5 nm slit width) and emission is detected at two wavelengths corresponding to CFP emission (474 nm, 2.5 nm slit width) and YFP emission (525 nm, 2.5 nm slit width). At 100 s, NaF and AlCl$_3$ are added to final concentrations of 20 m*M* and 30 μ*M*, respectively, to induce Gα activation, the subsequent binding of Gα$_{i1}$ to RGS4, and the resultant FRET between their conjoint CFP and YFP moieties.

Fluorescence measurements ($λ_{ex}$ = 485 nm, $λ_{em}$ = 530 nm) in the presence of only MPS-R14GL peptide (i.e., no added Gα$_{i1}$) and of only Gα$_{i1}$·GDP (i.e., no added peptide) are also shown. (B) As evaluated by the measurement of intrinsic Gα tryptophan fluorescence ($λ_{ex}$ = 292 nm; $λ_{em}$ = 342 nm), preincubation with 2 μ*M* of the palmitoylated AGS3 consensus GoLoco peptide (palmAGS3), but not the arginine-to-phenylalanine loss-of-function mutant version (palmAGS3^{R23F}), stabilizes Gα$_{i1}$ (200 nM) in its inactive, GDP-bound form and prevents activation by AlF$_4^-$ [as formed by the sequential addition of sodium fluoride (NaF; 20 m*M* final concentration) and aluminum chloride (AlCl$_3$; 30 μ*M* final concentration) at 150 seconds]. (C) Time course of AlF$_4^-$-mediated activation of Gα$_{i1}$ as measured by fluorescence resonance energy transfer (FRET) between 140 nM Gα$_{i1}$-CFP and 140 nM YFP-RGS4. FRET was measured, in real time, as the ratio of YFP emission intensity at 525 nm to CFP emission intensity at 474 nm (for more details, see Willard *et al.*, 2004b). Palmitoylated AGS3 consensus GoLoco peptides were solubilized with DMSO and preincubated at the indicated concentration with the FRET sensor protein pair for 5 min. At 100 s, the planar ion AlF$_4^-$ was created by the addition of NaF and AlCl$_3$ to final concentrations of 20 m*M* and 30 μ*M*, respectively. Data points are the mean of duplicate experiments.

Figure 5C illustrates the use of this FRET-based assay to measure the attenuation of $G\alpha_{i1}$ activation by the wild-type, palmitoylated AGS3 peptide and the lack of effect of the loss-of-function mutant palmAGS3^{R23F}. Data from Fig. 5C were fit to a single exponential association model using GraphPad Prism version 4.0 (Graph Pad Software, San Diego, CA). Rate constants were determined for FRET protein pair samples treated with DMSO control ($2.3 \times 10^{-2} \pm 1.1 \times 10^{-3}$ s^{-1}), 1 μM wild-type palmAGS3 peptide ($3.5 \times 10^{-3} \pm 3.7 \times 10^{-4}$ s^{-1}), and 1 μM palmAGS3^{R23F} peptide ($1.6 \times 10^{-2} \pm 1.6 \times 10^{-3}$ s^{-1}). Thus, a sevenfold molar excess of palmAGS3 peptide to $G\alpha_{i1}$-CFP protein resulted in a sevenfold reduction in $G\alpha_{i1}$ activation rate, whereas a sevenfold molar excess of the loss-of-function mutant peptide palmAGS3^{R23F} had only a minor effect on $G\alpha_{i1}$ activation, decreasing the rate by as little as 30%. We have found that this FRET-based assay is uniquely able to resolve the rapid kinetics of $G\alpha$ activation and its modulation by GoLoco peptides and is thus preferred over the use of intrinsic $G\alpha$ tryptophan fluorescence. A more detailed exposition of this FRET method can be found in Willard et al. (2004b).

Summary

We have described a protocol for the bacterial production of membrane-permeable GoLoco peptides as well as fluorescence-based biochemical assays for testing their ability to act as GDIs for $G\alpha$ subunits and inhibit aluminum tetrafluoride-mediated $G\alpha$ activation. These peptides should prove useful in understanding the cellular roles of GoLoco-containing proteins and heterotrimeric G proteins in directing asymmetric cell division and mitotic spindle forces. In addition, they have already proved useful in interrogating the specificity of receptor/G-protein/effector coupling (see Oxford and Webb, 2004).

Acknowledgments

This work was funded by National Institutes of Health Grants R01 GM062338 and P01 GM065533. R.J.K. gratefully acknowledges predoctoral fellowship support from the National Institute of Mental Health (F30 MH64319). F.S.W. is a postdoctoral fellow of the American Heart Association. D.P.S. is a recipient of the Burroughs-Wellcome Fund New Investigator Award in the Pharmacological Sciences.

References

Berman, D. M., Kozasa, T., and Gilman, A. G. (1996). The GTPase-activating protein RGS4 stabilizes the transition state for nucleotide hydrolysis. J. Biol. Chem. **271,** 27209–27212.

Colombo, K., Grill, S. W., Kimple, R. J., Willard, F. S., Siderovski, D. P., and Gönczy, P. (2003). Translation of polarity cues into asymmetric spindle positioning in *Caenorhabditis elegans* embryos. *Science* **300**, 1957–1961.

Covic, L., Gresser, A. L., Talavera, J., Swift, S., and Kuliopulos, A. (2002). Activation and inhibition of G protein-coupled receptors by cell-penetrating membrane-tethered peptides. *Proc. Natl. Acad. Sci. USA* **99**, 643–648.

De Vries, L., Fisher, T., Tronchere, H., Brothers, G. M., Strockbine, B., Siderovski, D. P., and Farquhar, M. G. (2000). Activator of G protein signaling (AGS3) is a guanine dissocation inhibitor for Galpha i subunits. *Proc. Natl. Acad. Sci. USA* **97**, 14364–14369.

Eichholtz, T., de Bont, D. B., de Widt, J., Liskamp, R. M., and Ploegh, H. L. (1993). A myristoylated pseudosubstrate peptide, a novel protein kinase C inhibitor. *J. Biol. Chem.* **268**, 1982–1986.

Elliott, G., and O'Hare, P. (1997). Intercellular trafficking and protein delivery by a herpesvirus structural protein. *Cell* **88**, 223–233.

Faurobert, E., Otto-Bruc, A., Chardin, P., and Chabre, M. (1993). Tryptophan W207 in transducin T alpha is the fluorescence sensor of the G protein activation switch and is involved in the effector binding. *EMBO J.* **12**, 4191–4198.

Frankel, A. D., and Pabo, C. O. (1988). Cellular uptake of the tat protein from human immunodeficiency virus. *Cell* **55**, 1189–1193.

Hamm, H. E. (1998). The many faces of G protein signaling. *J. Biol. Chem.* **273**, 669–672.

Hawiger, J. (1999). Noninvasive intracellular delivery of functional peptides and proteins. *Curr. Opin. Chem. Biol.* **3**, 89–94.

Higashijima, T., Ferguson, K. M., Sternweis, P. C., Smigel, M. D., and Gilman, A. G. (1987). Effects of Mg^{2+} and the beta gamma-subunit complex on the interactions of guanine nucleotides with G proteins. *J. Biol. Chem.* **262**, 762–766.

Joliot, A., Pernelle, C., Deagostini-Bazin, H., and Prochaintz, A. (1991). Antennapedia homeobox peptide regulates neural morphogenesis. *Proc. Natl. Acad. Sci. USA* **88**, 1864–1868.

Kimple, R. J., De Vries, L., Tronchere, H., Behe, C. I., Morris, R. A., Farquhar, M. G., and Siderovski, D. P. (2001). RGS12 and RGS14 GoLoco motifs are G alpha(i) interaction sites with guanine nucleotide dissociation inhibitor activity. *J. Biol. Chem.* **276**, 29275–29281.

Kimple, R. J., Jones, M. B., Shutes, A., Yerxa, B. R., Siderovski, D. P., and Willard, F. S. (2003). Established and emerging fluorescence-based assays for G-protein function: Heterotrimeric G-protein alpha subunits and regulator of G-protein signaling (RGS) proteins. *Comb. Chem. High Throughput Screen.* **6**, 399–407.

Kimple, R. J., Kimple, M. E., Betts, L., Sondek, J., and Siderovski, D. P. (2002a). Structural determinants for GoLoco-induced inhibition of nucleotide release by G alpha subunits. *Nature* **416**, 878–881.

Kimple, R. J., Willard, F. S., Hains, M. D., Jones, M. D., Nweke, G. K., and Siderovski, D. P. (2004). Guanine nucleotide dissociation inhibitor activity of the triple GoLoco motif protein G18: Alanine to aspartate mutation restores function to an inactive second GoLoco motif. *Biochem. J.* **378**, 801–808.

Kimple, R. J., Willard, F. S., and Siderovski, D. P. (2002b). The GoLoco motif: Heralding a new tango between G protein signaling and cell division. *Mol. Interv.* **2**, 88–100.

Kuliopulos, A., and Walsh, C. T. (1994). Production, purification, and cleavage of tandem repeats of recombinant proteins. *J. Am. Chem. Soc.* **116**, 4599–4607.

Lan, K. L., Remmers, A. E., and Neubig, R. R. (1998). Roles of G(o)alpha tryptophans in GTP hydrolysis, GDP release, and fluorescence signals. *Biochemistry* **37**, 837–843.

Lin, Y.-Z., Yao, S. Y., Veach, R. A., Torgerson, T. R., and Hawiger, J. (1995). Inhibition of nuclear translocation of transcription factor NF-kappa B by a synthetic peptide containing a cell membrane-permeable motif and nuclear localization sequence. *J. Biol. Chem.* **270,** 14255–14258.

McEwen, D. P., Gee, K. R., Kang, H. C., and Neubig, R. R. (2001). Fluorescent BODIPY-GTP analogs: Real-time measurement of nucleotide binding to G proteins. *Anal. Biochem.* **291,** 109–117.

Natochin, M., Gasimov, K. G., and Artemyev, N. O. (2001). Inhibition of GDP/GTP exchange on G alpha subunits by proteins containing G-protein regulatory motifs. *Biochemistry* **40,** 5322–5328.

Oxford, G. S., and Webb, C. K. (2004). GoLoco motif peptides as probes of Gα subunit specificity in coupling of G-protein-coupled receptors to ion channels. *Methods Enzymol.* **390,** 437–450.

Peterson, Y. K., Bernard, M. L., Ma, H., Hazard, S., Graber, S. G., and Lanier, S. M. (2000). Stabilization of the GDP-bound conformation of Gialpha by a peptide derived from the G-protein regulatory motif of AGS3. *J. Biol. Chem.* **275,** 33193–33196.

Peterson, Y. K., Hazard, S., Graber, S. G., and Lanier, S. M. (2002). Identification of structural features in the G-protein regulatory motif required for regulation of heterotrimeric G-proteins. *J. Biol. Chem.* **277,** 6767–6770.

Phillips, W. J., and Cerione, R. A. (1988). The intrinsic fluorescence of the alpha subunit of transducin: Measurement of receptor-dependent guanine nucleotide exchange. *J. Biol. Chem.* **263,** 15498–15505.

Schwarze, S. R., Hruska, K. A., and Dowdy, S. F. (2000). Protein transduction: Unrestricted delivery into all cells? *Trends Cell. Biol.* **10,** 290–295.

Siderovski, D. P., Diverse-Pierluissi, M., and De Vries, L. (1999). The GoLoco motif: A G-alpha-i/o binding motif and potential guanine-nucleotide exchange factor. *Trends Biochem. Sci.* **24,** 340–341.

Snow, B. E., Antonio, L., Suggs, S., Gutstein, H. B., and Siderovski, D. P. (1997). Molecular cloning and expression analysis of rat Rgs12 and Rgs14. *Biochem. Biophys. Res. Commun.* **233,** 770–777.

Snow, B. E., Brothers, G. M., and Siderovski, D. P. (2002). Molecular cloning of novel regulators of G-protein signaling (RGS) family members and characterization of the binding specificity of the RGS12 PDZ domain. *Methods Enzymol.* **344,** 740–761.

Webb, C. K., Kimple, R. J., Siderovski, D. P., and Oxford, G. S. (2004). D2 dopamine receptor activation of potassium channels is selectively decoupled by Gαi-specific GoLoco motif peptides. Submitted for publication.

Willard, F. S., Kimple, R. J., Kimple, A. J., Johnston, C. A., and Siderovski, D. P. (2004b). Fluorescence-based assays for RGS box function. *Methods Enzymol.* **389,** 56–71.

Willard, F. S., Kimple, R. J., and Siderovski, D. P. (2004a). Return of the GDI: The GoLoco motif in cell division. *Annu. Rev. Biochem.* **73,** 925–951.

Zhang, L., Torgerson, T. R., Liu, X. Y., Timmons, S., Colosia, A. D., Hawiger, J., and Tam, J. P. (1998). Preparation of functionally active cell-permeable peptides by single-step ligation of two peptide modules. *Proc. Natl. Acad. Sci. USA* **95,** 9184–9189.

[27] GoLoco Motif Peptides as Probes of Gα Subunit Specificity in Coupling of G-Protein-Coupled Receptors to Ion Channels

By Gerry S. Oxford and Christina K. Webb

Abstract

Biochemical and structural studies of signaling proteins have revealed critical features of peptide motifs at the interaction surfaces between proteins. Such information can be used to design small peptides that can be used as functional probes of specific interactions in signaling cascades. This article describes the use of a novel domain (the GoLoco motif) found in several members of the regulators of G-protein signaling (RGS) protein family to probe the specificity of Gα subunit involvement in the coupling of dopamine and somatostatin receptors to ion channels in the AtT20 neuro-endocrine cell line. Peptides encoding the GoLoco motifs of RGS12 and AGS3 were perfused into single cells during electrical recording of agonist-induced current responses by whole cell patch clamp methods. The particular sequences chosen have been demonstrated to bind selectively to the GDP-bound form of $G\alpha_i$, but not $G\alpha_o$, and preclude association of $G\beta\gamma$ and $G\alpha_i$ subunits. A functional manifestation of this property is observed in the progressive uncoupling of D2 dopamine receptors and Kir3.1/3.2 channels with repeated agonist application. Similar uncoupling is not observed with somatostatin receptors nor with D2 receptors coupling to calcium channels, suggesting Gα subunit specificity in these signaling pathways. Motifs found in other proteins in the GPCR signaling machinery may also prove useful in assessing G-protein signaling specificity and complexity in single cells in the future.

Introduction

The remarkable specificity of signaling pathways linking the G-protein-coupled superfamily of receptors (GPCRs) to a vast array of effector proteins is commonly thought to be related to the large number of molecular species directly in the signaling pathways, those modulating the pathways indirectly, and those providing compartmentalization of elements of the pathways. There is ample evidence for subclasses of signaling cascades largely segregated by the class of G-protein α subunit involved (e.g., Gs, Gi/o, Gq/11). However, despite some circumstantial evidence

METHODS IN ENZYMOLOGY, VOL. 390

that specificity in a given signaling cascade in each subclass resides in one or more of these elements, little experimental evidence exists linking a particular GPCR effector pathway to specific cascade elements such as specific $G\alpha$, $G\beta$, or $G\gamma$ subunits. For example, the coupling of GPCRs through Gi/o proteins to effectors such as GIRK and calcium channels occurs through $G\beta\gamma$ subunits (Herlitze et al., 1996; Ikeda, 1996; Logothetis et al., 1987) but is largely independent of the particular $G\beta$ or $G\gamma$ subtype (Wickman et al., 1994; but see Kleuss et al., 1992, 1993). Evidence for $G\alpha$ subunit specificity has to this point derived from three types of experiments: reconstitution of signaling blocked by pertussis toxin using intracellular dialysis of purified or recombinant $G\alpha$ subunits (Degtiar et al., 1997; Ewald et al., 1989; but see McGehee and Oxford, 1991), block of function with $G\alpha$-specific antibodies (Menon-Johansson et al., 1993; Takano et al., 1997), maintenance of signaling under PTX treatment by expression of mutant Pertussis toxin (PTX)-resistant $G\alpha_{i/o}$ subunits (Kammermeier et al., 2003; Leaney and Tinker, 2000; see also Chen et al., 2004; Ikeda and Jeong, 2004), or the absence of signaling during the antisense knockdown of $G\alpha_{i/o}$ subunits (Degtiar et al., 1996; Delmas et al., 1999; Kleuss et al., 1991; Takano et al., 1997). All of these approaches, however, involve either overexpression of a $G\alpha$ subunit, which alters normal signaling stoichiometry, or incomplete reduction in specific $G\alpha$ subunit expression with the attendant ambiguity of sufficient amplification for signaling from the remaining subunits of that species.

Regulator of G-protein signaling (RGS) proteins are a large family of multidomain proteins that modulate GPCR signaling in a variety of ways. The namesake domain of this family, the RGS box, has been shown to accelerate the intrinsic GTPase activity of certain $G\alpha$ subunits and termination of signaling. Another 19 amino acid domain, found in several RGS proteins, as well as other unrelated proteins, is the $G\alpha_{i/o}$–Loco interaction or GoLoco motif (Siderovski et al., 1999). We have taken advantage of three features of the GoLoco motifs found in RGS12, RGS14, and AGS3 proteins to develop an alternative approach to defining GPCR-to-G-protein coupling specificity. The key features of these particular GoLoco motifs are their guanine nucleotide dissociation inhibitor (GDI) activity, which essentially locks $G\alpha_i$ subunits in their inactive GDP-bound form; their binding to a site on $G\alpha_i$ that precludes association of the cognate $G\beta\gamma$ subunits, leaving $G\alpha_i$ in its monomeric inactive form; and the unique specificity of this binding for $G\alpha_i$ but not $G\alpha_o$ proteins (Kimple et al., 2001, 2002). These features suggest the possibility that synthetic peptides encoding these GoLoco motifs could interrupt signaling in pathways preferentially utilizing $G\alpha_i$ over $G\alpha_o$ subunits, thus discriminating signaling specificity at this level. Such peptides would represent relatively small, but specific probes of the $G\alpha_i$–$G\beta\gamma$ interface, reducing ambiguity associated with overexpression of signaling components or with specificity of antibody block. By perfusing synthetic GoLoco motif peptides into AtT20 cells, we have demonstrated interference with the coupling of D2 receptors to Kir3.1/3.2 channels and the absence of such interference in somatostatin (SST) receptor coupling to these channels and to D2 receptor coupling to P/Q-type calcium channels in the same cells, suggesting differential $G\alpha$ subunit signaling for each pathway.

Selection and Synthesis of GoLoco Motif Peptides

While GoLoco motifs have been found in a variety of proteins, those that bind $G\alpha_i$ subunits, but not $G\alpha_o$ subunits, preferentially are found in RGS12, RGS14, and AGS3. Thus it was decided to synthesize GoLoco motif peptides derived from RGS12 ("R12GL," EAEEFFELISKAQSN-RADDQRGLLRKEDLVLPEFLR; amino acids 1186–1221 of SwissProt accession number O08774) and RGS14 ("R14GL," DIEGLVELLNRVQ-SSGAHDQRGLLRKEDLVLPEFLQ; amino acids 496–531 of SwissProt accession number O08773) representing the minimal fragment necessary for full activity (Kimple et al., 2001). The peptide AGS3Con represents the consensus GoLoco motif sequence from AGS3 (TMGEEDFFDLLAK-SQSKRMDDQRVDLAG). Control peptides consisted of a scrambled version of the R12GL peptide ("R12Scr," AQLRFISAEAREDNGSF-KDEQ) or an arginine-23 to phenylalanine loss-of-function mutant (Peterson et al., 2000) of the AGS3 consensus peptide ("AGS3[R>F]"). Peptide sources have been described previously (Kimple et al., 2001; see also Kimple et al., 2004) and were kindly provided by Drs. David Siderovski and Randall Kimple (University of North Carolina, Department of Pharmacology).

Choice of Cell for Evaluation of GPCR-to-Ion Channel Coupling

A variety of cells and cell lines have been employed in the study of GPCR signaling. In the case of GPCR-to-GIRK channel coupling, most studies have utilized native tissues (e.g., atrial muscle cells or neurons) with a fixed complement of all signaling partners, heterologous expression of receptors and channels in a "null background" cell [e.g., Xenopus oocytes, Chinese hamster ovary (CHO) cells, or human embryonic kidney (HEK) cells], or cell lines natively expressing some or all components. We have taken the latter approach using a clonal mouse anterior pituitary cortico-troph cell line (AtT20) that exhibits morphological and functional proper-ties of a differentiated neuroendocrine cell. AtT20 cells endogenously express Gi/o-linked SST receptors and m4 muscarinic receptors (Jones, 1992; Patel *et al.*, 1994) that couple to native GIRK channels formed from only Kir3.1 and Kir3.2 subunits and to calcium channels of the $Ca_v2.0$ (P/Q-type) family (Kuzhikandathil and Oxford, 1999; Kuzhikandathil *et al.*, 1998; Surprenant *et al.*, 1992). We have also heterologously expressed either human D2 or D3 dopamine receptors in these cells—neurotransmitter receptors that are also Gi/o-linked and elicit GIRK currents upon agonist stimulation (Kuzhikandathil *et al.*, 1998). The native SST and m4 receptors serve as good controls for signaling of such heterologously expressed receptors. AtT20 cells can be obtained from American Type Culture Collection (ATCC, Manassas, VA) as line CCL89.

Cell Culture, Plating, and Transfection Procedures

Our culture procedures for AtT20 cells involve propagating and maintaining them in Ham's F10 culture medium supplemented with 5% fetal bovine serum, 10% heat-inactivated horse serum, 2 mM L-glutamine, and 1 mg/ml gentamycin antibiotic. Cells are maintained at 37° in 5% CO_2 in 25-cm plastic tissue culture flasks and passaged every 4 to 5 days at a 1:5 dilution. With this regimen, cells do not reach confluence in the flasks; rather, they occupy roughly 50% of the surface. A standard trypsin-EDTA buffer is incubated with the cells for about 2 min to loosen adherent cells prior to passaging. At the time of passage, small droplets of cell suspension are plated onto glass coverslips (12- or 15-mm round, #1 thickness) pre-coated with 40 μg/ml poly-D-lysine or poly-L-lysine, placed in 12-well culture plates (Costar), and allowed to sit for 15 min in an incubator before adding 1 ml of culture medium per well.

In order to express human dopamine receptors in AtT20 cells, we employ lipid-mediated transfer of plasmid DNA expression vectors. The coding sequences for human receptors are subcloned into the pcDNA3.1 expression vector (Invitrogen), which we cotransfect in fivefold excess of an enhanced green fluorescent protein marker plasmid (pEGFP, Clontech) used to evaluate transfection efficiency and visualize expressing cells by fluorescence microscopy. We have had great success transfecting expression plasmids into CHO and HEK cells using the popular Lipofectamine reagent and its newer variants (Invitrogen); however, AtT20 cells have proven quite resistant to these reagents. We initially had a great deal of success using one of the PfX reagents (PfX-2, Invitrogen), but it is no longer available. We now use DMRIE-C (Invitrogen) and routinely achieve expression efficiencies between 30 and 40% following the manufacturer's protocol. The ratio of lipid reagent to DNA is 1.0:1.4 (μg:μg).

Total DNA (1.4 μg) is added to each well in a volume of 1 ml. A noncoding DNA (pUC19 or pcDNA3.1) is used to maintain a constant amount of DNA per well. Experiments are done 24 to 48 h after transfections.

Electrophysiological Methods to Measure GIRK Channel Currents

We employ standard whole cell patch-clamp methods (Hamill *et al.*, 1981) to record agonist-induced GIRK or calcium channel currents from individual AtT20 cells under voltage clamp conditions. Coverslips containing cells are placed in a custom acrylic chamber of 0.5 ml volume with a round glass coverslip (25 mm) forming the bottom. The chamber is then transferred to the stage of an inverted microscope (Nikon Diaphot TMD) outfitted for Hoffman modulation contrast and fluorescence optics. Cells are visualized for recording at 400× (40×, 0.85 NA objective) and assessed for EGFP expression using a standard FITC filter cube set.

Patch electrodes (typically 1 to 3 MΩ resistance) are pulled from cleaned N51A borosilicate glass tubing (Drummond Scientific, Broomall, PA) on a PP-83 puller (Narashige, Tokyo). To reduce capacitance of the electrode, tips are coated with dental wax (Kerr Sticky wax, Romulus, MI) by dipping in warm molten wax and allowing to harden. Electrode tips are then fire polished on a custom microforge at 600X magnification such that the wax is effused from the tip and the glass is smoothed.

The standard external solution (SES) contains (in mM) 145 NaCl, 5 KCl, 2 CaCl$_2$, 1 MgCl$_2$, 10 HEPES, and 10 glucose adjusted to pH 7.4 and osmolarity of 295 mOsM. For experiments measuring GIRK currents, the electrodes are filled with a solution containing (in mM) 130 K-aspartate, 20 NaCl, 1 MgCl$_2$, 1 EGTA, 10 glucose, 10 HEPES, 5 Mg-ATP, and 0.1 GTP, adjusted to pH 7.4 and osmolarity of 305 mOsM. The intracellular sodium is necessary to stabilize activated GIRK channels (Kuzhikandathil *et al.*, 1998; Sui *et al.*, 1996), while GTP is provided to support G-protein signaling. In order to amplify the GIRK currents that are inwardly rectifying, we routinely increase the ionic driving force by elevating the extracellular K$^+$ concentration to either 30 or 75 mM by equimolar substitution of KCl for NaCl in the external solution.

For recording barium currents through calcium channels, the external solution contains (in mM) 10 BaCl$_2$, 125 NaCl, 10 TEA-Cl, 1 MgCl$_2$, 10 HEPES, and 10 glucose. To block sodium channels, 1 μM tetrodotoxin (TTX) is added. The electrodes are dipped into a CsCl solution (in mM: 150 CsCl, 1 MgCl$_2$, 10 HEPES, pH 7.4) and then backfilled with (in mM) 50 CsCl, 90 *N*-methyl-D-glucamine aspartate, 5 NaCl, 1 MgCl$_2$, 5 EGTA, 10 glucose, and 10 HEPES, pH 7.4.

Introduction of Peptides into the Cell Interior

GoLoco motif peptides are added to the internal pipette solution at a final concentration of 10 μM and diffuse into the cell interior following attainment of the whole cell recording configuration. To first determine whether molecules the size of these peptides (3 to 4 kDa) could diffuse into the cell within the time frame of a typical experiment, we assessed the rate of diffusion of 10 kDa rhodamine dextran B (10 μM, Molecular Probes) from the patch electrode into the cell. In such tests we determined that, within 2 min, the entire cell was filled with rhodamine dextran, as assessed by fluorescence microscopy (Fig. 1). As the GoLoco peptides are smaller than the rhodamine dextran used, we concluded that the peptides likely equilibrate throughout the cell interior in under 5 min of achieving whole cell configuration. Consequently, recordings always followed a 5-min delay after achieving whole cell recording configuration.

Local Application of Agonists to Single Cells

To limit and control application of agonists to the cells of interest during an experiment we employ a planar multibarrel pipette array. The array is constructed from polyimide-coated quartz capillaries (Polymicro Technologies, Phoenix, AZ) attached together with cyanoacrylate glue. We use five to seven capillaries with an inside diameter of 250 μm (Polymicro Technologies) that are cut to the desired length with scissors and

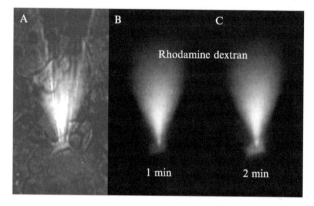

FIG. 1. Diffusion of rhodamine-conjugated dextran (10 kDa, 10 μM) into an AtT20 cell. Interference contrast image (A) of AtT20 cells during initial stages of whole cell recording from the cell at the end of the patch electrode. Fluorescence images of the electrode and the recorded cell at 1-min (B) and 2-min (C) intervals following rupture of the membrane patch to gain intracellular access to the electrode contents.

filed to a square tip by circular polishing on a fine jeweler's stone. Each barrel is connected to a reservoir of drug or control solution (5-ml syringe) via Tygon microbore tubing and an independent plastic stopcock. The array is positioned near the recorded cells using a three-dimensional fluid drive manipulator (MO103, Narashige, Tokyo, Japan), and active barrels are micropositioned under visual control such that agonist application can be achieved and terminated within 1 s. For more rapid changes, we occasionally employ a custom-designed piezoelectric lateral translator constructed from a piezoelectric bimorph (T215-H4CL-103X, Piezo Systems, Cambridge, MA) driven by a computer-controlled stimulus isolation unit (S940/910, Dagan Corp., Minneapolis, MN). This system permits selection and switching among any of five barrels in less than 5 ms. For the results presented here, the slower manual system was employed to deliver the dopamine receptor agonist quinpirole (QPRL; Sigma-RBI) and somatostatin (Sigma-RBI) each at a concentration of 100 nM and perfused over the cells at a rate of <0.5 ml/min.

Data Acquisition and Analysis

Stimulation protocols are generated and data are acquired and digitized using an Axopatch 200B patch clamp in combination with pClamp/Clampex software (v8.2; Axon Instruments, Burlingame, CA). To monitor GIRK currents, a step-ramp protocol consisting of a step from a holding potential of −60 to −100 mV followed by a 0.4-mV/ms ramp from −120 to 40 mV (Fig. 2B) is repeated every 2 to 5 s during changes in external solution. Currents at the end of the step are averaged and plotted to monitor current changes during agonist changes, whereas current during the voltage ramp serves as a monitor of seal integrity and cell health. To measure barium currents through calcium channels, voltage steps from a holding potential of −80 mV to voltages between −60 and 80 mV are applied to assess the voltage dependence of channel gating (Fig. 3B). All currents are normalized for cell size by dividing the absolute current by membrane capacitance.

GoLoco Peptides and GIRK Current Dynamics

Prior to activation of GPCRs by agonist application to a resting cell, the overwhelmingly dominant configuration of the pertussis toxin–sensitive Gi/o protein complement is the heterotrimeric form in which GDP-bound Gα subunits are complexed with Gβ and Gγ subunits. Upon agonist application, the Gα subunit exchanges GDP for GTP and the heterotrimeric complex dissociates. Following hydrolysis of GTP by the intrinsic

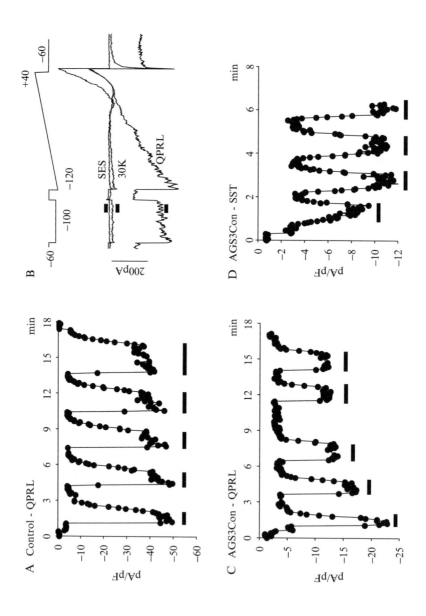

GTPase activity of the $G\alpha$ subunit, the free GDP-bound form can reassociate with the $G\beta\gamma$ complex. Given that the GoLoco motif of certain RGS proteins associates only with the GDP-bound form of $G\alpha_i$ (Kimple et al., 2001), it is in this latter time window that a GoLoco peptide has the opportunity to bind to $G\alpha_i$ and preclude association of $G\beta\gamma$ to reform the heterotrimer. The relatively high stability of the GoLoco–$G\alpha_i \cdot$ GDP complex suggests that, following agonist removal, a fraction of the $G\alpha_i$ pool will be bound by the GoLoco peptide rather than $G\beta\gamma$ and that subsequent receptor activation of the remaining heterotrimer pool will result in a progressive shift from heterotrimers to inactive GoLoco–$G\alpha_i \cdot$ GDP complexes. Consequently, the coupling of activated receptors to their effectors through $G\alpha_i$ subunits would be progressively weaker, whereas coupling through $G\alpha_o$ subunits, which have low affinity for the GoLoco peptides employed here, would be unaffected. Thus one would be able to discriminate preferential coupling of GPCRs to effectors through $G\alpha_i$ from coupling through $G\alpha_o$. A corollary prediction is that the initial responses of effectors to agonist activation of distinct GPCRs should not differ in the presence of GoLoco motif peptides, as the probability of association between peptides and $G\alpha_i \cdot$GDP is insignificant without heterotrimer dissociation.

Both of these predictions are manifest in the dynamics of GIRK current responses to the D2/D3 receptor agonist, quinpirole, and to somatostatin, which activates endogenous SST receptors in AtT20 cells. As can be seen in Fig. 2A, under control conditions, GIRK currents induced by repetitive application of quinpirole to AtT20 cells expressing human D2 receptors are relatively stable with only a small desensitization during each agonist challenge. In contrast, when a peptide encoding the consensus GoLoco motif sequence of AGS3 (AGS3Con) is exposed to the cytoplasmic compartment, repetitive responses to quinpirole are progressively smaller (Fig. 2C). In contrast, SST responses during internal dialysis of AGS3Con are not diminished (Fig. 2D). This result is consistent with a significant fraction of the D2 receptor–GIRK coupling occurring through G proteins

---

Fig. 2. Interference with D2 receptor-GIRK channel coupling by GoLoco peptides. Constitutive and quinpirole activated GIRK currents were monitored at 5-s intervals from two AtT20 cells at -100 mV in the absence (A) or presence (C) of 10 μM AGS3Con peptide added to the normal internal solution. Quinpirole was applied repetitively during the periods marked by solid bars in each case. Note the progressive decline of quinpirole responses in the presence of AGS3Con. Typical current responses to the step-ramp protocol employed are shown (B) for external perfusion with SES, elevated K^+ (30K), or 30K + quinpirole (QPRL). Data points in A, C, and D were derived from averages of the currents during the period marked by solid bars. Internal perfusion with the AGS3Con peptide did not progressively inhibit GIRK responses of another AtT20 cell to repeated applications of 100 nM somatostatin (D).

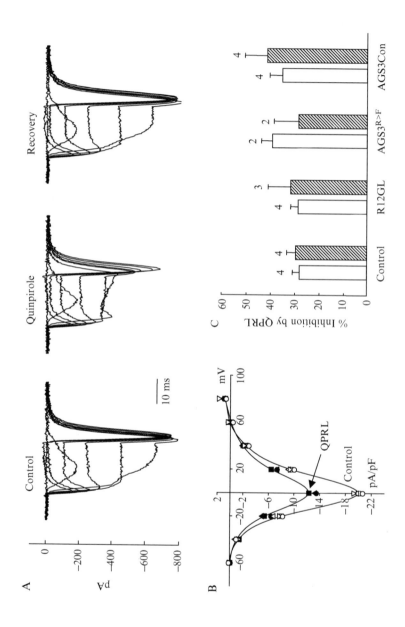

composed of $G\alpha_i$ subunits, whereas SSTR–GIRK coupling, which is also pertussis toxin sensitive, may occur exclusively through $G\alpha_o$ subunits.

GoLoco Peptides and Calcium Channel Dynamics

Several GPCRs couple to the inhibition of calcium channels in neurons (Dolphin, 2003), including D2 receptors, which reversibly inhibit P/Q-type calcium channels in AtT20 cells (Fig. 3A; Kuzhikandathil and Oxford, 1999). In several experiments in which GoLoco peptides were perfused into AtT20 cells expressing D2 receptors, during whole cell recording of barium currents neither R12GL nor AGS3Con peptides altered the degree of inhibition of currents by quinpirole (Fig. 3C, open bars). Furthermore, repetitive application of quinpirole also failed to change the degree of inhibition in the presence of these peptides (Fig. 3C, hatched bars). These observations suggest that, in contrast to D2 receptor coupling to GIRK channels, coupling of D2 receptors to calcium channels in AtT20 cells occurs predominantly via pertussis toxin-sensitive G-protein heterotrimers that include $G\alpha_o$ subunits.

Conclusions

Investigations of the molecular specificity of G-protein-coupled receptor signaling have employed a variety of approaches during the past decade, including the use of antibodies, antisense knockdown methods, and engineered mutants that are constitutively active, nonfunctional, or dominant negative. Each of these approaches has its advantages and limitations, thus the search for new tools to address these important issues of specificity continues. One of the latest and increasingly popular approaches involves

---

Fig. 3. Inhibition of P/Q-type calcium channels is not altered by GoLoco motif peptides. (A) Families of barium currents from a single AtT20 cell at membrane potentials ranging from −60 to 40 mV (30-ms steps) during external perfusion with control SES or 100 nM quinpirole as indicated. The electrode contained only control cesium-NMG internal solution without a GoLoco peptide. Note that the late steady (predominantly P/Q-type) currents, but not the early transient (T-type) currents, are reduced by quinpirole. (B) Current vs voltage relationship for another cell dialyzed internally with the R12GL peptide challenged repetitively with either control SES (open symbols) or 100 nM quinpirole (filled symbols). Note that no progressive change in the degree of inhibition of currents by quinpirole occurs (different symbols represent distinct repetitive agonist applications). (C) Percentage of barium current inhibition at 0 mV for the first (open bars) and third (hatched bars) quinpirole applications to cells dialyzed with control internal solution, active R12GL or AGS3Con peptides, and the inactive mutant AGS3$^{R>F}$ peptide as indicated. Data are mean values ± SEM for the indicated number of cells.

harnessing a natural process of interference of protein translation by small RNA sequences (siRNA; e.g., Shi, 2003), but design criteria and specificity of these functional probes are still not optimized universally (Scacheri et al., 2004; Ui-Tei et al., 2004). The use of small peptides with highly specific and well-characterized interactions with elements of the GPCR signaling pathways, such as the particular GoLoco motif peptides described here, offers another tool for such studies. As molecular and structural information about the peptide interfaces governing protein–protein interactions between signaling partners in GPCR cascades emerges, it is likely that other highly specific small interfering peptide probes will be developed.

Acknowledgments

The authors express their appreciation to Drs. David Siderovski and Randall Kimple for helpful discussions, for initially suggesting the use of GoLoco peptides as probes of GPCR/ion channel coupling, and for providing the peptides used in these experiments. This work was supported by NIH Grants NS18788 to G.S.O. and NS18788-S1 to C.K.W.

References

Chen, H., Clark, M. A. and Lambert, N. A. (2004). Endogenous regulators of G-protein signaling (RGS) proteins regulate presynaptic and postsymptic function: Functional expression of RGS-insensitive Gα subunits in central nervous system neurons. *Methods Enzymol.* **389**, 190–204.

Degtiar, V. E., Harhammer, R., and Nurnberg, B. (1997). Receptors couple to L-type calcium channels via distinct Go proteins in rat neuroendocrine cell lines. *J. Physiol.* **502**, 321–333.

Degtiar, V. E., Wittig, B., Schultz, G., and Kalkbrenner, F. (1996). A specific G_0 heterotrimer couples somatostatin receptors to voltage-gated calcium channels in RINm5F cells. *FEBS Lett.* **380**, 137–141.

Delmas, P., Abogadie, F. C., Milligan, G., Buckley, N. J., and Brown, D. A. (1999). Beta gamma dimers derived from G(o) and G(i) proteins contribute different components of adrenergic inhibition of Ca^{2+} channels in rat sympathetic neurons. *J. Physiol. (Lond)* **518**, 23–36.

Dolphin, A. C. (2003). G protein modulation of voltage-gated calcium channels. *Pharmacol Rev.* **55**, 607–627.

Ewald, D. A., Pang, I. H., Sternweis, P. C., and Miller, R. J. (1989). Differential G protein-mediated coupling of neurotransmitter receptors to Ca^{2+} channels in rat dorsal root ganglion neurons *in vitro*. *Neuron* **2**, 1185–1193.

Hamill, O. P., Marty, A., Neher, E., Sakmann, B., and Sigworth, F. J. (1981). Improved patch-clamp techniques for high resolution current recording from cells and cell-free membrane patches. *Pflüg. Arch.* **391**, 85–100.

Herlitze, S., Garcia, D. E., Mackie, K., Hille, B., Scheuer, T., and Catterall, W. A. (1996). Modulation of Ca^{2+} channels by G-protein β gamma subunits. *Nature* **380**, 258–262.

Ikeda, S. R. (1996). Voltage-dependent modulation of N-type calcium channels by G protein βgamma subunits. *Nature* **380**, 255–258.

Ikeda, S. R., and Jeong, S.-W. (2004). Use of regulator G-protein signaling (RGS)-insensitive Gα subunits to study endogenous RGS protein action on G-protein modulation of N-type calcium channels in sympathetic neurons. *Method. Enzymol.* **389,** 170–189.

Jones, S. V. (1992). m4 muscarinic receptor subtype activates an inwardly rectifying potassium conductance in AtT20 cells. *Neurosci. Lett.* **147,** 125–130.

Kammermeier, P. J., Davis, M. I., and Ikeda, S. R. (2003). Specificity of metabotropic glutamate receptor 2 coupling to G proteins. *Mol. Pharmacol.* **63,** 183–191.

Kimple, R. J., De Vries, L., Tronchere, H., Behe, C. I., Morris, R. A., Farquhar, M. G., and Siderovski, D. P. (2001). RGS12 and RGS14 GoLoco motifs are Gα$_i$ interaction sites with guanine nucleotide dissociation inhibitor activity. *J. Biol. Chem.* **276,** 29275–29281.

Kimple, R. J., Kimple, M. E., Betts, L., Sondek, J., and Siderovski, D. P. (2002). Structural determinants for GoLoco-induced inhibition of nucleotide release by Gα subunits. *Nature* **416,** 878–881.

Kimple, R. J., Willard, F. S., and Siderovski, D. P. (2004). Purification and *in vitro* functional analyses of RGS12 and RGS14 GoLoco motif peptides. *Methods Enzymol.* **390,** 419–436.

Kleuss, C., Hescheler, J., Ewel, C., Rosenthal, W., Schultz, G., and Wittig, B. (1991). Assignment of G-protein subtypes to specific receptors inducing inhibition of calcium currents. *Nature* **353,** 43–48.

Kleuss, C., Scherubl, H., Hescheler, J., Schultz, G., and Wittig, B. (1992). Different beta-subunits determine G-protein interaction with transmembrane receptors. *Nature* **358,** 424–426.

Kleuss, C., Scherubl, H., Hescheler, J., Schultz, G., and Wittig, B. (1993). Selectivity in signal transduction determined by gamma subunits of heterotrimeric G proteins. *Science* **259,** 832–834.

Kuzhikandathil, E. V., Yu, W., and Oxford, G. S. (1998). Human dopamine D3 and D2L receptors couple to inward rectifier potassium channels in mammalian cell lines. *Mol. Cell. Neurosci.* **12,** 390–402.

Kuzhikandathil, E. V., and Oxford, G. S. (1999). Activation of human D3 dopamine receptor inhibits P/Q-type calcium channels and secretory activity in AtT-20 cells. *J. Neurosci.* **19,** 1698–1707.

Leaney, J. L., and Tinker, A. (2000). The role of members of the pertussis toxin-sensitive family of G proteins in coupling receptors to the activation of the G protein-gated inwardly rectifying potassium channel. *Proc. Natl. Acad Sci. USA* **97,** 5651–5656.

Logothetis, D. E., Kurachi, Y., Galper, J., Neer, E. J., and Clapham, D. E. (1987). The β gamma subunits of GTP-binding proteins activate the muscarinic K channel in heart. *Nature* **325,** 321–326.

Menon-Johansson, A. S., Berrow, N., and Dolphin, A. C. (1993). G(o) transduces GABAB-receptor modulation of N-type calcium channels in cultured dorsal root ganglion neurons. *Pflüg. Arch.* **425,** 335–343.

McGehee, D. S., and Oxford, G. S. (1991). Bradykinin modulates the electrophysiology of cultured rat sensory neurons through a pertussis toxin-insensitive G protein. *Mol. Cell. Neurosci.* **2,** 21–30.

Patel, Y. C., Panetta, R., Escher, E., Greenwood, M., and Srikant, C. B. (1994). Expression of multiple somatostatin receptor genes in AtT20 cells: Evidence of a novel somatostatin-28 selective receptor subtype. *J. Biol. Chem.* **269,** 1506–1509.

Peterson, Y. K., Bernard, M. L., Ma, H., Hazard, S., III, Graber, S. G., and Lanier, S. M. (2000). Stabilization of the GDP-bound conformation of Gα$_i$ by a peptide derived from the G-protein regulatory motif of AGS3. *J. Biol. Chem.* **275,** 33193–33196.

Ruiz-Velasco, V., and Ikeda, S. R. (2000). Multiple G-protein betagamma combinations produce voltage-dependent inhibition of N-type calcium channels in rat superior cervical ganglion neurons. *J. Neurosci.* **20,** 2183–2191.

Scacheri, P. C., Rozenblatt-Rosen, O., Caplen, N. J., Wolfsberg, T. G., Umayam, L., Lee, J. C., Hughes, C. M., Shanmugam, K. S., Bhattacharjee, A., Meyerson, M., and Collins, F. S. (2004). Short interfering RNAs can induce unexpected and divergent changes in the levels of untargeted proteins in mammalian cells. *Proc. Natl. Acad. Sci. USA* **101,** 1892–1897.

Shi, Y. (2003). Mammalian RNAi for the masses. *Trends Genet.* **19,** 9–12.

Siderovski, D. P., Diversé-Pierluissi, M. A., and DeVries, L. (1999). The GoLoco motif: A $G\alpha_{i/o}$ binding motif and potential guanine-nucleotide-exchange factor. *Trends Biochem. Sci.* **24,** 340–341.

Sui, J. L., Chan, K. W., and Logothetis, D. E. (1996). Na^+ activation of the muscarinic K^+ channel by a G-protein-independent mechanism. *J. Gen. Physiol.* **108,** 381–391.

Surprenant, A., Horstman, D. A., Akbarali, H., and Limbird, L. E. (1992). A point mutation of the alpha 2-adrenoceptor that blocks coupling to potassium but not calcium currents. *Science* **257,** 977–980.

Takano, K., Yasufuku-Takano, J., Kozasa, T., Nakajima, S., and Nakajima, Y. (1997). Different G proteins mediate somatostatin-induced inward rectifier K+ currents in murine brain and endocrine cells. *J. Physiol.* **502,** 559–567.

Ui-Tei, K., Naito, Y., Takahashi, F., Haraguchi, T., Ohki-Hamazaki, H., Juni, A., Ueda, R., and Saigo, K. (2004). Guidelines for the selection of highly effective siRNA sequences for mammalian and chick RNA interference. *Nucleic Acids Res.* **32,** 936–948.

Wickman, K., Iniguez-Lluhl, J. A., Davenport, P. A., Taussig, R., Krapivinsky, G. B., Linder, M. E., Gilman, A. G., and Clapham, D. E. (1994). Recombinant G protein βgamma subunits activate the muscarinic-gated atrial potassium channel. *Nature* **368,** 255–257.

Subsection C

Other Modulators

[28]   Experimental Systems for Studying the Role of
G-Protein-Coupled Receptors in Receptor Tyrosine
Kinase Signal Transduction

By Nigel J. Pyne, Catherine Waters, Noreen Akhtar Moughal,
Balwinder Sambi, Michelle Connell, and Susan Pyne

Abstract

Early conception of G-protein-coupled receptor (GPCR) and receptor tyrosine kinase (RTK) signaling pathways was that each represented distinct and linear modules that converged on downstream targets, such as p42/p44 mitogen-activated protein kinase (MAPK). It has now become clear that this is not the case and that multiple levels of cross-talk exist between both receptor systems at early points during signaling events. In recent years, it has become apparent that transactivation of receptor tyrosine kinases by GPCR agonists is a general phenomenon that has been demonstrated for many unrelated GPCRs and receptor tyrosine kinases. In this case, GPCR/G-protein participation is *upstream* of the receptor tyrosine kinase. However, evidence now demonstrates that numerous growth factors use G proteins and associated signaling molecules such as β-arrestins that participate *downstream* of the receptor tyrosine kinase to signal to effectors, such as p42/p44 MAPK. This article highlights experimental approaches used to investigate this novel mechanism of cross-talk between receptor tyrosine kinases and GPCRs.

G-Protein-Coupled Receptor (GPCR)-Mediated Transactivation of
Receptor Tyrosine Kinases

The mechanism of GPCR-induced transactivation of the epidermal growth factor (EGF) receptor involves the intrinsic catalytic tyrosine kinase activity of this receptor, which is activated in response to heparin-binding EGF (HB-EGF). The latter is synthesized as a pro-ligand (proHB-EGF) that is anchored to the plasma membrane and released upon proteolytic cleavage (Daub *et al.*, 1996, 1997; Prenzel *et al.*, 1999) (see Scheme 1, A) Shedding of HB-EGF occurs in the presence of a GPCR

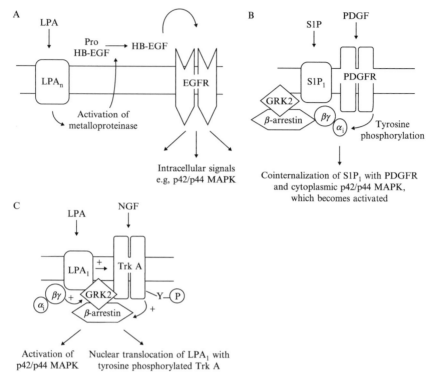

SCHEME 1. Different modes of GPCR-dependent regulation of receptor tyrosine kinase signaling. (A) GPCR transactivation of the EGF receptor via HB-EGF release. (B) PDGFβ receptor/S1P$_1$ receptor signal integration involving PDGFβ receptor-catalyzed tyrosine phosphorylation of G$_i$α. (C) Mixed transactivation/signal integration for Trk A/LPA$_1$ receptor signaling.

agonist, such as lysophosphatidic acid (LPA) (Hirata *et al.*, 2001) and involves a metalloproteinase. The major characteristic of the GPCR-dependent transactivation of receptor tyrosine kinases is the requirement for tyrosine phosphorylation of the growth factor receptor, which then acts as an acceptor site for scaffold protein/adaptors that initiate intracellular signals. Moreover, receptor tyrosine kinase inhibitors can be used to block GPCR-stimulated growth factor receptor transactivation. Indeed, there are numerous studies where receptor tyrosine kinase transactivation by GPCRs has been invoked on this basis. However, in some cases, interpretations should be re-evaluated for the following reasons. First, these inhibitors do not discriminate between GPCR-dependent transactivation of receptor tyrosine kinases and GPCR/receptor tyrosine kinase signal integration, where GPCR/G-protein-dependent signals are integrated downstream of receptor tyrosine kinase activity (see Scheme 1, B and C). In the integrative signaling model, efficient GPCR-dependent stimulation of effector pathways has an obligate requirement for receptor tyrosine kinase activity. Second, there is a need to demonstrate the role of the G protein upstream of the transactivated receptor tyrosine kinase. These experimental approaches, more often than not, have not been performed to validate transactivation mechanisms.

Platelet-Derived Growth Factor (PDGF)β Receptor and GPCR Signal Integration

Evidence shows a requirement for G protein in enabling efficient PDGF receptor signal transmission in mammalian cells. For instance, in airway smooth muscle (ASM) cells and PDGFβ receptor-transfected HEK 293 cells, pertussis toxin reduces the PDGF-dependent activation of c-Src and p42/p44 MAPK, yet has no effect on PDGF-stimulated autophosphorylation of tyrosine residues on the PDGFβ receptor (Alderton et al., 2001; Conway et al., 1999; Waters et al., 2003). Rosenfeldt and colleagues (2001) have also shown that the PDGF-dependent activation of c-Src in fibroblasts is inhibited by pretreatment of these cells with pertussis toxin. Moreover, Freedman and colleagues (2002) showed that PDGF increases GTPγS binding to $G_i\alpha$ and that pertussis toxin partially inhibits p42/p44 MAPK activation by PDGF in vascular smooth muscle cells. Others have reported that PDGF induces the formation of reactive oxygen species via agonist-dependent coupling to $G_i\alpha1$ and 2 (but not other subtypes) to the PDGFα receptor (Kreuzer et al., 2003).

Pertussis Toxin Sensitivity

The ability of pertussis toxin to ablate PDGF receptor-dependent stimulation of p42/p44 mitogen-activated protein kinase (MAPK) implicitly suggests the involvement of a GPCR in regulating PDGF receptor signal transmission. This toxin functions to uncouple G proteins from their respective GPCR by ADP ribosylation of the C-terminal cysteine of $G_i\alpha$, preventing interaction with the third cytoplasmic loop of seven transmembrane G-protein-coupled receptors. Therefore, it is likely that certain GPCRs are either bound with agonist released from cells in an autocrine loop or display some degree of tonicity in their activation state (e.g., partially constitutively active). Therefore, activation of a GPCR leads to the release of G-protein α and $\beta\gamma$ subunit dimers that appear to be used subsequently by the PDGF receptor tyrosine kinase to initiate activation of the p42/p44 MAPK pathway. This process is facilitated by close proximity between, for instance, the PDGFβ receptor and specific GPCRs (Alderton et al., 2001; Waters et al., 2003) (see Scheme 1, B).

For the purposes of this article, we focus on experimental studies using HEK 293 cells and ASM cells (Alderton et al., 2001; Conway et al., 1999; Waters et al., 2003). HEK 293 cells are maintained in minimal essential medium (MEM) containing 20% (v/v) fetal calf serum (FCS) containing L-glutamine (2 mM), penicillin (100 units/ml), streptomycin (100 μg/ml), and nonessential amino acids (Invitrogen). These cells are placed routinely in serum-free MEM for 24 h before experimentation. Airway smooth muscle cells are derived from guinea pig trachea (Pyne and Pyne, 1993) and are maintained in Dulbecco's modified Eagle's medium (DMEM) supplemented with 10% (v/v) FCS and 10% (v/v) horse serum. Primary ASM cells are used routinely at passage 3–4 and placed in serum-free DMEM for 24 h before experimentation. Their identity was confirmed to be smooth muscle by the presence of α-actin using smooth muscle-specific mouse anti-α-actin monoclonal antibodies (Sigma, UK) by western blotting. Typically, cells are pretreated with pertussis toxin (0.1 μg/ml) for 18 h in serum-free medium as required. Cell lysates are prepared by adding 0.25 ml of boiling Laemmli buffer [0.125 M Tris–HCl, pH 6.7, 0.5 mM Na$_4$P$_2$O$_7$, 1.25 mM EDTA, 2.5% (v/v) glycerol, 0.5% (w/v) sodium dodecyl sulfate (SDS), 25 mM dithiothreitol (DTT), 1% (w/v) bromphenol blue] to 0.5 $\times$ 10^6 cells/sample and passing the lysate three times through a 0.24-mm gauge needle to shear DNA prior to analysis of phosphoproteins by SDS–PAGE.

The phosphorylated forms of p42/p44 MAPK, PKB, and p38 MAPK are detected by western blotting cell lysates with respective antiphospho-specific antibodies. These antibodies exhibit good specificity and are available commercially. Antiphosphorylated p42/p44 MAPK, PKB, and p38 MAPK are from New England Biolabs (UK). Anti-p42 MAPK (BD Biosciences, UK), PKB (New England Biolabs, UK), and p38 MAPK (New England Biolabs, UK) antibodies are also used for western blotting to establish equal loading of the protein in each sample. Transfer of proteins from SDS–PAGE to nitrocellulose is at performed 100 V, 0.6 mA for 1 h at room temperature. After transfer of the proteins from SDS–PAGE, the nitrocelluose sheets are blocked in 3% (w/v) bovine serum albumin (BSA) or 1% (w/v) nonfat milk [used when probing with PY20 HRP-linked antiphosphotyrosine antibodies (BD Biosciences, UK)] in 10 mM Tris–HCl, pH 8, and 0.15 M NaCl (TBS) plus 0.1% (v/v) Tween 20, pH 7.5 for 1 h at 23°. The sheets are then probed with antibodies in TBS containing 1% BSA (w/v) plus 0.1% (v/v) Tween 20, pH 7.5, overnight at 4°. After this time, the nitrocellulose sheets are washed three times in TBS plus 0.1% (v/v) Tween 20. Detection of immunoreactivity is performed by incubating nitrocellulose for 1–2 h at 23° with reporter HRP-linked antimouse/rabbit

IgG antibodies in TBS containing 1% BSA (w/v) plus 0.1% (v/v) Tween 20, pH 7.5. After washing the blots as described earlier (to remove excess reporter antibody), immunoreactive bands are revealed using an enhanced chemiluminescence (ECL, Amersham Bioscience) detection kit.

Pertussis toxin induces an approximate 70–90% inhibition of PDGF-stimulated p42/p44 MAPK activation in PDGFβ receptor-transfected HEK 293 cells (Alderton *et al.*, 2001) (Fig. 1) and 30–50% inhibition of both PDGF-stimulated c-Src and p42/p44 MAPK in ASM cells (Conway *et al.*, 1999). The effect of pertussis toxin on PDGF-stimulated c-Src activity can be established by assaying c-Src activity in anti-c-Src immunoprecipitates. Anti-c-Src immunoprecipitates are prepared according to the following protocol. Cells (10^6 per sample) are lysed in ice-cold immuno-precipitation buffer (1 ml) containing 20 mM Tris–HCl, 137 mM NaCl, 2.7 mM KCl, 1 mM MgCl$_2$, 1 mM CaCl$_2$, 1% (v/v) Nonidet P-40 (NP-40), 10% (v/v) glycerol, 1 mg/ml BSA, 0.5 mM sodium orthovanadate, 0.2 mM phenylmethylsulfonyl fluoride (PMSF), leupeptin, antipain, and aprotinin [all protease inhibitors are at 10 μg/ml (pH 8)] for 10 min at 4°. The material is harvested and centrifuged at 22,000g for 5 min at 4°, and 200 μl of the cell lysate supernatant (protein, 0.5–1 mg/ml) is taken for immunoprecipitation with antibodies (5 μg of anti-c-Src antibodies/sample, Santa Cruz). Cell lysates are incubated with antibodies for 1 h at 4°, after

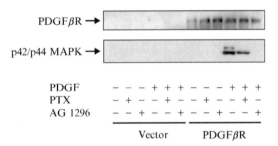

FIG. 1. The effect of pertussis toxin or tyrphostin AG 1296 on PDGF-stimulated p42/p44 MAPK activation in HEK 293 cells. HEK 293 cells were transfected transiently with vector or PDGFβ receptor plasmid construct. Transfected cells were pretreated with and without pertussis toxin (0.1 μg/ml, 18 h) or tyrphostin AG 1296 (20 μM, 20 min) prior to stimulation with PDGF (10 ng/ml, 10 min). (Top) A western blot probed with anti-PDGFβ receptor antibodies showing the lack of effect of pertussis toxin or tyrphostin AG 1296 on PDGFβ receptor levels in PDGFβ receptor-transfected cells. (Bottom) A western blot probed with antiphospho-p42/p44 MAPK antibodies showing the inhibitory effect of pertussis toxin or tyrphostin AG 1296 on the stimulation of p42/p44 MAPK by PDGF in PDGFβ receptor-*versus* vector-transfected cells. Total p42 MAPK loading was equal for all the samples and was detected using an anti-p42 MAPK antibody (see Alderton *et al.*, 2001). These are representative results of an experiment performed three times.

which 30 μl of 1:1 (w/v) protein-Sepharose A is added. After agitation for 1 h at 4°, the immune complex is collected by centrifugation at 22,000g for 15 s at 4°. Immunoprecipitates are washed twice with buffer A containing 10 mM HEPES, pH 7, 100 mM NaCl, 0.2 mM PMSF, 10 μg/ml leupeptin, 20 μg/ml aprotinin, and 0.5% (v/v) NP-40 and twice in buffer A without NP-40. Anti-c-Src immunoprecipitates are then incubated with 1.6 μg of acid-denatured rabbit muscle enolase (Sigma), 20 mM HEPES, pH 7, 10 mM MnCl$_2$, 20 μg/ml aprotinin, and 5 μCi of [γ-^{32}P] ATP (3 μM) in a total reaction volume of 20 μl. The reaction is stopped by adding 2$\times$ sample buffer containing 200 mM Tris–HCl, pH 6.7, 10 mM EDTA, 20% (v/v) glycerol, 4% (w/v) SDS, 5.6 M 2-mercaptoethanol, and 1% (w/v) bromphenol blue. Samples are then boiled for 2 min, cooled, and subjected to SDS–PAGE. After drying the gel [after washing in 10% (v/v) propan-2-ol], phosphorylated enolase is visualized by autoradiography before excising ^{32}P-labeled protein bands from the gel and quantification by Cerenkov counting.

The *in vivo* endogenous ADP-ribosylation of G$_i\alpha$ is established by the inability of pertussis toxin to catalyze the *in vitro* [^{32}P]NAD$^+$-dependent ADP ribosylation of G$_i\alpha$ in isolated membranes (Pyne *et al.*, 1989). Validation of the effect of pertussis toxin is therefore assessed by [^{32}P]ADP ribosylation of the G$_i\alpha$ subunit in membranes prepared from cells treated with and without pertussis toxin. This involves isolation of cell membranes followed by an *in vitro* pertussis toxin-catalyzed ADP-ribosylation of the G$_i\alpha$ subunit with [^{32}P]NAD$^+$. Cells are harvested in 0.25 M sucrose/10 mM Tris–HCl (pH 7.4)/1 mM EDTA/0.1 mM PMSF/2 mM benzamidine and homogenized by repeatedly (three times) passing through a 0.24-mm gauge needle. Membranes are collected by centrifugation at 22,000g for 10 min at 4° and are subjected to *in vitro* pertussis toxin-catalyzed ADP-ribosylation. This involves combining membranes (20 μg of protein) with buffer containing (final concentrations) 30 mM thymidine, 1 mM ATP, 80 mM potassium phosphate, pH 7.5, 20 mM arginine, 1 mM MgCl$_2$, and 7.5 μCi of 20 μM [^{32}P]NAD$^+$. Pertussis toxin (1.33 μg/assay) is added to each assay mixture and incubated at 37° for 1 h (pertussis toxin was preactivated by combining it with an equal volume of 50 mM DTT). Pertussis toxin-catalyzed ADP-ribosylation reactions are then stopped by adding an equal volume of 2$\times$ sample buffer containing 0.125 M Tris–HCl, pH 6.7, 2.5% (v/v) glycerol, 0.5% (w/v) SDS, 5.6 M 2-mercaptoethanol, and 1% (w/v) bromphenol blue and are then subjected to SDS–PAGE.

As a point of caution, pertussis toxin can abrogate tonic inhibition of adenylyl cyclase by G$_i$, thereby causing a rise in intracellular cAMP. In certain cases, cAMP can interrupt signaling from growth factor receptors to effectors such as p42/p44 MAPK. Therefore, in order to validate that the effects of pertussis toxin reflect abrogation of the G_i-mediated activation of the p42/p44 MAPK and not an indirect effect via cAMP, it is important to establish that this toxin does not modulate intracellular cAMP levels or that any increase in cAMP is insufficient to dampen the growth factor-dependent regulation of kinase modules, such as p42/p44 MAPK, p38 MAPK, and JNK. Intracellular cAMP is determined using a radioligand-binding assay with ^3H cyclic AMP (Amersham Biosciences) and a partially purified preparation of the catalytic subunit of PKA, derived from bovine heart (Rubin *et al.*, 1974). Briefly, after pretreatment of cells with pertussis toxin (0.25 × 10^6 cells/sample), the medium is removed and 500 μl of ice-cold 80% (v/v) ethanol is applied to the cells. The cAMP extract is harvested into a centrifuge tube and cellular debris is removed by centrifugation (14,000g, 10 min, 4°) before the supernatants are lyophilized. These are reconstituted directly into the assay incubation buffer [50 mM Tris–HCl (pH 7.4), 4 mM EDTA] to which [^3H]cAMP (50,000 dpm) and binding protein are added. After incubation on ice for 1 h, a mixture of activated charcoal (2%, w/v)/BSA (1% w/v) in incubation buffer is added (to absorb unbound [^3H]cAMP), the samples are centrifuged (14,000g, 10 min, 4°), and radioactivity associated with the supernatant (containing [^3H]cAMP associated with the binding protein) is determined by scintillation counting. A standard curve using known amounts of cAMP (0–125 pmol; nonspecific binding measured at 250 pmol cAMP) is conducted in parallel and the percentage specific binding is calculated for all samples. The standard curve is used to determine the cAMP concentration of the cell extracts (Pyne *et al.*, 1994).

Using these approaches, we have established that PDGF can use both G-protein-dependent and -independent routes to initiate activation of the p42/p44 MAPK pathway (Alderton *et al.*, 2001; Conway *et al.*, 1999). Two possible signaling pathways initiated by the ligand-bound PDGF receptor have been proposed: one involving G_i released from a GPCR that is closely associated with the PDGFβ receptor (see Scheme 1, B) and the other involving growth factor receptor autophosphorylation on tyrosine residues followed by the recruitment of adaptor/signaling molecules to the receptor. Interestingly, overexpression of $G_i\alpha_2$ reduces PDGF-stimulated autophosphorylation of the PDGFβ receptor (Alderton *et al.*, 2001). Therefore, the G protein may function to shift the balance between the two signaling routes in favor of G_i-mediated signaling. It is also *very* important to note that pertussis toxin sensitivity of any given cellular system appears to be dependent on the PDGFβ receptor:GPCR ratio. Therefore, if the PDGFβ receptor:GPCR ratio were 1:1, PDGF stimulation of p42/p44 MAPK would be expected to be blocked entirely by pertussis toxin. Alternatively, if the ratio were 10:1 in favor of the PDGFβ receptor, PDGF-stimulated p42/p44 MAPK activation would be expected to be virtually insensitive to pertussis toxin and routed almost entirely through the conventional mechanism involving the recruitment of signaling proteins to tyrosine phosphates on the PDGFβ receptor. Therefore, cell-specific stoichiometric expression of the PDGFβ receptor and sphingosine 1-phosphate receptor-1 (S1P$_1$) may define the extent of pertussis toxin sensitivity. This is not without precedent, as Luttrell and colleagues (1995) demonstrated that pertussis toxin sensitivity in IGF-1 signaling to p42/p44 MAPK is lost when high levels of the IGF-1 receptor are expressed. Thus, a high receptor tyrosine kinase density may bypass or mask the requirement for G-protein involvement by enabling conventional signaling from growth factor receptor phosphotyrosine sites to predominate. This phenomenon is not restricted to the IGF-1 receptor. Indeed, high concentrations of nerve growth factor (NGF) also override the G-protein input associated with Trk A receptor signaling in PC12 cells (Rakhit *et al.*, 2001). We routinely use 3–10 ng/ml of PDGF in order to stimulate cells. ASM have an intermediate expression level of PDGFβ and PDGFα receptors, whereas overexpression studies with the PDGFβ receptor in HEK 293 cells involve cotransfection with the S1P$_1$ plasmid construct (see later) (Alderton *et al.*, 2001; Waters *et al.*, 2003).

G-Protein Tyrosine Phosphorylation by PDGFβ Receptor Kinase

The overexpression of G$_i\alpha_2$ in HEK 293 cells markedly augments PDGF-stimulated activation of p42/p44 MAPK (Alderton *et al.*, 2001). Intriguingly, the overexpression of G$_i\alpha_2$ also induces an increased PDGF-dependent tyrosine phosphorylation of G$_i\alpha_2$ due to enhanced substrate availability (Alderton *et al.*, 2001). In order to establish the effect of G$_i\alpha_2$ on PDGFβ receptor signaling, HEK 293 cells (0.5×10^6 cells) at 70% confluence are placed in MEM and 2% (v/v) FCS and are cotransfected transiently with 1.5 μg G$_i\alpha_2$- and 1.5 μg PDGFβ receptor-pcDNA3.1 plasmid constructs/3 μl of LipofectAMINE 2000 according to the manufacturer's instructions. The G$_i\alpha_2$-pcDNA3.1 plasmid construct is from Dr. A. Wise (Glaxo-Smith Kline, UK) and the PDGFβ receptor-pcDNA3.1 plasmid construct is from Professor C.-H. Heldin (Ludwig Institute for Cancer Research, Uppsala, Sweden). After incubation for 24 h at 37°, cDNA-containing media are removed and the cells are incubated for a further 24 h in serum-free medium prior to agonist addition. The effect of G$_i\alpha_2$ on PDGF receptor signaling is determined by analysis of the activation state of p42/p44 MAPK (as assessed by western blotting of phosphorylated p42/p44 MAPK with specific antibodies). These studies have led us to conclude that the phosphorylated version of G$_i\alpha_2$ might be the "active" intermediate, which transmits signals from the PDGFβ receptor to p42/p44 MAPK. Only a small fraction of the recombinant $G_i\alpha_2$ that is overexpressed in the cells is apparently accessible for tyrosine phosphorylation by the PDGFβ receptor kinase. This is established by comparing the relative immunoreactivity of $G_i\alpha_2$ in vector-*control versus* $G_i\alpha_2$ plasmid construct-transfected cells, with the corresponding increase in the PDGF-stimulated tyrosine phosphorylation of both endogenous and recombinant $G_i\alpha_2$ (representing approximately 10% of the total recombinant $G_i\alpha_2$ expressed) (Alderton *et al.*, 2001).

Tyrosine-phosphorylated $G_i\alpha_2$ is immunoprecipitated from cell lysates (0.5×10^6 cells per sample) using anti-$G_i\alpha$ antibodies (CN Bioscience, UK). This is achieved using an immunoprecipitation procedure in which cells are lysed in ice-cold immunoprecipitation buffer (1 ml) containing 20 mM Tris–HCl, 137 mM NaCl, 2.7 mM KCl, 1 mM MgCl$_2$, 1 mM CaCl$_2$, 1% (v/v) NP-40, 10% (v/v) glycerol, 1 mg/ml BSA, 0.5 mM sodium orthovanadate, 0.2 mM PMSF, leupeptin, antipain, and aprotinin [all protese inhibitors are at 10 μg/ml (pH 8)] for 10 min at 4°. The material is harvested and centrifuged at 22,000g for 5 min at 4°, and 200 μl of the cell lysate supernatant (protein, 0.5–1 mg/ml) is taken for immunoprecipitation with antibodies (5 μg of antibody/sample). Cell lysates are incubated with antibody for 1 h at 4°, after which 30 μl of 1:1 (w/v) protein A-Sepharose is added. After agitation for 1 h at 4°, the immune complex is collected by centrifugation at 22,000g for 15 s at 4°. Immunoprecipitates are washed twice with buffer A containing 10 mM HEPES, pH 7, 100 mM NaCl, 0.2 mM PMSF, 10 μg/ml leupeptin, 20 μg/ml aprotinin, and 0.5% (v/v) NP-40 and twice in buffer A without NP-40. The immunoprecipitates are then combined with boiling sample buffer containing 0.125 M Tris–HCl, pH 6.7, 0.5 mM Na$_4$P$_2$O$_7$, 1.25 mM EDTA, 2.5% (v/v) glycerol, 0.5% (w/v) SDS, 25 mM DTT, and 1% (w/v) bromphenol blue. Samples are then subjected to SDS–PAGE and Western blotting with anti-$G_i\alpha_2$ antibodies (CN Bioscience, UK) and PY20 HRP-linked antiphosphotyrosine antibodies (New England Biolabs, UK). Pretreating cells with tyrphostin AG 1296 (1–20 μM, 20 min) blocks the PDGF-dependent tyrosine phosphorylation of $G_i\alpha_2$, suggesting that the PDGFβ receptor kinase may catalyze tyrosine phosphorylation of $G_i\alpha_2$ (Alderton *et al.*, 2001).

In terms of G-protein subunit organization, the functional consequence of tyrosine phosphorylation is unknown. However, it is possible that tyrosine phosphorylation of the α subunit might inhibit the GTPase activity of the G protein, thereby preventing reassociation of $G_i\alpha_2$ with $\beta\gamma$, thus prolonging the lifetime of both. Under these conditions, GTP-bound $G_i\alpha_2$ will accumulate in response to PDGF stimulation due to constitutive activation of the GPCR releasing $G_i\alpha_2$ and attendant phosphorylation of

$G_i\alpha_2$ by the PDGFβ receptor kinase. To all intents and purposes, this would appear as if PDGF was activating G-protein heterotrimeric subunit dissociation in a manner similar to GPCR agonists. Caution should therefore be exercised here in stating that certain growth factors activate G proteins in the classical manner of GPCR agonists. Indeed, increasing evidence shows that growth factors can modulate the GTP-bound state of G proteins. For instance, Freedman and colleagues (2002) showed that PDGF increases GTPγS binding to $G_i\alpha_2$ and that pertussis toxin partially inhibits p42/p44 MAPK activation by PDGF in vascular smooth muscle cells. Others have reported that FGF-2 promotes an apparent dissociation of the G_s $\alpha\beta\gamma$ heterotrimer and that FGF-2 stimulation of adenylyl cyclase proceeds via the α subunit, whereas NADPH oxidase inhibition involves the $G_s\beta\gamma$ subunits (Krieger-Brauer et al., 2000). Finally, Liebmann and colleagues (1996) have reported that EGF induces the tyrosine/threonine phosphorylation of $G_s\alpha$ in A431 cells. Subsequently, the activated G_s becomes refractory to further activation by bradykinin receptor stimulation, suggesting that the receptor tyrosine kinase can, via an unidentified mechanism, uncouple G_s from the BK receptor (Liebmann et al., 1996). Others have reported that the EGF receptor kinase catalyzes the phosphorylation of amino acid residues in the C- and N-terminal regions of $G_s\alpha$ that are involved in receptor interaction and nucleotide exchange (Moyers et al., 1995).

GRK2/β-Arrestin I Association with the PDGFβ Receptor

A great deal of evidence now shows that receptor tyrosine kinases signal via an endocytic pathway and that the signals generated within endosomes are distinct from those originating at the cell surface. We have proposed that the receptor tyrosine kinase can use the GPCR-dependent endocytic signaling machinery and *vice versa* (Waters et al., 2003). In this regard, we showed that GRK2/β-arrestin I are constitutively associated with the PDGFβ receptor in transfected HEK 293 cells (Alderton et al., 2001). These finding have been confirmed by studies from Freedman et al. (2002). These results were obtained from experiments in which HEK 293 cells were transfected with PDGFβ receptor and GRK2 or β-arrestin I plasmid constructs and where GRK2 and β-arrestin I were coimmunoprecipitated with the PDGFβ receptor using anti-PDGFβ receptor antibodies (Alderton et al., 2001). GRK2 is not coimmunoprecipitated with the PDGFβ receptor from cells that have not been transfected with GRK2 plasmid constructs (Fig. 2; Alderton et al., 2001) or the PDGFβ receptor-pcDNA3.1 plasmid construct (Fig. 2). These data thereby validate the specific coimmunoprecipitation of GRK2 with the PDGFβ receptor (all

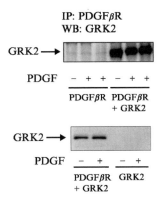

IP: PDGFβR
WB: GRK2

GRK2 →

PDGF   − + +   − + +

PDGFβR   PDGFβR
          + GRK2

GRK2 →

PDGF   − +   − +

PDGFβR   GRK2
+ GRK2

FIG. 2. Association of GRK2 with the PDGFβ receptor in HEK 293 cells. HEK 293 cells were transfected transiently with GRK2 and PDGFβ receptor plasmid constructs. Transfected HEK 293 cells were stimulated with PDGF (10 ng/ml, 10 min). A western blot showing the association of GRK2 with the PDGFβ receptor in anti-PDGFβ receptor immunoprecipitates is shown. The blot also shows endogenous GRK2 in anti-PDGFβ receptor immunoprecipitates from cells where the recombinant GRK2 plasmid construct had been omitted from the cotransfection. The PDGFβ receptor is coimmunoprecipitated with GRK2 (Alderton *et al.*, 2001). Also shown is the absence of GRK2 in anti-PDGFβ receptor immunoprecipitates isolated from vector-transfected cells. These are representative results of an experiment performed three times. IP, immunoprecipitation; WB, western blot.

antibodies from Santa Cruz) and suggest that GRK2 might interact with the GPCR associated with the PDGFβ receptor.

Internalization of the PDGFβ receptor–GPCR complex requires the presence of PDGF to induce β-arrestin/GRK2-mediated signaling. In this regard, it is well established that c-Src is involved in β-arrestin I-mediated signaling (Miller and Lefkowitz, 2001), and we have speculated that PDGF is required to induce recruitment of c-Src to the PDGFβ receptor–GPCR complex. Upon association, c-Src might then be stimulated by G protein (possibly Gβγ) released as a consequence of tonic GPCR activation or addition of the GPCR ligand. Indeed, as described earlier, the PDGF-dependent activation of c-Src is also sensitive to inhibition by pertussis toxin and c-Src does associate with the PDGFβ receptor–GPCR complex (unpublished data). Thus, β-arrestin I-mediated signaling to p42/p44 MAPK is dependent on both PDGF receptor tyrosine kinase activity and G protein and is a point of signal integration.

Downstream Signaling via PDGFβ Receptor–GPCR Complexes

As part of the signal integration, PDGF stimulates the G_i-dependent tyrosine phosphorylation of Gab1 in airway smooth muscle cells (Rakhit *et al.*, 2000). Grb-2 associated binder 1 (Gab1) contains several potential tyrosine sites for phosphorylation, which can form acceptor sites for SH2 containing adaptor-signaling molecules. Gab1 is constitutively associated with growth factor receptor binder-2 (Grb-2) and can bind phosphoinositide 3-kinase (PI3K). PDGF-stimulated tyrosine phosphorylated Gab1 is isolated in anti-Gab1 and anti-Grb-2 immunoprecipitates (Rakhit *et al.*, 2000). The consequence of the tyrosine phosphorylation of Gab1 is the binding of PI3K1a to Gab1, which is an essential step in the subsequent association of dynamin to a complex of PI3K1a-Gab1-Grb-2 (Rakhit *et al.*, 2000). The p85 regulatory subunit of PI3K1a can be detected in both anti-Gab1 and anti-Grb-2 immunoprecipitates by probing western blots with anti-PI3K1a antibodies (all antibodies from Santa Cruz) (Rakhit *et al.*, 2000). Indeed, anti-Grb-2 immunoprecipitates also contain a PDGF-sensitive PI3K activity, consistent with the presence of the 110-kDa catalytic unit of this kinase (Rakhit *et al.*, 1999).

In order to measure Grb-2-associated PI3K activity, resuspended anti-Grb-2 immunoprecipitates (40 μl) are each combined with 20 μl of phosphatidylinositol (3 mg/ml) in incubation buffer containing 1% cholate, 20 mM β-glycerophosphtate, 5 mM sodium pyrophosphate, 30 mM NaCl, and 1 mM DTT (pH 7.2). To each, 40 μl of [^{32}P]ATP (3 μM Na$_2$ATP, 7.5 mM MgCl$_2$, and 0.25 mCi of [^{32}P]ATP/ml) is added. The reaction is performed at 37° for 15 min and is terminated by adding 450 μl of chloroform/methanol (1:2, v/v). Organic and aqueous phases are resolved by adding 150 μl of chloroform and 150 μl of 1 M HCl. Samples are mixed and centrifuged (4200g for 10 min). This is repeated and the lower phase is harvested, dried, and [^{32}P]phosphatidylinositol 3-phosphate ([^{32}P]PI3P) resolved by TLC [Merck Kieselgel 60 plates, presoaked in 1:1 (v:v) methanol:1% (w/v) potassium oxalate/2 mM EDTA, air dried, and baked at 150° for 2 h] using choloroform/methanol/ammonia/water (14:20:3:5, by volume) in parallel with a phosphatidylinositol 3-phosphate nonradioactive standard (visualised by iodine staining). Radioactive bands are revealed by autoradiography, and samples corresponding to [^{32}P]PI3P are scraped from the plate and subjected to Cerenkov counting. The PI3K present in anti-Grb-2 immunoprecipitates is regulated by the G protein (Rakhit *et al.*, 2000, 2001) and is differentially controlled compared with the PI3K that associates directly with the PDGF receptor. The latter is not subject to regulation by the G protein. Therefore, PI3K associated with the PDGF receptor is insensitive to pertussis toxin and is most likely involved in conventional signaling from this receptor and/or Akt activation. This regulated form of PI3K is isolated with the PDGF receptor in antiphosphotyrosine immunoprecipitates and is stimulated up to 30-fold by treating ASM cells with PDGF (Rakhit *et al.*, 1999).

Dynamin is a GTPase (Schmid *et al.*, 1998) that functions to catalyze pinching off of endocytic vesicles containing GPCR-signaling complexes (that include Raf-MEK-1) for subsequent relocalization with and activation of cytoplasmic p42/p44 MAPK. Therefore, the recruitment of dynamin to the PI3K1a-Gab1-Grb-2 complex is an essential step in PDGFβ receptor/GPCR-mediated signaling to p42/p44 MAPK (Rakhit *et al.*, 2000). The PDGF-dependent association of dynamin to the PI3K1a-Gab1-Grb-2 complex is detected by western blotting of anti-Grb-2 immunoprecipitates probed with respective antibodies (Rakhit *et al.*, 2000). Essentially, the immunoprecipitation procedure used is identical to that described previously, except that anti-Grb-2 and anti-Gab1 antibodies (5 µg/10^6 cells/sample) were used to isolate protein signaling complexes. p85 PI3K1a and dynamin II are detected in these immunoprecipitates from cells treated with PDGF (10 ng/ml, 10 min) by western blotting with anti-p85 PI3K1a and anti-dynamin II antibodies (Santa Cruz) (Rakhit *et al.*, 2000).

PDGFβ Receptor–S1P₁ Receptor Complexes

These findings have led us to suggest that there are specific examples of complex formation between the PDGFβ receptor and GPCRs (Alderton *et al.*, 2001; Waters *et al.*, 2003). The best example describes the interaction between the receptors for PDGF and S1P, a bioactive sphingolipid that acts via several G-protein-coupled receptors, including the S1P₁ receptor (Chun *et al.*, 2002). In this context, it has been shown that the PDGFβ receptor ($M_r = 180$ kDa) forms a complex with the S1P₁ receptor (45 kDa) in PDGFβ receptor- and myc epitope-tagged S1P₁ receptor-cotransfected HEK 293 cells (Alderton *et al.*, 2001). This was demonstrated by the coimmunoprecipitation of the PDGFβ receptor–S1P₁ receptor complex from cell lysates using anti-PDGFβ receptor and anti-myc epitope tag antibodies (Santa Cruz). The methodology involves cotransfection of HEK 293 cells with 1.5 µg of the myc-tagged (C-terminal) S1P₁ receptor–pcDNA3.1 plasmid construct (from Dr. Palmer, Glasgow University, UK) and 1.5 µg of the PDGFβ receptor–pcDNA3.1 plasmid construct 3 µl of LipofectAMINE 2000 according to the manufacturer's instructions. There is no significant effect on the expression of each receptor by cotransfecting cells with both plasmid constructs (Alderton *et al.*, 2001). Immunoprecipitation is performed as described previously, except that 5 µg of either the anti-PDGFβ receptor or the anti-myc epitope tag antibody/sample is used. Stimulation of cells with either PDGF (10 ng/ml, 5–10 min) and/or S1P (5 µM, 10 min) did not increase the amount of each receptor in the complex, suggesting that it is preformed. No immunoreactive PDGFβ receptor (M_r = 180 kDa) or myc-tagged S1P$_1$ (M_r = 45 kDa) is detected when the antibody is omitted from the immunoprecipitation procedure or when immunoprecipitates are western blotted with secondary antibody alone (Alderton et al., 2001; Waters et al., 2003). Likewise, neither the PDGFβ receptor nor the myc epitope-tagged S1P$_1$ receptor is coimmuno-precipitated with respective antibodies unless cells are transfected with plasmid constructs (Alderton et al., 2001). Typically, GPCRs can migrate on SDS–PAGE as dimeric proteins with an approximate mass of 80–90 kDa and are difficult to resolve as a monomer on SDS–PAGE due to pronounced oligomerization. However, we have found that the use of sample buffer containing 0.125 M Tris–HCl, pH 6.7, 0.5 mM Na$_4$P$_2$O$_7$, 1.25 mM EDTA, 2.5% (v/v) glycerol, 0.5% (w/v) SDS, 25 mM DTT, and 1% (w/v) bromo-phenol blue at room temperature (rather than boiling) is sufficient to re-move the PDGFβ receptor and the myc epitope-tagged S1P$_1$ receptor from protein A-Sepharose beads and to reduce oligomerization of the GPCR on SDS–PAGE. This allows resolution of the protein as a monomeric 45-kDa species. In addition, it is clear from some studies that epitope tagging of the GPCR at the N terminus can prevent correct insertion of the receptor into plasma membrane and can abrogate trafficking of the receptor from the endoplasmic reticulum (Kohno et al., 2003). Therefore, we have used con-structs in which the epitope is placed at the C-terminal end of the GPCR. However, it should also be noted that C-terminal tagging might abrogate the binding of interacting signaling proteins with the PDZ domain docking sites present in the C terminus of some GPCRs. We were also able to show complex formation between the S1P$_1$ receptor and PDGFβ receptor in ASM cells using 5 μg of each of either anti-PDGFβ receptor or anti-S1P$_1$ receptor antibodies (Santa Cruz) (Waters et al., 2003).

It remains to be determined whether the PDGFβ receptor interacts directly with the S1P$_1$ receptor or whether there is a functional tethering protein. To establish the presence and identity of tethering proteins one could use the following approaches.

    i. Coimmunoprecipitation of potential tethering proteins with growth factor receptor–GPCR complexes and identification by mass spectroscopic analysis. Alternatively, copurification techniques can be employed, including streptavidin agarose and gel filtration chromatography.

    ii. Overexpression of potential tethering protein to establish whether this increases complex formation between GPCR and growth factor receptor. This may require coexpression of at least one of the component receptors if, for example, the endogenous form of this receptor is limiting.

iii. Deletion of candidate interaction site on the tethering protein to establish whether this causes disruption of the growth factor receptor–GPCR complex.

S1P can function as a co-mitogen with PDGF in several cell types (Conway *et al.*, 1999). The model of integrative signaling proposed here can explain these data and predicts that PDGF stimulation of p42/p44 MAPK will be dependent on the $S1P_1$ receptor, whereas S1P stimulation will require PDGFβ receptor kinase activity. Indeed, overexpression of the $S1P_1$ receptor (partially constitutively active) in HEK 293 cells increases the PDGF-dependent activation of p42/p44 MAPK, whereas overexpression of the PDGFβ receptor (which exhibits high basal tyrosine kinase activity) enhances S1P stimulation of the p42/p44 MAPK pathway in HEK 293 cells (Alderton *et al.*, 2001) (Fig. 3). Additional evidence supporting the integrative model stems from data showing that the treatment of ASM cells with the antisense $S1P_1$ receptor-pcDNA-3.1 plasmid construct reduced the PDGF- and S1P-dependent activation of p42/p44 MAPK (Waters *et al.*, 2003) (Fig. 4). The antisense construct contains the entire open reading frame of the $S1P_1$ receptor cDNA, which is inserted into the vector in the antisense orientation. This type of antisense construct is particularly effective at inducing the depletion of $S1P_1$ receptor expression levels (Waters *et al.*, 2003). Cells (0.5×10^6/sample) at 90% confluence are placed in DMEM containing 2% FCS and are transfected with 2–4 μg plasmid construct following complex formation with LipofectAMINE 2000 according to the manufacturer's instructions (Waters *et al.*, 2003). cDNA-containing media are removed after incubation for 24 h at $37°$, and the cells are incubated for a further 24 h in serum-free medium prior to agonist addition. Higher amounts of plasmid construct (>4 μg) tend to induce cell

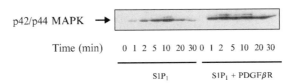

p42/p44 MAPK →

Time (min)   0 1 2 5 10 20 30   0 1 2 5 10 20 30

                    $S1P_1$ $S1P_1$ + PDGFβR

FIG. 3. The effect of overexpression of the PDGFβ receptor on S1P-stimulated activation of p42/p44 MAPK in HEK 293 cells. HEK 293 cells were transfected transiently with vector and/or plasmid constructs for $S1P_1$ and PDGFβ receptor as indicated. Transfected cells were then stimulated with S1P (5 μM, indicated times). A western blot probed with antiphospho-p42/p44 MAPK shows the effect of PDGFβ receptor on $S1P_1$-mediated activation of p42/p44 MAPK. Total p42 MAPK loading was equal for all the samples and was detected using an anti-p42 MAPK antibody (see Pyne and Pyne, 2002). These are representative results of an experiment performed three times.

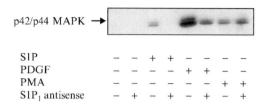

S1P	−	−	+	+	−	−	−	−
PDGF	−	−	−	−	+	+	−	−
PMA	−	−	−	−	−	−	+	+
S1P$_1$ antisense	−	+	−	+	−	+	−	+

Fig. 4. The effect of the S1P$_1$ receptor antisense plasmid construct on PDGF- and S1P-stimulated activation of p42/p44 MAPK activation in ASM cells. ASM cells were pretreated with the antisense S1P$_1$ plasmid construct (2–4 μg, 24 h) prior to stimulation with PDGF (10 ng/ml, 10 min), S1P (5 μM, 10 min), or phorbol ester (PMA; 1 μM, 10 min). Western blots show the effect of an antisense S1P$_1$ plasmid construct on PDGF, S1P- or PMA-dependent activation of p42/p44 MAPK. The antisense S1P$_1$ plasmid construct reduced S1P$_1$ receptor expression (Waters *et al.*, 2003). Total p42 MAPK loading was equal for all the samples and was detected using an anti-p42 MAPK antibody (see Waters *et al.*, 2003). These are representative results of an experiment performed three to four times.

death, possibly as complete depletion of the receptor might compromise cell survival.

The interaction between the PDGFβ receptor and the S1P$_1$ receptor in ASM and HEK 293 cells does not involve the PDGF-dependent release of S1P, which would have the potential to act in an autocrine manner (Alderton *et al.*, 2001; Waters *et al.*, 2003). This has been established using three approaches: (i) inhibitors of sphingosine kinase (e.g., DL-*threo*-dihydrosphingosine and *N*,*N*-dimethysphingosine at concentrations of 1–10 μM, 15-minute pretreatment), the enzyme that catalyzes the phosphorylation of sphingosine to S1P, fail to modulate activation of p42/p44 MAPK by PDGF; (ii) overexpression of point-mutated, catalytically inactive sphingosine kinase (Pitson *et al.*, 2000) or wild-type sphingosine kinase fails to modulate activation of p42/p44 MAPK by PDGF; and (iii) S1P is not released from cells (Waters *et al.*, 2003). S1P can be measured using [^{3}H]sphingosine labeling of cells. ASM cells or PDGFβ receptor-transfected HEK 293 cells (10^6/sample) are quiesced overnight before preincubation for 1 h in serum-free medium containing 2 mg/ml fatty acid-free BSA. Incubations, in the presence and absence of PDGF (10 ng/ml), are initiated by replacing the medium with that containing [^{3}H]sphingosine (Tocris Cookson, UK) (2.22 × 10^5 dpm/10^6 cells). After 3 min, the medium is removed and the cells are harvested into 0.5 ml ice-cold methanol. Lipids are extracted by vortexing in the presence of an equal volume of chloroform. [^{3}H]S1P is isolated by thin-layer chromatography on silica gel G60 plates developed with chloroform:methanol:acetic acid:H$_2$O (25:10:1:2, v/v) (Meyer Zu Heringdorf *et al.*, 1998). A [^{32}P]S1P standard (approximately 20,000 dpm) is run in parallel. This is prepared by a diacylglycerol kinase-catalyzed phosphorylation of ceramide in the presence of $[\gamma\text{-}^{32}P]ATP$ (Desai et al., 1992), followed by deacylation using 6 M HCl/butan-1-ol (1:1, v/v; 1 h at 100°) and purification by TLC, as described earlier. $[^{3}H]S1P$, which comigrates with the $[^{32}P]S1P$ standard, is measured by scintillation counting of appropriate sections of silica that are excised from the TLC plate.

S1P and/or PDGF also promotes the co-internalization of the PDGFβ receptor and S1P$_1$ receptor in the same endocytic vesicles (Fig. 5) (Waters et al., 2003). These vesicles colocalize with p42/p44 MAPK, whereupon the kinase appears to be activated (Waters and Pyne, unpublished), as imaged by epifluorescence and confocal microscopy. Furthermore, inhibitors of clathrin-mediated endocytosis (e.g., monodansylcadaverine, hyperosmolar sucrose, concanavalin A) prevent p42/p44 MAPK activation by both S1P and PDGF (Rakhit et al., 2000). These data validate the possibility that internalization of the PDGFβ receptor requires GPCR endocytic components and vice versa. PDGFβ receptor endocytic signaling appears to be initiated by GRK2/β-arrestin I that has been recruited to the PDGFβ receptor by association of the latter to the S1P$_1$ receptor, and can be increased by S1P (Alderton et al., 2001).

Epifluorescence microscopy and confocal microscopy are some of the most commonly used imaging techniques. By taking advantage of fluorescent-labeled antibodies or proteins, the spatial regulation of signaling pathways can be explored. Two or more proteins can be detected simultaneously if the primary antibodies are raised in different species, e.g., anti-myc epitope-tagged GPCR (mouse) and anti-PDGFβ receptor (rabbit). Colocalization of these two proteins can be seen when each primary antibody is detected using a secondary antibody conjugated to a different fluorophore, e.g., myc epitope tag detection using antimouse IgG TRITC (red) and PDGFβ receptor detection using antirabbit IgG FITC (green).

For both epifluorescence and confocal microscopy, cells are grown on glass coverslips and are fixed using one of several fixing protocols, which must be determined for each antigen of interest. Fixing for 10 min at room temperature in phosphate-buffered saline (PBS) containing 3.7% paraformaldehyde is suitable for the majority of antigens. Alternatively, cells can be fixed in 100% ice-cold methanol for 5 min at minus 20°. Following paraformaldehyde fixation, the cells must be permeabilized to allow penetration of the antibodies into the interior of the cell. This is achieved by incubating the fixed cells with 0.1% Triton X-100 in phosphate-buffered saline for 1 min. Methanol-fixed cells do not require further permeabilization. To prevent nonspecific binding of antibodies, the cells are blocked by incubation for 1 h at room temperature in PBS containing 5% serum

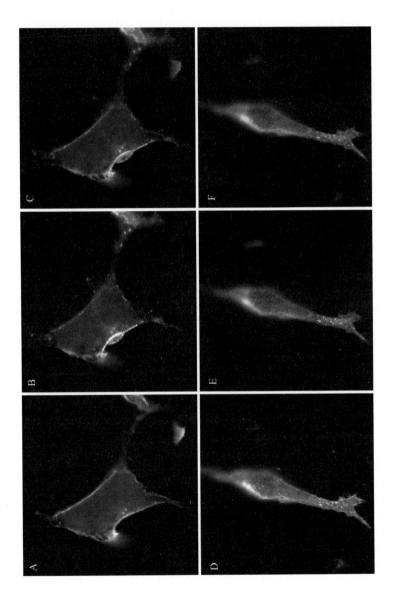

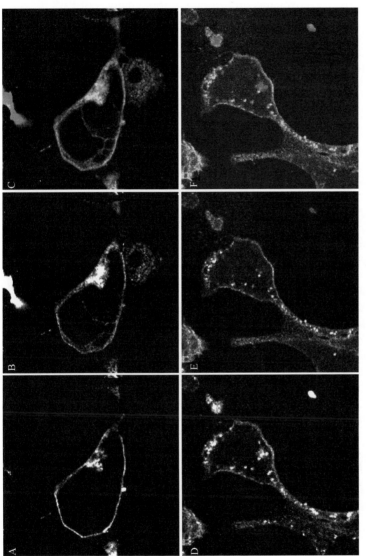

Fig. 5. Co-internalization of PDGFβ receptor–S1P$_1$ receptor complexes. HEK 293 cotransfected with the myc epitope-tagged S1P$_1$ and PDGFβ receptor were quiesced overnight and then stimulated with S1P (5 μM, 5 min). (Top) Epifluorescence images and (bottom) confocal fluorescence images. Images A and D are stained using the anti-myc epitope primary antibody followed by an antimouse secondary antibody conjugated to TRITC (red). Images B and E are stained using the anti-PDGFβ receptor primary antibody followed by an antirabbit secondary antibody conjugated to FITC (green). Images C and F show merged red and green channels where colocalization is visible as yellow. (A–C) Controls. (D–F) S1P stimulated. (See color insert.)

(ideally from the species in which the primary antibody is raised) and 1% (w/v) BSA. Primary antibodies are diluted in blocking solution and are incubated for 1 h at room temperature or overnight at 4°. To remove unbound antibody, the coverslips are washed in four changes of PBS.

Secondary antibodies are incubated and washed as described earlier. For enhanced detection, biotinylated primary or secondary antibodies followed by fluorophore-conjugated avidin can be used. Four biotin molecules bind one molecule of streptavidin. Thus, the signal is amplified substantially. The coverslips are then mounted onto glass slides on a medium such as Vectashield (Vector Laboratories) that contains an antifade component to prevent photobleaching and fading. For short- or long-term storage of slides, the coverslips can be sealed around the edges with nail varnish or a commercially available sealant. Slides should be stored in the dark at 4°. Images of fluorescence signals within defined wavelengths are acquired and manipulated using image analysis software such as Metamorph (Meta Imaging). The acquired images are then given false color (e.g., red or green) and overlayed using the software to generate images where colocalization will be displayed as yellow. In addition, information such as intensity measurements, distance measurements, and many more complex analyses can be carried out.

A major drawback of epifluorescence microscopy is the unwanted "glow" from structures above and below the plane of focus due to blanket illumination of the sample (see Fig. 5, top). This can degrade the image quality and often obscure the signals of interest. This is less of a problem in very flat cells, but in rounder cells, such as HEK 293 cells, this can be avoided by the use of confocal microscopes. Confocal microscopy differs in that the microscope optics are set up to focus on a single plane through the cell and to exclude light from other planes. The sample is scanned by a laser light source in the X and Y directions. The Z depth can be set by the user. This results in an optical sectioning of the sample and allows superior image quality and resolution (Fig. 5, bottom). Another advantage of confocal imaging is that multiple optical Z sections can be taken and "stacked" to produce a three-dimensional representation of the sample.

Nerve Growth Factor Receptor/GPCR Signal Integration

Nerve growth factor promotes the survival and differentiation of sensory and sympathetic neurons and activates p42/p44 MAPK via a B-Raf-dependent mechanism (Segal et al., 1996). We have focused on the potential role of G-proteins in regulating signaling from the Trk A receptor (which binds NGF) in PC12 cells (Moughal et al., 2004) (see Scheme 1, C). PC12 cells are maintained in Dulbecco's modified Eagle's medium supplemented with 10% (v/v) fetal calf serum. Cells are placed routinely in DMEM supplemented with 0.1% (v/v) FCS in the presence of pertussis toxin (0.1 μg/ml) as required for 24 h before experimentation.

Intriguingly, NGF reduces intracellular cAMP via an archetypical G-protein-dependent mechanism in PC12 cells (Rakhit *et al.*, 2001). Moreover, the NGF-dependent activation of p42/p44 MAPK is reduced by pretreating PC12 cells with pertussis toxin (Rakhit *et al.*, 2001). We routinely use between 1 and 5 ng/ml NGF in order to observe G-protein input. At concentrations of 50 ng/ml NGF, the effect of pertussis toxin on NGF-dependent activation of p44 MAPK, but not p42 MAPK, is diminished (Rakhit *et al.*, 2001). The effect of pertussis toxin is only observed between 0 and 10 min of stimulation with NGF and is not evident after this time (Rakhit *et al.*, 2001), suggesting that regulation of the p42/p44 MAPK cascade by NGF is regulated by at least two mechanisms that require differential G-protein input. In this regard, GRK2 is constitutively associated with the Trk A receptor and NGF promotes binding of β-arrestin I to the GRK2–Trk A receptor complex in a pertussis toxin-sensitive manner within the 10-min cell stimulation time frame (Rakhit *et al.*, 2001). This was further supported by results showing that the overexpression of recombinant GRK2 or β-arrestin I increased the NGF-dependent activation of p42/p44 MAPK in PC12 cells (Rakhit *et al.*, 2001). These data suggest that there may be a close functional association between the Trk A receptor and a GPCR.

To support a role of a GPCR in Trk A receptor signaling, we have obtained evidence to suggest that the LPA_1 receptor (also termed Edg-2), which binds lysophosphatidic acid, forms a complex with the Trk A receptor in PC12 cells (Moughal *et al.*, 2004). This was evidenced, in part, by the separate coimmunoprecipitation of Trk A receptor–LPA_1 receptor complexes from lysates of PC12 cells with anti-Trk A antibodies (Santa Cruz) and anti-LPA_1 receptor antibodies (Santa Cruz) (Moughal and Pyne, unpublished results). Verification of LPA_1 receptor expression was established by detection of the mRNA transcript (826 bp) (Fig. 6). Cells are harvested, homogenized, and RNA isolated using the RNeasy minikit (Qiagen, UK). RNA is reverse transcribed with the Superscript II RNase H$^-$ reverse transcriptase (Invitrogen) RT-PCR kit. The LPA_1 forward primer is 5′-TGGCTGCCATCTCTACTTCC-3′, and the reverse primer is 5′-AACCAATCCAGGAGTCCAGC-3′. Amplification involves preincubation at 94° for 2 min followed by 25–35 cycles of 1-min denaturation at 94°, 1-min annealing at 50°, and 2-min polymerization at 72°. The reaction products are extended at 72° for 10 min. The identity of the amplicon should be verified by nucleotide sequencing.

The Trk A receptor–LPA_1 receptor complex functions to regulate the p42/p44 MAPK pathway in PC12 cells (Moughal *et al.*, 2004). Evidence for

Base
pairs

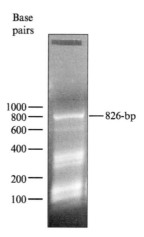

FIG. 6. RT-PCR of LPA$_1$ receptor transcript from PC12 cells. The 826-bp product was sequenced and exhibited 100% homology with the rat LPA$_1$ receptor.

functional interaction between Trk A receptor and LPA$_1$ receptor signaling systems is based on several observations. First, the treatment of PC12 cells with a submaximal concentration of LPA and NGF induced synergistic activation of p42/p44 MAPK at early time points (within 10 min of cell stimulation). Second, the transfection of PC12 cells with the LPA$_1$ receptor antisense construct, which reduced the expression of LPA$_1$, abrogated *both* LPA- and NGF-stimulated activation of p42/p44 MAPK. Third, overexpression of the recombinant LPA$_1$ receptor potentiated LPA- and NGF-dependent activation of p42/p44 MAPK. These data are compatible with a model in which Trk A receptor signaling is LPA$_1$ receptor dependent. Fourth, the treatment of cells with LPA induced transactivation of the Trk A receptor. Thus, LPA$_1$ receptor signaling appears to be Trk A receptor dependent. Transactivation of the Trk A receptor was assessed by western blotting anti-Trk A receptor immunoprecipitates (from 0.5×10^6 cells/sample) with PY20 HRP-linked antiphosphotyrosine antibodies. Transactivated Trk A receptors can also be detected by western blotting cell lysates with antiphosphotyrosine-490 Trk A receptor antibodies (New England Biolabs, UK) (PTyr-490 represents an adaptor site for FRS linked to the activation of p42/p44 MAPK by NGF).

Future Studies

It is necessary to more fully investigate the dynamics of the interaction between GPCRs/G proteins and receptor tyrosine kinases (e.g., identification of interaction sites, tethering proteins). In this manner, one can then attempt to disrupt complex formation to examine effects on signaling as proof of concept. The downstream signaling initiated by receptor tyrosine kinase/GPCR complexes involving GRK2/β-arrestin I is well established, as is its role in regulating endocytosis of the active receptor signalosomes. However, more research is required to define precisely the regulatory steps involved (e.g., the mechanistic activation of GRK2, the role of c-Src in β-arrestin I-mediated endocytic signaling, and the involvement of RGS proteins in growth factor signaling). There is also a need to establish the vectoral and spatial arrangement of these proteins in complexes with receptor tyrosine kinase/GPCRs in order to understand how distinct signaling pathways from receptor tyrosine kinases (e.g., G-protein-mediated *versus* conventional receptor tyrosine phosphate-mediated signaling) are arranged within the cellular context. In this regard, stoichiometric analysis of interacting partners is essential in identifying the extent to which each signaling pathway can be activated (amplification) to define the physiological reason as to why cells need to integrate GPCR and receptor tyrosine kinase signaling. Using well-defined strategies for modulating G-protein input into receptor tyrosine kinase signaling, one should be able to define the precise genetic programs activated (by gene array analysis). This will enable their comparison with gene induction patterns activated by conventional receptor tyrosine kinase signaling and by GPCR-mediated transactivation mechanisms.

Acknowledgments

This work was supported by the Welcome Trust and the BBSRC.

References

Alderton, F., Rakhit, S., Kong, K. C., Palmer, T., Sambi, B., Pyne, S., and Pyne, N. J. (2001). Tethering of the platelet-derived growth factor beta receptor to G-protein-coupled receptors: A novel platform for integrative signaling by these receptor classes in mammalian cells. *J. Biol. Chem.* **276**, 28578–28585.

Chun, J., Goetzl, E. J., Hla, T., Igarashi, Y., Lynch, K. R., Moolenaar, W. H., Pyne, S., and Tigyi, G. (2002). International Union of Pharmacology XXXIV: Lysophospholipid receptor nomenclature. *Pharmacol. Rev.* **54**, 265–269.

Conway, A. M., Rakhit, S., Pyne, S., and Pyne, N. J. (1999). Platelet-derived-growth-factor stimulation of the p42/p44 mitogen-activated protein kinase pathway in airway smooth muscle: Role of pertussis-toxin-sensitive G-proteins, c-Src tyrosine kinases and phosphoinositide 3-kinase. *Biochem. J.* **337**, 171–177.

Daub, H., Wallasch, C., Lankenau, A., Herrlich, A., and Ullrich, A. (1997). Signal characteristics of G protein-transactivated EGF receptor. *EMBO J.* **16**, 7032–7044.

Daub, H., Weiss, F. U., Wallasch, C., and Ullrich, A. (1996). Role of transactivation of the EGF receptor in signalling by G-protein-coupled receptors. *Nature* **379**, 557–560.

Desai, N. N., Zhang, H., Olivera, A., Mattie, M. E., and Spiegel, S. (1992). Sphingosine-1-phosphate, a metabolite of sphingosine, increases phosphatidic acid levels by phospholipase D activation. *J. Biol. Chem.* **267**, 23122–23128.

Freedman, N. J., Kim, L. K., Murray, J. P., Exum, S. T., Brian, L., Wu, J. H., and Peppel, K. (2002). Phosphorylation of the platelet-derived growth factor receptor-beta and epidermal growth factor receptor by G protein-coupled receptor kinase-2: Mechanisms for selectivity of desensitization. *J. Biol. Chem.* **277**, 48261–48269.

Hirata, M., Umata, T., Takahashi, T., Ohnuma, M., Miura, Y., Iwamoto, R., and Mekada, E. (2001). Identification of serum factor inducing ectodomain shedding of proHB-EGF and studies of noncleavable mutants of proHB-EGF. *Biochem. Biophys. Res. Commun.* **283**, 915–922.

Kohno, T., Matsuyuki, H., Inagaki, Y., and Igarashi, Y. (2003). Sphingosine 1-phosphate promotes cell migration through the activation of Cdc42 in Edg-6/S1P4-expressing cells. *Genes Cells* **8**, 685–697.

Kreuzer, J., Viedt, C., Brandes, R. P., Seeger, F., Rosenkranz, A. S., Sauer, H., Babich, A., Nurnberg, B., Kather, H., and Krieger-Brauer, H. I. (2003). Platelet-derived growth factor activates production of reactive oxygen species by NAD(P)H oxidase in smooth muscle cells through Gi1,2. *FASEB J.* **17**, 38–40.

Krieger-Brauer, H. I., Medda, P., and Kather, H. (2000). Basic fibroblast growth factor utilizes both types of component subunits of Gs for dual signaling in human adipocytes: Stimulation of adenylyl cyclase via Galpha(s) and inhibition of NADPH oxidase by Gbeta gamma(s). *J. Biol. Chem.* **275**, 35920–35925.

Liebmann, C., Graness, A., Boehmer, A., Kovalenko, M., Adomeit, A., Steinmetzer, T., Nurnberg, B., Wetzker, R., and Boehmer, F. D. (1996). Tyrosine phosphorylation of Gsalpha and inhibition of bradykinin-induced activation of the cyclic AMP pathway in A431 cells by epidermal growth factor receptor. *J. Biol. Chem.* **271**, 31098–31105.

Luttrell, L. M., van Biesen, T., Hawes, B. E., Koch, W. J., Touhara, K., and Lefkowitz, R. J. (1995). G beta gamma subunits mediate mitogen-activated protein kinase activation by the tyrosine kinase insulin-like growth factor 1 receptor. *J. Biol. Chem.* **270**, 16495–16498.

Meyer zu Heringdorf, D., Lass, H., Alemany, R., Laser, K. T., Neumann, E., Zhang, C., Schmidt, M., Ruan, U., Jakobs, J. H., and Van Koppen, C. J. (1998). Sphingosine kinase-mediated calcium signalling by G protein-coupled receptors. *EMBO J.* **17**, 2830–2837.

Miller, W. E., and Lefkowitz, R. J. (2001). Expanding roles for beta-arrestins as scaffolds and adapters in GPCR signaling and trafficking. *Curr. Opin. Cell. Biol.* **13**, 139–145.

Moughal, N., Waters, C., Pyne, S., and Pyne, N. J. (2004). Nerve growth factor signalling involves interaction between the Trk A receptor and lysophosphatidate receptor 1 systems: Nuclear translocation of the lysophosphatidate receptor 1 and Trk A receptors in pheochromocytoma 12 cells. *Cell. Signal.* **16**, 127–136.

Moyers, J. S., Linder, M. E., Shannon, J. D., and Parsons, S. J. (1995). Identification of the in vitro phosphorylation sites on Gs alpha mediated by pp60c-src. *Biochem. J.* **305**, 411–417.

Pitson, S. M., Moretti, P. A. B., Zebol, J. R., Xia, P., Gamble, J. R., Vadas, M. A., D'Andrea, R. J., and Wattenberg, B. W. (2000). Expression of a catalytically inactive sphingosine kinase mutant blocks agonist-induced sphingosine kinase activation: A dominant-negative sphingosine kinase. *J. Biol. Chem.* **275**, 33945–33950.

Prenzel, N., Zwick, E., Daub, H., Leserer, M., Abraham, R., Wallasch, C., and Ullrich, A. (1999). EGF receptor transactivation by G-protein-coupled receptors requires metalloproteinase cleavage of proHB-EGF. *Nature* **402**, 884–888.

Pyne, N. J., Moughal, N., Stevens, P. A., Tolan, D., and Pyne, S. (1994). Protein kinase C-dependent cyclic AMP formation in airway smooth muscle: The role of type II adenylate cyclase and the blockade of extracellular-signal-regulated kinase-2 (ERK-2) activation. *Biochem. J.* **304,** 611–616.

Pyne, N. J., Murphy, G. J., Milligan, G., and Houslay, M. D. (1989). Treatment of intact hepatocytes with either the phorbol ester TPA or glucagon elicits the phosphorylation and functional inactivation of the inhibitory guanine nucleotide regulatory protein Gi. *FEBS Lett.* **243,** 77–82.

Pyne, S., and Pyne, N. J. (1993). Bradykinin stimulates phospholipase D in primary cultures of guinea-pig tracheal smooth muscle. *Biochem. Pharmacol.* **45,** 593–603.

Pyne, S., and Pyne, N. J. (2002). Sphingosine 1-phosphate signalling and termination at lipid phosphate receptors. *Biochim. Biophys. Acta* **1582,** 121–131.

Rakhit, S., Conway, A.-M., Tate, R., Bower, T., Pyne, N. J., and Pyne, S. (1999). Sphingosine 1-phosphate stimulation of the p42/p44 mitogen-activated protein kinase pathway in airway smooth muscle: Role of endothelial differentiation gene 1, c-Src tyrosine kinase and phosphoinositide 3-kinase. *Biochem. J.* **338,** 643–649.

Rakhit, S., Pyne, S., and Pyne, N. J. (2000). The platelet-derived growth factor receptor stimulation of p42/p44 mitogen-activated protein kinase in airway smooth muscle involves a G-protein-mediated tyrosine phosphorylation of Gab1. *Mol. Pharmacol.* **58,** 413–420.

Rakhit, S., Pyne, S., and Pyne, N. J. (2001). Nerve growth factor stimulation of p42/p44 mitogen-activated protein kinase in PC12 cells: Role of G(i/o), G protein-coupled receptor kinase 2, beta-arrestin I and endocytic processing. *Mol. Pharmacol.* **60,** 63–70.

Rosenfeldt, H. M., Hobson, J. P., Maceyka, M., Olivera, A., Nava, V. E., Milstien, S., and Spiegel, S. (2001). EDG-1 links the PDGF receptor to Src and focal adhesion kinase activation leading to lamellipodia formation and cell migration. *FASEB J.* **15,** 2649–2659.

Rubin, C. S., Erlichman, J., and Rosen, O. M. (1974). Cyclic AMP-dependent protein kinase from bovine heart muscle. *Methods Enzymol.* **38,** 308–315.

Schmid, S. L., McNiven, M. A., and De Camilli, P. (1998). Dynamin and its partners: A progress report. *Curr. Opin. Cell Biol.* **10,** 504–512.

Segal, R., Bhattacharyya, A., Rua, L., Alberta, J., Stephens, R., Kaplan, D., and Stiles, C. (1996). Differential utilization of Trk autophosphorylation sites. *J. Biol. Chem.* **271,** 20175–20181.

Waters, C., Sambi, B., Kong, K. C., Thompson, D., Pitson, S. M., Pyne, S., and Pyne, N. J. (2003). Sphingosine 1-phosphate and platelet-derived growth factor (PDGF) act via PDGF beta receptor-sphingosine 1-phosphate receptor complexes in airway smooth muscle cells. *J. Biol. Chem.* **278,** 6282–6290.

[29] Identification and Biochemical Analysis of GRIN1 and GRIN2

By Naoyuki Iida and Tohru Kozasa

Abstract

We have identified the novel Gα$_z$-binding protein, which is referred to as the G-protein-regulated inducer of neurite outgrowth (GRIN1) using the far-western method. GRIN1 is expressed specifically in brain and binds preferentially to the activated form of α subunits of G$_z$, G$_i$, and G$_o$.

METHODS IN ENZYMOLOGY, VOL. 390

Coexpression of GRIN1 and the activated form of $G\alpha_o$ induce neurite outgrowth in Neuro2a cells. We have further identified two human GRIN1 homologs, GRIN2 and GRIN3, in the database. This article shows that GRIN2 can also bind to the GTP-bound form of $G\alpha_o$. These findings suggest that the GRIN1 family may function as a downstream effector for $G\alpha_o$ to regulate neurite growth.

Introduction

G-protein-coupled receptors (GPCRs) form the largest family of integral membrane proteins involved in various signal transductions, ranging from sensory transduction to cell proliferation and differentiation (Gilman, 1987). Ligand binding to a GPCR catalyzes the GDP-GTP exchange of the heterotrimeric G protein. The GTP-bound form of G-protein α subunits and, in some cases, released $\beta\gamma$ subunits initiate a broad range of cellular signaling events.

To elucidate the signaling pathway coupled to the $G\alpha_i$ subfamily, various binding proteins to $G\alpha_i$ subfamily α subunits have been identified. Yeast two-hybrid screening using $G\alpha_i$ as a bait has revealed RGS-GAIP, rap1GAPII, Ric-8A, and Eya2 as partners for $G\alpha_i$ (De Vries et al., 1995; Fan et al., 2000; Mochizuki et al., 1999; Tall et al., 2003). AGS3 was identified by a functional assay based on the pheromone response pathway in *Saccharomyces cerevisiae* (Takesono et al., 1999). We have identified a novel $G\alpha_z$-binding protein, the G-protein-regulated inducer of neurite outgrowth (GRIN1) using the far-western method (Chen et al., 1999). ^{32}P-phosphorylated, GTPγS-bound $G\alpha_z$ was used as a probe to screen a mouse embryo expression phage library. This article describes the methods used for the identification of this $G\alpha_z$-binding protein and the method used to study the interaction of GRIN1 with G-protein α subunits.

Preparation of Recombinant Proteins

Purification of $G\alpha_z$

Recombinant $G\alpha_z$ protein is purified from Sf9 cells infected with baculoviruses of $G\alpha_z$, $G\beta_1$, and His-tagged $G\gamma_2$ subunits (Kozasa and Gilman, 1995, 1996). Sf9 cells (Invitrogen) are grown in suspension in IPL-41 medium (Life Technology) containing 1% Pluronic F68, 10% heat-inactivated fetal calf serum (FCS), and 50 μg/ml gentamicin at 27° with constant shaking (125 rpm). For large-scale cultures, the concentration of FCS is reduced to 2% and 1% lipid mix (Chemically Defined Lipid Concentrate, GIBCO BRL) is supplemented. Sf9 cells (4-liter culture; 1.5–2.0 $\times$ 10^6

cells/ml) are infected with amplified recombinant baculoviruses encoding the G$_z$ α subunit, the β_1 subunit, and the His-tagged γ_2 subunit. Usually, 15 ml of α, 10 ml of β, and 7.5 ml of γ viruses are infected to 1 liter of Sf9 cell culture. Cells are harvested 48 h after infection by centrifugation at 1000g for 10 min and are suspended in 600 ml of ice-cold lysis buffer [50 mM NaHEPES (pH 8.0), 0.1 mM EDTA, 3 mM MgCl$_2$, 10 mM 2-mercaptoethanol, 100 mM NaCl, 10 μM GDP, 0.02 mg/ml phenylmethyl-sulfonyl fluoride (PMSF), 0.03 mg/ml leupeptin, 0.02 mg/ml 1-chloro-3-tosylamido-7-amino-2-heptanone, 0.02 mg/ml L-1-p-tosylamino-2-pheny-lethyl chloromethyl ketone (TPCK), and 0.03 mg/ml lima bean trypsin inhibitor]. Cells are lysed by nitrogen cavitation (Parr bomb) at 500 psi for 30 min. Cell lysates are centrifuged at 750g for 10 min to remove intact cells and nuclei. The supernatants are centrifuged at 100,000g for 30 min, and the resultant pellets are suspended in 300 ml of wash buffer [50 mM NaHEPES (pH 8.0), 3 mM MgCl$_2$, 10 mM 2-mercaptoethanol, 50 mM NaCl, 10 μM GDP, and proteinase inhibitors as described earlier] using a Potter–Elvehjem homogenizer and centrifuged as described earlier. The pellets are resuspended in 200 ml of wash buffer, and the protein concentration is determined. This suspension (cell membranes) can be frozen in liquid nitrogen and stored at $-80°$.

Cell membranes are diluted to 5 mg/ml with wash buffer containing fresh proteinase inhibitors. Sodium cholate is added to a final concentration of 1% (w/v), and the mixture is stirred on ice for 1 h prior to centrifugation at 100,000g for 30 min. The supernatants (membrane extract) are collected, diluted fourfold with buffer A [20 mM NaHEPES (pH 8.0), 100 mM NaCl, 1 mM MgCl$_2$, 10 mM 2-mercaptoethanol, 10 μM GDP, and 0.5% C$_{12}$E$_{10}$], and loaded onto a 4-ml Ni-NTA (Qiagen) column, which has been equilibrated with buffer A. The column is washed with 100 ml of buffer B (buffer A containing 300 mM NaCl and 10 mM imidazole) and 12 ml of buffer C (buffer A containing 0.2% sodium cholate) at 4°. The column is then warmed to room temperature. The column is washed with 12 ml of buffer C at 30° and 32 ml of buffer D (buffer A containing 30 μM AlCl$_3$, 50 mM MgCl$_2$, 10 mM NaF, and 5 mM imidazole) at 30°. Finally, the column is washed with 12 ml of buffer E (buffer A containing 150 mM imidazole). Fractions (4 ml) are collected during washing with buffers D and E. Gα_z or β_1His-γ_2 is eluted during the wash with buffer D or buffer E, respectively. Fractions are analyzed by silver staining and western blotting. Peak fractions of Gα_z from the column are diluted threefold with buffer F [20 mM NaHEPES (pH 7.4), 1 mM EDTA, 3 mM MgCl$_2$, 3 mM dithiothreitol (DTT), 0.7% CHAPS]. This solution is then applied to a Mono S HR 5/5 cation-exchange column for FPLC (Pharmacia Biotech Inc.), which has been equilibrated with buffer F. Gα_z is eluted with a 25-ml gradient of NaCl (0–550 mM) and normally appears in fractions with 400–450 mM NaCl. GDP (1 μl of a 5 mM solution) is put in the tubes for fractions (0.5 ml). Fractions are assayed by silver staining and by measurement of [^{35}S]GTPγS-binding activity. Peak fractions are concentrated, and the buffer is changed to buffer F containing 100 mM NaCl and 0.5 μM GDP using a Centricon 30 (Amicon). Samples are frozen in liquid nitrogen and stored at $-80°$.

Purification of PKCα

Sf9 cells (1 liter, 1.5 × 10^6 cells/ml) are infected with a recombinant baculovirus encoding rabbit PKCα (Fujise *et al.*, 1994). Cells are harvested 48 h after infection and suspended in 120 ml of ice-cold lysis buffer [20 mM Tris–HCl (pH 7.5), 5 mM EGTA, 1 mM EDTA, 10 mM 2-mercaptoethanol, and protease inhibitors]. After cell lysis (nitrogen cavitation at 500 psi for 30 min) and centrifugation (100,000g for 30 min), PKCα is purified chromatographically using DEAE-Sephacel, hydroxyapatite, phenyl-Superose HR10/10, and Mono Q HR5/5. The yield is 1 mg from a 1-liter culture, the specific activity is 800 units/mg, and the protein is more than 90% pure based on silver staining after gel electrophoresis.

Preparation of Phosphorylated Gα_z Probe

Purified Gα_z (250 μg) is incubated with GTPγS at 30° for 60 min in the presence of 5 mM EDTA and 2 mM MgSO$_4$. Phosphorylation of GTPγS-bound forms of Gα_z with PKCα (4 μg) is performed in buffer containing 25 mM Tris–HCl (pH 7.5), 5 mM MgSO$_4$, 0.125 mM CaCl$_2$, 1 mM DTT, 3 μM [γ-^{32}P]ATP (40,000 cpm/pmol), and 20 μg/ml phosphatidylserine-diolein (Sigma) for 30 min at 30°. After phosphorylation, free [γ-^{32}P]ATP is removed by gel filtration (PD-10 column; Amersham Pharmacia Biotech) in buffer G [20 mM NaHEPES (pH 7.4), 100 mM NaCl, 3 mM MgCl$_2$, 1 mM EDTA, 0.05% C$_{12}$E$_{10}$ (polyoxyethlene 10-lauryl ether), 1 mM DTT, and 10 mM β-glycerophosphate]. The labeled protein is stored at 4° and is used as a probe for screening cDNA expression libraries. The specific radioactivity of the probe obtained by this method is in the range 4–9 × 10^7 cpm/μg protein.

Isolation of Gα_z-Binding Proteins

A 16-day mouse embryo λEXlox expression library (Novagen) is screened for the expression of Gα_z-binding proteins. Approximately 3 × 10^5 clones are screened using the BL21(DE3)pLysE strain as the recipient *Escherichia coli* and phosphorylated Gα_z-GTPγS as a probe.

Phages are plated at a density of 3–5×10^4 per 150-mm plate and incubated at $37°$ for 7 h. The plates are overlaid with Hybond-C filters (Amersham Pharmacia Biotech) that have been saturated with 10 mM isopropyl-β-D-thiogalactopyranoside. Plates are further incubated at $37°$ for 3.5 h to induce expression of proteins encoded by cDNAs. The filters are then rinsed with buffer G at room temperature and blocked with buffer G containing 5% dry milk at $4°$ overnight. The filters are probed with 50 nM phosphorylated GTPγS-bound Gα_z in buffer G with 5% dry milk for 4 h at $4°$. The filters are washed three times with buffer A for 5 min at $4°$, air dried, and exposed to film at $-70°$ for 2 days.

Phage particles are eluted from positive plaques into SM medium [50 mM Tris-HCl (pH 7.8), 100 mM NaCl, 8 mM MgSO$_4$, 0.1% gelatin] and replated at 3000 plaques per 90-mm plate. Positive clones are subjected to successive cycles of eluting, replating, and screening as described earlier until phage clones are purified to homogeneity. The cloned DNA inserts in λEXphage are converted into pEXlox plasmids by infecting the E. coli strain BM25.8 that contains P1 Cre recombinase. Host strain BM25.8 E. coli are cultured in LB medium containing 10 mM MgSO$_4$, 0.2% maltose, 50 μg/ml kanamycin, and 34 μg/ml chloramphenicol. Fifty microliters of the diluted phage is mixed with 50 ml of BM25.8 host cells. The mixtures are spread on an LB plate containing 100 μg/ml ampicillin. Plasmids prepared from colonies in BM25.8 are retransformed into DH5α E. coli cells. Plasmid DNA are prepared from transformed DH5α E. coli cells for DNA sequencing.

We have isolated two clones, designated Z-13 and Z-16, by this method. As shown in Fig. 1, the product of the Z-13 cDNA bound GDP-Gα_z and GTPγS-Gα_z equally well, whereas the product of the Z-16 cDNA bound GTPγS-Gαz selectively. The Z-13 clone encodes nucleobindin, a protein also isolated as a binding protein for Gα_{i2} using the yeast two-hybrid method (Mochizuki et al., 1995). The Z-16 clone encodes a protein that has an open reading frame of 273 amino acid residues. We further extended the 5'-end sequence of this clone by 5'-RACE and isolated a full-length cDNA with an open reading frame of 827 amino acid residues and a calculated molecular mass of 84,700 Da (Frohman, 1993). We refer to this protein as GRIN1. The homolog of GRIN1 in chicken was also identified by the yeast two-hybrid screening using activated Gα_o as bait (Jordan and Iyengar, 2002).

Identification of GRIN1 Homologs in the Database

We performed a sequence database search to identify GRIN1 homologs using the C-terminal GRIN1 amino acid sequence and identified related clones, which are referred to as GRIN2 and GRIN3 (Chen et al.,

GDP   GTPγS

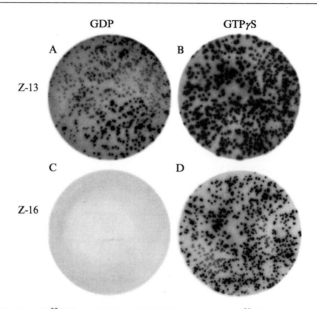

FIG. 1. Binding of [^{32}P]Gα$_z$ to Z-13 and Z-16. Interactions of [^{32}P]Gα$_z$ with purified phage expressing Z-13 (A and B) or Z-16 (C and D) are shown. Phosphorylated Gα$_z$ was tested in either its GDP-bound (A and C) or its GTPγS-bound (B and D) form. From Chen *et al.* (1999), with permission.

1999; Nagase *et al.*, 1998). Human GRIN1, GRIN2, and GRIN3 are composed of 856, 461, and 776 amino acids, respectively. Comparison of their amino acid sequences revealed that the GRIN family shares high homology in their C-terminal region (Fig. 2). GRIN homologs are conserved in the pufferfish *Fugu rubripes*; however, no GRIN homologs are found in the nematode *C. elegans* or fruit fly *Drosophila melanogaster* genomes.

Binding of GRIN1 and GRIN2 to G-Protein Subunits

Purification of GRIN1 and GRIN2

GRIN1 and GRIN2 cDNA are subcloned into the pFastBacHTb vector (Life Technologies, Inc.), and recombinant baculoviruses encoding His$_6$-GRIN1 or His$_6$-GRIN2 are generated. Membranes from Sf9 cells infected with baculoviruses encoding His$_6$-GRIN1 or His$_6$-GRIN2 are prepared and extracted with 1% C$_{12}$E$_{10}$. His$_6$-GRIN1 and His$_6$-GRIN2 are then purified from these extracts using Ni-NTA (Qiagen) chromatography. The proteins

```
GRIN1 693: APRAPSPPSRRDAGLQVSLGAAETRSVATGPMTPQAAAPPAFPEVRVRPGSALAAAVAPP 752
            ::    :: : ::: :    :     ::: :     ::    ::::    :: : :
GRIN2 326: APAEASPLSAQDAGVQAA-PVAACKAVATSPSLEAPAALHVFPEVTL--GSSLEEA---- 378
             :                     :              :::             :
GRIN3 616: SPGSGKKTPSRSVKASPRRPSRVSEFLKEQKLNVTAAAAQVGLTPGDKKKQLGADSKLQL 675


GRIN1 753: EPAEPVRDVSWDEKGMTWEVYGAAMEVEVLGMAIQKHLERQIEEHGRQGAPAPPPAARAG 812
            : ::::::  ::  ::::::::::::   ::::  ::::::::  : :    :   :::
GRIN2 379: -PS-PVRDVRWDAEGMTWEVYGAAVDPEVLGVAIQKHLEMQFEQLQR--APASED----- 429
            ::::: ::: ::::::::::   : :: ::: :: ::: ::           :
GRIN3 676: KQSKRVRDVVWDEQGMTWEVYGASLDAESLGIAIQNHLQRQIREH-EKLIKTQNSQTR-- 732


GRIN1 813: PGRSGSVRTAPPDGAAKRPPGLFRALLQSVRRPRCCSRAGPTAE    856
            : ::        :   : :     ::   :: ::: ::   :  :
GRIN2 430: ---SLSV-----EG--RRGP--LRAVMQSLRRPSCCGCSGAAPE    461
            :: : :.       :       :     :   :  :     :
GRIN3 733: --RSISSDTSSNKKLRGRQHSVFQSMLQNFRRPNCCVRPAPSSVLD 776
```

FIG. 2. Alignment of the amino acid sequences of human GRIN1, GRIN2, and GRIN3. The C-terminal amino acid sequences of human GRIN1, GRIN2, and GRIN3 are aligned. Identical residues are shown as a colon.

are purified further by Mono Q and Superdex 200 column chromatography in 20 mM NaHEPES (pH 8.0), 5 mM MgCl$_2$, 2 mM EDTA, 50 mM NaCl, 1 mM DTT, and 0.5% C$_{12}$E$_{10}$.

Binding of GRIN1 and GRIN2 to G-Protein Subunits

Hexa-histidine-tagged GRIN1 or GRIN2 purified from Sf9 cells (1.5 μg) is mixed with 2 μg of recombinant Gα_o or Gα_z bound with GDP, GTPγS, or GDP·AlF$_4^-$ in 100 μl of buffer G [50 mM NaHEPES (pH 8.0), 5 mM MgCl$_2$, 10 mM 2-mercaptoethanol, 0.1% C$_{12}$E$_{10}$] and incubated on ice for 30 min. NaF (10 mM) and AlCl$_3$ (30 μM) are included in the buffer to maintain Gα·GDP·AlF$_4^-$. Ni-NTA resin (25 μl), equilibrated with buffer G, is added to the mixture of proteins, followed by further incubation on ice for 5 min. The resin is collected by brief centrifugation, and the supernatant is saved as the flow-through fraction. The resin is washed twice with 150 μl of buffer G containing 500 mM NaCl and 10 mM imidazole. NaF and AlCl$_3$ are included in the wash buffer when they are present initially. The resin is finally eluted twice with 50 μl of buffer G containing 200 mM imidazole. Fractions are analyzed by immunoblotting after SDS–PAGE. In Fig. 3, both purified GRIN1 and GRIN2 interacted with the GTPγS- or GDP·AlF$_4^-$-bound form of Gα_o and Gα_z specifically. Interaction with the GDP-bound forms of these two proteins was not detected.

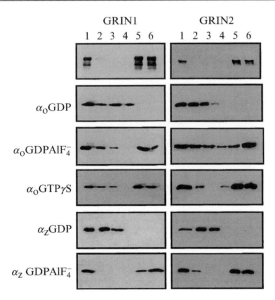

FIG. 3. Interactions of GRIN1 and GRIN2 with G-protein subunits. Purified recombinant (Sf9 cell-derived) hexahistidine-tagged GRIN1 (left) or hexahistidine-tagged GRIN2 (right) (1.5 μg of each) was mixed with 2 μg of recombinant Gα_o or Gα_z bound with GDP, GTPγS, or GDP·AlF$_4^-$ as indicated. Ni-NTA resin was added to the mixture, and precipitation of protein complexes was performed as described in the text. Fractions were resolved by SDS–PAGE, and immunoblots were developed with antibodies reactive with GRIN1 or GRIN2 (top, left and right, respectively) or with Gα_o or Gα_z, as appropriate. The applied proteins are shown in lane 1 of each panel; lane 2, flow through; lanes 3 and 4, sequential washes with 500 mM NaCl and 10 mM imidazole; and lanes 5 and 6, sequential elutions with 200 mM imidazole–HCl. From Chen *et al.* (1999), with permission.

Conclusion

We screened a mouse embryo cDNA expression library to search for proteins capable of interacting with GTPγS-Gα_z and identified a G$_i$ subfamily (G$_i$, G$_o$, G$_z$)-binding protein referred to as GRIN1.

GRIN family proteins have moderate homology in their C-terminal region that is probably responsible for binding to Gα_o. GRIN family proteins have no RGS domain and do not show GAP activity for Gα_i nor Gα_o (Nakata *et al.*, unpublished results). GRIN1 and GRIN2 interact preferentially with activated forms of α subunits of the G$_i$ subfamily (G$_i$, G$_o$, and G$_z$). Both GRIN1 and Gα_o are highly expressed in brain. In particular, they localize at the growth cone of neurons, and coexpression of GRIN1 with activated Gα_o induces neurite outgrowth (Chen *et al.*,

1999). We thus hypothesize that GRIN1 may function as a downstream effector for Gα_o to regulate neurite growth.

References

Chen, L. T., Gilman, A. G., and Kozasa, T. (1999). A candidate target for G protein action in brain. *J. Biol. Chem.* **274**, 26931–26938.

De Vries, L., Mousli, M., Wurmser, A., and Farquhar, M. G. (1995). GAIP, a protein that specifically interacts with the trimeric G protein G alpha i3, is a member of a protein family with a highly conserved core domain. *Proc. Natl. Acad. Sci. USA* **92**(25), 11916–11920.

Fan, X., Brass, L. F., Poncz, M., Spitz, F., Maire, P., and Manning, D. R. (2000). The alpha subunits of Gz and Gi interact with the eyes absent transcription cofactor Eya2, preventing its interaction with the six class of homeodomain-containing proteins. *J. Biol. Chem.* **275**(41), 32129–32134.

Frohman, M. A. (1993). Rapid amplification of complementary DNA ends for generation of full-length complementary DNAs: Thermal RACE. *Methods Enzymol.* **218**, 340–354.

Fujise, A., Mizuno, K., Ueda, Y., Osada, S., Hirai, S., Takayanagi, A., Shimizu, N., Owada, M. K., Nakajima, H., and Ohno, S. (1994). Specificity of the high affinity interaction of protein kinase C with a physiological substrate, myristoylated alanine-rich protein kinase C substrate. *J. Biol. Chem.* **269**(50), 31642–31648.

Gilman, A. G. (1987). G proteins: Transducers of receptor-generated signals. *Annu. Rev. Biochem.* **56**, 615–649.

Jordan, J. D., and Iyengar, R. (2002). Identification of putative direct effectors for G alpha o, using yeast two-hybrid method. *Methods Enzymol.* **345**, 140–149.

Kozasa, T., and Gilman, A. G. (1995). Purification of recombinant G proteins from Sf9 cells by hexahistidine tagging of associated subunits. *J. Biol. Chem.* **270**, 1734–1741.

Kozasa, T., and Gilman, A. G. (1996). Protein kinase C phosphorylates G12α and inhibits its interaction with G$\beta\gamma$. *J. Biol. Chem.* **271**, 12562–12567.

Mochizuki, N., Hibi, M., Kanai, Y., and Insel, P. A. (1995). Interaction of protein nucleobindin with G alpha i2, as revealed by the yeast two-hybrid system. *FEBS Lett.* **373**(2), 155–158.

Mochizuki, N., Ohba, Y., Kiyokawa, E., Kurata, T., Murakami, T., Ozaki, T., Kitabatake, A., Nagashima, K., and Matsuda, M. (1999). Activation of the ERK/MAPK pathway by an isoform of rap1GAP associated with G alpha(i). *Nature* **400**(6747), 891–894.

Nagase, T. *et al.* (1998). Prediction of the coding sequences of unidentified human genes. IX. The complete sequences of 100 new cDNA clones from brain which can code for large proteins *in vitro*. *DNA Res.* **5**, 31–39.

Takesono, A., Cismowski, M. J., Ribas, C., Bernard, M., Chung, P., Hazard, S., 3rd., Duzic, E., and Lanier, B. S. (1999). Receptor-independent activators of heterotrimeric G-protein signaling pathways. *J. Biol. Chem.* **274**(47), 33202–33205.

Tall, G. G., Krumins, A. M., and Gilman, A. G. (2003). Mammalian Ric-8A (synembryn) is a heterotrimeric G alpha protein guanine nucleotide exchange factor. *J. Biol. Chem.* **278**(10), 8356–8362.

Author Index

Subject Index

A

AC, *see* Adenylyl cyclase
Activator of G-protein-signaling-3
 intestinal cell expression, 25, 27
 macrophagy role, 27–28
 regulators, 28
 structure, 27
Adenylyl cyclase
 domains, 83–84
 isoforms, 84
 RGS2 interactions
  adenylyl cyclase activity assays
   cytoplasmic domains, 95
   membrane-bound enzyme, 95–96
  cyclic AMP accumulation assay in cells,
   91–94
  expression vector construction,
   84–86
  gel filtration profile studies, 96–97
  isoform specificity, 84
  pull-down assays
   assay I, 87–89
   assay II, 89–91
   overview, 86–87
α_{2A}-Adrenergic receptor, mitogen-activated
 protein kinase signaling inhibition by
 RGSZ1, 46–48, 50
AGS3, *see* Activator of G-protein-signaling-3
Autophagy, *see* Macrophagy

B

Baculovirus-Sf9 cell system
 Gβ5 recombinant protein preparation
  cell culture, 167
  immunodetection, 166–167
  infection, 167–168
  ion-exhange chromatography of RGS
   dimers, 169–171
  nickel affinity chromatography,
   168–169
  virus preparation, 164

regulators of G-protein signaling
 recombinant protein preparation
  cell culture, 167
  immunodetection, 166–167
  infection, 167–168
  ion-exhange chromatography of Gβ5
   dimers, 169–171
  nickel affinity chromatography,
   168–169
  virus preparation, 164
 RGS9-1 expression, 184–186
 Ric-8A expression, 378

C

Calcium channel
 GoLoco motif peptide studies of
  G-protein-coupled receptor/ion
  channel coupling
   agonist application, 442–443
   calcium channel current dynamics
    coupled to dopamine receptors, 447
   cell culture, plating and transfection,
    440–441
   cell lines, 439–440
   data acquisition and analysis, 443
   peptide design and synthesis, 439
   peptide injection, 442
   whole cell patch-clamp, 441
 RGS12–N-type calcium channel
  interaction studies
   affinity chromatography with channel
    peptides, 233
   dorsal root ganglion neuron culture
    feeding, 227
    isolation of neurons, 225–226
    materials, 225
    plating, 226–227
    trituration, 225
   electrophysiology studies
    data analysis, 236–237
    peptide introduction through
     recording pipette, 237–238

505

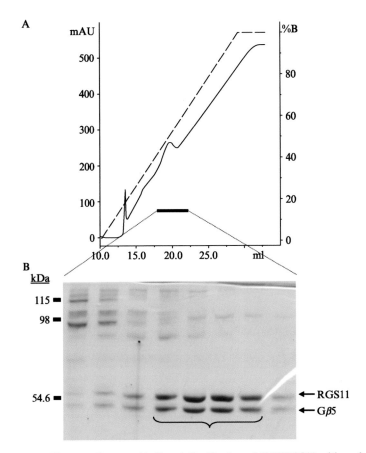

Hooks and Harden, Chapter 11, Fig. 4. Purification of Gβ5/RGS11 with a chelating FPLC column. (A) Elution of a protein peak (A_{280}) from a Ni-charged chelating column is shown. Imidazole, used in the elution gradient, also absorbs light at this wavelength and accounts for the linear increase in absorbance that is superimposed on the elution of protein. The left-hand y axis and solid line indicate mAU (A_{280}), and the right-hand y axis and hatched line indicate percentage of elution buffer (1 M NaCl). The bold line indicates fractions analyzed by SDS–PAGE. (B) The purity of the heterodimer eluting from the chelating column is shown on a Coomassie-stained SDS–PAGE gel. The bracket indicates which fractions were pooled prior to concentration.

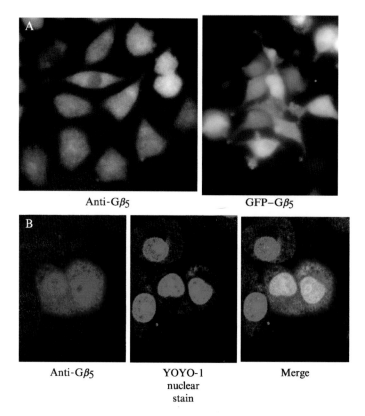

Anti-Gβ_5 GFP–Gβ_5

Anti-Gβ_5 YOYO-1
nuclear
stain   Merge

Sɪᴍᴏɴᴅs *ᴇᴛ ᴀʟ.*, Cʜᴀᴘᴛᴇʀ 14, Fɪɢ. 2. (A) Epifluorescence microscopic images (black and white) of rat pheochromocytoma PC12 cells. Untransfected PC12 cells processed for immunohistochemistry with affinity-purified ATDG anti-Gβ_5 primary antibody and fluorescein isothiocyanate (FITC)-labeled antirabbit secondary antibody (left). PC12 cells transfected with the pEGFP-C2 construct encoding a GFP–Gβ_5 fusion protein (right). Modified from Zhang *et al.* (2001), with permission. (B) Laser confocal immunofluorescence analysis of stably transfected PC12 cells. Cells transfected stably with HA epitope-tagged Gβ_5 were analyzed by laser confocal microscopy after dual staining with affinity-purified ATDG anti-Gβ_5 antibody (red signal) and the nuclear counterstain YOYO-1 (green signal). The yellow–orange signal in the merge image indicates colocalization of probes. Modified from Zhang *et al.* (2001), with permission.

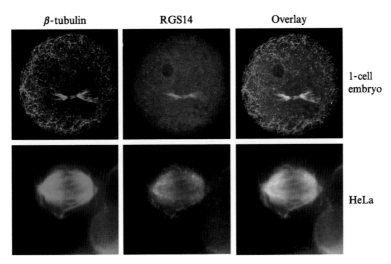

β-tubulin RGS14 Overlay

1-cell embryo

HeLa

MARTIN-MCCAFFREY *ET AL.*, CHAPTER 16, FIG. 2. Colocalization of RGS14 and β-tubulin in preimplantation mouse embryos (one-cell stage; top) and mitotic HeLa cells (bottom). Microtubules (green) were detected with a monoclonal antibody raised to β-tubulin and an FITC-conjugated goat anti-mouse secondary antibody. RGS14 (orange) was detected using an affinity-purified rabbit anti-RGS14-60 antiserum (antigen: rat RGS14 amino acids 60–544) and a Cy3-conjugated goat anti-rabbit secondary antibody. Colocalization between RGS14 and microtubules appears yellow in the overlay images because of the combination of the orange and green pseudocolors.

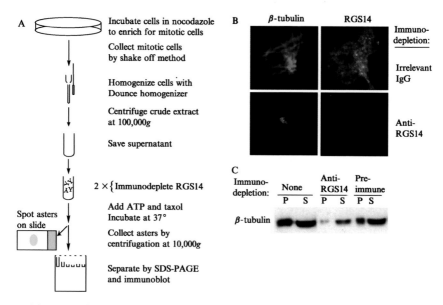

Martin-McCaffrey *et al.*, Chapter 16, Fig. 5. Immunodepletion of RGS14 from HeLa cell mitotic extracts disrupts aster formation. (A) Flow diagram of method for preparing and immunodepleting mitotic cell extracts. (B) RGS14 was immunodepleted from 50 μl of extract (3×10^7 cells/ml) twice with 100 μg affinity-purified anti-RGS14-263 antiserum (antigen: rat RGS14 amino acids 263–544) conjugated to a Seize-IP matrix (Pierce). Induction of mitotic aster formation on RGS14-depleted and control extracts (depleted with irrelevant IgG) was accomplished by adding 3 μM ATP and 10 μM taxol to the extracts and incubating at $30°$ for 1 h. RGS14-depleted asters are very small compared to control asters. (C) Asters were collected by centrifugation and analyzed by SDS–PAGE. The relative amounts of asters formed were compared by immunoblotting for β-tubulin. There was less tubulin present in the pellet ("P") formed by RGS14-immunodepleted mitotic extracts induced to form asters than those formed by mitotic extracts depleted with preimmune serum or nondepleted, untreated mitotic extracts, indicating that RGS14 is required for normal mitotic aster formation. The content of β-tubulin in the supernatent ("S") postimmunodepletion of RGS14 is not altered grossly.

VÁZQUEZ-PRADO *ET AL.*, CHAPTER 17, FIG. 3. The yeast two-hybrid system is a powerful tool used to reveal novel protein–protein interactions regulating RGS-containing RhoGEFs. This figure illustrates different steps involved in the cloning of cDNAs of PDZ-RhoGEF-interacting proteins. As explained in the text, a structural domain of PDZ-RhoGEF is used as bait to fish out interacting proteins expressed from a cDNA library. The interaction reconstitutes a transcription factor, which allows the yeast to grow under restrictive conditions (left) and turn blue in the presence of X-αGal (middle). To confirm the interactions, cDNA of the putative-interacting protein is isolated and transformed in yeast containing the bait or an unrelated control (right). Those yeast strains carrying the bait and a positive-interacting protein grow and turn blue under restrictive conditions (-AHLT + XαGal). The efficiency of the transformation is verified by the growth in medium only restricting for the presence of the plasmids (-LT).

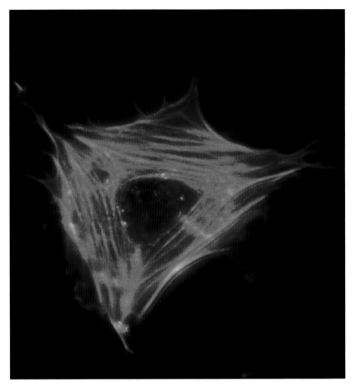

VÁZQUEZ-PRADO *ET AL.*, CHAPTER 17, FIG. 4. Texas Red-conjugated phalloidin was used to label polymerized actin stress fibers in this LPA-treated cell.

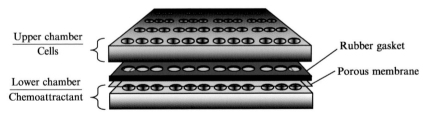

Upper chamber
Cells
Lower chamber
Chemoattractant

Rubber gasket
Porous membrane

VÁZQUEZ-PRADO *ET AL.*, CHAPTER 17, FIG. 5. The Boyden chamber illustrated here consists of two plexiglass members separated by an extracellular matrix protein-coated membrane. The upper portion contains 48 holes, arranged in four rows of threes that pass completely through the lid. The base contains the same pattern of holes closed at the bottom end to form wells capable of holding dilutions of the chemoattractant. The two members are secured together, usually by tightening small screws on the lid, after insertion of the rubber gasket between them to prevent leaks. A solution of cells is then added to the upper wells, directly on top of the exposed membrane. At the appropriate time, the chamber is disassembled and the membrane is processed as indicated in the text. The final stained membrane is digitized, and the intensity of staining at each spot is quantified with the NIH Image or other appropriate software. The intensity of spots is determined in triplicate and is illustrated numerically and as a bar graph.

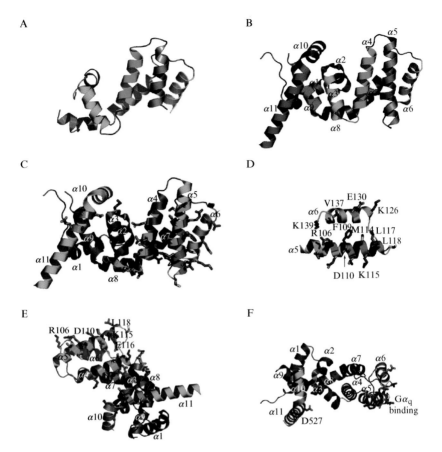

STERNE-MARR, ET AL., CHAPTER 20, FIG. 2. GRK2 RH domain structure. (A) Axin-based homology model of the GRK2 RH domain. The RGS box of GRK2 (amino acid residues 51–171) was aligned with RH domains of known structure, and a homology model was built manually as described in the text. Note that two regions, the α1 helix and the α5:α6 loop, bear little sequence homology to analogous regions of other RH domains. In our original model (Sterne-Marr et al., 2003), these regions were modeled after the axin RH domain but are omitted here. Helices are indicated by color: α2 is blue, α3 is turquoise, α4 is green, α5 is pink, α6 is yellow, α7 is orange, α8 is brown, and α9 is red. (B) The GRK2 RH domain determined from the GRK2:Gβγ crystal structure (Lodowski et al., 2003). The helix color scheme is the same as in A, with the addition of α1 in gray, α10 in gold, and α11 in magenta. The axin-based homology model was a good predictor of the bundle subdomain (α helices 4–7), but a poor predictor of the terminal domain (α helices 1–3 and 8–11). The terminal subdomain was a region of variability among known RH domain structures, with the RhoGEF (Chen et al., 2001; Longenecker et al., 2001) and GRK2 RH domains containing extra α helices in this subdomain. In the GRK2 primary sequence, α helices 10 and 11 are separated from α helices 1–9 by the kinase domain. Differences between the axin-based homology model and the crystal structure are outlined in the text and can be visualized here. (C) Traditional view of the

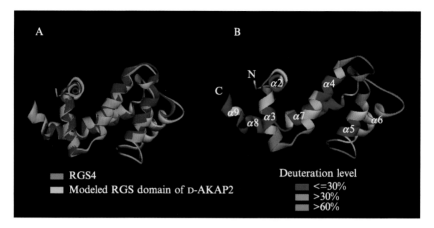

Burns-Hamuro *et al.*, Chapter 22, Fig. 2. (A) A backbone structural alignment of RGS4 and modeled RGS B domain of D-AKAP2. (B) Deuterium exchange data after 3000 s on exchange mapped onto the structure of the modeled RGS B domain. Reprinted from Hamuro *et al.* (2002), with permission.

GRK2 RH domain indicating residues that were mutagenized to test binding to either $G\alpha_{q/11}$ or mGluR1a. Mutated residues that perturbed $G\alpha_{q/11}$ binding are shown with magenta backbone, whereas mutated residues, which had no effect on $G\alpha_{q/11}$ binding, have green backbones. (D) $G\alpha_{q/11}$-binding site on GRK2. In the "traditional" RH domain view, the $G\alpha_{q/11}$-binding site faces the side of the structure. From the traditional view, the structure was rotated around the y axis $\sim90°$ to bring the $\alpha5$ and $\alpha6$ helices proximal to the viewer and was then rotated $\sim90°$ in the z axis to bring the $\alpha5$ and $\alpha6$ helices into a horizontal orientation. Color scheme is the same as in C, except that mutated residues that exhibit dramatic and mild defects in $G\alpha_{q/11}$ binding are shown in magenta and pink, respectively. (E) Membrane-proximal view of GRK2 RH domain showing location of the $G\alpha_{q/11}$-binding site with respect to the membrane. Because the structure of GRK2 was determined in complex with $G\beta\gamma$ subunits, the relative location of the plasma membrane can be predicted with a high degree of confidence. In this view, the GRK2 RH domain $\alpha11$ helix is juxtaposed to the plasma membrane and the $G\alpha_{q/11}$-binding site points toward the cytoplasm. Color scheme is the same as in C, except $\alpha11$ helix residues that were mutated to test binding to mGluR1a have an orange backbone if no effect was detected (T528A, A531E, E532A) or a blue backbone (D527A) if the mutation disrupted coimmunoprecipitation of GRK2 with mGluR1 (Dhami *et al.*, 2004). (F) The structure in E was rotated so as to visualize the D527 side chain. Although solvent exposed, D527 points toward the $\alpha3$ helix. Color scheme is the same as in C and E.

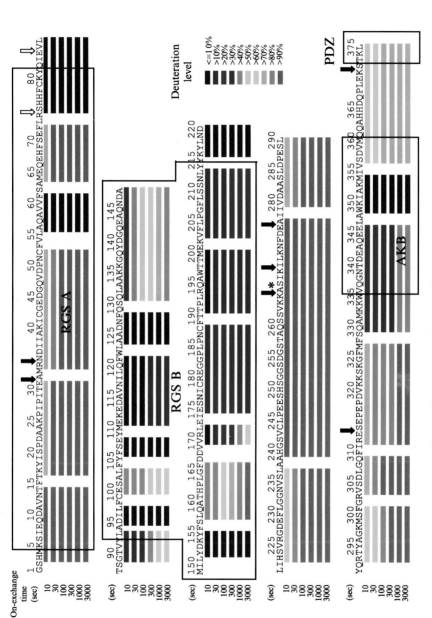

BURNS-HAMURO *ET AL.*, CHAPTER 22, FIG. 4. (*continued*)

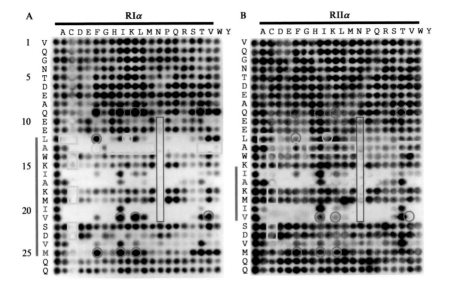

Burns-Hamuro *et al.*, Chapter 22, Fig. 5. A peptide substitution array of the 27 residue AKB domain of D-AKAP2 probed with either RIα (A) or RIIα (B) using an antibody sandwich detection system as described in the text. Each spot represents an immobilized peptide prepared by SPOT synthesis. All 20 amino acids (top) are substituted along each position of the AKB domain (left). Residues most sensitive to substitution are indicated in red. The bar to the left of each array highlights the region sensitive to substitution. Residues that are circled green represent peptides that show preferential binding to RIα. Residues that are boxed yellow represent peptides that show preferential binding to RIIα. Proline substitutions disrupt binding to both RI and RII along a defined sequence length (blue box). Reprinted from Burns-Hamuro *et al.* (2003), with permission.

Burns-Hamuro *et al.*, Chapter 22, Fig. 4. Domain organization of D-AKAP2 probed by deuterium exchange mass spectrometry. A soluble form of mouse D-AKAP2 (372 amino acids), which included the C-terminal half of RGS A, the full RGS B domain, the PKA-binding site, and the PDZ-binding motif, was used for DXMS experiments. The deuteration level of several D-AKAP2 peptides after on exchange for 10–3000 s is presented. The peptides are denoted under the sequence of D-AKAP2 as color bars, representing the extent of deuteration (see legend). The on-exchange time is listed to the left of the sequence. Blue represents heavily protected areas, suggesting an ordered region and red represents highly exchanging regions, suggesting a disordered region. Residues corresponding to the partial RGS A domain, the full RGS B domain, the AKB domain, and the PDZ-binding motif are indicated. Limited proteolysis cleavage sites by either trypsin or Glu-C are illustrated after 1 (black arrows) and 24 (white arrows) h of digestion. The asterisk above serine 267 is highlighting and is an *in vitro* PKA phosphorylation site. Reprinted from Hamuro *et al.* (2002), with permission.

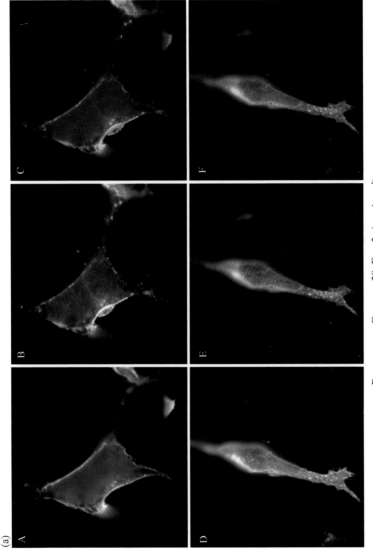

PYNE *ET AL.*, CHAPTER 28, FIG. 5. *(continued)*

PYNE *ET AL.*, CHAPTER 28, FIG. 5. Co-internalization of PDGFβ receptor–S1P₁ receptor complexes. HEK 293 cotransfected with the myc epitope-tagged S1P₁ and PDGFβ receptor were quiesced overnight and then stimulated with S1P (5 μM, 5 min). (a) Epifluorescence images and (b) confocal fluorescence images. Images A and D are stained using the anti-myc epitope primary antibody followed by an antimouse secondary antibody conjugated to TRITC (red). Images B and E are stained using the anti-PDGFβ receptor primary antibody followed by an antirabbit secondary antibody conjugated to FITC (green). Images C and F show merged red and green channels where colocalization is visible as yellow. (A–C) Controls. (D–F) S1P stimulated.